The Hot-Blooded Insects

The Hot-Blooded Insects

STRATEGIES AND
MECHANISMS OF
THERMOREGULATION

BERND HEINRICH

SPRINGER-VERLAG BERLIN HEIDELBERG GMBH

Originally published by Springer-Verlag Berlin Heidelberg in 1993
Softcover reprint of the hardcover 1st edition 1993

10 9 8 7 6 5 4 3 2 1

ISBN 978-3-662-10342-5   ISBN 978-3-662-10340-1 (eBook)
DOI 10.1007/978-3-662-10340-1

This book is printed on acid-free paper and its binding
materials have been chosen for strength and
durability.

Designed by Gwen Frankfeldt

Contents

Prologue

NO aspect of the physical environment is more important to insects than temperature. In most environments temperature fluctuates through time, but insects also experience extreme temperature variations in space. A large mass, such as a human body weighing 65 kg, would register no measurable temperature increase by stepping from shade to sunshine for several minutes; a 10 mg fly, however, heats up some 10°C in only 10 seconds when it lands in a sunfleck. Needless to say, the thermal environment faced by insects is potentially much more severe than it is to us or to any other vertebrate animal. And it is probably not an exaggeration to claim that insects have evolved some of the most amazing feats of thermal adaptation and thermoregulation in the entire animal kingdom. Yet, little over 20 years ago, that statement would have seemed eccentric.

As we sat one night in 1969 in front of a white sheet illuminated by a lantern on Mount Kainde in New Guinea among dark forest trees festooned with orchids, moss, and tangles of lianas, one after another beautiful moth, of dozens of different species, flew out of the darkness. I captured the arriving insects by net and thrust a thermocouple probe into each one to measure its body temperature. My host, Peter Shanahan, owner of the coffee plantation and of the white sheet, kindly tolerated my unusual routine.

I was grasping for any and all clues that might ultimately allow me to figure out what to me seemed a deeply puzzling question: How could moths, who have no sweat glands, no lungs, no capillary system for peripheral blood circulation, and no major muscles besides those used for flight—how could such creatures possibly regulate their body temperature and fly at the same time?

After two years of grappling with the problem for my Ph.D. thesis under George A. Bartholomew and Franz Engelmann at UCLA, I had only recently become convinced that some sphinx moths could, indeed, stabilize their thoracic temperature within 2–3 °C of 40 °C while in free flight over a wide range of air temperatures. But it was still a mystery to me how they did it. There were no clues to go on and no precedent. Except for work being done at one other laboratory, there was also little or no interest in the question. Almost nobody was thinking about "insect thermoregulation" then.

It had been difficult enough for me to trust my own previous data of a sphinx moth stabilizing its thoracic temperature in flight over a wide range of ambient temperature while its level of heat production appeared to remain totally unchanged. This result was either unprecedented or a big faux pas. I suspected the latter because a team of other researchers on the topic had just published not one but a whole series of papers with a conclusion entirely different from mine. But now in New Guinea I was pleasantly shocked to find that some of the smaller moths with large wings barely heated above air temperature. Some of the sphinx moths, however, maintained thoracic temperatures near the phenomenally high level of 46 °C. The night was chilly, and these large, narrow-winged moths, being heated some 9 °C above my own body temperature, were clearly not warm-blooded. They were hot-blooded.

As would later become apparent from comparative studies of many other insects described in this book, neither the "hot" nor the "cold" moths that night were an exception. Some moths (and other insects) even fly with body temperatures near 0 °C, the freezing point of water. In a few short years I've come to believe that certain insects are among the most highly evolved organisms on earth with respect to mastering temperature as a variable of the physical environment. In that and many other respects they are as adequate as any homeotherm.

The insects were the first animals on earth to evolve social systems. They were the first group of organisms in the history of life on earth to fly. Millions of years before the dinosaurs appeared, the invention of flight by insects had already made their diversification possible. And flight made many of them the first endothermic, or hot-blooded, and ultimately also thermoregulating animals on earth. (Later, endothermy became not only a consequence of flight, but also in some cases a necessity for it.)

Insects are delicate microscopic whiteflies and massively chunky goliath beetles. They are colorful butterflies flying lonely missions through the jungle, and they are teeming masses of blind termites building towering castles of clay taller than a two-story building. It is difficult to imagine a group of animals more different from us, or more varied. As Howard Ensign Evans so aptly pointed out, it is as if they were an independent form of life from another planet. Their riot of forms and exuberance of diverse life-styles make them ideal organisms for the comparative studies needed to unravel evolutionary mechanisms, especially those that relate to temperature adaptation. Despite their usefulness for illuminating general theoretical insights, insects have often been looked upon as if they were indeed from another planet. As I write, a just-published book on new directions in ecological physiology gives less than passing lip service to this, the most abundant, diverse, and species-rich group of animals on earth. Yet I believe (and this is a personal opinion that I hope will be supported in this book) that insects, perhaps more than any other animals, have much to teach us about how physiological adaptations contribute to behavior and ecology.

In insects the implications of thermoregulation have routinely been explored to the ecological levels of organization (particularly in bees, beetles, butterflies, and grasshoppers) because their thermoregulation is often closely and intimately related to ecology. In contrast to our close study of vertebrate animals, however, our ignorance of physiological mechanisms in insects still looms large, and it is still plagued by glaring controversies and disagreements. As one of the more prominent workers on insect thermoregulation has told me, "One of the things I most like about working with insects is that much physiology is so poorly understood that you constantly confront the sorts of basic questions that were solved in vertebrates a century or two ago. Vertebrate physiology seems bankrupt by comparison." I share this idea that, in comparison with vertebrate biology, the study of insects has yielded many important new insights, and there is more still to learn.

There are several reasons for the slow progress with insects until recently. Undoubtedly one impediment was the view of insects as primitive animals easily classified as "poikilotherms"—animals whose body temperature follows that of the environment. This notion was hardly one to spark exciting questions.

Another reason for slow progress on insect thermal physiology is their body size. Julian S. Huxley speculated in 1941 (in "The

Uniqueness of Man") that insects have been cut off from further "progress" by their breathing mechanism: "The land arthropods have adopted the method of air-tubes or tracheae, branching to microscopic size and conveying gases directly to and from the tissues, instead of using a dual mechanism of lungs and a bloodstream. The laws of gaseous diffusion are such that respiration by tracheae is extremely efficient for very small animals, but becomes rapidly less efficient with increasing size, until it ceases to be of use at a bulk below that of a mouse. It is for this reason that no insect has ever become, by vertebrate standards, even moderately large." He goes on to argue that it is also for this same reason that none has become even moderately intelligent, because none has reached a size large enough to provide the minimum number of neurons required for "the multiple switchboards that underlie intelligence."

The small body size of insects that clamps a low ceiling on their intelligence also severely taxes human ingenuity to study their physiology of thermoregulation. Thermoregulation can be accomplished only by studying intact organ systems, and it usually involves the simultaneous and smooth operation of many organ systems, those involving locomotion, water balance, blood circulation, and gas exchange. All must often be operating simultaneously before the phenomenon under study can be observed. And instrumental observations are difficult and often nearly impossible to make without doing gross damage to the system when that system is of minute dimensions. Many aspects of normal experimentation that are easily possible with vertebrate animals are difficult with insects. But this is part of their charm.

The available evidence for some physiological mechanisms in insects is sometimes still more circumstantial than one might wish. Since the insects' small size often places a large burden on an experimentalist's skill and the instruments used, it is sometimes debatable whether or not the data on the simplest of measurements, such as body temperature, are reliable. This is a legitimate concern, and I shall point out repeatedly in this book where I believe studies have gone astray. I hope here to provide not only a review and a synthesis of over 1,000 technical research papers from the scientific literature, which has been growing steadily since 1965 (see Figure P.1), but also a critique of the available work and suggestions for how it may be improved. Most of these articles are each as long as some of the following chapters, and I've been forced to reduce many of them to one-liners. For these

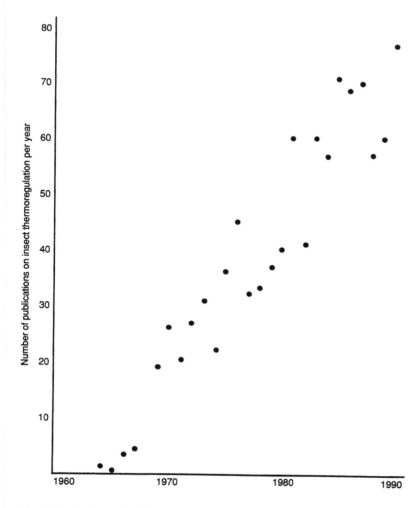

Fig. P.1 Number of publications on insect thermoregulation per year from 1965 to 1990, as retrieved by a computer search (BIOSIS database, cross-indexing *body temperature/thermoregulation and insects/Insecta*) by Craig A. Robertson.

perhaps unavoidable indignities I apologize, nevertheless. Much of the work I explore has not been previously consolidated and critically digested so as to make it accessible even to a scientific audience.

If the data are taken into consideration, then we have come a remarkably long way in understanding insect thermoregulation since a thermometer was first laid alongside the abdomen of insect resting in a little closed container over a century ago, even though only a tiny fraction of the species which might reasonably be expected to thermoregulate has been examined. Despite the million or so insect species that exist, it is reasonable to expect that mechanisms of thermoregulation are relatively conservative because research so far has been consistent with this supposition. Patterns are emerging and major puzzles are being solved, and I felt that it was now appropriate to consolidate our information.

No generalities may be made until accurate empirical information is established. In insect biology, because of the very small size, the complexity, and the variety of animals, and also the ease of getting (some) data (and ease of misapplying it), there is much imperfect knowledge of how different physiological systems interact. Furthermore, the same sets of data often have entirely different relevance when viewed from the ecological rather than the physiological perspective. Both views are necessary, but both are often not available from the same data set. It is difficult for me to imagine a scientific field at times so muddled with controversies as a result of imperfect information or the complexities of getting good information and applying it critically to the appropriate perspective. These are problems for any young field, and to counter them I describe experiments in those cases where there is reasonable doubt, one way or the other, on how to interpret them. Indeed, I feel that the designs of the experiments used to answer questions are fascinating in their own right, and try to give at least a feel for how new knowledge in the field is acquired.

This book is based on the primary literature. Nevertheless, there are numerous reviews of insect thermoregulation and particular details of insect thermoregulation that the reader may wish to consult for other coverage, for historical knowledge, and for other points of view. Biochemical considerations (Hochachka and Somero, 1984), the effects of energetics (Heinrich, 1979) and locomotion on thermoregulation (Kammer and Heinrich, 1978; Heath and Heath, 1982), and aspects of thermal energy budgets and the thermal environment affecting body temperature are found elsewhere (Parry, 1951; Gates, 1980; Gates and Schmerl, 1975; Porter and Gates, 1969; Willmer, 1982; Cossins and Bowler, 1987). Other selected aspects of invertebrate (Wieser, 1973) and insect thermoregulation have been previously reviewed (Cloudsley-Thomson, 1970; Heinrich, 1974, 1981; May, 1979, 1985; Casey, 1988).

In organizing this book I faced the choice of highlighting topics such as shivering, heat-loss mechanisms, and basking in separate chapters. I felt, however, that this approach would inevitably result in lumping together the great diversity of taxa into a clumsy *Gemisch* in which important details would be lost. Instead, I cover the major taxa separately, thereby showing sometimes parallel patterns as well as diverse patterns. A summary is provided at the end of each chapter and at the end of the book to draw together the main conclusions and bring parallel threads into line. Thus, the text provides access to the taxa, but the index provides the main access to the cross-taxonomic topics. The down-side of this organization is that all of the supporting data for some theories are not organized all in one place.

Peter D. Medawar, a Nobel laureate in experimental medicine, has said: "Quantification as such has no merit except insofar as it helps to solve problems." So far, most of the solutions to the problems of how and why insects regulate body temperature have required large amounts of common sense, not excessive amounts of statistical manipulations. I have therefore tried in this book to cover those approaches which so far have provided the most direct access to an understanding of the insects' thermal problems and solutions to them. I refrain from descriptive "models" unless they reveal new discoveries of mechanisms.

A biologist may take many potential approaches, and I need to specify also what this book is *not* and where alternative information can be found. I concentrate on thermoregulation (behavioral and physiological mechanisms of altering body temperature), not thermal balance, as such. Thermal balance is contingent on physical constraints of the environment (Gates, 1980; Gates and Schmerl, 1975) and on size (Bartholomew, 1981; Casey, 1981). Furthermore, activity at a given body temperature depends on the biochemical substrate (Hochachka and Somero, 1984). And this biochemical substrate may itself adapt in response to temperature stress (Alahiotus, 1983) and seasonal changes (Lee and Denlinger, 1991). In turn, biochemistry affects the geographical distribution of active insects (Parsons, 1978; Kimura, 1988) and the ability of insects to survive over the winter in an inactive state (Lee and Denlinger, 1991). In addition, temperature sensitivity can change remarkably rapidly to selection (White, deBach, and Garber, 1970; Bennett, Dao, and Lenski, 1990; Huey et al., 1991), presumably because of underlying biochemical adaptations.

All of these considerations are essential components for a complete understanding of the thermobiology of insects. All are "relevant" and fascinating topics in their own right. But for the most part they are not unique to insects, and they are not covered in depth in this book. I decided instead to take an ecological-evolutionary perspective and to concentrate on the behavioral and physiological mechanisms of the insects themselves that directly affect body temperature.

This book is meant generally for an academic audience, but I have tried to make both the general principles and the great diversity of details accessible to an interested wider audience without sacrificing substance.

Finally, I have disagreed at one point or another with almost everyone who has made significant and lasting contributions to the field. This is to be expected. Many concepts are not yet frozen into place. They are still being discussed, and I confront them.

A Few Words on Terminology

A variety of terms are used to describe the thermal responses of animals, but in practice almost none are exclusive and most are highly misleading. Each term merely provides a short-hand description of a specific aspect of the thermal response that one may wish to acknowledge.

In general, birds and mammals that maintain high and stable body temperatures at all times were, and still are, called *warm-blooded* or *homeothermic* (from the Greek *homoios* = similar), while most fish, amphibians, and insects were traditionally called *cold-blooded* or *poikilothermic* (from the Greek *poikilos* = changeable) because their body temperature changes along with that of their environment. These terms are broad generalizations that say nothing about the specific body temperatures maintained or how they are maintained. For example, a butterfly basking on a rock with a thoracic temperature of perhaps 42 °C maintains a much higher body temperature than a hibernating mammal, such as a bat or a ground squirrel, which may cool to near 0 °C.

Other sets of terms are used to describe temporal aspects of body temperature, as well as the source of the heat used to elevate body temperature. A bat, while hibernating, is poikilothermic over broad ranges of ambient temperature. It only becomes homeothermic after arousal. Animals such as bats that change from an unregulated to a regulated state are called *heterotherms* (*heteros* = different). Animals can be *endothermic*, by elevating their body temperature using internally generated heat (from muscular activity, for example); or they can be (like the butterfly basking on a rock) primarily *ectothermic*, relying on an external heat source (such as solar radiation).

All of the above terms are descriptive aids that can direct our attention to specific aspects of the thermal biology we wish to emphasize. But they are not descriptive of the whole animal. For example, we humans can generate our own body heat but we may also bask in the sun. Our body temperature declines at night. Strictly speaking, we are thus warm-blooded homeothermic endotherms who are also to a small degree poikilothermic, heterothermic, and ectothermic. The same six terms might all apply to some insect, but each to a different degree and each to a degree that changes almost constantly throughout the day at a more rapid rate than in a human.

Nevertheless, according to the definitions of a few researchers, certain animals are not "endothermic" even if they have a high body temperature from their own metabolism, provided they exceed certain size limits. By this rationale insects would not be considered endotherms, either, because they exceed certain activity levels that result in inevitable heating. Therefore, to call them "hot-blooded" might be even more objectionable. The title of this book notwithstanding, I hope here to avoid categorization wherever possible. From a limited perspective it would be possible to classify different insects in any of a variety of terms—as poikilotherms, cold-blooded, warm-blooded, heterotherms, endotherms, heliotherms, and even homeotherms. Terms tend to categorize, however, and categorization is not helpful in understanding thermal mechanisms in insects because many assumptions are associated with the terms. I therefore use these terms loosely, for the sake of convenience and only to focus our attention on specific aspects of behavior or physiology.

What is "thermoregulation"? As indicated in detail by Huey, Pianka, and Hoffman (1977), thermoregulatory behavior is often not easily documented in the field. For a simple working definition, I define thermoregulation as the maintenance of a specific body temperature (or temperature range) independent of *passive* processes, such as radiation, convection, evaporation, and the body's metabolism during different activities. Any or all of these processes and others may be exploited to achieve body temperatures that fluctuate about some specific setpoints, and as indicated in this book, a variety of approaches are used to measure or infer thermoregulation.

Numerous specific terms and abbreviations are relevant to insect thermoregulation. Three temperatures are often used as convenient baselines upon which to gauge whether or not the animal is endothermic and/or regulating its body temperature. One common term is ambient, environmental, or air temperature (T_a). To be meaningful for physiological studies, T_a must be measured within a centimeter or so of the insect.

A more inclusive term for environmental temperature in answering behavioral and/or ecological rather than physiological questions is effective environmental temperature (T_e). This term is useful if one wishes to combine the effects of all nonphysiological sources of heat gain and loss on the animal's body temperature *without* differentiating the effects from air temperature, as such. (T_e is determined by allowing dead animals or models thereof to come to temperature equilibrium under specific field conditions where T_a, solar radiation, convection, and other factors are variable.) T_e is often called operative temperature, T_o.

In the absence of direct solar radiation, T_a approximates T_e, but if solar radiation is present then T_e will be greater than T_a, although it will still converge on T_a if wind speed (and hence convection heat loss) increases. For a really minute insect, T_a should equal T_e irrespective of solar radiation. Thus, T_a suffices in many instances except where solar radiation is involved.

The various terms for body temperature (T_b) include thoracic temperature (T_{thx} in this book, in other sources occasionally T_{th}); head temperature (T_{hd} here, or T_h); and abdominal temperature (T_{abd}, or T_{ab}). The difference between body temperature and ambient temperature is called the *temperature excess*, and it may be expressed as ($T_b - T_a$).

Units of Measure

In various parts of the literature one encounters a confusing welter of terms for measuring energy. Historically, the unit of thermal energy is the *calorie* (cal), which is defined as the amount of heat required to raise the temperature of 1 g of water from 14.5°C to 15.5°C at an atmospheric pressure of 760 mm Hg. (The specific temperature interval is necessary for measurements involving questions of physics, but it is of no practical importance for resolving behavioral and physiological questions of animal biology.)

Other units of energy include the *joule* (J). One calorie equals 4.184 J. The rate of energy transfer can be expressed as calories per unit time, or in terms of *watts* (W), where 1 W = 1 J/s. Biology deals largely with aqueous systems, so the calorie has an obvious and direct meaning that is easily interpretable when it is applied to temperature. Because of its obvious utility and its historical use in biology, the calorie is the most commonly used energy unit in the literature on insect energetics. In this book, however, I adhere to the suggestions of numerous colleagues who have urged me to convert to the SI (Système Internationale) or international units of joules or watts now required in many journals.

Another useful measure is *specific heat capacity*, which is the amount of heat required to change the temperature of a specific quantity of substance. The specific heat of water is 4.18 J/g/°C. The specific heat capacity of muscle tissue is generally assumed to be 80 percent that of water, close to 3.34 J/g/°C. Thus, to raise the temperature (in the absence of heat loss) of a 2.0 g moth from 20°C to 40°C requires (3.34 J/g/°C) × (2 g) × (20°C temperature change) = 133.6 J.

The specific heat capacity of water (and tissue containing mostly water) is very high, some eleven times higher than that of copper. That of air is only 0.24 cal/g/°C (1.0 J/g/°C), and since the density of air is 1.2 g/l, the heat capacity of 1 liter of air is only 1.2 J/°C.

The amount of heat loss from the evaporation of water can be calculated from measurements of the rate of water loss, since we know the heat of vaporization of water. In physiology it is customary to use the value of 584 cal/g H_2O, or 2,443 J/g H_2O, although the value shifts slightly with temperature.

The calorie is also a biologically useful term for converting between measurements of different forms of energy. For most practical purposes, an animal aerobically metabolizing 1 mg of sugar gains a yield of 3.7 calories or 15.5 J, which is sufficient to raise the body temperature 4.6°C in a 1 g bee: (15.5 J) × (1 g) ÷ (3.34 J/g/°C) = 4.6°C.

In practical terms, most physiological research measuring energy metabolism has traditionally relied on the measurement of oxygen consumption, and it still does, mainly because it is easy.

According to the following equation describing the combustion of one mole of glucose:

$$C_6H_{12}O_6 + 6O_2 \rightarrow 6CO_2 + 6H_2O$$

$$(180 \text{ g}) + (192 \text{ g}) \rightarrow (264 \text{ g}) + (108 \text{ g})$$

Six moles of oxygen are consumed whenever 1 mole of glucose (or similar carbohydrate) is completely metabolized as an energy source. Since 1 mole of an ideal gas occupies 22.4 l under standard conditions (0°C, 760 mm Hg), the 1 mole (180 g) of sugar requires 134.4 l of O_2 [(6 moles O_2) × (22.4 l O_2/mole O_2) = 134.4 l O_2]. Converting liters to milliliters and grams to milligrams, each mg of sugar requires 0.74 ml of oxygen [134.4 ml O_2/180 mg sugar = 0.75 ml O_2/mg sugar] for complete combustion. At 3.7 cal or 15.5 J/mg sugar, each ml of O_2 consumed yields 4.93 cal or 20.7 J (3.7 cal/mg ÷ 0.75 ml O_2/mg = 4.93 cal/ml O_2, or 15.5 J/mg ÷ 0.75 ml O_2/mg = 20.7 J/ml O_2).

The chemical reactions for utilizing lipid and protein are more complex. Similar logic applies, however, and the corresponding caloric equivalents per ml O_2 are generally close to 4.7 cal (= 19.7 J) and 4.5 cal (= 18.8 J), respectively.

Rates of energy expenditure can be expressed in many ways. Most vertebrate physiologists have an inordinate fondness of expressing rates of energy expenditure (or heat production) on the basis of ml O_2 per gram or per kilogram of whole-body weight per hour. The intent is to "factor out" size, thereby permitting comparisons of metabolic rates of animals of different size. In insects, however, the thorax (which can weigh less than a third as much as the rest of the body but which contains its flight muscles) may alone be responsible for over 99 percent of the metabolism during flight activity. It is thus legitimate and often appropriate to express metabolism in terms of thoracic weight. Various researchers have used, for insects, units other than grams, including µg, mg, and kg. Since most insects weigh in the range of grams or milligrams, the gram may perhaps be the most appropriate unit to use, unless comparison with vertebrates are desired. Rate functions in the literature range from seconds to hours. Life would be simpler if everyone used watts, based on seconds, but the hour seems to have acquired general usage. Nevertheless, for metabolic rate alone, there are 54 possible permutations of the units of measure (such as µl per g body weight per second, ml per kg thoracic weight per hour, etc.), and most of them have probably been used. Energy has been measured in rates of oxygen consumption, calories, joules, and watts. Wherever possible (and sometimes it is not), I try here to follow arbitrary restrictions and use those units that are the closest to what was actually measured and that require the least conversion.

Notes on Methodology

Near the end of the heyday of the "noose 'em and goose 'em" era of thermoregulation studies on lizards, when students were scouring the desert with a monofilament noose secured to a pole in one hand and a Schultheiss thermometer in the other, James H. Heath (1964) at the University of Illinois cautioned in the prestigious journal *Science* about the pitfalls of interpreting data collected in this way. He said he could show "thermoregulation" of beer cans (filled with water) in the desert using methods commonly applied to show thermoregulation in lizards. We are now witnessing the heyday of the "grab-and-stab" era of insect-thermoregulation studies in the field: nets are used instead of noose poles, and thermocouples inside hypodermic needles with temperatures read on digital read-out meters. Heath's cautions apply to insects a hundred-fold: at least the lizards were large enough so that the temperatures read were accurate; the same can't be said for insects, and there are several other potential sources of error.

The first major consideration is the disparity in size between probe and object measured. Temperature probes must be small enough relative to the object measured so that the heat they absorb does not seriously affect the temperature of the animal. This is not a problem when hypodermic-needle probes (0.3 mm in diameter) are used on insects weighing over 200 mg, but large errors are evident and corrections need to be made when animals smaller than 50 mg are measured (Heinrich and Pantle, 1975).

A second major potential problem is body-temperature lability. The body temperature of a small insect can change (either fall or rise) several degrees between the time the animal is captured and before its temperature is measured. In the grab-and-stab technique, the insect is caught between gloved fingers (if possible) so that a measurement may be made within 2–3 seconds. If hand-capture is not possible then net-capture is required, which doubles or triples the time (and the possible temperature change) between the insect's activity of interest and the measurement.

The third and perhaps most common—and also perhaps the more serious—source of error concerns not precision but interpretation of the data for biological relevance. For example, the takeoff temperatures of butterflies can be easily measured, but they say little about flight temperatures in the field, as is sometimes assumed in some studies. The thoracic temperature of an insect confined in a respirometer may reflect alarm and escape responses, not necessarily its temperature normally maintained under biologically relevant circumstances in the field. A bee in a respirometer may have an entirely different thermal regime from one buzzing from flower to flower. The precisely measured thoracic temperature of a moth suspended in flight may have almost no bearing on the body temperature the animal normally maintains in free flight.

Caging, suspending, restraining, and skewering insects are often necessary evils for the scientist asking physiological questions. It may be necessary to have continuous temperature records simultaneously in several body parts in order to determine their *interrelationships* or their relationships to other physiological variables. For such studies, however, the emphasis is on the interrelationships, not on their absolute values relative to ecological questions.

A new technology has become available, thermal imaging, that can be used to measure body-surface temperatures of unrestrained insects. As yet this technique is difficult to use in the field, and because of the high cost and the considerable technical expertise required to use it, it is generally impractical for most researchers.

Thermocouples, in contrast, are easy to make and read-out meters (Physitemp, 154 Huron Avenue, Clifton, NJ 07013) relatively inexpensive. Thermocouples are nothing more than electrical junctions between two wires of dissimilar metal, generally copper and constantan. Thermocouples installed inside hypodermic needles for "grab-and-stab" measurement are commercially available (Bailey Instruments, 515 Victor Street, Saddle Brook, NJ 07662). Such probes were used in many of the studies cited in this book. However, they are useless for studies where the animal must be given some freedom of movement while its temperature is continually monitored. In the past, commercially available thermistors have been used. But they tend to be bulky because they require a coating. Thermocouples are recommended, but they must be made by the researcher.

Constructing thermocouples from wire is ridiculously simple.

The insulation from the tips of two thermocouple wires is scraped off and the tips soldered together. It is critical that the solder junction is tiny, so that temperature will be measured at only one point. (The recorded temperature is an average of all of the temperatures of the wire wherever there are electrical contacts.) If the junction is as much as a millimeter in diameter then it is already too large for many measurements with small insects.

Frustrations can arise in making thermocouples (and in most studies where they are needed one is faced with making them almost continuously) when the thermocouple wires are so thin that they are barely visible with the unaided eye (these are available from Omega Engineering, Box 4047, Stamford, CT 06907). I offer the following simple tips that, as far as I know, are not published elsewhere. They will permit making even the smallest thermocouples a simple routine. First scrape the insulating enamel off about a millimeter from the ends of the two wires. Then hold the wires between thumb and forefinger so that they are lying side by side with the uninsulated tips touching. Dip the tip into soldering flux, and then dip the tip into solder that has just melted and rapidly pull it back out. (If the solder is so hot that the soldering flux smokes, then it will likely burn off more insulation from the thermocouple wires. Therefore, avoid very hot solder and long immersion in it.) The soldering flux reduces the surface tension of the solder, and with any luck some solder will have joined the two wires at the tip. Now examine the connection under a microscope and with sharp scissors cut off any surplus tip, so that the thermojunction is as small as possible. Rinse in soap water to remove any excess acid from the soldering flux. The thermocouple, when connected to leads going to a read-out meter (possibly equipped with recorder), is now ready to use. As indicated in this book, in some studies 4 or more temperatures in different parts of the body have been recorded simultaneously.

For the time being the grab-and-stab techniques and continuously implanted thermocouples, despite their shortcomings, still offer the most for the money when it comes to answering specifically defined questions. Potential problems and ways to surmount them have been discussed (Heinrich and Pantle, 1975; Heinrich, 1975; Stone and Willmer, 1989). There are, unfortunately, no hard-and-fast rules in distinguishing a "good" data point from a meaningless one. The best choice depends on a number of factors that have to be worked out in each case. I have here singled out the difficulties in measuring the presumably most simple datum of all: temperature. Measuring physiological variables, of course, presents much greater challenges.

Finally, behavior in the field often varies over time periods that, though very short, are often sufficiently long to affect body temperatures. As a consequence, random measurements of individuals in the field are apt to obscure much of potential interest as regards mechanisms of thermoregulation. Nevertheless, the data can have value for answering ecological questions where average performance is of interest.

Night-Flying Moths

A
S anyone who has ever passed a summer evening on a country porch will attest, moths are a highly varied group. The order Lepidoptera, besides butterflies, includes over 10,000 primarily nocturnal species of moths in North America and Mexico alone. Despite this great variety, a little over half of the 50 existing publications on moth thermoregulation concern just one family, the Sphingidae (commonly called "sphinx" or "hawk moths"), and 10 of these papers are on a single species, the common tobacco hornworm moth, *Manduca sexta* (formerly *Protoparce*). At a weight of 2 to 3 grams, *M. sexta* is one of the relatively large sphingids, a distinct advantage for a biologist seeking information on the physiology of body-temperature regulation of an insect.

Most sphinx moths are so large, in fact, that they superficially resemble hummingbirds (Fig. 1.1). Roger Tory Peterson even depicts one alongside an Anna's hummingbird in the 1990 edition of his *Field Guide to Western Birds*. The smallest hummingbirds weigh near 3 grams, whereas sphinx moths may range in mass from a little under 300 milligrams to over 6 grams. And these generally nocturnal (some also fly in the daytime) moths are very rapid and adroit flyers that hover, as hummingbirds do, in front of flowers and sip nectar, although some species do not feed at all, relying instead on energy reserves accumulated during the larval stage.

Sphingids are of particular interest to students of insect thermoregulation and energetics because of their historical importance. They were the first insects from which individual measurements of body temperature were taken; it was from them

Fig. 1.1 "Hummingbird" moth, *Hemaris* sp., sipping nectar from a flower. Unlike most sphinx moths, *Hemaris* is diurnal, and unlike many it uses its first pair of legs for partial support while hovering.

we learned that insects were not necessarily all poikilothermic. Furthermore, they were the subject of stark controversies on the mechanism of insect thermoregulation, controversies that eventually stimulated productive research and sharpened our focus. More is known about thermoregulation in moths than in any other group of insects, except possibly bees, and they are now a model of many of the principles and mechanisms of insect thermoregulation in general. For these reasons I have chosen to examine them in detail both to illustrate general principles of thermoregulation in insects and to provide a historical perspective of how the insights were derived.

The Physiology of Pre-Flight Warm-Up

The first person to measure the temperature of individual insects was the noted geologist Johann F. Hausmann, who in 1803 (quoted in Porfirij Bachmetjew, 1899) reported an increase of 2° C in the air temperature in a small vial containing a *Sphinx convolvuli*. Still using a mercury thermometer, but laying it against the insect directly, George Newport (1837) of the Entomological Society of London and the Royal College of Surgeons published temperatures of 5.5°F and 0.6°F above air temperature for a *Sphinx ligustri* female and male, respectively. From these and other measurements he concluded that flying insects had higher body temperatures than crawling ones.

Mercury thermometers were far too large to measure the body temperature of insects, and a major breakthrough for further studies was achieved in 1831 when Leopoldo Nobili and Macedonio

Melloni first used thermocouples to measure the temperature of insects (caterpillars and pupae). Using thermocouples to measure internal temperatures, Bachmetjew (1899), from the Physics Institute of the Hochschule in Sophia, Bulgaria, went on to show (in a male *Saturnia pyri* moth) the then-surprising result that the insect could vary its own body temperature (T_b) over fluctuations of at least 7°C. He determined, further, that these fluctuations are correlated with wing movements. Forty years later, the South African researcher M. J. Oosthuizen (1939), while working at the University of Minnesota, noted that "From a review of the voluminous literature on the body temperature of insects, it is apparent that the available data are rather fragmentary and in some cases inexact." He then went on to report body temperatures for the saturniid moth, *Samia cecropia* as determined with implanted thermocouples. He showed that changes of thoracic temperature (T_{thx}) are linked to activity of the flight muscles (Fig. 1.2).

The functional significance of periodic thoracic warming had already been noted by Heinz Dotterweich (1928) from the Zoological Institute at Kiel. He showed that moths belonging to the families Noctuidae, Bombycidae, and Sphingidae are incapable of flight until the temperature of the flight muscles has been raised through shivering or "wing whirring" (Fig. 1.3). The oleander hawk moth *Deilephila nerii* raised its thoracic temperature to 32–36°C before it could fly, and in continuous flight it could reach 41.5°C. Moths heated to 35°C in an oven flew without prior wing-whirring, proving that wing-whirring had some function

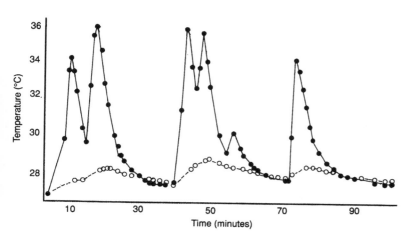

Fig. 1.2 Thoracic (*filled circles*) and abdominal temperatures (*open circles*) of a tethered, intermittently active, 4-day-old female *Samia (Hyalophora) cecropia* moth. Thoracic temperature rose as a function of muscular activity (that is, during periods of continuous wing movements); T_{thx} dropped when the animal was at rest. (From Oosthuizen, 1939.)

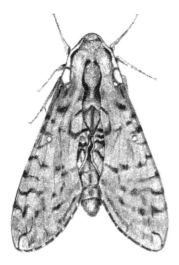

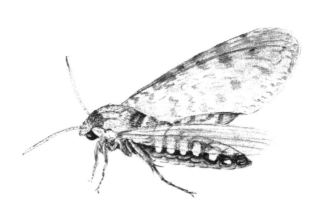

Fig. 1.3 The sphinx moth *Manduca sexta,* perched at rest *(left)* and beating its wings during pre-flight warm-up *(right).*

other than pumping the moth up with air as had previously been suggested. Dotterweich thus showed that wing vibrations prior to flight serve solely to raise muscle temperature.

Little was known about the underlying physiology of the "wing-whirring" of moths first described by Dotterweich (1928) except that it raised $T_{t;ix}$ sufficiently for flight. By 1968, however, Ann E. Kammer, now at the University of Arizona at Tempe, published a paper in which she compared the neural activation of flight in different moths during warm-up and during flight.

Lepidoptera are neurogenic flyers (Moran and Ewer, 1966); each muscle contraction is stimulated by one or several impulses from the central nervous system. Kammer (1968) made simultaneous recordings of the electrical activity of a number of upstroke and downstroke muscles of the wings and showed, as expected, that during flight these muscles are activated out of phase with each other (Fig. 1.4), so that the muscles contract alternately. During warm-up, however, the upstroke and downstroke muscles are fired nearly synchronously, rather than alternately. That is, the muscles are caused to contract isometrically against each other so that only a little wing movement results. The synchrony isn't perfect, hence some vibration of the wings, the "wing whirring," is still visible to the naked eye.

At a given muscle temperature, full-amplitude wing beats yield amounts of heat per muscle contraction closely similar or identical

Warm-up

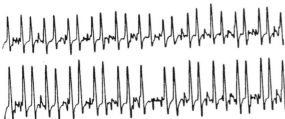

Warm-up

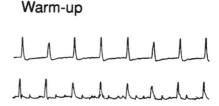

Flight

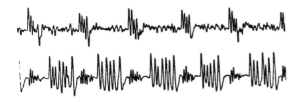

Flight

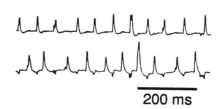

200 ms

Fig. 1.4 Muscle and nerve activity is electrical in nature—more specifically, it depends on changes in the concentrations of positive and negative ions on either side (inside or outside) of a cell's membrane. The difference in electrical charges across the membrane creates an electrical potential, which can be measured with an electrode. When a potential is made more negative by the movement of ions, the change is called "depolarization."

The graphs shown here (from Kammer and Rheuben, 1976; Kammer, 1968) record the changes in potential in opposing flight-muscle groups of a saturniid *(left)* and a sphingid *(right)* moth during warm-up and during flight. The top trace in each group was recorded from the dorsal longitudinal muscles, the bottom from a dorso-ventral muscle. (These are extracellular recordings, for which the electrode was placed outside the membrane of the muscle cell.) In both moths, the two muscle groups are activated synchronously during warm-up and alternately during flight. One difference between the two species is that in the saturniid there is multiple activation per muscle unit per wing stroke whereas in the sphingid there is single activation.

"Activation" is indicated by the peaks and valleys of the traces, which are often called "spikes" or "impulses" or "action potentials." Although traces are usually not labeled—we are usually interested only in the frequency and relative amplitude of the spikes—most researchers graph electrical potential in millivolts (mV) over time in milliseconds (ms).

those produced by the wing vibrations during warm-up (Heinrich and Bartholomew, 1971). The advantage of a slight temporal shift in the neural activation of the muscles to produce the warm-up pattern is therefore not to produce more heat. Instead, by reducing wing movements while contracting its flight muscles the moth (1) reduces convective cooling; (2) reduces body movements that could attract predators; and (3) avoids potential damage to its wings by not flopping around.

The neural-activation pattern for warm-up movements in moths is different not only in degree but in kind from the flight pattern. For example, in saturniid moths, which unlike sphingids have relatively large wings that beat slowly, the dorsal longitudinal muscles (wing depressors) and the dorso-ventral muscles (wing elevators) are both activated by *bursts* of several impulses per muscle contraction in flight, whereas during warm-up (when wing-beat frequency is greater than in flight) they are activated by a single impulse rather than by bursts of impulses (Fig. 1.4).

Phase relationships may vary as well as the number of impulses per wing beat. The muscles that provide the main power for either the upstroke or the downstroke of the wings are each activated simultaneously or in phase with each other, respectively, in all moths during both warm-up and flight. Additional flight muscles are recruited for heat production during warm-up, however, and within several species of sphingids the phase of some of these smaller flight muscles, which function primarily in flight control, may vary arbitrarily in relation to the other flight muscles during warm-up. These phase characteristics are species-specific (Kammer, 1970), suggesting that the neurophysiology of warm-up involves specialized, evolved motor patterns; they are not just immediate or proximate attempts to hold the wings steady, for a number of different possible motor patterns can do the job equally well.

The distinct motor patterns for pre-flight warm-up and flight are already apparent in young moths prior to their emergence from the pupal stage. Recording from wires implanted in the developing flight muscles of the pupae of the saturniid moths *Antheraea polyphemus* and *A. perny* and the sphinx moth *Manduca sexta*, Kammer and Rheuben (1976) and Kammer and Kinnamon (1979) found the same saturniid- and sphingid-specific warm-up patterns in the pupae as in the emerged adults (Fig. 1.5). Thus, the specific saturniid vs. sphingid motor patterns correlated with wing size in adults are clearly controlled not only by sensory

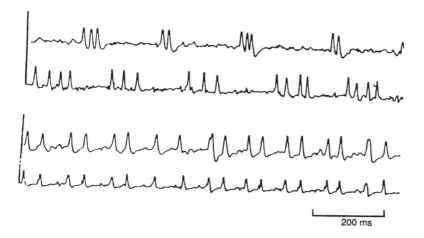

Fig. 1.5 The activity of antagonistic (upstroke and downstroke) flight muscles of an *Antheraea polyphemus* pupa, recorded extracellularly 3 days prior to its emergence from the pupal stage (eclosion). Note the typical saturniid activation patterns found both for flight *(top set of traces)* and for warm-up *(bottom set)*. (From Kammer and Rheuben, 1976.)

200 ms

feedback from the moving wings. Instead, motor patterns are hard-wired.

Nevertheless, sensory input from the wing movements modifies the existing motor patterns. In saturniid moths the muscles moving the relatively large wings are usually activated 3–4 times per contraction (Fig. 1.4), but if the wings must move against great resistance (as when the moth is forcibly held stationary), the upstroke and downstroke muscles are activated at 9–10 times per contraction (Fig. 1.6). Presumably, greater force of muscle contraction per wing beat is achieved by activating the power-producing muscle with greater frequency per contraction. (See Chapter 4 for similar results in grasshoppers.)

During warm-up, when the muscles contract isometrically, the resistance to movement that any contracting muscles experience is nearly equal to the force exerted by the opposing muscles. As in the example above of a moth whose wings are held still, in warm-up a moth's wings are also prevented from beating. Rather than activating the flight muscles in bursts, though, the saturniid as well as sphinx moths instead activate their muscles with single spikes during warm-up (Fig. 1.4). Presumably bursts of spikes should also keep the wings of shivering moths stationary while still producing the same amount of heat, and it is therefore not yet clear why saturniid moths do not activate their flight muscles during warm-up in bursts, even though they normally do so in flight.

Although the motor patterns of the flight muscles have received

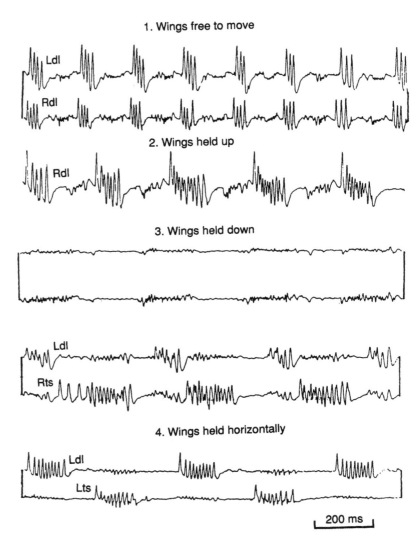

1. Wings free to move

Ldl

Rdl

2. Wings held up

Rdl

3. Wings held down

Ldl

Rts

4. Wings held horizontally

Ldl

Lts

200 ms

Fig. 1.6 Muscle potentials from adult *Antheraea polyphemus.* When the animals attempt to move the wings against great force, more action potentials per wing-beat cycle are recorded in the appropriate muscles. *(1)* In a moth whose wings are free to move, the left and right dorsal longitudinal wing-depressor muscles are activated synchronously by 3–4 action potentials. *(2)* When the wings are forcibly held up, the same wing depressors now fire at 7–10 times per wing-beat cycle. *(3)* When the wings are held down, wing-depressor muscle activity ceases or decreases, but the activity of the wing elevators *(Rts,* right tergosternal muscles) greatly increases. *(4)* When the wings are pinned horizontally, alternate groups of 9–10 action potentials per wing-beat cycle in the left wing depressor *(Ldl,* left dorsal longitudinal) and elevator *(Lts)* muscles are recorded. (From Kammer and Rheuben, 1976.)

THE HOT-BLOODED INSECTS

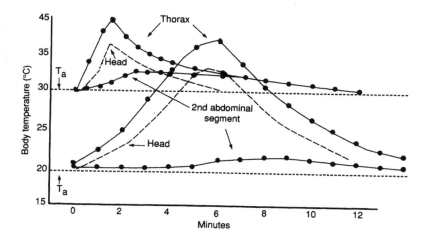

Fig. 1.7 Temperatures at different parts of the body during warm-up and subsequent cool-down in *Manduca sexta* at 30°C *(top)* and 19.5°C *(bottom)*. (T_{thx} and T_{abd} from Heinrich and Bartholomew, 1971; T_{hd} interpolated from data in Hegel and Casey, 1982.)

a great deal of research attention, other aspects of the warm-up physiology have been little studied. For example, in all pre-flight warm-up so far investigated, abdominal temperature (T_{abd}) remains low but it immediately shoots up noticeably if for some reason the warm-up is aborted and shivering stops (Fig. 1.7). These observations suggest that perhaps the animals are controlling heat distribution during shivering. In other words, they are actively preventing heat flow to the abdomen. I examine the evidence for this sort of thermoregulation in the next section.

A Debate on Thermoregulation

The elaborate pre-flight warm-up behavior as first demonstrated in moths indicates that specific thoracic temperatures are necessary for flight. This raises several questions: Why is a particular temperature required? Is body temperature regulated in flight? Why do some species fly at one body temperature while others fly at very different temperatures?

Having examined T_{thx} during warm-up, D. A. Dorsett (1962), from the University College, Ibadan, Nigeria, noted that "The need for preliminary warming suggests that the power that can be produced by the flight muscles at the normal environmental temperatures is insufficient to raise the insect from the ground or sustain it in controlled flight." He speculated that a large moth with small wings would have to work harder (by generating more frequent or more forceful wing strokes) than a smaller moth with large wings; consequently, this increase in work should be reflected in a greater quantity of heat liberated in the muscle and hence in higher T_{thx}. In other words, the moths did not regulate their T_{thx} but instead T_{thx} in flight was a direct function of the normal work output.

Dorsett tested his ideas on the numerous species of hawk moths common at most times of the year in Nigeria, although his most detailed observations were made on 45 specimens of *Deilephila nerii*, a species which also occurs in Europe. Results of two experiments on the same individual moth showed warm-up rates of 4.25°C/min at 20°C and 6.45°C/min at 27°C. As expected and confirmed in numerous subsequent studies (Fig. 1.8), warm-up duration and rate are inversely related to ambient temperature (T_a); at 20°C the *Deilephila* moths warmed up in 5 minutes, but at 27°C they warmed up in half that time. Dorsett (1962) also showed that the frequency of the wing beats throughout warm-up increased directly with thoracic temperature, which is also true for all other moths examined so far (Fig. 1.9).

The observation that the rate of pre-flight warm-up is very slow at low ambient temperatures and then rises steeply at higher ambient temperatures is now well documented in many other insects as well. Wing-beat frequencies, however, are independent of ambient temperature; they vary strictly as a function of T_{thx} during warm-up (Fig. 1.9). Wing-beat frequency in any one warm-up thus rises gradually with increasing T_{thx} until the muscles are warm enough to achieve the minimum wing-beat frequency and power output necessary to permit liftoff and flight. The shivering moths were producing heat in direct relation to muscle temperature, but *not* in relation to *ambient* temperature. These results gave no hint of possible stabilization of T_{thx}. However, Dorsett's work extended the idea already becoming established for other insects—namely, that vigorous muscle activity results in heat production and that the very vigorous activity of insect flight can result in astonishing elevations of body temperature.

The tantalizing (and for this book critical) question had not yet been raised: Is thoracic temperature stabilized—more precisely, regulated—over a range of ambient temperatures? The body temperature of an animal at any one time is due to the balance of rates of heat production and rates of heat loss. Thermoregulation implies the maintenance of a *specific* body temperature *independent* of passive processes such as radiation, convection, evaporation, and the body's metabolism during different activities. Thermoreg-

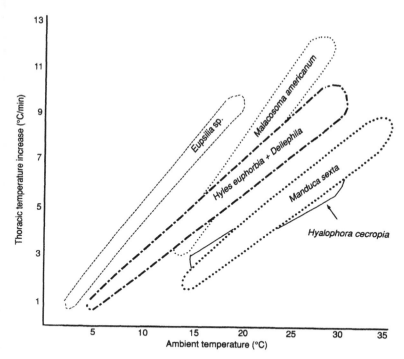

Fig. 1.8 The initial rates of pre-flight warm-up, measured as the increase in T_{thx} per minute, at different ambient temperatures. Included are data from the sphinx moth *Manduca sexta* (Heinrich and Bartholomew, 1971), the sphinx moths *Hyles euphorbia* and *Deilephila elpenor* (Heinrich and Casey, 1973), the lasiocampid moth *Malacosoma americanum* (Casey, Hegel, and Buser, 1981), the saturniid moth *Hyalophora cecropia* (Bartholomew and Casey, 1973), and the winter noctuid moths *Eupsilia* sp. (Heinrich, 1987a). For most moths the warm-up rate remains almost the same throughout the entire warm-up, even though it is almost strictly a function of ambient temperature.

The areas outlined enclose most of the data points in each of the studies. They are meant to indicate the pattern of response, not all the values measured. If one makes extrapolations of the values, one arrives at the lower limits of ambient temperature from which warm-up can occur in the different species.

ulation implies the ability to vary either heat input or heat loss to counteract the effects of the environment on body temperature, or a combination of regulation of both heat production and loss to reach the same end. Theoretically, any of the above avenues of heat flux, or a combination thereof, could be used to maintain body temperature within specific bounds. Undoubtedly it seemed

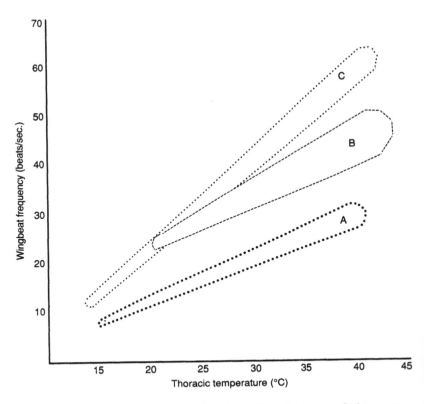

Fig. 1.9 Wing-beat frequency as a function of T_{thx} during pre-flight warm-up in *(A)* the sphinx moths *Manduca sexta* (Heinrich and Bartholomew, 1971), *(B) Euchloron magacra, Pseudoclanis postica,* and *Deilephila nerii* (Dorsett, 1962), and *(C)* the lasiocampid moth *Malacosoma americanum* (Casey, Hegel, and Buser, 1981).

preposterous at the time that individual insects, which until very recently had all been lumped together as "poikilotherms," could do what the "advanced" endotherms were capable of. But these assumptions were challenged in 1964; the question was then deemed appropriate and it precipitated much research.

In 1964 Phillip A. Adams and James E. Heath published a short paper on the white-lined sphinx moth *Celerio* (now *Hyles*) *lineata*. *Hyles lineata* has a Holarctic distribution and a very wide latitudinal and altitudinal range. Unlike most other sphingids, *H. lineata* flies both at night and in the daytime, possibly depending on ambient thermal conditions. It is a very rapid and agile flyer that also resembles a small hummingbird when in flight (Fig. 1.10).

Adams and Heath (1964) inserted bead-thermistor probes into the moths' thoraxes and attached meter-long leads to the thermistors. These leads served as a leash or tether, but they allowed the animals considerable movement, including flight. Like other moths, *H. lineata* remain in torpor with body temperature close to ambient temperature until they shiver (Fig. 1.2) in preparation for takeoff. When the tethered moths studied by Adams and Heath were not flying continuously, they sometimes alternated between shivering (when T_{thx} declined to 34.8°C) and cooling (when T_{thx} reached 37.9°C). Moths artificially heated with a heat lamp crawled away from the heat on reaching a mean T_{thx} of 37.7°C. The authors concluded: "We believe these results establish that *Celerio lineata* is not only capable of raising the thoracic temperature to consistent levels prior to flight, but that it is also competent to maintain this temperature throughout its activity period." This was indeed a landmark paper. Unfortunately, their simple statement, though correct, sowed the seeds for a misunderstanding on the mechanism: under natural conditions, "continuous activity" for a sphinx moth is *uninterrupted flight*, not alternations between shivering and resting (or walking between sun and shade) as they had been limited to in the laboratory study.

The pioneering paper by Adams and Heath for the field of insect thermoregulation was followed by a flurry of others by these authors and by Heath's students at the University of Illinois at Urbana. Their combined work suggested not only that *H. lineata* regulates its T_{thx} (Adams and Heath, 1964; Adams, 1969; Heath and Adams, 1965, 1967), but also that the sphinx moth *Manduca*

Fig. 1.10 The white-lined sphinx moth *Hyles lineata*, sipping nectar.

sexta (McCrea and Heath, 1971; Jacobs and Heath, 1967) and the saturniid moth *Hyalophora cecropia* (Hanegan and Heath, 1970a,b) all did so as well. All of these papers maintained that T_{thx} is regulated in "flight" by the regulation of heat production. No underlying physiological mechanism was provided, although, given Kammer's (1968) work, it might be presumed that regulation could potentially involve increased activation of the flight muscles. No such evidence was found, however, and instead a flurry of concurrent and subsequent publications on *M. sexta* (Heinrich, 1970a, 1970b, 1971a, 1971b; Heinrich and Bartholomew, 1971), *H. lineata* (Casey, 1976a,b), *H. cecropia* (Bartholomew and Casey, 1973), and other moths (Casey, 1980, 1981a,b; Heinrich and Casey, 1973; Heinrich, 1987a) found *no* evidence for any regulation of heat production whatsoever in flight. How, then, was body temperature regulated in flight, given that at least 8 publications claimed it was accomplished by varying heat production, while just as many publications claimed it was explicable by varying heat loss only?

It is probably not proper, nor is it my intention, to delineate here all the data, arguments, and ideas that could lead to such diametrically opposed views of how moths regulate their thoracic temperature. As discussed in detail elsewhere (Heinrich, 1974), the main problem was simple: it lay in not differentiating between "flight" and "continuous activity." The two are not identical, and "flight" and "continuous activity" cannot be interchangeably used to include on-off wing movements (or amputated wing-stub movements), bouts of shivering, bouts of flight, continuous shivering, continuous flight, or various combinations of shivering and flight.

A Thermoregulatory Mechanism in Flight

Vertebrate homeotherms regulate body temperature roughly along the "thermostat" analogy. Internal and external thermosensors feed information on body temperature into an integrative center in the hypothalamus that has temperature setpoints and that then regulates muscle activity for heat production and controls avenues of heat loss about those setpoints. As indicated, sphinx moths require a high (above 35 °C) T_{thx} before they take off for flight, but once in flight, I shall argue, they do *not* regulate their T_{thx} about some setpoint by the thermostat analogy. Before takeoff, moths indeed shiver if T_{thx} is below some specific temperature, which is probably determined by a lower temperature setpoint of the neural "thermostat" in the thoracic ganglia. Once in flight, however, they no longer attempt to maintain T_{thx} above some minimum. Instead, T_{thx} either falls or rises because of the heat produced as a by-product of the flight metabolism alone. However, at some point during flight at high T_a, potential overheating occurs and a heat-loss mechanism is activated as a high-temperature setpoint is finally encountered. These somewhat surprising conclusions were first worked out as follows in the sphinx moth *Manduca sexta* (Fig. 1.3).

Like other sphinx moths, *M. sexta* are hoverers, and in the field they fly continuously without stopping whenever they are active. Previous studies claiming that sphinx moths regulated their T_{thx} "in flight" by regulating heat production had, for reasons of experimental utility, used moths that were tethered by short thermocouple leads or moths that had their wings cut off so that they might more easily be accommodated into very small flasks during measurements. Such moths could indeed turn heat production on and off (by beating their wings in shivering or flight modes or by not beating them). A moth in the field, however, must produce heat continuously to fly. It cannot stop beating its wings and still be "continuously active."

In order to eliminate behavioral on-off heat production resulting as an experimental artifact, it was necessary to provide an experimental protocol in which the effect of flight duration on T_{thx} was controlled. Following the ideas of the researchers on locust physiology (see Chapter 4), I built a flight mill on which moths could be flown in circles while at the same time T_{thx} could be *continuously* recorded. With this device it was easy to show that the moths indeed heated up in flight, as expected. But the suspended moths already initiated flight (as opposed to warm-up) behavior at T_{thx} less than 20°C. Furthermore, there was very little indication that T_{thx} was stabilized at any one specific temperature setpoint, as the numerous papers had reported. In other words, these data (as far as they went) indicated that the moths did not thermoregulate at all in flight. I was almost ready to give up the seemingly wild-goose chase on "the moth problem." But I still wondered if the results from the flight mill might, like those of some previous studies, *also* be an experimental artifact. It was obviously necessary to measure the T_{thx} of continuously active moths in the field to find out if thermoregulation was a real biological phenomenon or only a laboratory artifact.

Sphinx moths are usually not common nor easy to capture, but I was fortunate to "grab and stab" several white-lined sphinx moths, *Hyles lineata*, in the Anza Borrego desert of southern California. My few measurements were exciting: T_{thx} of animals in the field were even higher than any other previously measured in the laboratory. Animals at ambient temperatures of 16°C at night already had T_{thx} of about 41°C. I subsequently measured nearly identical T_{thx} of other *H. lineata* flying in sunshine at higher ambient temperatures. These data revived in me the idea that maybe the moths could, indeed, regulate their T_{thx} in *free* flight, where rates of energy expenditure (it turned out) were nearly double those of moths flying while suspended by a tether. In effect, the moths on the flight mill had been flying under conditions where the effects of gravity had been reduced (see also May et al., 1980), which meant that they did not have to work to stay aloft.

Measurements of T_{thx} by the perhaps crude (but telling) "grab-and-stab" technique (Fig. 1.11) then confirmed that the sphinx moth *Manduca sexta* also regulates its T_{thx} in *free* continuous flight (in a temperature-controlled room) where work output remained high. Strangely, however, there was no evidence for any regulation of heat production; the metabolic rate of flying moths was, at about 45 ml O_2/g body weight/hr, the same at all air temperatures (Fig. 1.12). The independence of metabolic rate (i.e., heat production) as a function of ambient temperature during flight

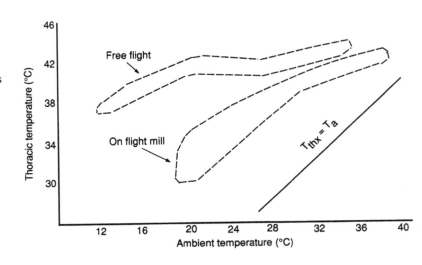

Fig. 1.11 Thoracic temperature in *Manduca sexta* as a function of ambient temperature. The upper sausage shows T_{thx} in free flight while the lower sausage shows the T_{thx} of animals flying in circles while suspended on a flight mill (from Heinrich, 1971a).

THE HOT-BLOODED INSECTS

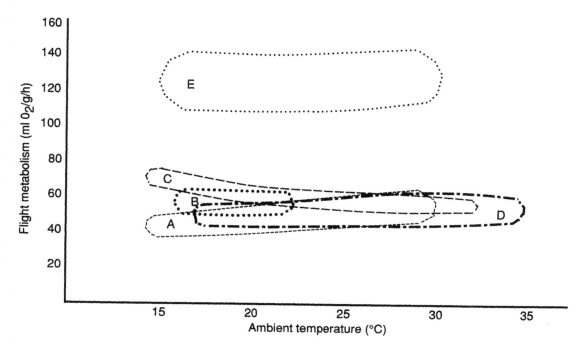

Fig. 1.12 Energy expenditure during free flight of *(A) Manduca sexta* (Heinrich, 1971a), *(B) Deilephila elpenor* and *Hyles euphorbia* (Heinrich and Casey, 1973), *(C) Hyles lineata* (Casey, 1976a,b), *(D)* male gypsy moths (Lymantriidae) *Lymantria dispar* (Casey, 1980), and *(E)* tent caterpillar moths (Lasiocampidae) *Malacosoma americanum* (Casey, Hegel, and Buser, 1981).

has now been confirmed for a great range of different moths and other insects that thermoregulate during free flight.

Despite their constant heat production, the *Manduca* moths maintained a difference of 26°C between T_{thx} and ambient temperature at 12°C and a difference of only half that (or 13°C) at 30°C. The greater temperature gradient between T_{thx} and the ambient air temperature at the low as opposed to the high air temperatures meant that there is greater passive-convective heat loss (in this case, two times greater) from the thorax at the low than at the high temperatures. But the *total* rates of heat loss from the thorax are necessarily identical and equivalent. This observation isn't parenthetical or trivial: at high ambient temperatures T_{thx} rises only slightly whereas rates of heat production remain high, the same as rates at low air temperature. It follows that the balance of the heat loss from the thorax at high air temperatures must be accounted for by some mechanism besides passive convection, and this mechanism apparently becomes activated at increasing body or air temperatures. The mystery to be solved was this: What is the mechanism of heat loss by which body temperature is regulated during flight?

As far as was known, insects could not vary the insulating capacity of their "fur" on the thorax by piloerection, thus probably precluding a major mechanism of thermoregulation used by vertebrate animals to vary heat loss. Regulation of blood flow to vary heat loss and evaporative cooling are other potential mechanisms, but Church (1959b) in his analysis of various endothermic insects had (prematurely) argued that "circulation of the haemolymph during flight contributes little to heat flow."

Morphology is the handmaiden of physiology, and some of the emerging physiological principles of sphinx moths were illuminated by the anatomical studies of Franck Brocher in France. Brocher (1920) had made a detailed study of blood circulation in the sphinx moth *Sphinx convolvuli*. By injecting India ink into the dorsal side of the abdomen near the heart (Fig. 1.13) of moths that were depilated (so that he could see the fate of the dye through the transparent cuticle), he showed that the heart pumped blood anteriorly into the thorax. The heart, he showed, makes a loop through the center of the thorax (Fig. 1.14). In view of the recent knowledge that the moths in flight must somehow transfer heat away from their overheating thoracic musculature, Brocher's studies suggested to me the hypothesis that the aortic loop in the thorax could serve as a cooling coil. As the blood is heated in the thorax and pumped throughout the body it could, at least potentially, dissipate the heat through the head or the abdomen, the latter of which could act both as a heat sink and as a heat radiator. The abdomen in insects does not contain any major muscle mass capable of appreciable heat production. Abdominal temperature should thus remain low during flight, as it does during warm-up (Figs. 1.2 and 1.7). The idea that the moths use the abdomen as a heat radiator seemed on balance the most reasonable hypothesis—the one most worth investing in to test.

If heat is dissipated into the abdomen via the blood, then temperature in the ventrum of the abdomen, where blood from the thorax enters, should rise. The ventral diaphragm in the abdomen pushes blood posteriorly and to the sides, which means that the T_{abd} should decline posteriorly while most of the heat is lost anteriorly. After having been cooled, the blood enters the heart and is then pumped into the thorax; therefore dorsal abdominal temperatures should be relatively low and uniform. These temperature profiles of the abdomen were observed in flying moths

THE HOT-BLOODED INSECTS

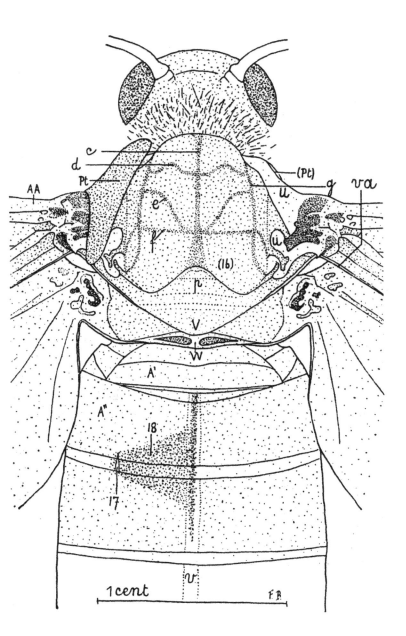

Fig. 1.13 Dorsal view of a depilated *Sphinx convolvuli.* The number 17 indicates the site of injection of India ink, and stippling traces the path of the black ink particles. (From Brocher, 1920.)

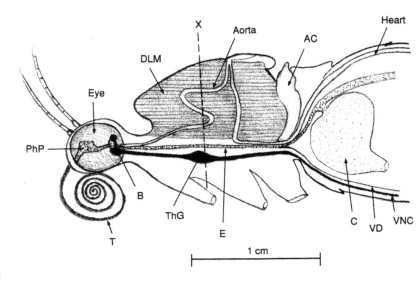

Fig. 1.14A Sagittal section through the thorax and the anterior portion of the abdomen of *Sphinx convolvuli*, showing the heart, aorta, dorsal longitudinal muscles *(DLM)*, ventral diaphragm *(VD)*, and air chamber *(AC)*. The other structures illustrated here are not directly related to thermoregulation (*T* = tongue, *PhP* = pharyngeal pump, *B* = brain, *ThG* = thoracic ganglion, *E* = esophagus, *C* = crop, and *VNC* = ventral nerve cord). The dashed line labeled *X* indicates the position of the transverse section illustrated in Fig. 1.14B. (From Brocher, 1920.)

Fig. 1.14B Cross section of *S. convolvuli* thorax (see Fig. 1.14A for position of section and identification of labels). The aorta, because of its loops, is exposed three times in this view. The dorso-ventral muscles *(DVM)*, not visible in the sagittal section of Fig. 1.14A, act as wing elevators. (From Brocher, 1920.)

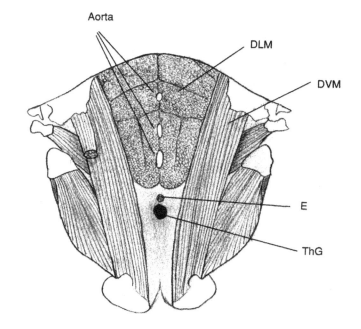

THE HOT-BLOODED INSECTS

(Heinrich, 1971b) as well as in tethered moths heated on the thorax to mimic the heat generation that normally occurs in flight (Fig. 1.15). As T_{thx} approached 44 °C, T_{abd} began to rise, first in the ventrum of the abdomen and then also dorsally. All of the dorsum of the abdomen heated to a uniform temperature, thus excluding the possibility that all of the T_{abd} rise was due to passive heat conduction. These experiments showed that, in a stationary moth, the abdomen could indeed serve as a radiator by dissipating heat from the thorax via the blood circulatory system. Furthermore, whereas the thorax is heavily insulated, the abdomen is very lightly insulated (Heinrich, 1971b; Casey, 1976b), so it can serve as a thermal window.

Heat transfer measured in a stationary moth, as such, did not yet imply similar thermoregulation in flight. The key to the problem of thermoregulation still lay in examining what happened in continuous *free* flight. And that presented what seemed an insurmountable problem: how do you measure blood flow and heat transfer in a sphinx moth while it is flitting about like a hummingbird? While I was still struggling with this problem, I heard William A. Calder III give a seminar on the mechanics of canary song. He used the toy of the day, an "impedance pneumograph," which converted mechanical movement occurring between two electrodes to electrical signals. Might one be able to measure both the amplitude and frequency of a moth's heart by placing one

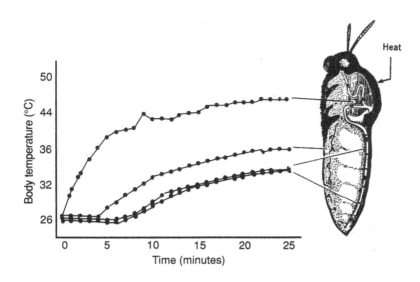

Fig. 1.15 Continually recorded body temperatures of a tethered *M. sexta* moth as heat was applied to the thorax to mimic the heating normally resulting from the flight metabolism. Lines on the sagittal view of the moth point to the positions where the 4 thermocouples were implanted. Not shown here are data for a moth killed by injection of ethyl acetate and then heated as before. The initial rate of T_{thx} increase was the same as in the live moth, but after 12 minutes the dead moth had a T_{thx} of 53 °C and T_{abd} was still at 26 °C (from Heinrich,

electrode on each side of the heart, hitching the electrodes to the impedance converter, and picking up the signal on a multichannel chart recorder into which one would simultaneously feed signals from various thermocouples in different portions of the moth? If that were possible, one could then test if the moths thermoregulate by transferring hot blood to the abdomen. One could not do this in a free-flying moth. But by heating a stationary moth on the thorax, one would be doing even better: one could actually separate flight as such from the overheating response resulting from flight.

Not being an electrophysiologist, I had my doubts in setting up the experiment. But I knew that if it went as I did not even dare to predict, then it would prove something very novel. It would be a startling result that would have to cause a re-evaluation of well-established ideas in the field. It would also come as an intense relief to me after two long years of a frustrating search for a novel thesis problem. The stakes were more personal, too. I had just learned that another student, at another university, had been working also for two years on the same problem with the same species; I might get scooped.

With some anxiety I thus watched the blue ink flow onto the green-lined chart as the signals—thoracic and 3 abdominal temperatures (Fig. 1.15) and heart rate and heart amplitude (Fig. 1.16)—were all being simultaneously recorded as I heated the moth's thorax with the narrow beam of a heat lamp. I could also visually observe the beats of the greenish-colored heart, because I had scraped the scales from the transparent cuticle of the dorsum of the abdomen. The data came out precisely as I had hardly dared to imagine. I was electrified with excitement, but still I refused to believe what seemed too good to be true: the moth appeared to prevent itself from overheating by pumping heat with its blood to the abdomen. This particular moth, I knew, would now change my life. Hundreds of disparate details suddenly clicked together into an unexpected pattern of reality, no longer a figment of mere fancy. (After the run I carefully killed the moth by an injection of ethyl acetate, and I left all the wire thermocouples and electrodes in place in its body. I then cut the wire leads and tied them together with a small red thread. I knew I'd keep this historic moth for the rest of my life). A few more repeat runs that day and the next on other moths showed that the first one was not a fluke, so I consolidated the results in a paper I sent to *Science*

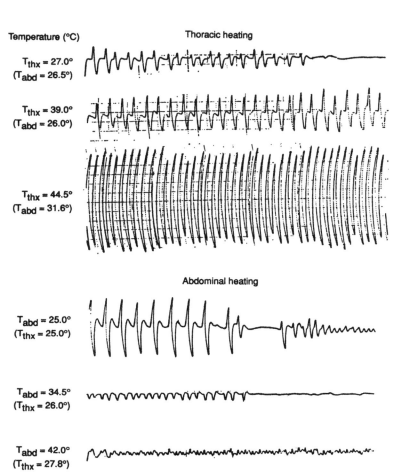

Temperature (°C)

Thoracic heating

$T_{thx} = 27.0°$
$(T_{abd} = 26.5°)$

$T_{thx} = 39.0°$
$(T_{abd} = 26.0°)$

$T_{thx} = 44.5°$
$(T_{abd} = 31.6°)$

Abdominal heating

$T_{abd} = 25.0°$
$(T_{thx} = 25.0°)$

$T_{abd} = 34.5°$
$(T_{thx} = 26.0°)$

$T_{abd} = 42.0°$
$(T_{thx} = 27.8°)$

Fig. 1.16   Records of the heart beats of a tethered *M. sexta* being heated on the thorax *(top)* or on the abdomen *(bottom)*. Unlike previous graphs of muscle activity, which recorded electrical activity, these traces are records of *mechanical* activity, or heart beats. (From Heinrich, 1971a).

(Heinrich, 1970b) to proudly announce what to me to this day is still my most memorable "Eureka" experience.

Despite what seemed to me overwhelming evidence for the new mechanism, Professor Franz Englemann, co-chairman of my thesis committee, remained unconvinced when I enthusiastically thrust the discovery upon him. "*If* the circulatory system is causing the heat transfer and thermoregulation," he said, "then you have to eliminate that system and demonstrate that heat transfer and thermoregulation are both abolished." Franz was known to be hard to please, and this was a tall order, because the circulatory system obviously has numerous other vital functions besides the one I might want to test. If you destroy those you likely destroy the whole animal. And dead or dying animals don't thermoregulate. Nevertheless, I reasoned that perhaps one could *temporarily* abolish the blood circulation at precisely that time when it is needed to distribute excess heat, simply by tying the heart shut just before flying a moth.

The results of flying moths under the above condition, an induced "cardiac arrest," were dramatic: those that flew sufficiently long for T_{thx} to stabilize (at least 2 minutes), showed no thermoregulation at all; their temperature excess was approximately 24°C at all air temperatures. Consequently, they overheated to near lethal (near 46°C) temperatures already at relatively low ambient temperatures. It could be proven that the moths stopped flying because of an inability to dissipate excess heat: when their "fur coat" was removed from the thorax, the cardiac-arrest moths could again fly, but they did so with a temperature excess of only 16°C rather than 24°C (Fig. 1.17). These experiments showed that the dorsal aorta is the critical organ for regulating T_{thx} and for allowing flight at ambient temperatures from at least 12°C to 35°C. Without their thermoregulatory capacity, sphinx moths (most species are of tropical distribution) would be restricted to flying only at relatively low ambient temperatures. The experiment was not unlike pinching shut the radiator hose of a car and then trying to drive it through the desert. The car *could* run, but only a short distance.

The volumes of blood pumped to transfer heat have so far not been measured, but calculations of the minimum volumes required suggest that the circulatory system could indeed be adequate to accomplish most of the observed thermoregulation in flight (Fig. 1.18). Relatively modest volumes of blood can transfer enough heat to account for the observed thermoregulation, provided the temperature gradient between T_{thx} and T_a remains sufficiently high (at least 8°C).

The regulation of T_{thx} could theoretically be achieved automatically. As ambient temperature increases, the abdomen necessarily gets warmer, and the warmed heart could simply pump more rapidly to transfer more heat, and then it would become warmer still and transfer even more heat, and so on. Such automatic "regulation" following the Q_{10} response (the rate increase produced by a temperature increase of 10°C) likely occurs to a limited extent, because the pumping rate of the heart of resting moths is strongly dependent on temperature (Heinrich, 1971b). However,

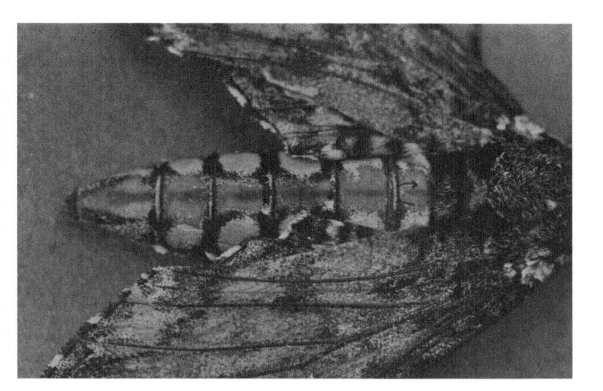

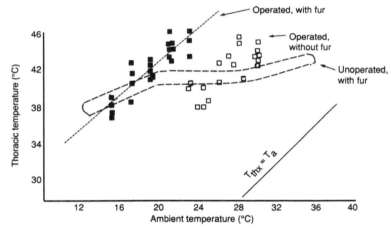

Fig. 1.17A The scales have been removed from the top of this *M. sexta* moth to show the heart through the transparent cuticle. The curved arrow at the first abdominal segment shows where the heart was ligated to produce "cardiac arrest."

Fig. 1.17B Thoracic temperatures of *M. sexta* in free flight as a function of ambient temperature for moths in which the aorta was tied off with a fine thread. The 23 filled squares plot the data for animals that have been operated on in this way but whose fur was left intact; the 14 open squares plot data for moths that were operated on and whose thoracic insulation was removed (from Heinrich, 1970a). For comparison, the range of thoracic temperatures for unaltered moths in free flight is also included.

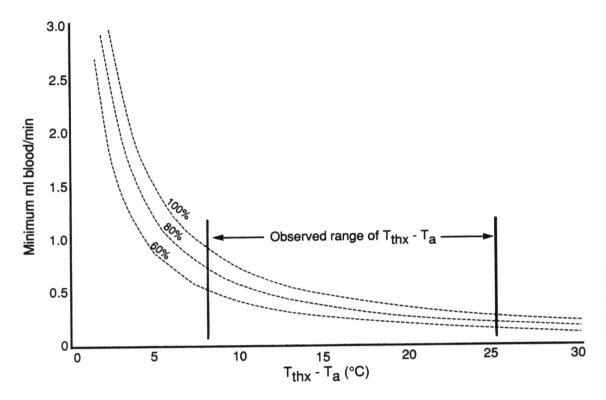

Fig. 1.18 Minimum calculated volumes of blood that must be pumped through the thorax at different levels of temperature excess (thoracic minus ambient temperature) if stabilization of T_{thx} were by blood circulation alone. The three lines represent the volumes required if 100, 80, or 60 percent of the total energy expended for flight were converted to heat (from Heinrich, 1971b).

the abdominal heart starts vigorous, high-amplitude pumping movements even if abdominal temperature remains low, provided T_{thx} exceeds 42°C (Fig. 1.16), and the heart stops pumping if it is heated (that is, if T_{abd} is raised) and T_{thx} remains low. Furthermore, the abdominal heart ceases to respond to heating in the thorax if the ventral nerve cord between thorax and abdomen is severed (Heinrich, 1970b; Hanegan, 1973). Therefore, in addition to the "automatic" temperature response of the heart there is also a neural control mechanism that depends on the temperature of the *thorax.*

The thoracic ganglia control the coordination of the various flight muscles, but the temperature sensors that affect both the shivering vs. flight response of these muscles and the pulsations of the heart in the abdomen to prevent overheating of the thorax could reside also in the thoracic ganglia, or they may be located elsewhere and send their input to the thorax. Experiments by James L. Hanegan (1973) and Hanegan and James E. Heath

(1970c) suggest, however, that in the cecropia silk moth, *Hyalo-phora cecropia*, the temperature sensor(s) that affect both heart activity and shivering vs. flight responses of the muscles reside in the thoracic ganglia.

Hanegan and Heath heated the moths' thoracic ganglia by placing them in contact with an insulated thermistor probe that was in turn heated by driving electric current through it. Alternatively, the ganglia were cooled (or heated) by touching them with the tip of a silver wire that was cooled (or heated) by pumping water through a tube containing all but the tip of the wire. By the above methods Hanegan and Heath were able to heat or cool the ganglia independently of thoracic-muscle temperature, and they demonstrated that the transition from the warm-up motor pattern (synchronous activation of the antagonistic flight muscles) to the motor pattern (Fig. 1.4) characteristic of flight (alternate activation, resulting in wing beats rather than shivering) depends on the temperature of the thoracic ganglia independently of thoracic-muscle temperature (Hanegan and Heath, 1970c). Of course, in intact moths the temperature of the muscles and the ganglion normally vary in parallel, and ganglion temperature is normally a reliable indicator of muscle temperature.

Mechanical activity of the heart was not measured, but electrical potentials were recorded instead (Hanegan, 1973) and their frequency described as "heart rate." However, unlike flight-muscle potentials (which are reliable records of flight-muscle activity), electrical potentials from a heart do not accurately depict heart rate (because the heart is controlled by several sets of different muscles affecting constriction, dilation, and elongation). "Heart rates" in the cecropia moths using electrical potentials as a criterion were erratic and correlated neither with T_{thx} nor T_{abd}, nor did external heating of the thorax increase heart rate (i.e., frequency of electrical potentials). Moreover, the thoracic heating experiments were restricted to 25–40°C, and these temperatures are unlikely to be high enough to cause a heat-dissipation response. On the other hand, heating of the thermistor (and the ganglion) to 48°C resulted in a maximum response from the heart—up to 180 potentials per minute (Hanegan, 1973). Cooling of the ganglia resulted in a reduction of the electrical heart response, but only if the neural connection between thorax and abdomen was left intact. (See the caption to Fig. 1.4 for a brief explanation of electrical potentials.)

It would be of interest to repeat the above experiments in a sphinx moth known to regulate its T_{thx} in flight, and to measure the *mechanical* heart activity (heart-beat amplitude and frequency) as well. However, rapid heart beat, as such, need not imply that any pumping occurs, because lateral heart beats as such cannot propel blood. Pumping requires both lateral and forward motions for the propagation of a (generally) posterior-to-anterior peristaltic contraction of large amplitude.

Many of the integrative functions in insects for controlling locomotion are regulated by the thoracic ganglia (Wilson, 1968), while the sensory neurons of the head lead to a head ganglion or "brain" that integrates the information from eyes, antennae, and mechanoreceptive facial hairs. The motor patterns resulting in flight control do not require signals from the brain (Kammer, 1971), but the sensory input that affects the flight movements probably do. The head therefore plays an important role in integrated flight.

Head temperature may also be important in neural coordination, but the head is potentially another thermal window for dissipation of excess heat from the thorax. Timothy M. Casey and students from Rutgers University examined this possibility. In both the sphinx moth *M. sexta* (Hegel and Casey, 1982) and the lasiocampid moth *Malacosoma americanum* (Casey, Hegel, and Buser, 1981), head temperature is relatively independent from ambient temperature (Fig. 1.19). Probably because of the close proximity of the head to the thorax, T_{hd} unavoidably follows T_{thx}. More than passive heat conduction is involved, however, because T_{hd} of live moths is consideraly higher than in dead moths heated to the same T_{thx} (Hegel and Casey, 1982). Furthermore, as the thorax in some (live) *M. sexta* is heated to near 40°C, cyclic fluctuations in head temperature result, and these fluctuations are concurrent with decreases of both T_{thx} and heart activity. Injections of ethyl acetate to kill the animal immediately stop both the temperature fluctuations and the heart activity and result in increases of T_{thx} and decreases of T_{hd}.

Given that moths transfer heat into the head from the thorax (which increases T_{hd}), do they also use the head as a thermal window to dissipate excess heat from the thorax? The evidence so far does not support this hypothesis: If the head acts as a thermal window to dissipate heat for thermoregulation, then the animals would need to dissipate *more* heat from it at high ambient temperatures than at low. But in fact the opposite occurs. Considerable heat *is* lost from the head—after all, it is small and only

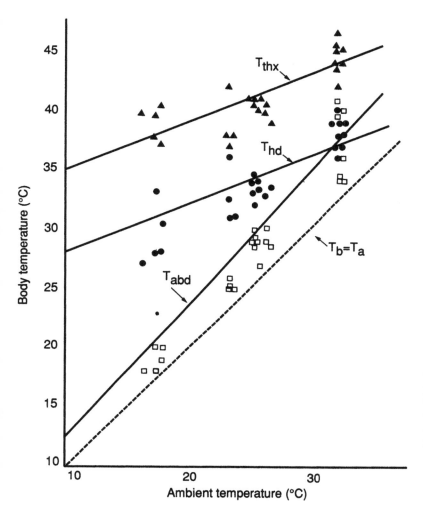

Fig. 1.19 Body temperatures as a function of ambient temperature in *Manduca sexta* in free flight. Note that head temperature is relatively independent of ambient temperature, as is T_{thx}, whereas the temperature excess of the abdomen increases with increasing ambient temperature. (From Hegel and Casey, 1982.)

lightly insulated and therefore it cools much more rapidly than the thorax (Hegel and Casey, 1982)—but there is no increased thermal gradient between T_{hd} and T_{thx} at higher than at low ambient temperatures, as there is between T_{abd} and T_{thx}. A higher gradient would be necessary to increase convective heat loss at high ambient temperatures. Instead, the gradient is in the opposite direction (Fig. 1.19). The head is therefore not used as a thermal window for thermoregulation (as the abdomen is).

The work of Casey (1976a,b) on the sphinx moth *Hyles lineata*, which was initially thought to regulate its T_{thx} in flight by adjust-

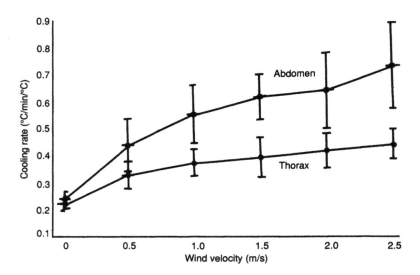

Fig. 1.20 Cooling rates of *Hyles lineata* body parts as a function of wind velocity. Note that the abdomen cools nearly twice as fast as the thorax does. Cooling rate is measured as the change in T_b per minute, per each degree difference between T_b and T_a. (From Casey, 1976b.)

ing heat production, confirms the idea that T_{thx} in this moth is regulated as it is in *M. sexta*. *Hyles* in free flight also has a metabolic rate nearly independent of ambient temperature (Fig. 1.12) and, as in *M. sexta*, when T_{thx} is regulated in flight the temperature excess of the abdomen increases at high ambient temperatures, as would be predicted if the abdomen serves as a heat dissipater. Also as in *M. sexta*, in *H. lineata* the abdomen is lightly insulated, and very small increases in wind speed have dramatic effects on abdominal cooling rates (Fig. 1.20). During flight, then, when air movement over the body is rapid, even modest increases of T_{abd} can result in large rates of heat loss.

At least theoretically sphinx moths could (as honeybees do, see Chapter 8) lose heat through evaporation from the mouthparts. Indeed, Adams and Heath (1964a) observed a resting sphinx moth *Pholus achemon* exuding a drop of fluid from its proboscis, and this stationary moth had a faster thoracic cooling rate than predicted by passive cooling. But this incident does not prove the moth cooled by this method. As we have just shown, sphinx moths have a physiologically facilitated mechanism of thoracic cooling unrelated to evaporative cooling. This mechanism is apparently employed to speed up cool-down immediately upon cessation of flight (Bartholomew and Epting, 1975b). Hegel and Casey (1982) at no time observed *M. sexta* regurgitate fluid, either while flying at high ambient temperatures or when grounded and externally heated.

The Energetics of Pre-Flight Warm-Up

Like other moths (Casey and Hegel-Little, 1987), *M. sexta* shiver to produce heat during warm-up at rates near the physiological maximum (Heinrich and Bartholomew, 1971). Since heat output at any moment depends strongly on muscle temperature (Fig. 1.21), which determines wing-beat frequency (Fig. 1.9), the initial rate of warm-up is greater the higher the ambient (and body) temperature.

As warm-up proceeds, the rate of heat production (and wing-beat frequency) increases sharply. Because of the increasing temperature gradient between T_{thx} and T_a, however, the rate of passive convective heat loss also increases. Obviously, the rate of heat production remains much greater than the rate of heat loss throughout a warm-up, hence T_{thx} continues to rise. That rate of T_{thx} increase continues to be much more rapid throughout any

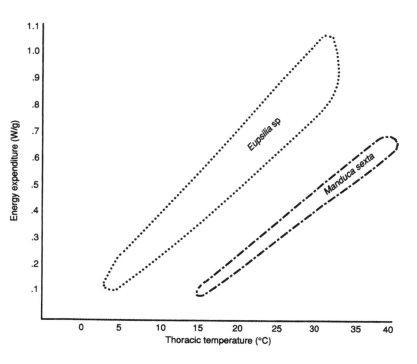

Fig. 1.21 Energy expenditure in watts / gram thorax of *Manduca sexta* (Heinrich and Bartholomew, 1971) and *Eupsilia* sp. (Heinrich, 1987a) during pre-flight warm-up at various temperatures from 0 °C to 30 °C. Energy expenditure is calculated from body-heat flux derived from body mass, specific heat, and curves of warm-up and passive cooling.

warm-up from high than from low T_a, because at a given T_{thx} the temperature gradient (and hence the rate of heat loss) is much greater at the lower T_a.

Heat production during warm-up must exceed heat loss, or else T_{thx} would not increase at all. Indeed, if heat production is not continually greater than heat loss, then all the energy expended in shivering is energy lost. It is apparent, therefore, that the moths "should" shiver at the maximum rates that they are capable of, and without pause, throughout any one warm-up cycle. They must do so to maximize heat storage and minimize heat loss, and thus to economize in the energy costs for any one warm-up.

The absolute energy costs of warm-up had originally been calculated for moths (Fig. 1.21) on the basis of heat storage in the body and heat loss to convection (Heinrich and Bartholomew, 1971). But such methods underestimate total heat exchange because, being derived from measurements of T_{thx}, they ignore heat loss and heat storage of the abdomen and the head. (These heat losses are low during warm-up, but they are not negligible.) With newer, improved instruments it is now possible to derive instantaneous measurements of energy expenditure during pre-flight warm-up (Bartholomew, Vleck, and Vleck, 1981; Casey and Hegel-Little, 1987). These measurements confirm previous findings that heat production during warm-up is unrelated to ambient temperature and strictly dependent on muscle temperature. Wing-beat frequency rises throughout warm-up as a strict function of T_{thx} (Fig. 1.9) and although the rate of heat production rises sharply throughout warm-up, the energy expenditure per muscle contraction remains relatively constant over wide ranges of muscle temperature. The total cost of warm-up may increase close to an order of magnitude from the highest to the lowest ambient temperature where the moths will shiver to warm up.

The large energy cost of warm-up from low ambient temperature is due primarily to convective heat loss, as the gradient between T_{thx} and T_a increases. All of the energy expended for a pre-flight warm-up would be lost in the insect that starts to shiver at a T_a that is too low for it to achieve a sufficiently high T_{thx} for flight. No wonder most insects are very reluctant to attempt warm-up at very low temperatures, even at temperatures above the point at which shivering becomes physiologically possible.

There is often considerable individual variation between animals' willingness to begin shivering warm-up. Some individuals can shiver more vigorously than others, and thus there is also variation in the capacity to achieve a sufficiently high T_{thx} for flight at low T_a. As far as I know, however, no study has attempted to answer whether or not there is a correlation between temperature threshold of warm-up, vigor of warm-up, and ability to achieve flight temperature. Do individual moths of different shivering capabilities "know" whether they are vigorous enough to warm up before they waste considerable amounts of energy in trying?

Abdominal temperature could also affect energy economy. Like many other moths (Dorsett, 1962; Bartholomew and Casey, 1973; Heinrich and Casey, 1973; Casey, Hegel, and Buser, 1981; Heinrich, 1987a,b), *M. sexta* shiver until their T_{thx} reaches 37–39°C (Fig. 1.7) when warming up at all air temperatures from which warm-up can occur (Heinrich and Bartholomew, 1971). Takeoff temperatures are generally several degrees above minimum flight temperature, as would be expected since the moths are required to have sufficient power to lift off vertically as opposed to merely having to maintain level flight.

Given the high energy cost of warm-up, energy economy during this process would be enhanced considerably if heat were sequestered in the thorax and prevented from leaking into the abdomen. Indeed, in almost all insects so far examined that have the capacity to shunt heat into the abdomen, T_{abd} remains low during preflight warm-up (Fig. 1.7). Since even very small quantities of blood leaking into the abdomen can result in considerable heat transfer, particularly at high T_{thx} (Heinrich, 1971b), and since resting moths that no longer show an elevated T_{thx} often have conspicuous heart pulsations in either direction (Wasserthal, 1981), it is likely that blood circulation is minimized during warm-up to reduce potential heat flow. Head temperature, in contrast, is elevated (Fig. 1.7) during warm-up by physiologically facilitated heat flow (Hegel and Casey, 1982) to the head, at some cost to thoracic temperature increase.

One of the simplest and most direct mechanisms of stopping heat flow to the abdomen is probably to shut off or greatly reduce the blood flow. So far only a few indirect suggestions that this occurs are available. But at least in *M. sexta* heart pulsations indeed appear to be suppressed during warm-up (Heinrich and Bartholomew, 1971).

The energy balance sheet changes instantly after the cessation of activity. At the end of flight a high T_{thx} no longer serves any purpose. Instead, a high T_{thx} now drains valuable energy resources because of the elevation in resting metabolism due to the Q_{10} effect. To be sure, passive-convective cooling is rapid. But many researchers have additionally reported physiologically facilitated cooling immediately upon cessation of flight or warm-up (see Heinrich and Bartholomew, 1971; Bartholomew and Epting, 1975a,b), suggesting either a passive relaxation of the heat-retention mechanism probably operative during warm-up (or during flight at low ambient temperatures) or an active mechanism of heat dissipation for energy economy (Bartholomew, Vleck, and Vleck, 1981), which could be based on at least two different mechanisms.

The moths' dissipation of heat into the abdomen at the end of flight could have another function besides reducing T_{thx} to capitalize on the Q_{10} effect for energy economy. For example, sphinx moths (whose primary fuel is lipid in flight; Zebe, 1954) consume oxygen at rates above resting levels at *given* T_{thx} during post-"flight" cooling, which is thought to indicate that they are paying off an oxygen debt (Bartholomew, Vleck, and Vleck, 1981). However, these studies refer to cool-downs after bouts of *shivering*, not bouts of flight. The difference could be significant.

Apparently the moths use that carbohydrate already within the thorax during warm-up, sparing extrathoracic stores (Joos, 1983). They should therefore, during *warm-up*, be able to reduce blood flow for fuel mobilization from the abdominal depots and thus help to economize warm-up costs *by reducing abdominal heat loss*. Thus, the post-warm-up abdominal heating could be a by-product of the circulation related to replenishing the thoracic fuel. If so, there should be no post-*flight* abdominal heating, because moths in flight are adequately ventilated by thoracic pumping and presumably have no oxygen debt. In this scenario involving post-flight cooling, the energy economy would accrue during the warm-up by minimizing shivering, rather than later during the cool-down after flight by minimizing the rate of passive metabolism.

Efficiency and Heat Production

Flight is an almost totally aerobic process in insects, and measurements of oxygen consumption or CO_2 production (Kalmus, 1929) during warm-up and flight provide a good measure of chemical-energy input. The energy output appears in heat ("inefficiency") and in mechanical work. Since the efficiency of insect muscle is likely to be relatively constant between different species and between different activities, measured rates of oxygen consumption, which have traditionally been used to measure energy expenditure, have also been legitimately used as direct indicators of rates of heat production to answer comparative physiological questions. But how much heat is actually produced in absolute terms?

If sphinx moths, and other insects, expend energy in flight only to fly, and if the work that the wings must do can be calculated from aerodynamic models, then the absolute amounts of heat produced can be calculated if measurements of oxygen consumption are available. Such measurements have suggested that the flight-muscle efficiency (external work done/energy expended) for sphingids is near 20 percent (Casey, 1981b,c). However, subsequent studies suggest that insect flight muscles convert metabolic energy into considerably less mechanical work than previously thought. Using improved aerodynamic models, Ellington (1984, 1985) calculates that for both fibrillar and non-fibrillar muscles, efficiency is likely only 5–10 percent. These latter values are also near those derived for the sphinx moth *M. sexta* during free hovering flight (Stevenson and Josephson, 1990) and for the work performed by isolated metathoracic tergosternal muscles of the locust *Schistocerca americana* (Josephson and Stevenson, 1991). Estimates for overall muscle efficiency for a number of other animals are also close to 10 percent (references in Stevenson and Josephson, 1990).

What happens during shivering warm-up? It should be kept in mind, I believe, that the above measurements of "efficiency" were derived relative to work actually produced in flight or in work mimicking flight behavior. The total inefficiency that was measured could thus be attributable to any of several steps. There could be mechanical inefficiency in muscle contraction, the interface between muscles and wings (the wing hinge), and the interface between wing and air (the aerodynamic flow as affected by wing shape, size, flexibility, and wing movement). The shivering moth (producing almost *no* net mechanical work against the air or against other media) is 100 percent "inefficient" as far as the production of external work is concerned. Since the *flying* moth is already 90–95 percent "inefficient," however, the shivering moth presumably produces only 5–10 percent more heat within the muscles per given volume of oxygen consumption than does a flying moth.

Size and Morphometrics

Moths fly with an amazing range of thoracic temperatures. Some species fly with a muscle temperature as low as 0°C and do not thermoregulate. Others generate T_{thx} near 46°C in flight (Fig. 1.22) and regulate T_{thx} within 2–3°C over wide ranges of ambient temperatures. This large range of response is related to the combination of body mass and wing loading of the insect and the thermal conditions of the animal's environment.

In most insects the limitation of body mass on endothermy is very pronounced. Moths (Heinrich and Bartholomew, 1971) and many other insects (Heinrich, 1974) having a body mass less than about 200 mg show markedly lower T_{thx} than do larger insects

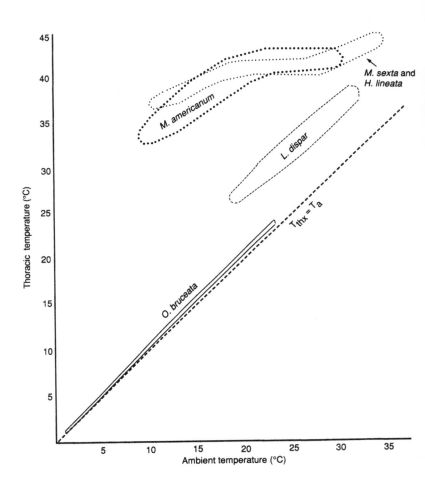

Fig. 1.22 Relation of thoracic temperatures to ambient temperature during fight for the sphinx moths *M. sexta* and *Hyles lineata* (Heinrich, 1971a; Casey, 1976b; Hegel and Casey, 1982); the gypsy moth *Lymantria dispar* (Casey, 1980); the geometrid winter moth *Operophtera bruceata* (Heinrich and Mommsen, 1985); and the lasiocampid moth *Malacosoma americanum* (Casey, 1981a).

THE HOT-BLOODED INSECTS

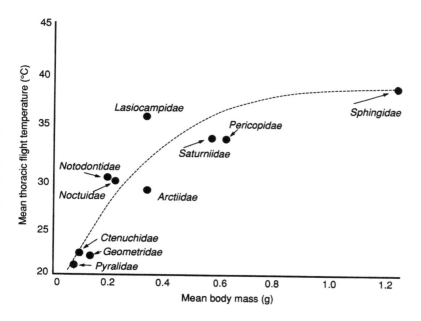

Fig. 1.23 The relation of mean T_{thx} during flight to mean body mass in ten families of moths from Costa Rica. Pyralids, notodontids, geometrids, periocopids, and lasiocampids were measured at ambient temperatures of 15–17°C; other families were measured at both 7 and 15–17°C. (From Bartholomew and Heinrich, 1973.)

and a very strong dependence of T_{thx} on ambient temperature (Figs. 1.22 and 1.23). At the extremes, a small (<15 mg) moth flies with T_{thx} within 1°C of T_a, whereas some sphinx moths weighing several grams fly with T_{thx} of more than 25°C above T_a. A comparison of the T_{thx} of some 10 families of moths from Costa Rica (Bartholomew and Heinrich, 1973) shows a strong correlation of T_{thx} with mass (Fig. 1.23). The same pattern emerges within two moth families at a temperate site (Casey and Joos, 1983). In both studies, however, there was also a strong tendency for the small-bodied moths to have low wing loading, which is the ratio of body weight to wing area (Fig. 1.24). This suggests that small moths, having difficulties maintaining a high T_{thx} in flight, may have evolved instead larger wings as a mechanism for flying with a low T_{thx} to compensate for their lower mass.

In sphinx moths with body mass in excess of 200 mg, T_{thx} reaches 40–45°C during flight. T_{thx} continues to be independent of mass in moths of this size up to the largest moths, those weighing 6 g (Fig. 1.25). In these animals body mass has little effect on T_{thx} presumably because the upper tolerable T_{thx} cannot be exceeded (Fig. 1.25). In them, the variation in T_{thx} is largely a function of wing loading alone (Fig. 1.26).

Large wings *per se* need not imply lower takeoff temperatures for flight (Marden, 1987), but large wings *permit* lower transport costs, although at the expense of lower maneuverability (which insects harvesting nectar by hovering over flowers obviously cannot afford). The very low average wing loading of many geometrids (Fig. 1.27) is achieved because some species do not feed at all (and the females who carry the eggs commonly have *no* wings). Thus all flying individuals of some species of this family may carry neither eggs nor gut. Similarly, saturniids, which have conspicuously larger wings than sphinx moths as well as lower T_{thx} and lower energy expenditures of flight (Bartholomew and Casey, 1978), also do not feed. The lower T_{thx} of the saturniids relative to the sphingids is observed despite their generally thicker insulation (Bartholomew and Epting, 1975a).

In the tent caterpillar moths *Malacosoma americanum* and male gypsy moths *Lyamantria dispar*, thoracic temperature in flight ex-

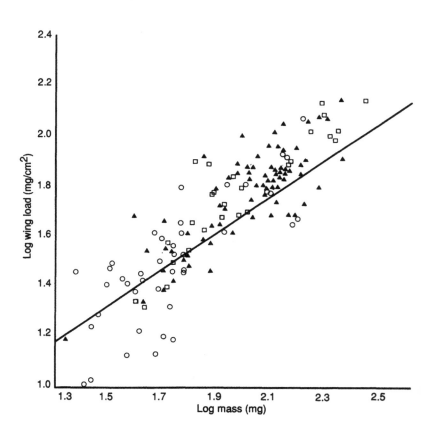

Fig. 1.24 Wing loading in relation to body mass in noctuids *(triangles)*, geometrids *(circles)*, and miscellaneous other *(squares)* moths. (From Casey and Joos, 1983.)

THE HOT-BLOODED INSECTS

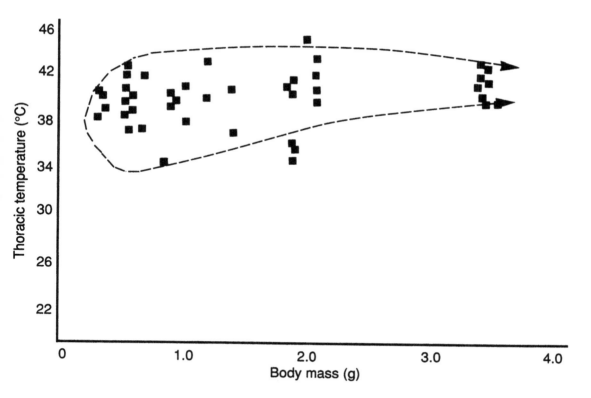

Fig. 1.25 Thoracic temperature *(squares)* of free-flying sphinx moths of 12 species (from New Guinea) in relation to mean body mass (from Heinrich and Casey, 1973) and of 12 species (enclosed by dashed line, no points) from Costa Rica (from Bartholomew and Heinrich, 1973).

ceeds ambient temperature by 27 °C and 7 °C, respectively (Casey, 1980, 1981a), even though both weigh near 100 mg. The male gypsy moths fly with a lower temperature excess because they have *both* less insulation as well as lower energy costs of flight than the smaller-winged *M. americanum;* they fly at half the wing-beat frequency but pay very nearly the same energy cost per wing stroke (Casey, 1981a).

Comparisons with Vertebrate Animals

Maximum body temperature is a function of rates of heat production and heat loss, and it can be predicted if both rates are known. Since insects are almost totally aerobic in flight, the rate of heat production can, as already mentioned, be determined relatively accurately from rates of oxygen consumption (or CO_2 production). Passive rates of heat loss can also be determined (Fig. 1.21) by a simple conversion using the animal's mass, specific

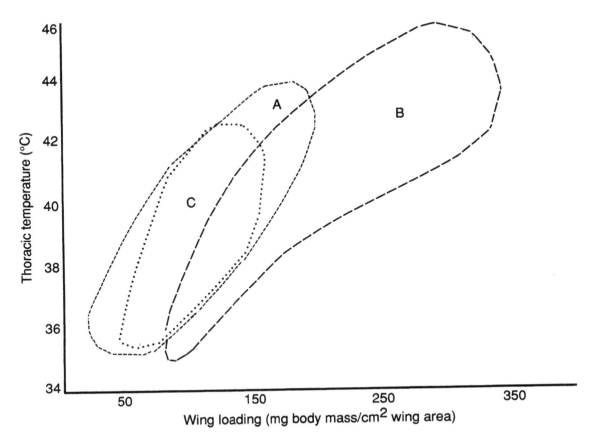

Fig. 1.26 Mean T_{thx} during free flight in relation to mean wing loading in (A) 3 genera of sphinx moths from Nigeria (Dorsett, 1962); (B) 12 species of sphinx moths from New Guinea (from Heinrich and Casey, 1973); and (C) 13 species of sphinx moths from Costa Rica (Bartholomew and Heinrich, 1973).

heat content, and passive cooling rates or conductance (°C temperature drop/minute/°C temperature difference). Given expected cooling rates at different masses, insulation can be compared for different kinds of animals.

In moths, only the thoracic weight is relevant in a comparison of conductance to evaluate insulation, because these animals heat up only the thorax whereas most vertebrate animals heat up all or most of their entire bodies. Taking this difference into consideration, Bartholomew and Epting found insulation values for both sphinx moths (1975a) and saturniid moths (1975b) similar to those predicted for homeothermic vertebrates of their size. Furthermore, rates of warm-up (Heinrich and Bartholomew, 1971), thoracic temperature in flight (Bartholomew and Heinrich, 1973; Bartholomew and Casey, 1978), and energy expenditure of flight in sphinx and saturniid moths (Bartholomew and Casey, 1978) also scale almost identically to those of bats and birds. The smallest bats, shrews, and hummingbirds have reached a lower size limit of near 3 g, and it was commonly supposed that this apparent limit was set by thermoregulatory constraints. Although this may be so for vertebrates, the data on sphinx and saturniid moths show that the same high body temperatures of 35–40°C can be regulated in animals weighing as little as 200 mg.

Evolution of Thermoregulation

Endothermy and thermoregulation have both likely evolved independently numerous times in the moths, because thoracic temperatures within the many families are similarly predictable according to common scaling parameters and independent of taxonomic affiliation. Throughout evolution, thermoregulation became necessary only in those species that, because of their scaling and flight characteristics, reached a T_{thx} close to the upper tolerable level of 45°C, which is also the upper limit of other insects and other animals. The solution to the problem of overheating then required only relatively minor modifications of the circulatory system that was already in place and used for other functions, such as fuel mobilization.

Fig. 1.27 The geometrid winter moth *Operophtera bruceata*. The male has large light wings and the female (perched below on a leaf) has no wings.

Pre-flight warm-up probably also evolved independently within the various groups. Pre-flight behavior is an obvious modification of flight in which most of the thoracic muscles are activated by the central nervous system relatively synchronously rather than alternately (Kammer, 1968).

The major indirect power-producing flight muscles—the dorsoventral (DVM) elevator muscles and the dorsal longitudinal (DLM) depressor muscles—are synchronously activated during warm-up (Kammer, 1967) in all species studied. However, some of the smaller flight muscles used in flight control show different, species-specific activations. For example, in the sphinx moth *Mimas tiliae* the subalar muscle typically fires with the DLM, both during warm-up and flight, while in *Hyles lineata*, another sphingid, these muscles are excited in antiphase (Kammer, 1968). Kammer (1970) found different species-specific patterns of activation of 8 different thoracic muscles in all 6 species of sphinx moth investigated. Function has not constrained evolution to just one pattern, and the variety of neural behavior that is possible for warm-up activity suggests that the species-specific patterns, even within the same moth family, are explicable by multiple independent evolutionary origins of warm-up (Kammer, 1970).

For the most part, heat production and endothermy are inevitable in any insect that contracts its muscles in flight and that reaches or exceeds a body mass of about 200 mg. The relevant question is not how or why insects have evolved to be endothermic. A potentially more interesting question is: Why has every insect (as far as we know) that is endothermic in flight, and that must often also dissipate heat, also evolved a warm-up mechanism? On first thought the answer seems simple enough: Insects need to warm up until the muscle-twitch durations are short enough to achieve the minimum wing-beat frequency and power output necessary to sustain flight. But why have the muscles not been modified instead to operate at a high power output at *low* temperature, in which case the costly pre-flight warm-up would not be necessary?

A simple hypothetical scenario will make the answer obvious. Suppose ambient temperatures average $20°C$ and the muscles of a large moth are adapted to operate optimally at, say, $22°C$. This moth should be able to take off instantly without any warm-up, but within several seconds of taking flight it would be subjected to temperatures many degrees above its optimum. The higher the T_{thx} the moth tolerates, the longer it could fly. Thus, if there is selective pressure for continuous flight, there is selective pressure for tolerating a high T_{thx} and/or maximizing power output at high T_{thx}. The higher the optimal T_{thx}, the less the problem of heat dissipation and the longer the possible duration of flight. In short, from the evolutionary perspective, the additional energy cost invested for pre-flight warm-up buys continuous flight.

Moths have evolved to fly with T_{thx} from $0°C$ to $46°C$. A lot is known about why one or another T_{thx} is ecologically relevant, but very little is known about the physiological mechanisms that have been selected. Several levels of organization could have been altered to allow moths to fly at different temperatures. One possibility is biochemical adaptation. For example, the thoracic muscles of winter-flying cuculiinid moths generate extracellular action potentials (Esch, 1988) at temperatures as low as $0°C$, and these insects can already start shivering at this body (and ambient) temperature (Heinrich, 1987a,b). No other moth has ever been seen to shiver at such low temperatures.

In conjunction with potential biochemical changes there are also likely adaptations of tissue morphology. For example, rates of muscle contraction are limited by calcium uptake during the contraction cycle, and that uptake occurs in the sarcoplasmic reticulum (SR). The more extensive the SR system, the more sites for calcium uptake and the more rapidly muscle contractions can be repeated. The problem with too much SR, however, is that it takes up space—it takes space that could be used for contractile proteins. In short, strength or power may be compromised for speed. In the highly endothermic sphinx moth *M. sexta*, which needs both speed and power for hovering, the sarcoplasmic reticulum is limited to the usual thin layer between the mitochondria and myofibrils, whereas the winter moth *Operophtera*, flying with a very low power output, nevertheless has *bundles* of SR (Fig. 1.28). Why the extensive SR? The geometrid moth has a much lower wing-beat frequency than the sphinx moth, so the elaborate SR system in *Operophtera* cannot be explained on the basis of rate of muscle contraction, per se. However, *Operophtera* flies with a substantially lower muscle temperature (15° to 0°C) than the sphinx moth does (38–42°C). Perhaps *Operophtera*'s highly elaborated SR system is related to the necessity to maintain muscle contractions *at very low temperature*; since calcium uptake per uptake site is low at low temperature, many slow sites compensate to maintain an overall high rate.

At the level of gross morphology, smaller moths tend to have lower wing loadings than do heavier moths (Heinrich and Bartholomew, 1971; Casey and Joos, 1983). Since lower wing loading is associated with lower energy expenditure in moths, perhaps one mechanism of flying at low muscle temperature (a necessity at small body size) is to reduce body weight or enlarge the wings, thereby enabling the moth to remain airborne, at the cost of maneuverability of flight. All of these factors, from subcellular mechanisms to body morphology, may be self-reinforcing: the reduced wight of a moth without a gut would automatically reduce the need for high T_{thx} while also reducing energy expenditure, which would in turn reduce the need for small wings for maneuverable flight useful in feeding, which would reduce the need for a high T_{thx}, and so on.

Steps in the evolutionary progression of changing wing morphology are still evident. For example, female gypsy moths in Europe and North America have wings but they do not fly. Within the Geometridae there are some species where both sexes have wings. In others the females have reduced wings, wing stubs, and in still other the females lack even wing stubs. Domesticated

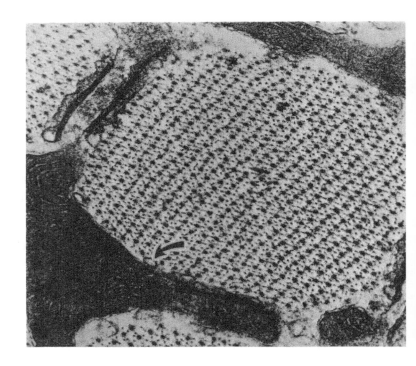

Fig. 1.28 Micrographs showing the muscle structure of
Manduca sexta (top) and *Operophtera bruceata (bottom)*. Arrows point to cross sections of
sarcoplasmic reticulum (SR)
between myofibrils and mitochondria. Note the typical pattern of a thin SR layer in *M.
sexta* vs. the large bundles of
SR in the winter moth. (Courtesy of J. H. Marden and R. L.
Anderson.)

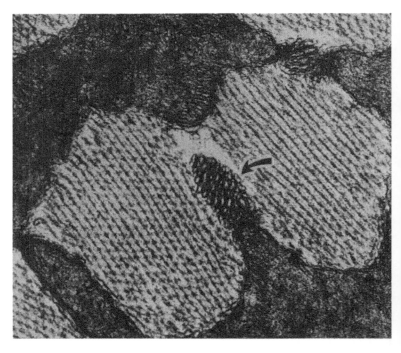

silkworm moths, *Bombyx mori*, have lost the ability to fly and the capacity (and necessity) to thermoregulate by shunting heat to the abdomen (Ploye, 1979), but they still show a limited ability to generate heat with their flight muscles.

As is true of the day-flying Lepidoptera (see next chapter), the largest species of moths are also from the lowland tropics. Apparently, Bergmann's rule (that congeners from warm areas are smaller than those from cold areas) doesn't apply to lepidopteran endotherms; if it does apply then the effect must be outweighed by other factors, making it invisible.

Unlike vertebrate animals, individual moths fly only for a few days or weeks in the year and then only for a very short time during any one day. Activity time may be an hour or less, and choice of activity time could play a large role in thermal balance. For example, in the Anza Borrego desert of southern California I have seen the white-lined sphinx moth *Hyles lineata* fly only once dusk had fallen, as the hot (>30°C) daytime temperatures were falling. In the much cooler mountains, however, this moth can commonly be observed flying in the daytime. Similarly, in the hot, lowland dry forests of Costa Rica, the huge (up to 6 g) sphinx moths *Cocytius* (3 species) restrict their flight activity to the early morning hours before sunup, when it is cooler than usual (Daniel Janzen, personal communication). Do the smallest summer moths fly on the warmest nights? How is activity spread out through the season on the basis of thermal balance? Clearly, the numerous behavioral options that moths have for activity can result in a blurring of possible physiological adaptations for specific climatic regimes. But sometimes, as the winter moths demonstrate, physiological adaptations allow activity to occur at times most suitable for predator avoidance and energy balance.

Winter-Flying Moths

Sphinx moths, most of which are at home in tropical lowlands, represent perhaps an extreme for a thermoregulatory challenge: They are heavy-bodied, they fly continuously with very high rates of wing beats and heat production, and they are active at high temperatures. All four factors act to raise the thorax to potentially intolerable temperatures. At the other extreme are several groups of small-bodied moths from north-temperate regions that have adapted to be active from 0°C to 10°C in the winter, thereby reducing predation from numerous bats and birds. These moths

Fig. 1.29 A winter cuculiinid moth (Noctuidae), shivering.

include Cuculiinae (a subfamily of the Noctuidae, or owlet moths) and a number of Geometridae. Both have different behavioral and physiological adaptations that provide an instructive contrast to the ways sphinx moths have evolved.

The Cuculiinae (Fig. 1.29) are dominantly of north-temperate distribution in Europe, Asia, and America and contain nearly 50 species in the northeastern United States (Franklemont, 1954). Many of these moths spend the summer in pupal diapause or estivation. They emerge in the fall, develop their eggs in the winter, and then oviposit in early spring before the buds of the trees open. They fly in any month of the winter, provided it warms up to above 0°C, and they emerge (as after the snow melts) from under the leaves of the forest floor where they spend their time when not active (Schweitzer, 1974).

Summer-flying moths characteristically roost on bark or on some other background material that matches their own coloration (Sargent, 1966). The winter noctuids of different species have a diversity of colors—grey, black, white, tan, brown, sienna—that could be used (and originally may have been used) to match the bark of a variety of trees. At the present time of evolution they have the habit of hiding under leaves on the forest floor, a habit that probably evolved independently among the different species as an adaptation to avoid freezing when perching on bark became thermally infeasible in the winter.

The winter noctuids are very sensitive to freezing. Their blood has a freezing point of about −2°C (close to freezing point for summer-active insects); hence, they lack antifreeze compounds (Heinrich, 1987a,b), although they supercool to −6°C and sometimes to near −15°C. Ambient temperatures under snow and under leaves where moths are found usually range near −2–0°C (Schweitzer, 1974), which is a moderate temperature relative to the thermal environment above the snow, so antifreeze compounds are not essential. Furthermore, it is probably the lack of antifreeze compounds in the blood that allows the moths almost immediately to fly on the first warm night of late winter or after being taken out of a refrigerator where they have been kept at T_a < −5°C for weeks; if antifreeze compounds were present, there would be a prolonged lag of inactivity before they could be degraded and cleared from the body.

The winter-flying cuculiinids are unusual in that they can begin to shiver from temperatures as low as 0°C and in that, despite their small mass (about 200 mg), they fly with T_{thx} near 30–35°

C even at T_a of 5°C (Fig. 1.30). As far as I am aware, no other insects can shiver and warm up from such low temperatures, nor are there other insects so small that are able to maintain a 30°C difference between internal and external temperatures.

FLIGHT TEMPERATURES

Cuculiinid moths can also fly at −3°C, but at such low ambient temperatures they cannot continuously sustain the 33–38°C temperature gradient necessary to maintain thoracic temperatures suitable for flight, because they then lose heat convectively faster than they produce it. As a result they must stop and shiver (heat storage occurs during shivering because convection is minimized) before resuming another short flight.

During flight at 4–13°C, T_{thx} is nearly identical to that at the end of warm-up. At ambient temperatures of 22°C and above, however, all the T_{thx} of flying moths is above takeoff T_{thx}, suggesting that overheating is occurring. Flights are brief at high ambient temperatures (>26°C), and as T_{thx} approaches 39.5°C some of the moths stop flying, presumably because they are unable to cool themselves. Apparently the moths are very good at retaining heat, but they are not capable of getting rid of it to prevent overheating at even modest (for us) ambient temperatures. In the field, however, they are seldom subjected to temperatures above 15°C. Instead, they must usually fly at temperatures less than 10°C, and it is an interesting problem how they maintain a high T_{thx} at such low ambient temperatures.

HEAT RETENTION IN THE THORAX

The thoracic pile of winter moths is thick (though not unusually thick in comparison with some other moths from temperate regions) and it serves as an effective barrier that retards the rate of heat loss in moving air. Cooling rates of moths with and without pile are near 0.6°C/min/°C difference between T_{thx} and T_a in still air, but at wind speeds of 5–6 m/s (near the moth's flight speed) depilated moths cool about twice as fast (or maintain half the temperature excess) as those with their thoracic pile intact (Fig. 1.31).

Apart from the heat lost from the thorax directly through the thoracic pile, considerable amounts of heat could also be lost through the head. The head necessarily receives heat from the thorax by conduction. Is additional heat shunted to the head to

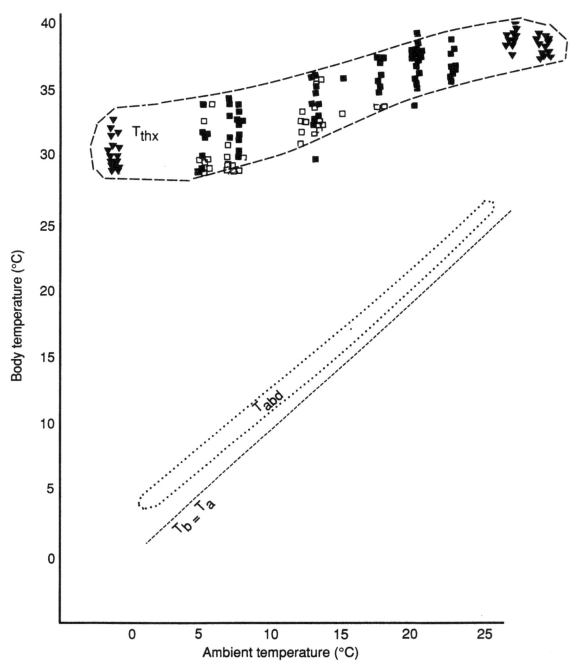

Fig. 1.30 Thoracic and abdominal temperatures of cuculiinid moths in free flight as a function of ambient temperature. Triangles = moths that stopped and restarted flight. Filled squares = *Eupsilia* sp. in continuous (>2 min) flight. Open squares = *Lithophane* sp. in continuous flight. (From Heinrich, 1987a.)

THE HOT-BLOODED INSECTS

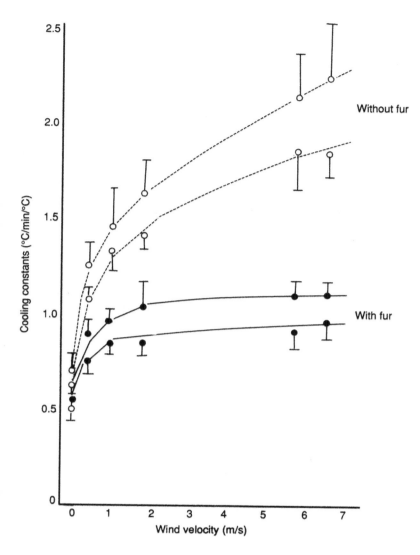

Fig. 1.31A A *Eupsilia* winter moth perched on a beech bud. Note the thick "fur coat" on the thorax.

Fig. 1.31B Cooling rates of the thorax in winter moths (*Eupsilia* and *Lithophane*), with intact thoracic fur (pile) and with thoracic fur removed, as a function of wind velocity. The top line of each pair plots data from *Eupsilia*, the bottom line represents *Lithophane*. (From Heinrich, 1987a.)

raise its temperature at low ambient temperatures? Head temperature in moths is tightly coupled with T_{thx}, averaging about 40 percent of T_{thx} excess in both live, stationary moths and dead moths. Therefore, heat reaches the head through passive conduction, and little or no physiological heat transfer to the head occurs.

During flight, head temperature immediately plummets to about 15 percent of T_{thx} excess, because the small head (5 mg) has a much higher cooling rate than the thorax (40 mg). In the summer-flying tent caterpillar moth *Malacosoma americanum*, which is also endothermic and of similar mass to most winter noctuids, head-temperature excess averages 64 percent of T_{thx} excess in flight (Casey, Hegel, and Buser, 1981). The head-temperature excess of winter moths is some 4 times less. These data suggest that, in contrast to the tent caterpillar moth, the winter moths have less heat leakage to the head, which (as I will show later) is due to a counter-current heat exchanger.

As previously demonstrated for sphinx moths, the abdomen is another major potential avenue of heat loss in winter owlet moths. But direct temperature measurements with thermocouples and thermovision imaging, both during pre-flight warm-up as well as during continuous flight, show that T_{abd} remains close to T_a at all T_a (Figs. 1.30 and 1.32). Furthermore, there is no increase of T_{abd} over T_a, even when the moths stop flying because of heat prostration. These observations are highly unusual in comparison to the data from sphinx moths, which use their abdomen as a major heat sink.

Part of the winter moths' capacity to retard heat flow to the abdomen is due to the thoracic and abdominal air sacs that insulate the abdomen from the thorax, which are also found in most other moths. However, major heat retention is likely also accomplished by two separate counter-current heat exchangers that retard heat loss to the head and abdomen, respectively (Fig. 1.33).

Within the thorax the ascending and descending loops of the aorta are closely appressed against each other. Therefore, cool

Fig. 1.32 The winter moth *Eupsilia morrisoni*, with its thick layer of pile on the thorax, and "thermal contour" drawings made from 3 successive thermovision pictures during pre-flight warm-up (all dorsal views). While the outermost areas remain at about 11–14°C during warm-up, the central thoracic area rises in temperature (as indicated by the pattern of contour lines, or isotherms) to 27–31°C. (From Heinrich, 1987a.)

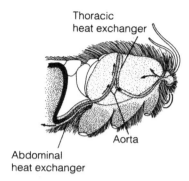

Thoracic
heat exchanger

Aorta

Abdominal
heat exchanger

Heart

Aorta

0 1 2 3 4 5
mm

Fig. 1.33 Diagrammatic representation of the head, thorax, and anterior portion of the abdomen of a sphinx moth the size of *M. sexta (bottom)* and a cuculiinid winter moth *(top)*. The direction of blood flow is indicated by the arrows. Counter-current heat exchangers occur where cold *(dark)* blood goes in one direction while warmer blood *(more lightly stippled)* flows in the opposite direction. Two of these areas of counter-current heat exchange are highly developed in the cuculiinid moth, but they are poorly or not at all developed in the sphinx moth. The latter can use the abdomen as a thermal window to get rid of excess thoracic heat. (From Heinrich, 1987a.)

blood from the abdomen ascends the aorta and is warmed by the thoracic muscles during flight. The temperature in the descending loop should then be higher than that in the ascending portion of the loop and heat, following the temperature gradient, would be recovered by the ascending loop (rather than being passed on to the head, where it would be immediately lost by convection).

Some direct measurements are in accord with this model: during pre-flight warm-up in *Eupsilia morrisoni* the temperature in the top of the heat exchanger is only 0.5 °C lower than at the thoracic surface. But during flight (fixed, with thermocouples still in place), the temperature at the top of the heat exchanger decreases up to 2.0 °C below surface T_{thx}, as would be predicted if cool blood flowed anteriorly in the aorta.

While the anterior counter-current in the thorax may help to account for the low T_{hd}, another counter-current exchanger in the anterior ventrum of the abdomen probably helps explain the low T_{abd}. All the blood pumped from the cool abdomen is confined in the aorta, which makes a loop in the ventral portion of the abdomen, and all the heated blood from the thorax returning to the abdomen must flow around or under this loop with cool blood. Counter-current heat exchange is therefore inevitable in that loop.

FLIGHT ACTIVITY AND ENERGETICS

On late evenings at the end of October the moths feed from the yellow flowers of witch hazel *(Hamamelis virginiana)*. This small tree is the last plant of the year to flower, producing splashes of yellow long after the leaves have fallen from other trees. The moths' primary food, however, is probably the sap oozing from taps created by red squirrels on sugar maple trees (Heinrich, 1992). Even in February, on the first day of a thaw, the cuculiinids often come in droves to maple sugar or almost any sugary, fermented concoctions smeared onto tree stems, where they tank up to near double their body weight. Afterward they are often forced to fast after crawling under leaves and being covered periodically under snowstorms until mid-April. By early May, when the first migrant warblers return and the leaves start to unfurl, the moths have already disappeared for another year.

Moths that have been kept for weeks in a refrigerator at temperatures below the freezing point of water *can* immediately warm up and fly when taken out, but despite that physiological capacity warm-ups from 0 °C are rare. On the other hand, at temperatures of near 10 °C almost all moths warm up upon only slight provocation. Warm-up durations are about 20–25 minutes at 0 °C and approximately 1 minute at 20 °C. The rate of heat production during warm-up is a direct function of muscle temperature, independent of ambient temperature, as in other insects (Fig. 1.21). Winter moths do not defend a lower lethal body temperature; they do not attempt to shiver when cooled to their freezing (and death) point ($-6\,^{\circ}$C to $-15\,^{\circ}$C). Such a response would not be expected, since the moths cannot shiver at $T_b < -2\,^{\circ}$C, and they likely spend long portions of the winter at T_b several degrees lower than $-2\,^{\circ}$C. If they shivered every time they cooled to near lethal temperatures they would soon exhaust their energy supplies and die.

Different rates of warm-up from different ambient temperatures are due to physical parameters of heat storage and loss. For example, during warm-up from $0\,^{\circ}$C, moths shivering with a T_{thx} of $5\,^{\circ}$C produce approximately 9.2 J/g thorax/min (= 0.15 W), but since they are losing 7.5 J of this heat to convection their warm-up rate is slow. During warm-up from $20\,^{\circ}$C, in contrast, moths with a similar body-temperature excess but with T_{thx} now at $25\,^{\circ}$C produce 46 J/g thorax/min, but they are still only losing 7.5 J of this to convection and hence warm-up rate is rapid.

We can extrapolate the rate of energy expenditure on T_{thx} to zero at T_a near $-2\,^{\circ}$C, confirming that the lowest T_a (and T_{thx}) at which the warm-up behavior has been observed is indeed the lowest temperature at which the moths can physiologically warm up. The minimum warm-up temperature coincides with the lowest at which action potentials can be recorded from the flight muscles (Esch, 1988).

After flying and subsequent feeding from sugar solutions painted onto trees, the moths have the option of cooling to ambient temperature or of shivering and remaining flight-ready. In general, they shiver and remain flight-ready at high ambient temperatures ($13\,^{\circ}$C and $17\,^{\circ}$C) but become torpid while feeding at low temperatures ($6\,^{\circ}$C). By not shivering, the moths save considerable energy, because at low temperatures the cost of shivering to maintain an elevated T_{thx} is especially high and the resting metabolic rate is low (Fig. 1.34). The resting metabolic rates of winter moths at specific T_{thx} are approximately 30 percent lower than those reported for tropical moths (Bartholomew and Casey, 1978). It is thus not surprising that the increase in metabolism from rest to flight at low ambient temperatures is enormous: flying metabolism is approximately 8,000 times resting metabolism at $-3\,^{\circ}$C, 1,600 times at $0\,^{\circ}$C. These increases are probably records of metabolic scope in the animal world.

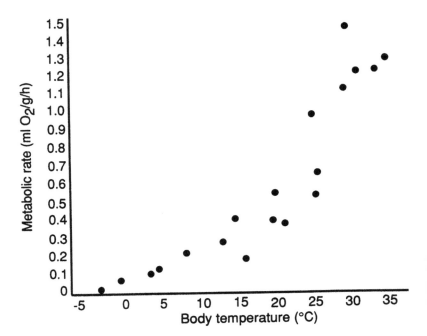

Fig. 1.34 Metabolic rates of *Eupsilia* sp. at rest as a function of ambient (and body) temperature. (From Heinrich, 1987a.)

THE COLD GEOMETRIDS

As already indicated, geometrids (the adults of "inchworm" caterpillars) are light-bodied, large-winged moths occurring in both tropical and north-temperate regions. A number of species typically fly in late fall and early winter, and of these late-flying forms some of the females are flightless, having either vestigial wings or no wings.

One of the species perhaps most adapted to cold is *Operophtera bruceata* (Fig. 1.29). Females lack wings and the males fly in November in the northeastern United States. In the early part of November *O. bruceata* flies mostly at night, but as the seasonal temperatures decline to <0°C at night the moths then fly only in the daytime. Finally, near the end of November when even daytime temperatures are <0°C, this moth flies only on sunlit slopes near noon (Heinrich, unpublished observations).

Not surprisingly, these moths (males) never bask; presumably because at their small mass (9.9 mg), they would again cool to near ambient temperature within about 5 seconds of flight. (As in other Lepidoptera, there is no need for them to be warm when they remain at rest. Heating occurs only to permit flight.) Since

THE HOT-BLOODED INSECTS

their small mass precludes any appreciable heat storage in flight (Fig. 1.22), they have had little or no advantage to evolve either shivering or basking behavior. Instead, they have become capable of takeoff with a T_{thx} of 0°C.

How can an insect fly with a T_{thx} at 0°? First, the moths are able to fly with a very low energy expenditure. The geometrid moths do not carry eggs in flight, nor do they have a digestive tract. Furthermore, their wing area relative to body weight is exceptionally large, and they undoubtedly give up maneuverability in favor of energetically cheap long-distance travel that often involves "sailing" along on a slight breeze. Energy expenditure of flight is related to wing-beat frequency and amplitude, which are a function of wing size. While endothermic winter moths require a wing-beat frequency near 60 Hz before they can fly, the geometrids can fly with a wing-beat frequency of only 4 Hz.

A second reasonable hypothesis is that the biochemical machinery is adapted to operate at those temperatures to which the animal is necessarily exposed. Surprisingly, however, the catalytic capacities of citrate synthase and pyruvate kinase from extracts of the muscles of this and another highly poikilothermal geometrid, *Alsophila pometaria*, show identical temperature dependence (Fig. 1.35) to those of the endothermic cuculiinid moths and highly endothermic sphinx moths, *Manduca sexta* (Heinrich and Mommsen, 1985). If these results apply to other moths as well, then it means that the structural enzyme units are extremely conservative. Nevertheless, the above results for isolated enzymes do not preclude the possibility that temperature-dependence is built into the *in vivo* system; the enzymes could be attached within the mitochondrial membrane or along the cytoplasmic reticulum in such a way that they work better at one or another temperature. Or, as mentioned, the sarcoplasmic reticulum (Fig. 1.28) and other micro-anatomical features could be adapted to permit activity at low muscle temperatures (Bennett, 1985).

James Marden from the University of Vermont has shown that there are considerable differences in flight-muscle performance between different moths. Marden used a force transducer to record the lift generated during flight attempts while simultaneously monitoring muscle temperature with a thermocouple. In a comparison of lift-generating capabilities between the winter-flying geometrid *Operophtera bruceata* and the sphinx moth *M. sexta*, he showed that the cold-bodied geometrid has a relatively low thermal dependence of contraction and relaxation rates. It can generate lift at muscle temperatures from 0°C to 33°C, with the muscles best adapted to operate in the range of 5–25°C. However, the tradeoff for that broad thermal breadth is decreased power at *all* temperatures. The normally hot-bodied sphinx moth *M. sexta*, in contrast, has a very narrow thermal breadth for flight-muscle operation, but its muscles are much faster and more powerful at temperatures of 35–41°C (Fig. 1.36). The basis for these different muscle performances is unknown, but the implications are clear: As elaborated in detail elsewhere (Heinrich, 1977), an insect can adapt its flight machinery to operate over a wide range of temperatures, but for maximum power output it must specialize to operate at a specific temperature.

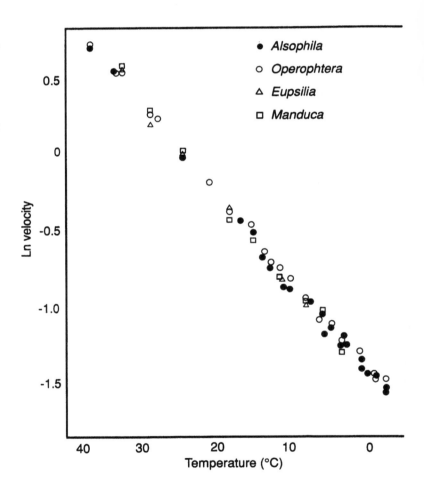

Fig. 1.35 Temperature dependence of the citrate-synthase reaction in 4 species of moths. The slopes of the regression lines are identical for the enzyme over a temperature range from below 0°C to over 35°C. (From Heinrich and Mommsen, 1985.)

THE HOT-BLOODED INSECTS

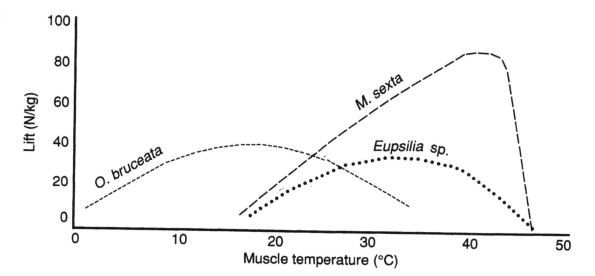

Fig. 1.36 Maximum lift (measured in newtons) per muscle mass as a function of muscle temperature in three moths displaying different thermal strategies. *O. bruceata* is poikilothermic and flies at muscle temperatures down to 0°C and up to 25°C. *Eupsilia*, an endothermic winter moth, flies with muscle temperatures from 30°C to 37°C. *Manduca sexta* regulates its T_{thx} in flight at 40°C. (Courtesy of J. H. Marden.)

Summary

Moths are a diverse group of insects that live in a broad range of thermal environments. Their thermal physiology has been explored in greater depth than has that of any other group. As a consequence, moths are now a window to many of the main features and principles of insect thermoregulation.

At one extreme are poikilothermic moths less than 10 mg in mass. Although too small to retain endothermically produced heat, some of these insects are nevertheless able to fly at ambient temperatures near 0°C. Flight at such low muscle temperatures involves a number of adaptations that reduce energy expenditure (mainly reduced body weight and increased wing size) and adaptations of muscle morphology, including probably a much-elaborated sarcoplasmic reticulum system.

Moths that have evolved to fly rapidly (with a high energy expenditure) have adapted their muscles to operate in a relatively narrow range of muscle temperatures. Endothermic moths differ in this from poikilothermic animals, which can operate over a broad range of temperature but can generate less force at any one temperature. In general, the larger the moth, the higher and the more stable the T_{thx} it maintains in flight. Abdominal temperature is not regulated.

Large, fast-flying moths, such as sphinx moths and possibly also saturniid moths, inevitably produce heat from their flight metabolism and they regulate T_{thx} at about 40°C over an ambient-temperature range of some 20°C while in flight. (Head temperature is maintained at several degrees below T_{thx}.) All of the heat raising T_b in flight is a by-product of the flight metabolism. Throughout flight, heat production is not regulated; instead, thermoregulation is accomplished through increased blood circulation, which shunts excess heat to the abdomen. When the thorax heats up, the thoracic ganglia stimulate the heart, in the abdomen, to increase circulation. Given the option of flying with a low energy expenditure, moths do not compensate by producing more heat because of the declining T_{thx}. Instead, heat production in flight is varied with respect to load, and it remains independent of temperature regulation.

The large moths that inevitably heat up in flight have evolved to also *require* a high T_{thx} in order to produce sufficient power to support flight. Prior to flight they warm up to near the temperatures normally generated in flight by shivering. Shivering involves the nearly synchronous contraction of the up- and downstroke muscles of the wings, which in flight contract alternately. Wing-vibration frequency and rates of heat production during warm-up are strictly a function of muscle temperature, and they are independent of ambient temperature. Heat production per wing stroke is relatively constant over wide ranges of T_{thx}, and thus heat production rises almost in direct proportion to wing-vibration frequency throughout any one warm-up.

Endothermic moths begin cool-down within a wing beat of stopping flight, and they then are poikilothermic, as are all resting moths. Some moths, principally sphinx moths, show physiologically facilitated cool-down after shivering and possibly after flight cessation. Cool-down may be facilitated by a mechanism functioning to conserve energy resources, or it could be related to transfer of fuel from abdominal to thoracic depots.

Large sphingid and saturniid moths have an aortic loop sandwiched in between the right and left dorsal longitudinal muscles, and this loop acts as a cooling coil to pick up heat from the muscles. Winter-flying noctuidae lack this loop and instead the aorta is folded into a counter-current heat-exchanger arrangement that retards heat flow to the head. The small, endothermic winter moths capable of flying at 0°C with a T_{thx} above 30°C also retard heat flow from the thorax to the abdomen by a second counter-current heat exchanger located in the anterior of the abdomen.

There is little evidence that moths have evolved to vary either body size or body temperature in relation to climate. Instead, physiology and behavior have compensated for environmental changes.

Remaining Problems

1. What are the cellular and biochemical differences between moths flying with a T_{thx} of $0°C$ and those flying with a T_{thx} of $45°C$? (Why can some moths fly with a low T_{thx} while others need a high T_{thx}?)
2. How do large, tropical sphinx moths manage to prevent overheating?
3. What are the neurological mechanisms for body-temperature control?
4. What is the significance of the apparent thoracic-temperature limit near $45°C$?
5. How much of the heat-sequestering mechanism of winter moths can be accounted for by the counter-current heat exchangers, and how much is due to reduction of blood flow?
6. Do those moths with inefficient counter-current heat exchangers shut off blood supply to the abdomen during warm-up and thereby incur an oxygen debt?
7. Why does the Bergmann rule not apply to moth endotherms?
8. Do the day-flying moths that superficially mimic butterflies (family Uraniidae) thermoregulate in the same way as endothermic moths do, or are they more like heliothermic butterflies?

TWO   *Butterflies and Wings*

O N A N Y walk in a hot, lowland, tropical jungle you
will meet brilliant butterflies on their sojourns amongst
the foliage. You will see them on mountain meadows
covered with carpets of colorful flowers and on the tundra above
the Arctic circle. Butterflies are found in an extraordinary range
of geographical, ecological, and thermal environments, and wherever they live their activity is strongly affected by thermoregulation. In the northern hemisphere two of the more eye-catching
examples are the European peacock *Inachis io* (Fig. 2.1) and the
Holarctic mourning cloak *Nymphalis antiopa*. Both butterflies hi-
bernate as adults. In Vermont and Maine the mourning cloak is
already active in late March, while the ground is still covered in
deep snow. It flies long before any leaf or flower buds have
opened, stopping periodically to feed on sap oozing from tree
wounds and to expose its dark "cloak" to the warming sun.

Thermoregulation is a crucial feature of the biology of butterflies, because their fecundity may relate directly to body temperature (Stern and Smith, 1960; Gossard and Jones, 1977;
Kingsolver, 1983). But just as important, foraging, the seeking of
mates and oviposition sites, and the avoidance of predators all
depend on constant or intermittent flight. And flight requires the
maintenance of a high muscle temperature. All butterflies so far
examined, whether they are active at air temperatures of 34°C in
a tropical lowland or at 4°C on the Arctic tundra, fly with an
elevated and as a group in a remarkably narrow (relative to other
insect orders) range of thoracic temperatures. The ability to regulate a high and narrowly constrained thoracic temperature while
subjected to wide ranges of air temperature is in itself noteworthy.

But what makes it even more surprising is that different species of butterflies have accomplished it despite a wide range of body masses (from less than 10 mg to at least 200 times greater than that in some of the Australasian ornithopteran birdwing butterflies) and hence in huge differences in passive cooling rates.

Butterflies and moths belong to the same order (Lepidoptera), but their principles of thermoregulation differ greatly. In general butterflies, unlike moths, derive most of their heat from solar radiation, while they are perched. In order to bask they require *thin* insulation or no insulation at all. As a result they cool convectively in flight, especially if they are small-bodied. In addition, wing-beat frequency and rate of endothermic heat production are lower in butterflies than in moths, because wing loading is low since the wings are large.

In the large volume of literature on thermoregulation in butterflies one will find that general trends and principles have been proposed. Many of the individual studies, however, have limited

Fig. 2.1 The peacock butterfly from northern Europe in dorsal-basking position.

value in elucidating the mechanisms of thermoregulation. It is often the case that T_b taken in the field will be reported without a description of what the butterfly has been doing at the time of the measurement to achieve that T_b. Sometimes T_b is given in relation to "flight activity" (which is not the same as T_b after a sufficiently long flight duration to stabilize T_b), to "activity" (which could involve basking, flight, or any combination thereof), or to "basking" (without reference to the beginning, middle, or end of the basking bout). In addition, necessary distinctions are commonly not drawn between T_b taken of animals that are free, tethered by thermocouple wires, or dead. Often the mass of the animals is not given.

Differences in mass are physiologically, behaviorally, and ecologically highly relevant because they greatly amplify the effect that the ambient thermal conditions can exert on body temperature. For example, a small New Guinea butterfly *(Orsotriaena medus)* heats from 30°C to lethal temperatures of 55°C in 2 minutes when forcibly placed into tropical sunshine, while a large animal from the same area (930 mg, *Papilio aegeus*) requires a minute longer to reach the same (lethal) temperature (Heinrich, 1972). A small New England butterfly (24 mg, *Coenonympha inornata*) cools convectively from 34°C to 29°C in 30 seconds of flight at 23°C (Heinrich, 1986a) despite its flight metabolism, whereas a much larger north-temperate butterfly (313 mg, *Nymphalis antiopa*) can fly and intermittently soar even at 5–7°C lower ambient temperature while stabilizing its thoracic temperature near 35°C (Heinrich, 1986b) while being heated by its flight metabolism alone. As these examples suggest, and as I will here explore, the key to butterfly thermoregulation resides in an integration of basking (heating) and flight (generally cooling) behavior. Butterfly thermoregulation therefore raises interesting questions not only on the mechanisms but also on the ecology and evolution of thermoregulation.

Behavioral Mechanisms of Thermoregulation

Most butterflies close their wings dorsally when resting, so that only the highly cryptic wing undersides are exposed (Findlay, Young, and Findlay, 1983). When active in sunshine, however, butterflies reveal the often brilliantly colored dorsal surfaces of their wings, which may be opened and closed rhythmically. Since a butterfly's visual conspicuousness from colorful dorsal wing surfaces functions in sexual signaling (see for example, Tinbergen, 1942; Magnus, 1958; Brower, van Zandt Brower, and Cranston, 1965; Hidaka, 1973; Hidaka and Yamashita, 1975), it is perhaps understandable that the wing opening of butterflies in sunshine was originally considered primarily a display behavior (Parker, 1903) with wing closing functioning for concealment (Longstaff, 1905a,b, 1906; Tongue, 1909).

Superimposed on the reasons for displays in many species is thermoregulation. Basking in sunlight not only shows the wings, it also raises body temperature and, depending on the species, it can involve either opened or closed wings. Cooper (1874) wrote that the sulphur butterfly *Colias philodice* leaned sideways with closed wings when resting "as if to receive the warmth of the sun." Other early observers who also speculated that different postures in butterflies related to thermoregulation include Radl (1903), Pictet (1915), Longstaff (1906), and Winn (1916).

Fig. 2.2 The comma butterfly *Polygonia comma*, hugging the ground in dorsal-basking position.

Following the above and other (Herter, 1953; Digby, 1955) early descriptions of the behavior of butterflies and other insects in sunshine, there appeared studies on behavioral thermoregulation in reptiles (Cowles and Bogert, 1944; Bogert, 1949, 1959; Bartlett and Gates, 1967) as well as pioneering experiments on butterfly thermoregulation (Vielmetter 1954, 1958). Apparently unaware of much of the earlier insect work, Clench (1966) acknowledged the studies on reptile thermoregulation and discussed potential behavioral thermoregulation in butterflies. Although he wrongly concluded that behavioral thermoregulation for butterflies had not been previously known, his paper nevertheless set the stage for many subsequent studies.

Clench (1966) classified four kinds of behavioral thermoregulation in butterflies: *dorsal basking* (common in papilionid, danaid, and nymphalid butterflies), *lateral basking* (common in many lycaenids and in the subfamily Coliadinae of the pierids), *body basking*, and *ground hugging* (Fig. 2.2). He also claimed that the circulating blood (Arnold, 1964) makes the wings "effective heat exchangers."

While "dorsal basking," as first described by Parker (1903), a butterfly holds its wings open and perpendicular to the incident radiation while the head is pointed away from the sun. On the other hand, during "lateral basking," as first described by Cooper (1874) for *C. philodice*, the butterfly closes its wings dorsally and tilts sideways to present the underside of its wings at right angles to the sun (Fig. 2.3). In "body basking," which Clench (1966)

Fig. 2.3 Lateral basking by the orange sulphur butterfly *Colias eurytheme*.

describes only for the cabbage butterfly *Pieris rapae* L., the wings are held open just enough to direct the solar radiation onto the body (Fig. 2.4). "Ground hugging" refers to butterflies heating themselves by appressing themselves to the warm substrate. Most of the ideas presented by Clench (1966) were speculative, however, and he admitted to "the necessity for obtaining actual body temperatures to amplify the observational data."

Perhaps unknown to Clench, Vielmetter (1954, 1958) had already reported that the *Argynnis paphia* maintains a body temperature near 34°C ± 1.5°C by dorsal basking. Body temperatures of these butterflies rise quickly when their wings are opened to the sun, and when a butterfly that is perched on the ground lifts its wings upwards the body-temperature excess is lowered by about 50 percent. Closing the wings results in a further lowering of body-temperature excess by 20 percent.

Clench (1966) supposed that the wings of dorsal baskers act as solar panels that intercept the sun's rays and then send the heat to the thorax by blood circulation (Arnold, 1964). He made this argument primarily by pointing out that many butterflies have dark pigmentation directly on the veins of the wings, which should aid in the absorption of radiant heat at the precise site of the major hemolymph flow. However, the heating rate of butterflies in sunshine is nearly identical whether they are dead or alive (Heinrich, 1972). As revealed in numerous later studies, the wings are indeed involved in thermoregulation, but the blood-circulation mechanism proposed by Clench (as well as a second mechanism that was subsequently proposed) turned out to be wrong.

Kammer and Bracchi (1973) enlarged on the experiments comparing dead *vs.* live butterflies (Heinrich, 1972). By cutting through the wings near the base and physically replacing the cut wings near (<1 mm) the cut, they showed that the heating rates in butterflies with cut and uncut wings were also identical. Together, these experiments decisively confirmed that blood circulation is not involved in heat transfer. These conclusions were corroborated with subsequent measurements (Wasserthal, 1983) of the hemolymph flow in the wings of living *Pieris rapae* and *P. brassicae*. There was no circular flow of hemolymph in the wing, and it required about 12 hours for stained hemolymph to traverse the wing membranes. This flow rate is much too slow to account for significant heat transfer.

Fig. 2.4 Dorsal body-basking behavior (Clench, 1966) or "reflectance basking" (Kingsolver, 1985a) of the eastern tailed blue, *Everes comyntas*.

The actual role of the wings in thermoregulation in dorsal baskers was first clarified in an elegant set of experiments by Lutz Wasserthal (1975) at the Ruhr University in Germany. In controlled laboratory experiments Wasserthal shaded the wings, or different portions of them, and showed that in both live and dead butterflies there is only a 30 percent reduction of the temperature excess (in butterflies heated with a halogen lamp) if the wings are shaded. Thus, 70 percent of the temperature excess in dorsal baskers is due to direct heating of the body itself. Although in some then-unknown way the wings had *some* effect on body heating, heating of only 15 percent of the area near the wing base resulted in most of the thoracic-temperature excess (Figs. 2.5 and 2.6). Therefore, the wing surface does not act as a solar panel as Clench (1966) had supposed.

Wasserthal's (1975) primary contribution was demonstrating that the wings are instrumental in the control of convection. In sunshine horizontal wings trap warm air underneath them if the animal is ground hugging upon a flat surface (Fig. 2.2), and this warm, still air causes body heating and additionally retards convective cooling. Wasserthal heated dried butterflies and held them in various orientations while keeping the direction of incident solar radiation constant (at right angles to the dorsal wing surface). The highest thoracic-temperature excess was achieved when the wing plane was horizontal; when the plane of the wing was vertical, the rising warm air from the heated wings was no longer trapped and temperature excess was lowered by 20 percent (Fig. 2.7).

Wasserthal's experiments neatly demonstrate that the wings act as convection baffles for thermoregulation. Dorsal baskers in the field vary wing angle accordingly, closing the wings at high T_{thx} to shade the thorax and opening the wings to the sunshine (Fig. 2.8) and simultaneously trapping warm air underneath them at low air temperatures.

In view of the importance of convection in the elevation and maintenance of a high T_{thx} in butterflies, it is perhaps not surprising that body positioning with respect to *wind* could also be crucial, especially for dorsal baskers (Polcyn and Chappell, 1986; Heinrich, 1990). Heating in the sunshine and therefore proper orientation to sunshine is the primary response in all butterflies so far examined, however, and no orientation with respect to wind (for convective cooling) has yet been demonstrated for any dorsal-basking butterfly. This makes sense for a heliotherm, because it is pointless for an animal to achieve a position that minimizes convective heat loss if it thereby does not gain heat! On the other hand, basking location, such as the ground vs. the foliage, is likely a very important variable (Fig. 2.9) that can be and is exploited to maximize heating and minimize convective cooling.

Lateral baskers (Fig. 2.3) do not require warm horizontal surfaces, such as the ground or broad leaves, to trap warm air. In lateral baskers the body is shielded by the basal portions of the

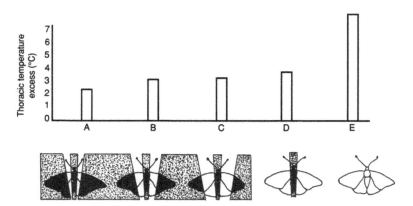

Fig. 2.5 The effect of heating the wings on thoracic-temperature excess in *Apatura ilia* during irradiation with a heat lamp. Note that the contribution of heat from the wings is mainly from the basal portion of the wings, as shown in the different shading patterns illustrated in *A–D*. The control *(E)* shows the result of whole-animal irradiation. (From Wasserthal, 1975.)

THE HOT-BLOODED INSECTS

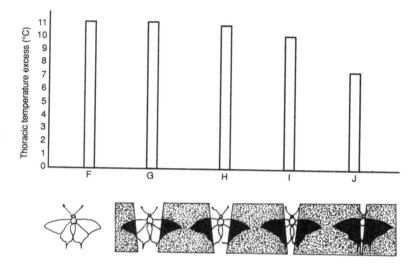

Fig. 2.6 The effect of heating the wings on thoracic-temperature excess in *Papilio machaon* during irradiation with a heat lamp. Note that this experiment is the reverse of Fig. 2.5: at first the whole body is heated and then increasingly large portions of the wings are shaded. The contribution to T_{thx} of the distal portion of the wings *(G–I)* is minimal. (From Wasserthal, 1975.)

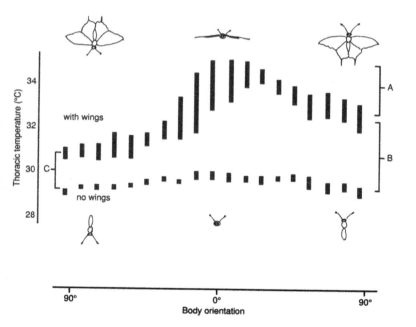

Fig. 2.7 The effect of spatial orientation on T_{thx} of totally irradiated butterflies, *Papilio machaon*, with fully open wings *(top bars)* or with no wings. The wings of the specimens were dried in the natural basking position, and the heat source was always held at right angles to the body (and/or wings). Ambient temperature was 22°C. The data show that the wings affect thoracic heating by reducing convective cooling of rising warm air *(A)* and, to a small degree, by heating by conduction *(B* and *C)*. (From Wasserthal, 1975.)

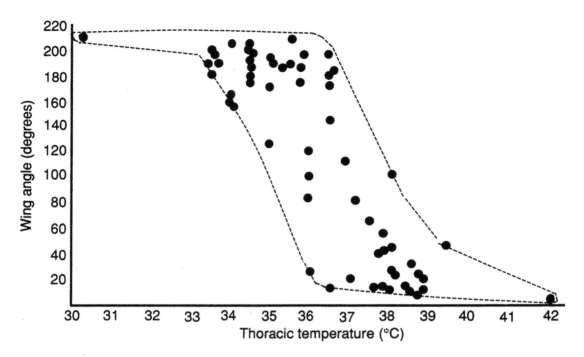

Fig. 2.8 Wing angles of perching *Euphydryas (Occidryas) editha* as a function of thoracic temperature in the field. Cool butterflies open their wings while hot individuals close them dorsally. (From Heinrich, 1986b.)

wings. The basal portions of the upper wing surfaces touch the thorax, and the upper wing surfaces are in *direct* contact with the body. Thus, when the ventral wing surface is heated by solar radiation (Kingsolver and Moffat, 1982), this heat is conducted through the thin wing surfaces directly into the body. Since the body is heated through direct contact with the wings, the animals are more liberated from basking above a heated substrate than are dorsal baskers.

A third type of basking behavior—Clench's "body basking," as seen in the European cabbage butterfly *Pieris rapae* as well as in the whites *(Pontia)*, blues *(Everes* and *Glaucopsyche)*, orange-tips *(Anthocharis)*, and coppers *(Lycaena)*—consists of only partially opening the wings to the sun (Fig. 2.4). Unlike most dorsal baskers (which are often also ground huggers when basking), body baskers are typically small and frail and they perch on grass and other vegetation where they would not likely be able to trap warm air under their wings. Clearly, since these butterflies bask with their wings partially raised, they are not heated by the dorsal-basking technique of trapping warmed air *under* the wings. Instead, in part because their wings are generally predominantly white or

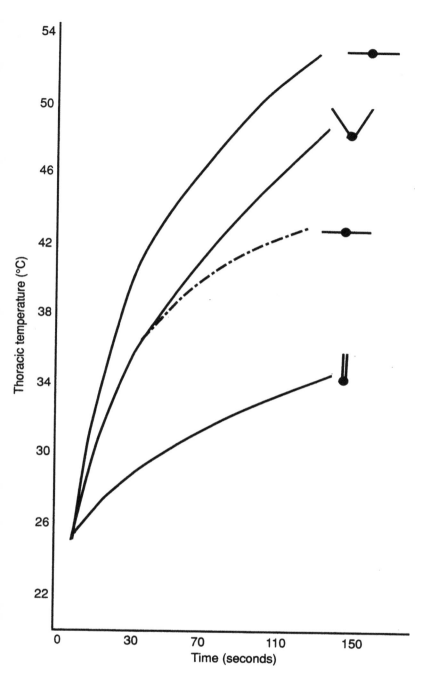

Fig. 2.9 The increase in T_{thx} of dead *Euphydryas (Occidryas) editha* in sunshine as a function of wing angle and substrate. The wings were held fully open *(top line)*, partially open *(second from top)*, or closed *(bottom line)*. The butterflies were on solid ground *(solid lines)* or on vegetation *(dashed line)*. (From Heinrich, 1986b.)

light-colored, it has been proposed that their wings act as solar reflectors (Clark, Cena, and Mills, 1973); this is known as the "reflectance-basking" hypothesis.

According to this hypothesis, the wings act as reflectance panels that focus heat onto the body. This hypothesis has been extensively promoted by Joel G. Kingsolver at the University of Washington, and it is elaborated on in numerous publications in sometimes extensive and perhaps elegant modeling (Kingsolver, 1985a,b,c, 1987, 1988; Kingsolver and Wiernasz, 1987).

I would love to believe that "reflectance basking" is a real phenomenon that makes an effective difference in the T_{thx} of a butterfly, because it would increase the number of wonderful and seemingly clever ways that insects have evolved for thermoregulation. Unfortunately, as I have discussed elsewhere (Heinrich, 1990), the empirical data used to support the hypothesis are grossly inadequate, and I have so far seen no critical experiments that support it. Instead, the data indicate that "reflectance-basking" behavior is fully explicable by another thermoregulatory mechanism (Heinrich, 1990).

The primary prediction of the "reflectance hypothesis" is that basking with the wings held open at an angle heats the butterfly up to a higher temperature than basking with the wings fully open. In order to test this prediction I compared the T_{thx} of a sample of 60 yellow butterflies with wings at 90° (open) with the T_{thx} of equal samples of butterflies with wings at 45° and at 22–23° to the sunshine. (The dead butterflies were held on a frame and there was full air circulation beneath them and around the wings.) The total sample size of 180 butterflies subjected to different treatments run concurrently should have given a powerful enough statistical test to detect any differences in heating, if the butterflies with their wings slanted to the sunshine heated to higher temperatures. They didn't (at least not in the direction predicted). Mean T_{thx} of butterflies with wings at 90° to sunshine was 39.90°C. Butterflies with wings at an angle to the sunshine heated concurrently did not have higher T_{thx}, but instead the mean T_{thx} was nearly identical (39.49°C). The thoracic heating obviously results from the effect of the sun's rays *directly* on the body itself. Presumed reflectance from the wings adds nothing to the temperature increase. Nevertheless, wing angle can still have a powerful effect on temperature excess and hence be a significant factor in thermoregulatory behavior.

The above experiments were conducted in wind-still conditions to eliminate forced convection, which normally occurs in the field. In the field weak air currents are ubiquitous and they have major effects on cooling in small butterflies (Heinrich, 1986a). In order to test the effect of wing angle on heat *loss*, butterflies were subjected to radiant heating from above at different wing angles, as before. This time, however, the animals were also subjected to slight air movements. In this situation body temperature could be made to rise or fall simply by raising or lowering the wings, provided the air currents were perpendicular to the body. When air currents were parallel to the body axis, raising or lowering the wings had almost no effect on T_{thx} (Fig. 2.10).

The first sets of experiments showed that raising the wings does not increase thoracic heating, and the second set of experiments showed that raised wings have a profound effect on raising T_{thx} because the wings act as convection baffles to *retard cooling*. "Reflectance basking" is thus a variation of dorsal or body basking, where an animal maximally exposes its thorax to heating while at the same time minimizing convective cooling.

Most so-called dorsal baskers with dark wings (*Nymphalis* sp., *Polygonia* sp., *Euphydryas* sp., *Boloria* sp., *Speyeria* sp.) also frequently raise their wings like "reflectance" baskers when they stop on thin vegetation (such as grass blades) where they cannot trap air under their wings (personal observations). Their *dark* dorsal wing surfaces are presumably also not effective solar reflectors, but raised wings can at least retard convective cooling, no matter where the animal perches or what the color of its wings. The light coloration of many small butterflies is not explained,

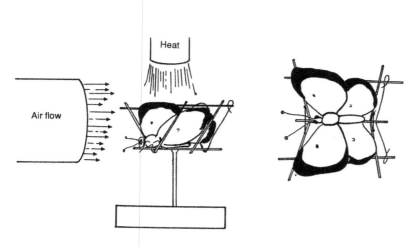

Fig. 2.10 Experimental set-up showing a dead butterfly whose wings can be raised or lowered on a rack while being simultaneously subjected to heat from above and to convective cooling from the left. The butterfly can be turned to face the air stream or to be lateral to it. Changing wing angle has little or no effect on T_{thx} when the butterfly is facing the air stream. But changing the wind direction has a profound effect on T_{thx}, especially when the wings are raised. (From Heinrich, 1990.)

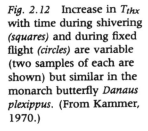

Fig. 2.11 Dorsal-basking behavior of a skipper butterfly, Hesperidae.

although it could function in sexual signaling, making the animals conspicuous from a distance.

Finally, in skippers (Hesperidae) the underwings are spread laterally during basking even though the forewings are raised (Fig. 2.11). These butterflies probably have it both ways (or maybe 3 ways—they also shiver; Kammer, 1968): they may reduce convective cooling across the dorsally exposed, heated thoracic surfaces, as indicated above, while at the same time their underwings reduce convection from below, as shown by Wasserthal (1975). However, the role of wing positioning in thermoregulation in skippers remains to be investigated.

Shivering is rare in butterflies. Aside from skippers, the monarch butterfly *Danaus plexippus* (Kammer, 1970, 1971; Masters, Malcolm, and Brower, 1988), and the red admiral *Vanessa atalanta* (Krogh and Zeuthen, 1941), several species of swallowtails *Papilio* sp. (Rawlins, 1980) and the New Guinea birdwing papilionid *Ornithoptera priamus poseidon* (Stone et al., 1988) also shiver. For theoretical reasons discussed in other chapters, shivering would likely evolve only in those insects that are large enough to show marked obligatory endothermy in flight. Rate of warm-up in shivering is similar to that during fixed flight (Fig. 2.12).

Some butterflies, such as *Danaus plexippus* (Gibo and Pallett,

Fig. 2.12 Increase in T_{thx} with time during shivering *(squares)* and during fixed flight *(circles)* are variable (two samples of each are shown) but similar in the monarch butterfly *Danaus plexippus*. (From Kammer, 1970.)

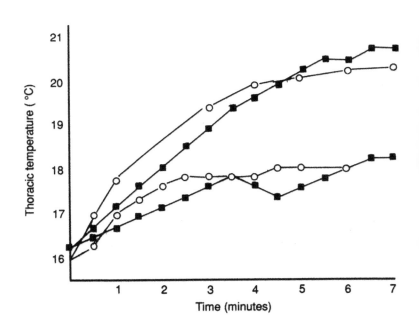

THE HOT-BLOODED INSECTS

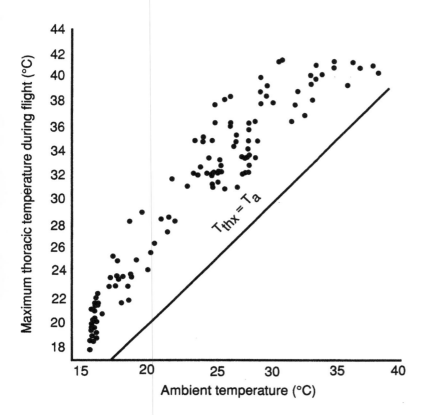

Fig. 2.13 Maximum T_{thx} attained in fixed flight of *D. plexippus* at various ambient temperatures. The butterflies do not thermoregulate T_{thx} in fixed flight. Instead, the higher the ambient temperature, the greater the wing-beat frequency and the greater the temperature excess. (From Kammer, 1970.)

1979) and birdwing butterflies (*Ornithoptera* sp.), may soar, thereby reducing the effect of metabolic heating. Monarch butterflies on migration may on occasion soar without a wing beat for at least 17 seconds (personal observation). But the use of soaring behavior for thermoregulation has not been investigated. So far there is no evidence that any butterfly can regulate its T_{thx} in flight, as many moths do. Flight generates heat that only modestly elevates T_{thx} even in the monarch butterfly (Fig. 2.13), and there is therefore little indication (or expectation) that heat is dissipated through a physiological mechanism, except possibly in very large butterflies during continuous free flight.

Ecology

A high T_{thx} is a prerequisite for flight in butterflies, and flight is required for finding mates, laying eggs, feeding, and avoiding predators. Since flight and thermoregulation are directly linked, the ecological significance of thermoregulation clearly concerns all of these essential functions, even though all may not be occurring at the same moment.

For example, predator avoidance is often an all-or-nothing event whereas flight for the other functions may be relatively continuous. Although the necessity to react quickly enough to avoid a predator may be a rare event, it could nevertheless be a potent selective force because of the severity of the consequences. Clearly, in many situations thermoregulation serves all or many of these functions simultaneously, inasmuch as an active butterfly feeds, chases mates, oviposits, and avoids predators as opportunity or necessity dictates. In other cases, however, a reasonable identification of thermoregulation with one or another function is feasible.

In some tropical butterflies it has been possible to identify thermoregulation with predator avoidance, because some of these butterflies are relatively well protected from bird predation. Birds have probably exerted a very potent selective force on numerous butterflies, and this selective pressure has had a strong effect on butterfly design (Chai and Srygley, 1989; Marden and Chai, 1989). In general, "hawking" birds such as jacamars, flycatchers, trogons, and motmots dive downward on potential butterfly prey. The butterflies they seek shift to an erratic, generally upward flight pattern. The net effect is a contest of flight performance between the two, where acceleration or speed is critical for turning maneuvers. In flying animals, acceleration is proportional to the relative amount of tissue mass devoted to flight muscle (Marden, 1987). Consequently, the outcome of aerial contests should be favored by the animal with a high relative flight-muscle mass. Indeed, of a sample of 124 species of neotropical butterflies, James H. Marden from the University of Vermont and Peng Chai from the University of Texas in Austin (1989) observed that more than 95 percent of butterflies palatable to birds have acceleration indexes (based on flight-muscle mass) greater than the maximum for the birds that chase them. In contrast, unpalatable butterflies have much lower flight-muscle mass relative to their predators. (These differences are independent of phylogenetic origins, strengthening the hypothesis that they are "adaptations." However, even if they were phylogenetically linked it appears to me that it would not disprove the hypothesis, because a "good" adaptation could lead to rapid speciation and radiation of the line carrying that trait.)

Given that flight performance is an important variable in predator avoidance, body temperature, which strongly affects flight performance in all insects, would be an equally important variable. As expected, Chai and Srygley (1986, 1989) found a direct correlation between thoracic temperature and palatability of birds in neotropical rainforest butterflies. The majority of butterflies that are unpalatable to the rufous-tailed jacamar *Galbula ruficauda* not only have low flight-muscle mass and long slender bodies, but they also fly with low T_{thx}. Slow, regular flight patterns apparently result from the combined effects of low T_{thx} and low flight-muscle mass. In contrast, the palatable butterflies with high flight-muscle mass and short, stout bodies are fast, erratic flyers with high thoracic temperatures. To maintain a high T_{thx}, the palatable butterflies bask and therefore prefer sunny edge habitat, whereas the unpalatable, slow-flying butterflies do not bask and instead seek cool, shady habitat and avoid the hot areas (Chai and Srygley, 1986; Srygley and Chai, 1989).

The ability to thermoregulate allows butterflies not only to escape predators. It also allows them to be active from the lowland tropics near the equator to the High Arctic. Few comparative studies of butterflies from different geographical areas are available. The primary study of Arctic butterflies is that by Kevan and Shorthouse (1970), who examined thermoregulatory behavior of five species found at Lake Hazen (81°49′N) on Ellesmere Island, Northwest Territories, Canada. These butterflies were active only on sunny days, and in 1968 they were seen only on 17 of the days between 8 July and 4 August.

During the 17 sunny days the butterflies further restricted their activity to gullies, depressions, and creekbeds. In these wind-protected areas substrate temperatures were near 25°C, as compared with 13°C outside them. Air temperatures 24 cm above ground were near 13°C within the protected areas, and 11°C outside. Air temperatures were highest close to the ground, and most butterflies flew no more than 45 cm above the ground. Dead butterflies *(Boloria chariclea)* in artificial basking postures in the field achieved T_{thx} between 28–32°C, presumably closely approximating the temperatures of live butterflies, in about 2 minutes. Heating rates vary depending on basking substrate, and the Arctic butterflies take abundant advantage of the warm substrate by ground hugging while they bask.

Of the five species studied by Kevan and Shorthouse (1970), the two fritillaries, *Boloria chariclea* and *B. polaris* (Nymphalidae), are dorsal baskers, the two Lycaenidae, *Lycaena feildeni* (a copper) and *Plebius aquilo* (a blue), are "body baskers" that hold their wings half-open, and the sulphur, *Colias hecla* Lef. (Pieridae), is a lateral basker.

No High Arctic butterflies have been observed to shiver, and like other small-bodied basking butterflies (I estimate 200 mg or less), they should cool rapidly when taking flight (see Heinrich, 1986a,b). Thus, even on the days that they are "active," they must spend a major portion of their time grounded and basking.

Prevailing temperatures in alpine environments above timberline tend to be low, approximating the Arctic environment, and a number of studies (Watt, 1968, 1969; Kingsolver and Watt, 1983, 1984; Kingsolver, 1983) have addressed thermoregulation and ecology in lateral-basking *Colias* species from the alpine environment of the mountains of western North America.

Watt (1968) first discussed how in *Colias* species reproductive success is a function of feeding, gamete production, mating, and female oviposition, all of which are governed to a large extent by body temperature and hence thermoregulation. In the alpine *Colias* tested (using implanted thermistors), no voluntary flight occurred at T_{thx} less than 28°C and the T_{thx} of "flying" butterflies was reported to be 30–40°C (Watt, 1968).

My own measurements of thoracic temperatures of *freely* flying *Colias* butterflies are considerably different from the above. From a sample of 47 *C. philodice, C. nastes,* and *C. hecla* in central Alaska at near 1,200 m elevation (in July 1989), I determined that the minimum flight T_{thx} is 19–20°C, not 28°C. Thoracic temperature of butterflies in continuous free flight increases in parallel with air temperature, averaging 14°C above it in sunshine and 6°C above it in shade (Heinrich, unpublished). However, few butterflies were in flight at ambient temperatures greater than 20°C in sunshine. (Thus, *overheating* may occur at T_{thx} near 34°C, and the presumed optimum 30–40°C previously reported for "flight" may, I suspect, reflect escape responses of the probed butterflies, rather than a true range for free flight.)

At low air temperatures the *Colias* butterflies I observed apparently attempted to minimize convective cooling by flying with the wind, and at "high" air temperatures (>17°C) they were instead almost always consistently flying against the wind. Some butterflies flew under overcast conditions when air temperature was as low as 13°C, presumably because they had initiated flight earlier or at some nearby sunny location where they could have warmed up to achieve the minimum flight temperature of 19–20°C to take off. As long as they did not stop flight they could thus continue activity. (*Colias* do not shiver.) Thoracic temperatures of *Colias* species in Vermont were identical to those in the Arctic, as predicted, since numerous species of *Colias* from a large altitudinal gradient and from diverse thermal environments have similar T_{thx} (Watt, 1969). Thus, the results I found from Alaskan *Colias* butterflies were not likely different from those of other researchers in different locations because of ecological variation. However, such variation could still exist but so far be obscured by methodological variables. As mentioned, I suspect that flight *initiation* in captive butterflies may occur when the animals are uncomfortable and attempt to escape, and as shown previously (Heinrich, 1986a), the T_{thx} at flight initiation is usually considerably higher than the T_{thx} in continuous flight.

Colias butterflies have presumably been under selective pressure to minimize warm-up time to increase flight time, because in cooler environments (higher latitudes, higher altitudes, and earlier in the season) the butterflies heat up faster at a given solar input because of darker pigmentation on the basal part of the underwings (Watt, 1968). Although the butterflies are grounded primarily because of their need to heat up, they sometimes experience overheating and have to stop flying even in the alpine environment (Watt, 1968; Kingsolver and Watt, 1983).

Colias butterflies do not shiver prior to flight (Watt, 1968), nor do they show physiological mechanisms of cooling during flight (Tsuji, 1980; Tsuji, Kingsolver, and Watt, 1986). Thus, dead butterflies or model butterflies made out of steel and paper (Kingsolver and Moffat, 1982) can be used to approximate body temperatures of live butterflies to determine when flight activity can occur. Given that butterflies will be active when they are heated in the sun, model butterflies, or "butterfly thermometers," were used to predict when the sun was shining, and this prediction, in turn, was then used as an accurate predictor of when the animals would be active in the field (Kingsolver, 1983). Theoretically, convection also plays a major role in pre-flight heating rates (Kingsolver and Watt, 1983), but in a windy habitat the animals can seek out relatively wind-free micro-sites for basking. However, they cannot avoid convection during flight.

Convection determines maximum flight duration and maximum flight durations can be predicted on the basis of heating and cooling rates (Heinrich, 1986a). In order to be active at a relatively low air temperature, a small butterfly must either be able to fly

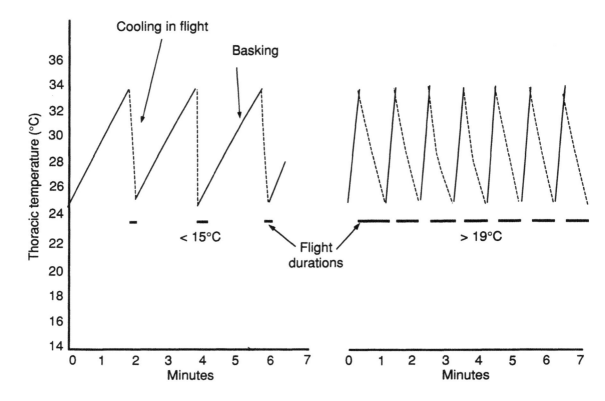

Fig. 2.14 Schematic representation of regulation of T_{thx} in *Agriades glandon* through activity regimens. Higher ambient temperature and more direct sun result in faster heating during basking and slower cooling in flight, and hence in longer flight durations, here depicted as solid horizontal lines. (Data from Heinrich, 1986b, figs. 1 and 2.)

with a low T_{thx} or, if it requires a high T_b, it must reduce convection and spend most of its time grounded and basking. A study of four butterflies active at 3,100 m in the Sierra Nevada mountains of California suggests that body temperatures are only weakly related to mass. Thus, as expected, activity regimens vary enormously as a function of mass (Heinrich, 1986b). The large effect of cooling in flight is well demonstrated by males of *Agriades glandon* (Fig. 2.14), the smallest (23 mg) butterflies examined. They spend 94 percent of their time in the morning grounded and basking, probably because they cool to suboptimal flight temperatures after only 10 seconds of flight. At noon, at some 4–5 °C higher air temperature, they are grounded only 25 percent of the time. They can then stay on the wing 75 percent of the time because they can maintain the same T_{thx} as that achieved by basking, since the convective cooling in flight is much reduced (owing to the reduced temperature gradient between thoracic and air temperature). The much larger (120 mg) *Colias eurytheme* can

THE HOT-BLOODED INSECTS

be and are nearly continuous fliers near 17–18°C at noon (Heinrich, 1986b) because their relatively large mass retards more heat loss in flight. The still larger (313 mg) *Nymphalis antiopa* generates a T_{thx} over 5°C higher than the *Colias* species in flight. The largest of the butterflies examined at the alpine site, *N. antiopa* (unlike the others) often lands in shade and avoids sunshine at noon. Furthermore, when flying at near 16–18°C, it proceeds at an apparently leisurely pace and unlike the smaller butterflies, it alternates bouts of wing beating with 1- to 5-second periods of soaring. These observations suggest that the very different activity regimens of butterflies are generated at least in part by thermoregulatory considerations.

A large portion of flying butterflies are often males in search of females. By staying grounded the males reduce convective cooling, and they then remain flight-ready at all times, able to pursue any females that might come by. (Presumably the females cannot afford this strategy of mating because they must also seek oviposition sites.) As temperatures increase males can adopt the alternative strategy of active wandering in search of females (Wickman, 1985; Heinrich, 1986a). However, the difference between basking and flight T_b may be profound, especially in small butterflies that cool within seconds after initiating flight (Fig. 2.15). The primary size disadvantage of the small butterflies in a cool environment is that they cool so rapidly in flight that they may have only a few seconds of flight before they need to bask again.

Several species, including *Agriades glandon* (Heinrich, 1986b) and *Coenonympha inornata* (Heinrich, 1986a), buy more time (and distance) per individual flight by storing heat in the thorax before each takeoff. They elevate T_{thx} during each basking bout to considerably higher levels than that needed for immediate flight (Fig. 2.16). *Agriades glandon*, for example, takes off after basking when T_{thx} has reached 33–34°C, and it lands when T_{thx} has cooled to 25–26°C (Heinrich, 1986b). Relatively coordinated (but not rapid) flight in this species, as in *Colias*, is possible with T_{thx} at least 20°C. In a checkerspot butterfly *Euphydryas (Occidryas) editha*, on the other hand, takeoff and post-flight T_{thx} are indistinguishable, and these butterflies appear to adopt a sit-and-wait strategy for intercepting females (Ehrlich and Wheye, 1984) rather than a search-and-pursue strategy. While perched in a sunfleck, the checkerspot butterflies regulate T_{thx} fairly precisely near 36–37°C by adjusting the position of their wings (Fig. 2.8), manip-

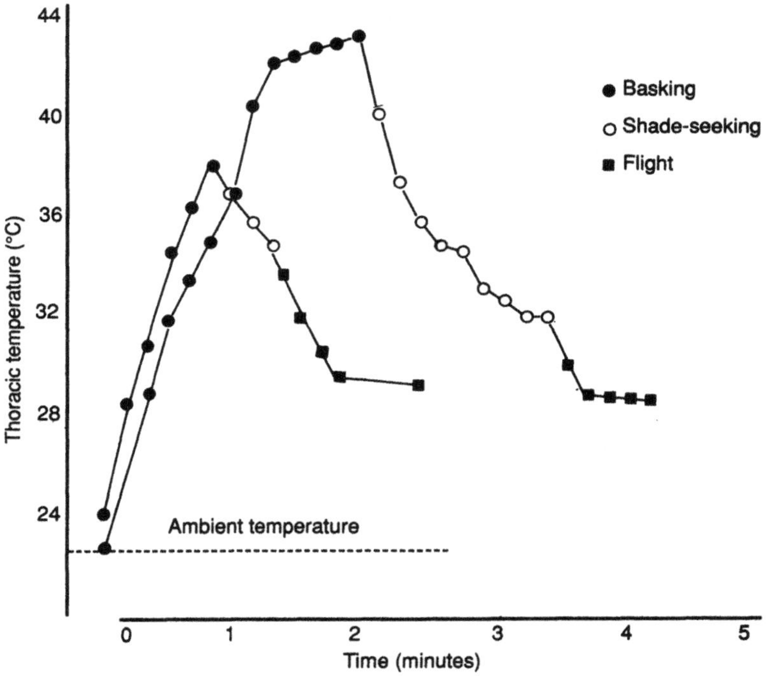

Fig. 2.15 Change in T_{thx} of two live *Coenonympha iornata* while basking, subsequent shade-seeking, and then in continuous flight. The butterflies basked and took off in flight at T_{thx} near 40°C, but flight T_{thx} stabilized near 28–29°C during flight, some 6°C above ambient temperature. (From Heinrich, 1986a.)

ulating solar heating and convection by the mechanisms previously indicated. (It is worthwhile noting that *E. editha* raise their wings in the apparent "reflectance-basking" posture to cool or prevent T_{thx} from rising further. Again, this behavior is explicable by other causes: they perch near the ground, and they are thus allowing the warm air underneath them to rise, not trapping it in under the wings.)

If even temperate and alpine butterflies can be subjected to potential overheating in sunshine (Kingsolver and Watt, 1983), then those species inhabiting the lowland tropics should face even greater thermal constraints. In the lowlands of New Guinea near the equator, the T_{thx} of butterflies rendered incapable of movement and placed into sunshine can rise to 44–52 °C in about one minute (Heinrich, 1972). Nevertheless, some butterflies (even black ones) are active there even in sunshine at midday, and they maintain thoracic temperatures similar to those butterflies from north-temperate regions.

Several strategies of achieving thermal balance were identified for the tropical species. Some very large butterflies (930 mg), like

Papilio aegeus, remain at all times in shade, and at air temperatures of 31°C their T_{thx} stay near 37°C almost exclusively from the heat produced during nearly continuous flight. The much smaller (60 mg) *Mycalesis elia* (Satyridae) also maintain T_{thx} approximately 6°C above air temperature, but their body heat is derived primarily from basking at sunflecks in shady forest. One species, *Precis (Junonia) villida*, with a distribution ranging into the cool coastal regions of Australia, appears unable (or unwilling) to enter shade, even when overheated. Generally perching only on the ground—where T_{thx} can exceed 50°C in one minute—the *Precis* butterflies maintain T_{thx} below 43°C by taking flight (and cooling convectively) whenever overheating is imminent (Heinrich, 1972). It has been reported (Stone et al., 1988) that giant ornithopteran birdwing (2–3 g) butterflies in New Guinea shiver as well as bask in the morning. Foraging peaks in the early morning and evening, and in the heat of the day the animals rest in the shade with their wings closed.

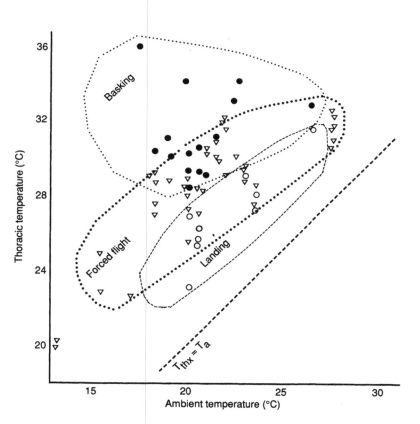

Fig. 2.16 Thoracic temperatures of *C. inornata* during basking (and takeoff), after landing (for undisturbed animals), and during continuous forced flight (for chased animals). These data show that at the lower ambient temperatures (<20°C) the animals cool in flight and need to stop to bask to resume flight at high T_{thx}. (From Heinrich, 1986a.)

Although it has not been critically investigated, I suspect that the large-bodied and thus potentially highly endothermic butter-flies in the tropics have two options. They may either fly very slowly and soar (like some of the birdwings) and thereby reduce the metabolic heat load, or they may fly very fast (like some of the other papilionids) to lose heat convectively despite high rates of metabolic heat production. In addition, diurnal shifts of both activity and habitat (Suzuki et al., 1985; Ohsaki, 1986; Stone et al., 1988), as well as of roosting sites (Rawlins and Lederhouse, 1978), have been identified as dependent on or strongly affected by the necessity to maintain specific body temperatures. When the butterflies are active (as a population), individual T_{thx} is also regulated by altering the relative proportions of basking vs. flight activity (Figs. 2.14–2.16).

A number of highly mathematical biophysical models have been published to predict the activity of certain butterflies. Unfortu-nately, they contain the assumptions that T_b of basking butterflies is within 1 °C of T_b of flying butterflies and that no reduction of T_{thx} occurs in flight. This may indeed sometimes be the case for a specific species, but if so then it is by lucky chance (see Fig. 2.15 and Fig. 1 in Heinrich, 1986b). What the models do show is, as one might predict intuitively, that *basking* butterflies are heated to different T_b by different solar-radiation intensities and different wind speeds. The hours of activity per day have been shown to coincide with the hours of sunshine needed to heat dead models of the butterflies with thermometers inside them. Unfortunately the activity predicted is measured in "hours/day" or "percent flight activity per day" and refers to the *population*. The models say little about whether an individual butterfly stays grounded during 1 percent or 99 percent of this time and is *ready* for instant takeoff and for flying 10 seconds, or whether it actually spends all of this activity time in the ecologically relevant activity, namely flight (Figs. 2.17 and 2.18). The difference between the innumer-able options is not distinguished in models that assume basking T_b is the same as flight T_b.

Geographical Distribution and Body Mass

As already mentioned, body mass is perhaps the major factor affecting flight activity at different ambient thermal conditions. Does mass, because of its thermoregulatory constraints, then vary ecologically? Apparently not. Small butterflies occur in the tropics,

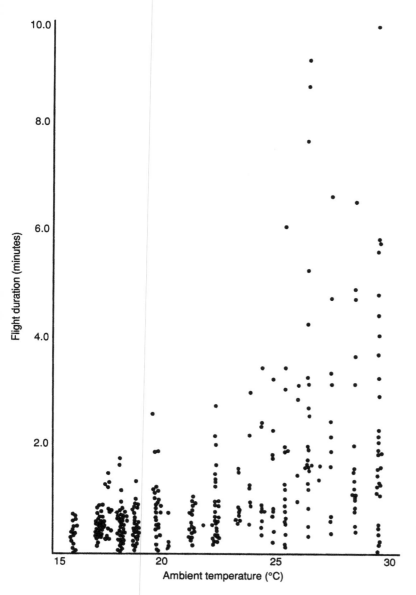

Fig. 2.17 Durations of flight for individual *Coenonympha inornata* as a function of ambient temperature. The butterflies were active only on sunny days. (From Heinrich, 1986a.)

in temperate alpine areas, and in Arctic regions. Furthermore, some of the largest butterflies, including ornithopterans that weigh up to 4 g, are from tropical regions.

Blau (1981) proposes that cool environments with variable sun promote small body size in butterflies because it "allows individuals to heat up rapidly and to make use of short intervals during

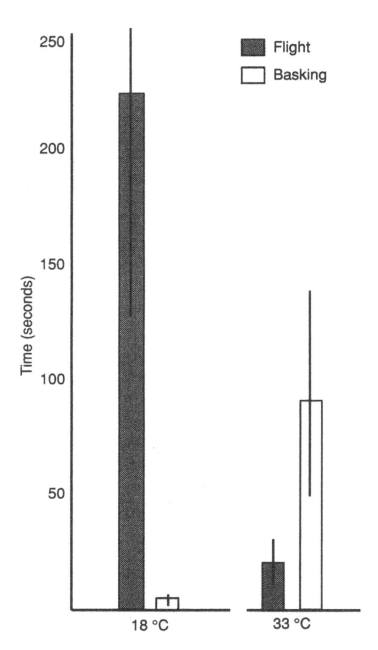

Fig. 2.18 Flight and basking durations of *Precis villida* at <18°C *(left)* and at 33°C *(right)*. (From Heinrich, 1972.)

which thermal conditions are favorable." Where the environment is warm but variable—for example, temperate summer—"larger individuals will more readily achieve optimal body temperatures, and will have the advantage of being better able to regulate their temperature to correct for either overheating or overcooling," therefore they will be more buffered against short-term environmental changes. Where the environment is warm and constant, "overheating will be the primary stress" and "convective cooling during flight in smaller individuals will be more rapid and effective."

This rationale, however, does not explain the available data on T_b, much less patterns of butterfly distribution. Smaller individuals indeed heat up slightly faster than larger ones. For example, under relatively constant conditions at the equator the 41 mg *Orsotriaena medus* heats from 30 to 45°C in 30 seconds of exposure to sunshine, while the 400 mg *Taenaris myops* heats to the same temperature in 70 seconds (Heinrich, 1972). But heating rates are impressively rapid even in the temperate regions; for example, a 26 mg *Caenonympha inornata* in Maine heated from 22.5 to 43°C in 80 seconds (Heinrich, 1986a).

In the field, pre-flight heating rate is not the crucial variable in activity regimens. The crucial variable with respect to mass is not warm-up time but it is, rather, the *flight* time that is bought with the warm-up. For example, moderately sized butterflies weighing near 300 mg (such as the mourning cloak *Nymphalis antiopa*) spend a half-minute longer to bask to reach flight temperature than a butterfly that weighs 10 times less (like *Agriades glandon*), but after having warmed up the mourning cloak can, at 20°C, fly *continuously*, while the smaller butterfly must spend 94 percent of its time grounded and basking at that temperature; individual flights are less than 10 seconds before cooling forces the smaller animal to stop flight and resume basking again. *Agriades glandon*, weighing 24 mg, buys only 8 seconds of flight time with each warm-up at 16°C (Heinrich, 1986b), whereas other Lepidoptera weighing as little as 200 mg are sufficiently large to remain heated by endothermy from *continuous* flight. Like most moths, they need not stop at all.

Large size, therefore, although it admittedly imposes a slight disadvantage in warm-up time by basking, nevertheless confers an enormous advantage in prolonging flight duration before convective heat loss forces another bout of basking (Heinrich, 1986a,b). Indeed, if the insect has very rapid wing beats and is

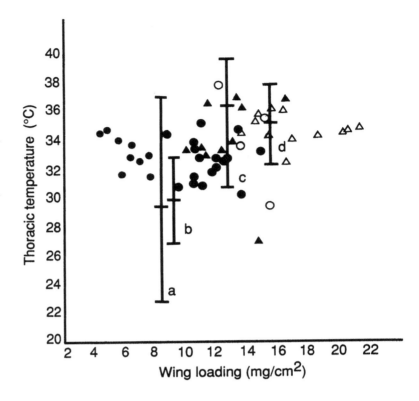

Fig. 2.19 Relationship between wing loading and T_{thx} of butterflies captured in the field from near sea level in Maine and Vermont *(points)* and from near 3,100 m in the Sierra Nevada mountains *(vertical lines)* of California. Four species from Maine and Vermont were sampled: *Vanessa atalanta (open triangles), V. virginiensis (filled triangles), Speyeria cybele (open circles),* and *Papilio glaucus (filled circles).* Vertical lines show mean T_{thx} and range of four species of butterflies sampled from 3,100 m in the Sierra Nevada of California: *a, Agriades glandon; b, Colias eurytheme; c, Occidryas editha; d, Nymphalis antiopa.* (From Heinrich, 1986b.)

large and insulated enough (as large moths are), it can fly without any basking or other intermittent warm-up behavior even at air temperatures near 10°C, or possibly even lower. Therefore, the small size commonly seen in Arctic butterflies, and in many other insects as well, is not likely to be explained by the immediate thermal advantage of a slightly quicker warm-up.

A possible counterbalance for the disadvantage of small size for heat balance in flight is low wing loading, which permits flight with lower T_{thx} (see Fig. 2.19 and Chapter 1). As indicated previously, some small moths can fly even with a T_{thx} close to 0°C, in part because of low wing loading. Similarly, males of the European skipper, *Thymelicus lineola,* can fly at cooler environmental conditions than females can because their lower wing loading permits flight at a 3°C lower T_b (Pivnick and McNeil, 1986, 1987).

Blau's (1981) argument that overheating is a potential problem for large butterflies where environmental temperatures are high and constant is supported. For example, in the lowland tropics,

THE HOT-BLOODED INSECTS

Papilio aegeus (weighing nearly 1 g) remains in shade and apparently never basks; it takes off with T_{thx} equaling ambient temperature (about 32°C), but it soon overheats to near lethal temperatures when flying in sunshine (Heinrich, 1972). A case for thermal constraints on body mass is thus made for *Papilio polyxenes*, which occurs from at least New York to Costa Rica, because the Costa Rican butterflies have thoracic diameters 43 percent lower than the northern ones (Blau, 1981). However, other selective pressures (Blau, 1981) must also be operating to affect body size, or else most of the world's largest moths and butterflies would occur in the Arctic, rather than the hot lowland tropics. Undoubtedly a very major constraint on adult size is larval development, which varies geographically (Ayres and MacLean, 1987).

Coloration

Coloration has only a very minor effect on thermal balance relative to the effect mass has, but the evidence is better that color has in some instances been influenced by selective pressure for thermo-regulation. In *Colias* species melanization of the basal portions of the underwings significantly accelerates basking warming rate (Watt, 1968). As predicted by the thermoregulatory hypothesis, *Colias* species at high elevations have greater melanization than those at lower elevations and lower latitudes both between species and within the same species (Watt, 1969; Roland, 1982). In *C. nastes* and *C. meadii* in the field in Alberta at high elevations (>2,200 m), the darker individuals travel a greater distance per day (Roland, 1982) than do the light ones (147 m vs. 109 m), presumably because of a thermal advantage.

Douglas and Grula (1978) propose that control of melanin deposition in the pierid butterfly *Nathalis iole* maintains sexual activity in cooler environments and facilitates northward range expansion. "Cold-weather" phenotypes in *Pieris* species (Shapiro, 1974, 1977) are largely determined by the photoperiod to which the larvae are subjected (Hoffmann, 1974, 1978; Douglas and Grula, 1978).

The melanization of the proximal portion of the hind wings in *Colias* may have direct fitness benefits. Why then are *Colias* primarily light-colored, both dorsally and on the distal portion of the hind wings? Why has evolution not fixed dark pigmentation? Overheating is not the likely explanation, because in flying butterflies the color of the wing *undersides* would not affect T_{thx}. And basking butterflies should *always* minimize basking time. I suspect the most likely explanation for selection for light coloration is to maximize visual contrast for sexual signaling.

Given the role of the wings previously described in dorsal baskers, it is not surprising that in general the pattern of melanization observed for lateral baskers does not hold for dorsal baskers and non-baskers. For example, many Papilionidae of both temperate and tropical regions are very darkly pigmented, yet many temperate species are primarily yellow. Furthermore, the alpine *Parnassus* butterflies (Papilionidae) are lightly pigmented, although their body and basal portions of the wings are dark and the dark pigmentation may affect T_{thx} and activity patterns (Guppy, 1986). In the Lycaenidae, northerly forms of both *Lycaena feildeni* and *Plebius aquilo* (a blue) show loss of color when compared with their more southerly counterparts (Downes, 1964, 1965), and some butterflies from near the equator in the lowlands of New Guinea are nearly jet black while others are reflective in both the visible and near infrared (Heinrich, 1972). The functional significance of all of these specific color patterns is unknown. But they do affect activity patterns. Are they the cause or the effect of specific activity patterns?

Thermoregulation and the Evolution of Wings

I shall here speculate about the evolution of thermoregulation as related to the evolution of wings. We have many fossils of wings, but none of endothermy and thermoregulation. Thus my attempt may seem to be foolhardy, especially since it is counter to all previously published theories. I'm going out on a limb, but not without some confidence.

It is easy to refuse to speculate, to say "we (who is *that*?) don't know" because whatever happened millions of years ago as regards the body temperature of a mayfly or a dragonfly is not preserved in the fossil record. The safe thing to do is to say nothing (you can't be wrong), or to say that numerous factors affected the evolution of endothermy of insects in complex ways, which is doing the same thing as saying nothing. But I believe there is much indirect evident that tells a story. It goes without saying that numerous interacting factors can and usually do *contribute* to the same end, and that most of them proceed in lock-step fashion (so that they become functionally related), rather than isolated one from another as we must often treat them in order to understand them. But I here speak to the main trend as derived from the combined evidence from observations of all presently existing insect orders. With new data perhaps another main trend will emerge (but I doubt it!).

All agree that endothermy and flight in insects are now related. But I shall show how use of the wings for thermoregulation is a derived rather than a primary function, unlike now commonly supposed. Even in a recent publication (Grodnitzky and Kozlov, 1991) it is assumed from the outset that the need for a high T_b came first; hence the curious conclusion that butterflies have large wings to gather more heat, and that moths have small wings to beat faster so that more energy is expended so that they can be endothermic. What was the driving force in evolution, the need for endothermy or the need for wings?

The insects' evolutionary progenitors were slow-moving creatures inhabiting fresh water, but upon coming onto land they diverged greatly and assumed a much more active life. What most distinguished their new way of life from that of their ancestors, and what probably ultimately caused their radical explosion of adaptation into hundreds of thousands of species of the most bizarre forms and habits, was flight.

The invention of insect flight in the second half of the Paleozoic Era, some 230 million years ago (Wootton, 1981), was not previously achieved by any other group of organisms in the history of life. Only one insect line out of the many then-existing forms evolved flight, and this single line soon underwent extensive radiation into a diversity of forms and species. In the initial "paleopterous" insects there was little differentiation between young and adult forms. But with the evolution of flight came metamorphosis, and that made specializations to radically different environments by larval and adult forms possible. It also made some of the larger insects endotherms.

The published scenario for why insects evolved endothermy and thermoregulation has, in my opinion, put the cart before the horse. And since much of the argument generally accepted until now on the evolution of insect thermoregulation is derived from modeling studies on butterflies, I therefore meet this issue in this chapter.

The predominant role of the wings in butterfly thermoregulation today is irrefutable. This sound experimental fact has many important implications, and it has also prompted a theory on insect-wing evolution that is based on elaborate modeling. This book is obviously not about the evolution of appendages as such, but in view of a now widely held hypothesis linking the evolution of insect *wings* to thermoregulation, the topic cannot rightly be here simply ignored.

The theory that wings evolved from thermoregulatory devices rests specifically on results of experiments by Mathew M. Douglas which repeat the elegant studies of Wasserthal (1975). Douglas (1981) continued work similar to that of Wasserthal, who showed that only the proximal portion of the wings accomplish most of the body warming during basking. Unlike Wasserthal (1975), who shaded the original wings of his experimental butterflies, Douglas instead removed the original wings and attached lobes. The lobes were constructed to mimic the thoracic lobes or "pro-wings" found in some fossil insects. Butterflies with the attached lobes achieved 55 percent greater temperature excess than those without them. Douglas then suggested from these experiments that, as Wasserthal had shown, the lobes trap warm air and simultaneously reduce convective heat loss. He concluded that since the lobes could have little aerodynamic function (Flower, 1964), they therefore could have functioned instead as thermoregulatory devices and later have led to the evolution of insect wings. This new theory added one more to many that had been proposed already (see review by Kukalova-Peck, 1978).

The same idea about the evolution of wings from thermoregulatory devices was later also adopted by Kingsolver and Koel (1985). These authors, using heat-flow dynamics and aerodynamic theory, "proved" that small, lateral projections could indeed pick up heat. Second, they similarly proved that these small lobes could have little aerodynamic function. Roger Lewin (1985) then devoted two pages in *Science* to these "elegant series of studies" and headlined his article with "Experimental data on thermoregulation and aerodynamics give the first quantitative test of a popular hypothesis for the evolution of flight in insects." Stephen J. Gould (1985) then picked up the banner, devoting an article in *Natural History* to the idea in which he credited Kingsolver and Koel for providing "compelling experimental evidence for . . . the origin of wings." Complex structures like wings obviously could not have evolved *de novo*. They must have evolved from some other previous structures that were later modified until they could gradually take on their new function. What were the pro-wings used for? Gould stated that from the experimental work on heating and aerodynamics of wing buds, we now have "the first hard evidence to support a shift from thermoregulation to flight as a scenario for the evolution of wings." And, Gould continued, "we could not hope for a more elegant experimental confirmation" than the work of Kingsolver and Koel (i.e., and Wasserthal and Douglas).

Despite such overwhelmingly strong consensus and enthusiasm for these ideas, I feel compelled to caution that the hypothesis is weak until the following points are answered. First, if wings evolved from thermoregulatory structures in non-flyers, then there should be intermediate stages. There are numerous species of insects with reduced or vestigial wings, but in not a single case are such extremely simple (and presumably easy to evolve) structures as lobes or similar devices used for thermoregulation. *Only* insects that already fly use their wings for heating. Also, to my knowledge, of all the millions of invertebrate animals that do *not* fly, there is not a single example of a lobe or bud or some other structure that functions solely for increasing body temperature. Could one not safely conclude that such lobes or heating structures are not useful now for non-flying animals? If so, why should they have been used for thermoregulation before flight evolved?

The reason that heating panels are not (and probably never were) needed by insects is quite obvious. Because of their very high surface-to-volume ratio, insects have no problem heating up quickly in sunshine. Even a large *Papilio* butterfly with its wings cut off can heat up at an impressive rate of $25\,^{\circ}C/min$ (Heinrich, 1972). Despite the rigor and mathematical sophistication that can decisively prove that a butterfly can heat up at a given, slightly faster rate when it possesses a laterally attached lobe, this result does not address the question of why the difference is important.

Physical models can determine the precise magnitude of response, but they do not necessarily address the biology and the evolutionary question. The evolutionary question does not concern heating rate, *per se*. Instead, the problem concerns whether or not heating should occur at all (in a *flightless* insect), given the costs. For example, if lateral lobes make a flightless animal more conspicuous to predators, require resources to produce, hinder locomotion in terrestrial escape, and increase T_b to a level that results in more rapid metabolism and depletion of energy reserves, then why produce them? Should the insect need to warm up to fight pathogens, increase growth rate, increase rates of egg production, accelerate crawling speed, or the like, then almost explosive heating rates are still possible without any lobes or pro-wings. Surface area is in an insect not likely to be a limiting factor during basking, as it may be in, for example, a stegosaur, whose spines perhaps were dorsal heating plates. But the availability of or capacity to reach sunflecks in a densely foliated swamp might be such a limiting factor.

Theories other than the wing-buds-for-thermoregulation hypothesis have been summarily dismissed as merely being "facile verbal arguments—little more than plausible 'just-so' stories." However, the "elegant" experimental models cited to evaluate the evolutionary question miss the point. The buds-for-thermoregulation hypothesis assumes some critical advantage of a very fast heating-up rate, or the maintenance of an elevated body temperature by basking, in a *non-flying insect*. This is not a good biological assumption. To the contrary, because of the Q_{10} effect it is generally energetically advantageous to maintain a *low* body temperature if at all feasible (that is, when the insect does not need to fly) to save energy resources. No butterfly, nor any other of the millions of extant flying insects, warms up by basking or shivering *except to fly*. Given our wealth of comparative data on insects in all orders, it is much more reasonable to speculate that the impetus for thermoregulation was flight, not vice versa as the buds-for-thermoregulation model a priori assumes. The need for pre-flight warm-up arose after insects became adapted to fly. As I have reasoned before (Heinrich, 1977), and as is indicated by examples throughout most chapters of this book, the only insects that have evolved warm-up mechanisms are those that *adapted* to the heating that they *generate* in flight or in other situations that result in body temperatures above normally encountered ambient temperatures (such as being in a dense crowd or living in an open, sunlit habitat).

The fossil evidence also does not favor the thermoregulatory hypothesis for wing evolution. Rather, it speaks against it. The detailed anatomical examinations of fossil insects by Jarmila Kukalova-Peck (1978) at Carleton University in Ottawa indicate that the pro-wings were not merely lobes such as would suffice to pick up heat. Instead, they were articulated at the base and they were likely used in rowing in water, as are the respiratory appendages of present-day aquatic mayfly larvae that resemble original insect stock. Indeed, the gills of lower Permian mayflies (Fig. 2.20) have an uncannily close resemblance to the pro-wings (Fig. 2.21) of contemporary mayfly nymphs (Kukalova, 1968), and with some

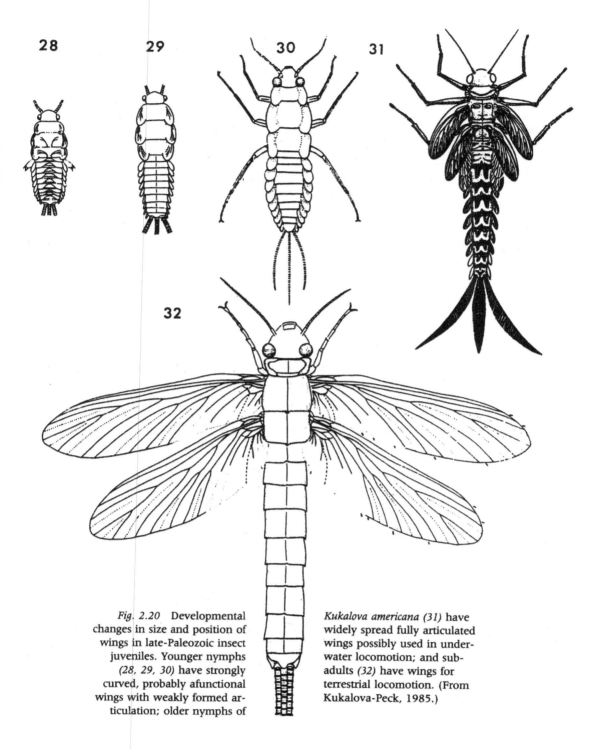

28　　**29**　　**30**　　**31**

32

Fig. 2.20 Developmental changes in size and position of wings in late-Paleozoic insect juveniles. Younger nymphs (28, 29, 30) have strongly curved, probably afunctional wings with weakly formed articulation; older nymphs of *Kukalova americana (31)* have widely spread fully articulated wings possibly used in underwater locomotion; and sub-adults (32) have wings for terrestrial locomotion. (From Kukalova-Peck, 1985.)

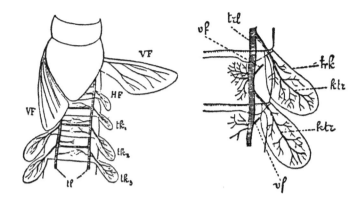

Fig. 2.21 Tracheal gills and rudiments of thoracic wings of the contemporary mayfly *(Cloeon dimidiatum)*, showing internal tracheal trunks and connections to gill trachea. The diagram at right shows the middle abdominal segment of a larva of *Batis binoculates* with details of tracheation of gills. (From Packard, 1898.)

caution (Wootton, 1981) there are good reasons to suppose that wings evolved from gills themselves (Wigglesworth, 1976) or from gill covers (Birket-Smith, 1984) used for locomotion. With respect to locomotion, this is a reverse scenario of the one where the "flippers" used by aquatic birds such as penguins are derived from wings.

In Paleozoic Ephemeroptera (and other insects), as in contemporary Ephemeroptera, each abdominal segment is equipped with a pair of paddle-like gills (Kukalova, 1968) that is essentially indistinguishable from pro-wings (Fig. 2.20). Indeed, Kukalova-Peck's (1978) detailed anatomical studies show that the three pairs of thoracic and nine pairs of abdominal gill plates of Paleozoic mayfly nymphs are strictly homologous to thoracic wings. She points out that the gill plates of mayfly nymphs are sequentially arranged abdominal appendages, always located above the spiracle, which evaginate and develop very much like the thoracic wings. The gills are also provided with true venation, are articulated between the subcoxa and tergum, and are moved by the subcoxo-coxae muscles like thoracic wings.

The pro-wings were richly veined, as would be expected if they were used for gas exchange. Throughout evolution, and especially in the adult stage, the number of winglets became reduced and the richly branched and fairly symmetrical venation became less branched and more asymmetrical (Kukalova-Peck, 1985), as would be expected if their function changed from respiration more to support for locomotion. Present-day ephemeropteran gills are still used for locomotion, and the use of wings (if they are wetted) in providing at least some gas exchange for respiration in present-day forms cannot be ruled out.

The Ephemeroptera are important insects in an analysis of wing evolution, because this group was the direct ancestral line to the Lepidoptera. The wing venation of primitive forms shows that the wings of Ephemeroptera were homologous with those of the odonates as well (Riek and Kukalova-Peck, 1984). Recent insects such as orthopteroids, hemipterans, coleopterans, and hymenopterans in turn also later evolved from the ephemeropteran-like derived stock (Kukalova-Peck, 1987), and it can be concluded that insect flight evolved only once, from the common ephemeropteran-like ancestor (Kukalova-Peck, 1978).

Further support for the idea that insect wings originated from movable appendages such as the mayfly-like gills that are distributed in pairs along the segments of both the thorax and the abdomen comes from neuroanatomy (Fig. 2.21) and neurophysiology (Fig. 2.22). Interneurons involved in the generation of motor activity in *flight* of the locust *Locusta migratoria* (Robertson, Pearson, and Reichert, 1982) and the cricket *Teleogryllus oceanicus* (Robertson, 1987) mirror the segmental repetition of articulated wing appendages on fossil mayfly nymphs (Dumont and Robert-

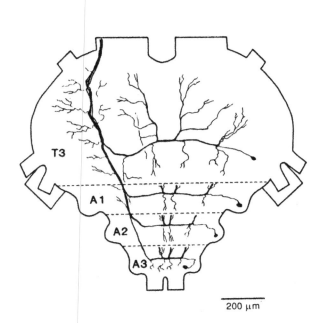

200 μm

Fig. 2.22 Segmental arrangement of flight interneurons in the locusts' metathoracic ganglion mirrors the repetition of articulated appendages of fossil mayfly nymphs. The adult locust *(Locusta migratoria)* ganglion is a fusion of four embryologically distinct ganglia *(dashed lines).* Only one neuron of each bilateral pair is drawn. The physiology of each of the drawn interneurons is identical; they have direct excitatory connections with wing-elevator motor neurons. (From Robertson, Pearson, and Reichert, 1982.)

son, 1986). In the locust, the flight motor neurons are distributed among at least six embryologically distinct ganglia, three of which are abdominal ganglia. The segmental nature of the flight motor neurons of these orthopterans serves no apparent function, but it reflects instead a prior evolutionary stage. The interneurons represent remnants of *previously* innervated lateral appendages along the *abdomen*, which were not wings. Thus the now-similar interneurons in the thorax (which now function in coordinating flight) were also not likely originally specifically adaptive for flight, but instead they represent some evolutionary stage, prior to flight but different from walking, in which the appendages on both the thorax and abdomen served a similar function.

While the remarkable uniformity of wing characteristics in insects (Kukalova-Peck, 1978) suggests that wings evolved only once, something as simple as mere projections or buds to speed up heating should have evolved numerous times, if there were any need for them. It is unlikely that there was need then, especially in the ancestors of present-day insects (which were semi-aquatic and living in warm swamps), any more than there is need for such devices now (in non-flying insects) despite the cooler environments and terrestrial habits of present-day insects.

In the gill theory (Wigglesworth, 1976), as in the thermoregulation theory, one function is presumed to have been taken over by another, namely by flight. Apparently the thermoregulation theory is now so vehemently supported because mathematical formulae and models "prove" that wing buds can heat the body yet not provide aerodynamic support. Would it bolster the argument for the gill theory if models, formulae, and detailed measurements were provided that conclusively proved that structures smaller than those suitable for flight can nevertheless effect some gas exchange? Hardly. However, that is the primary line of logic used to defend the main competing model.

How could gills enlarge to become wings? First, it is helpful to recognize that after the primitive Paleozoic Pterygota developed metamorphosis, the dispersal function of the animal shifted increasingly to the adult. In the larva, moving the gills resulted in ventilation and in locomotion and dispersal. Once the insects' ancestors left the water, the functional significance of the gills could shift to the second function, selecting for morphological changes that reduced the ventilatory and increased the locomotory function. For example, the anterior gill plates—those on the thorax—became enlarged along with their associated muscles, while the more posterior respiratory plates along the sides of the abdomen became reduced (Fig. 2.20).

Kukalova-Peck (1978) suggests that the first aerial excursions occurred as the aquatic insects sought resources on the vegetation above the water and then dropped back into the water to avoid predators. Competition may have driven the insects to crawl out of the water to feed on vegetation above the waterline. The shift to land could have resulted in greater exposure and vulnerability to terrestrial predators, resulting in selective pressure to regain the safety of the water when confronted by such predators. Ultimately, when the escape mechanisms (and flight) were perfected, the animals could exploit both above-water and submerged resources more effectively.

What other selective pressure could have promoted the evolution of wings from gills in semi-aquatic insects, which then, as now, used the gills for locomotion? We don't know. But it is a safe bet that *before flight evolved* there would have been even greater need for dispersal to find food by the nymphs themselves, hence selective pressure on the rowing organs, the gills, to be perfected for ever-better locomotory functions, such as increasing them in size and musculature so that they could alternately or additionally be used for short glides in air. Meeting less resistance in air than in water, startled animals might react by "swimming" even on dry land, similar to the typically rapid movements of wings (Birket-Smith, 1984). Since the gills were already articulated and supplied with muscles (to aid both in aeration and locomotion), they might also have been useful for control of body orientation while dropping and later for flapping flight.

The above scenario makes sense out of a vast array of data on insect thermoregulation. According to this scenario, after flapping flight (with its increasingly greater metabolic demands) had evolved, there then resulted inevitable increases of body temperature and the necessity to operate the muscles near those high temperatures inevitably generated by flight (Heinrich, 1977). The shift to higher muscle temperatures for optimal performance in turn necessitated the evolution of pre-flight warm-up to reach those temperatures encountered in flight using the wings themselves or the flight muscles that drive them. Obviously the evolution of wings is of utmost critical importance to any of a large variety of potential evolutionary considerations of endothermy and thermoregulation. But the role of the wings for thermoregulation is clearly a derived rather than a primary function.

Summary

Butterflies are found from above the High Arctic to the lowlands of the equator, yet in all areas active animals maintain similar and high (20–40°C) T_{thx}, regardless of body size. Since feeding, mate finding, oviposition, and escape from predators all depend on flight, thermoregulation assumes pivotal importance in all butterflies, and butterflies display an amazing array of thermoregulatory responses.

Only a few butterflies have been observed to shiver. The majority rely instead on basking to elevate T_{thx}. Different species exhibit different kinds of basking. Two kinds are observed: dorsal or body basking and lateral basking. During dorsal basking the animal exposes its dorsal body surfaces at right angles to the sun. A dorsal basker on warm substrate holds its wings down, trapping warm air and reducing convection. On cool substrate or up in vegetation, a dorsal basker holds its wings up at an angle to the sun, reducing convection across the dorsal body surface. Lateral-basking butterflies close their wings dorsally and tilt the underside of one set of wings to the sun. They are then heated directly by conduction through the heated wings. Butterfly wings act neither as solar panels to circulate heat to the body, nor as reflectance panels to heat the body by reflected light.

A major source of heat is that produced as a by-product of the flight metabolism. But relative to the heat-producing capacity of their nocturnal relatives, the moths, the endothermy of butterflies is modest because of their lower rates of energy expenditure in flight, their little or no insulation, and their often very much smaller size. Small butterflies cool rapidly by convection, and endothermic heat is of primary significance only to larger butterflies. Flight-generated elevation of T_{thx} is, however, of major importance in the overall activity patterns to some butterflies: At low ambient tenperature large butterflies may stay in flight nearly continuously because they remain heated by their flight metabolism, but small ones cool very quickly in fight and are able to fly for only a few seconds at a time, having to spend most of their time basking.

In lateral baskers melanization and furriness of the basal portions of the wings is of importance in promoting heating and thus increasing the time available for flight. Lateral-basking species from cooler climates are darker than those from warmer areas, and within some species there are seasonal morphs that are less melanized in the warmest part of the season. In dorsal baskers, on the other hand, there is in general no apparent correlation between melanization and ambient thermal conditions.

Body mass affects thermal balance at least an order of magnitude more than does coloration, but data so far do not indicate that body mass has been varied to affect thermal balance, presumably because of other constraints on body size (possibly larval biology). It can be predicted that many Arctic and north-temperate butterflies could fly nearly continuously if they were larger and thus did not cool in flight, and that many butterflies from the hot, tropical lowlands should be small-bodied, to prevent overheating in flight. Many of the larger butterflies occur in the tropics, however, and some of the smallest are at high latitudes and high altitudes. Overheating of the large tropical butterflies is inevitable if they fly in sunshine, and they must remain nearly permanently in shade. The north-temperate and Arctic forms, on the other hand, are obligate heliotherms and for the most part are totally dependent on direct sunshine for activity. Physiological considerations of thermoregulation dictate that flight patterns of butterflies should vary greatly as a function of body mass and climate, but behavioral thermoregulation affects such swift and powerful changes in body temperature at all body sizes that it may override other selective pressures.

The fact that butterflies rely on their wings for thermoregulation has generated a number of studies that appeared to prove that insects evolved wing-like structures that were first used for thermoregulation and were later modified to serve as wings. However, a synthesis of paleontological, physiological, morphological, and comparative evidence indicates instead that flight came first and then inevitably generated heating in the large-bodied insects, which secondarily led to thermoregulation. As first proposed by Sir Vincent Wigglesworth, the most likely scenario for the evolution of insect wings is that they evolved from rowing structures used also for gas exchange. The muscles used for warm-up likely evolved concomitantly to power flight, but flight provided the driving force. In butterflies, the wings then merely provided an economically cheap solar alternative to muscle power in body heating.

Remaining Problems

1. Are there differences in thermal reflectance between tropical, temperate, and Arctic butterflies?

2. What is the thermal significance of the unique basking posture of skippers?
3. What are the ecological correlates between flight patterns, mass, and color?
4. What is the role of predation in the evolution of flight-readiness and thermoregulation?
5. What are the relative advantages and disadvantages of dorsal vs. lateral basking?
6. Why are butterflies so conservative in body temperature despite their wide range into different thermal environments, whereas moths show a great range of T_{thx}?

Dragonflies Now and Then

I T I S difficult to look at a dragonfly today without being reminded of one of those paintings we've all seen of a prehistoric swamp scene, with a giant dragonfly soaring among lush cycads and perhaps an amphibian in the corner pulling itself up from the water onto land. With a wingspread of 70 cm (Wootton, 1981), one of those giant dragonflies *(Meganeura monyi)* that lived in the Permian or Caboniferous coal swamps of some 300 or 400 million years ago may or may not appear to us to be a scaled-up version of a present-day dragonfly. From considerations of insects in general, however, and from extrapolations of thermoregulation in dragonflies as a function of size in particular as reviewed in this chapter, it is almost certain that this winged monster was not only endothermic but also a good thermoregulator, as are many of its smaller cousins today. Insects were likely the first forms of life to have physiological mechanisms of thermoregulation, by at least 50 million years before the dinosaurs.

Dragonflies are fascinating because they exhibit a variety of thermoregulatory mechanisms in the contexts of different lifestyles (May, 1978). Some dragonflies thermoregulate behaviorally, some physiologically, and others do both. Still others show no overt evidence of thermoregulation. As is true of other insects, the differences reflect ecological adaptations that are constrained by body size and flight characteristics.

It had been assumed for a long time, even before measurements of body temperature were taken, that dragonflies thermoregulate. The British entomologist Philip S. Corbet (1963, pp. 125–133) discussed the possible role of various postures and behavior patterns for thermal balance, and he recognized two behavioral types of Odonata: "flyers" and "perchers." Flyers remain continuously on the wing, and perchers make only short flights to pursue prey, mates, and rivals. Michael L. May from Rutgers University (1976a,b,c) presented the first data on dragonfly body temperatures, and his data provided the first direct evidence that some of the perchers as well as some of the flyers regulate their thoracic temperature. The mechanisms of thermoregulation differ greatly between the two and they are still not fully understood.

Flyers

Flyer dragonflies typically cruise back and forth over a specific beat. They are active over a very wide range of air temperature and they are active largely independent of sunshine. Perhaps the best-studied dragonfly with regard to thermoregulation is the green darner *Anax junius*, which occurs over the whole American continent from Central America to Canada. In northern New England it flies throughout the summer, sometimes until late October, and in the summer it is frequently hunts insects on the wing late into the evening after the sun has set.

Like other flyers, *A. junius* warms up prior to flight by shivering (Figs. 3.1 and 3.2) until T_{thx} approaches 35 °C (May, 1979, 1986; Heinrich and Casey, 1978). T_{thx} is regulated in flight (May, 1986; Polcyn, 1989), because the difference between T_{thx} and air temperature declines at increasing air temperature (Fig. 3.3). The thoracic-temperature excess in flight may be as high as 25 °C at low ambient temperatures (May, 1986), as low as a reported 0 °C at high (Polcyn, 1989). (However, the T_{thx} reported by Polcyn were taken "during activity" or "during activity periods," raising the possibility that they do *not* represent flight temperatures. Polcyn's data could instead represent temperatures of animals that stopped flight at air temperatures higher than 40 °C because they were overheating.)

The mechanism whereby *A. junius* reduces its thoracic-temperature excess in continuous free flight most likely involves shunting excess heat from the thorax to the abdomen. The thorax is the site of heat production during flight, and the cylindrical abdomen, 4–5 cm long and some 3 mm wide, is a potential heat radiator for the dissipation of excess heat to the environment by convection. This possibility was experimentally investigated. Potential overheating in flight was mimicked by applying heat to the thorax with a focused beam of light from an incandescent lamp. When

Fig. 3.1 *Anax junius* during shivering warm-up.

T_{thx} was increased to near 40°C, T_{abd} increased by over 10°C, while in control (dead) animals T_{abd} increased by only 1°C. Most of the temperature increase in T_{abd} of live animals was not due to conduction, since the control animals had only a 1°C rise in T_{abd} near the thorax and since the *posterior* tip of the long abdomen in live animals increased in temperature almost as rapidly as that portion closest to the thorax (Heinrich and Casey, 1978). Furthermore, the ventral portion of the abdomen (where blood initially flows from the thorax) always warmed up to a higher temperature than the rest of the abdomen. The temperature gradient between the ventral and dorsal abdomen, a very short (3–4 mm) distance, is much greater than the gradient over the much longer (4–5 cm) distance between the anterior and posterior abdomen. This implies that very rapid heat transfer into the abdomen is initiated along the abdomen's ventrum. These results are all consistent with the hypothesis that blood circulation serves as the vehicle of heat transfer, but without further experiments they do not prove it.

Dragonflies exhibit conspicuous later-abdominal pumping ("breathing") movements that are potentially related to water loss by evaporation (Miller, 1962). When a dragonfly is heated on the thorax, these breathing movements are usually in synchrony with the heart pulsations (Heinrich and Casey, 1978). Therefore, both heart beats and abdominal breathing movements are correlated with abdominal heat transfer, and the T_{abd} profiles do not by themselves show what is driving the heat transfer.

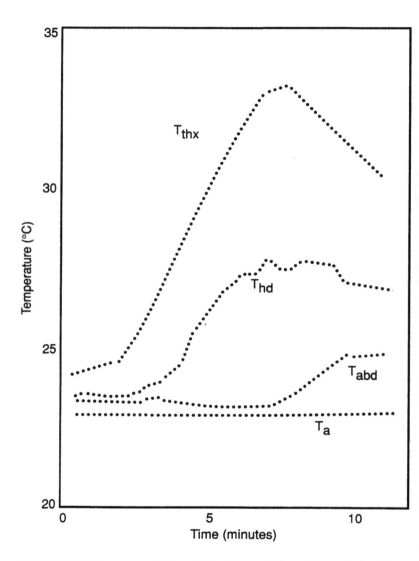

Fig. 3.2 Simultaneous variation of T_{thx}, T_{hd}, and T_{abd} of *Anax junius* during shivering warm-up and consequent cool-down. (From May, 1986.)

THE HOT-BLOODED INSECTS

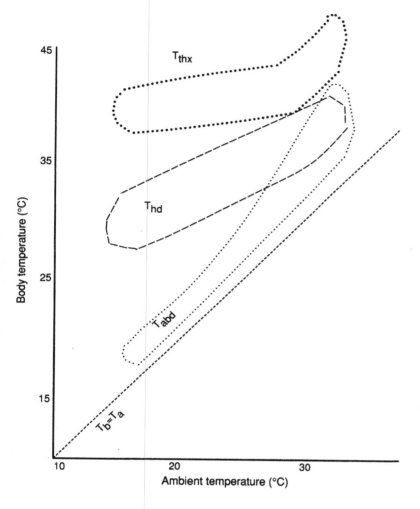

Fig. 3.3 Relationship of T_{thx}, T_{hd}, and T_{abd} to ambient temperature in *Anax junius* during free flight in the field (from May, 1986). The data of Polcyn (1989) for T_{thx} and T_{abd} of *A. junius* "during activity" are very much different (see text).

In an experiment to determine whether it is the abdominal pumping or the heart pumping that is functional in heat transfer (Heinrich and Casey, 1978), the abdominal pumping movements were abolished by waxing the ventral portions of the tergites of each abdominal segment together (Fig. 3.4). Heat transfer to the abdomen occurred before the operation and after it (Fig. 3.5), and again after the wax was removed. However, no abdominal heating occurred in dead animals and in those in which the heart was pinched shut (anywhere along its length). The fact that there was a decrement in heat transfer after breathing was stopped could suggest that the "breathing" movements help blood circulation, as in bumblebees (Heinrich, 1976), but this possibility has not yet been investigated. Even in *Libellula saturata*, which does not transfer heat into the abdomen, heart pulsations and abdominal breathing movements are usually synchronous (Fig. 3.6), suggesting that the coordination of the breathing and blood-circulation systems in insects is a general and not strictly a thermoregulatory phenomenon.

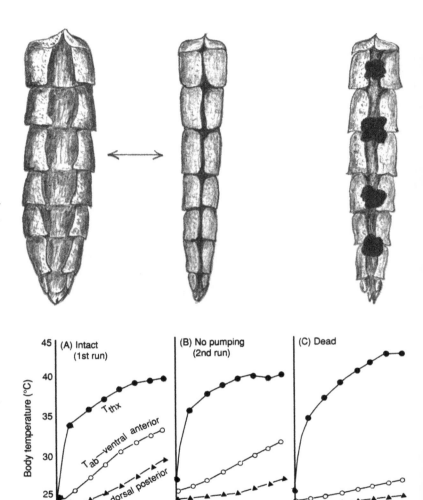

Fig. 3.4 Diagrams of the abdominal underside of a dragonfly showing expansion during inspiration *(left)* and contraction during expiration *(middle)*, when the air is pushed through the thorax. The diagram at far right shows how abdominal pumping motions were abolished by waxing the ventral portions of the tergites of each abdominal segment together. (From Heinrich and Casey, 1978.)

Fig. 3.5 Body-temperature changes during thoracic heating of the flyer dragonfly *Aeshna multicolor*, showing abdominal heating even when the abdominal pumping movements were abolished (from Heinrich and Casey, 1978).

THE HOT-BLOODED INSECTS

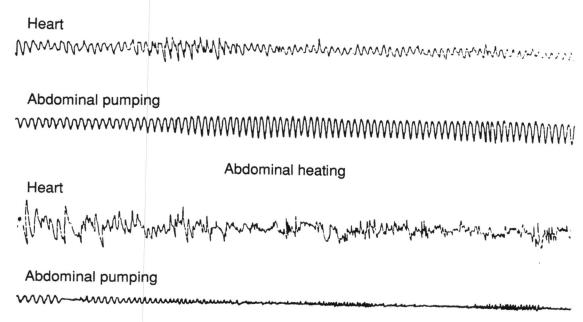

Thoracic heating

Heart

Abdominal pumping

Abdominal heating

Heart

Abdominal pumping

10 sec

Fig. 3.6 Mechanical activities of the heart and concurrent abdominal pumping during first thoracic and then abdominal heating. During the portions of the runs indicated, thoracic temperature increased from 29 to 39°C, and abdominal temperature increased from 26 to 32°C. Both heating experiments were performed on the same *Libellula saturata,* and the electrical leads remained in the same location. Note the amplitude increase of abdominal pumping and the synchrony of abdominal and heart pumping during thoracic heating. During abdominal heating, the two were not synchronous. (From Heinrich and Casey, 1978.)

When the heart is exposed (Fig. 3.7) and tied off, heat transfer is totally abolished (Fig. 3.8), although similarly operated animals in whom the heart is not pinched off transfer heat normally (Fig. 3.9). In animals which are heated on the thorax, pinching the heart shut results in an immediate increase to lethal T_{thx}, with a concomitant decrease in T_{abd}. These laboratory results show that the heart is responsible for the heat transfer by the blood from the thorax to the abdomen. The necessary experiments of measuring T_{thx} and T_{abd} of continually flying dragonflies with ligated hearts have not yet been done, but May (1986) measured in the field that the T_{abd} of flying *A. junius* are more elevated above air temperature at high rather than at low air temperature (Fig. 3.3). This observation is consistent with the idea that reductions of T_{thx} are achieved by dumping heat into the abdomen. Because head temperature varied in parallel with T_{thx} (May, 1986), we know the head does not serve as a heat radiator for stabilizing T_{thx}. The T_{abd} data of Polcyn (1989), in contrast, are inconsistent with the hypothesis that heat is dissipated through the abdomen possibly because his *A. junius* were "active" (flew sometimes?) rather than guaranteed in continuous flight.

Interestingly, although *A. junius* prevent heat flow into the abdomen during shivering warm-up (May, 1976; Heinrich and Casey, 1978), they initiate abdominal heating almost immediately

Fig. 3.7 Lateral view of the thorax and part of the abdomen of *Anax junius*, including an inside view of the thorax to show flight muscles as well as the dorsal aorta. A portion of the heart is lifted out with a pin to be tied shut to interrupt blood flow during thoracic heating (see Figs. 3.8 and 3.9).

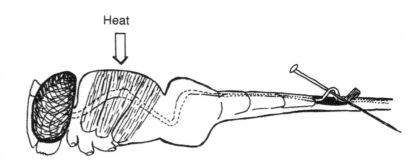

Fig. 3.8 Stabilization of T_{thx} by transfer of heat to the abdomen in an *A. junius* dragonfly heated on the thorax; after the heart was pinched to interrupt blood flow to the thorax, T_{thx} rapidly increased. (From Heinrich and Casey, 1978.)

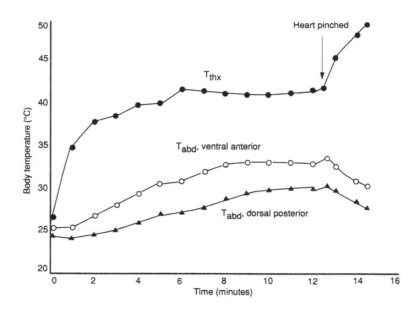

THE HOT-BLOODED INSECTS

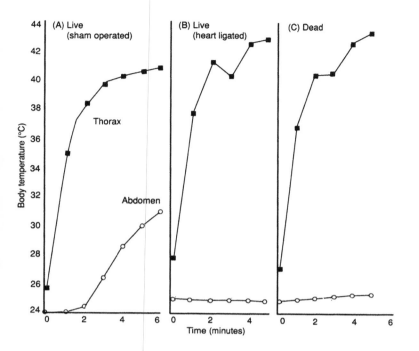

Fig. 3.9 The change in T_{thx} and T_{abd} (near the posterior end) of an *A. junius* while heat was focused onto the thorax. The sham-operated individual transfers heat to the abdomen, but when the heart is ligated (or when the dragonfly is dead), no abdominal heating occurs. (From Heinrich and Casey, 1978.)

when heated with a heat lamp in the laboratory. The reason for the latter response remains obscure. At least in some dragonflies the temperature of the abdomen itself has significance. In female *Erythemis simplicicollis*, increases in temperature from 23 to 33 °C of the tip of the abdomen increased egg-deposition rates 2.5-fold (McVey, 1984). So far, however, there is no evidence that T_{abd} is regulated in any dragonfly, possibly indicating that no studies have been done or that T_{abd} is indeed not regulated.

The same physiological response of heat dissipation from the thorax is found in another aeshnid dragonfly, *Aeshna multicolor* (Heinrich and Casey, 1978), but little is known about mechanisms of T_{thx} regulation in other flyers. For example, the relatively small-bodied flyer (~180 mg) *Tetragoneuria cynosura* regulates at least as well as *A. junius* in the field, though its T_{abd} closely tracks air temperature over the whole range of air temperature where it flies (May, 1987), making accelerated heat transfer to the abdomen unlikely as its primary mechanism of thermoregulation. May (1987) speculates how these dragonflies might regulate their T_{thx} in flight and suggests that the animals may vary their external heat load by shuttling between sunshine and shade. No single mechanism seems to account for the thermoregulation that is observed, however.

Another proposed mechanism for reducing the temperature excess in flight involves modulating the rate of heat production by varying flight activity. Dragonflies have a relatively large wing area for their body mass and are capable of gliding, and a number of observers (Hankin, 1921; Corbet, 1963; May, 1976a; Polcyn, 1989) report increased incidence of gliding rather than powered flight on hot sunny days in *Anax*, *Hemianax*, and various Trameinae. May (1976a) reports that at high ambient temperatures in sunny weather, *Tramea* and *Tauriphila* spend most of their time gliding, with occasional wing flaps. At dusk or during cloudy periods, their flight appeared to be more erratic, with little gliding. The idea that metabolic-heat input is decreased on hot days to prevent overheating needs additional testing. For example, perhaps prey availability also affects hunting techniques and flight behavior, and prey availability may vary as a function of weather.

If gliding is a mechanism for reducing temperature excess then it should be especially relevant to species that have no physiological control of heat dissipation. However, the behavior is not restricted to those species lacking physiological control of heat loss. May (1990) observed *Anax junius* one morning in Florida from 7 to 10 A.M. as air temperature increased from 24 to 31°C and radiation from 40 to 780 W/m². Initially flight was "rapid and powered by nearly continuous wingbeat," but by 10 A.M. "flight speed was reduced and long periods of gliding were interspersed with occasional shallow wingbeats." Similar observations were reported by Polcyn (1989) of *A. junius* in the hot environment of the Mojave Desert.

Dragonfly wing beats are shallow, and it is difficult to identify periods of gliding by eyeballing. Even then, relatively few quantitative observations are available, although May (1978, fig. 6, p. 38) published data showing *Tramea carolina* gliding approximately 20 percent of the time at 20°C, to about 50 percent of the time at 35°C. More than doubling the gliding time over this range of air temperature, where the temperature excess halved (May, 1976a), could potentially explain all of the thermoregulation observed in this species. May (1976a) also reports, however, that *Tramea carolina* "frequently" flies at high ambient temperatures with its abdomen depressed sometimes at angles 45° or more below the horizontal. This behavior may reduce exposure of the abdomen to the sun and probably also increase convective heat loss, but it is the *thorax* which would be overheating in flight, not the abdomen. Furthermore, flying with the abdomen at an angle of 45° would indeed increase convective heat loss (from the abdomen), but by reducing flight speed it would have the opposite effect on the thorax, where increased convective cooling would be useful. It is also not clear how the animals simultaneously glide at high air temperatures and hold the abdomen at an angle, as that would cause wind resistance. In short, the data so far available are intriguing, but they do not convincingly prove that gliding functions in thermoregulation.

Perchers

Despite sometimes similar body masses, flyers and perchers are widely different in morphology. In two comparative studies, perchers were found to have a much higher ratio of thoracic to total body mass: the thorax of flyers is about 44 percent of total body mass, that of perchers is near 65 percent (Heinrich and Casey, 1978; Marden, 1989). In perchers abdominal mass is near 16 percent of body mass, in contrast to 31–35 percent in the flyers. The abdomen of perchers is generally flattened dorso-ventrally, while it is cylindrical in flyers.

Marden (1989) indicates that the high flight-muscle ratio (flight-muscle mass/body mass) is especially pronounced in male perchers and that it serves them for rapid acceleration in aerial contests over females. If flight performance is optimized, then it can be predicted that thoracic temperature should be optimized as well, to take advantage of the existing high muscle ratio, but this possibility has not yet been investigated. However, as in other insects, the T_{thx} of dragonflies can vary depending on the task at hand: in *Sympetrum obtrusum*, males that carry females during mate guarding (by grasping them by the neck) have consistently higher T_{thx} than males that guard without contact (by hovering or perching nearby) (Singer, 1987).

Percher dragonflies are the phylogenetically more "advanced" dragonflies, but their circulatory system is smaller and blood volumes are 2–4 times lower than in flyers. In *Libellula saturata* (Heinrich and Casey, 1978) and in *L. pulchella, L. luctuosa, Plathemis lydia*, and *Pachydiplax longipennis* (M. May, personal communication), there is little if any evidence for physiologically facilitated heat transfer between thorax and abdomen. It is not clear what the penalty would be to these advanced dragonflies

Fig. 3.10 Basking posture of a percher dragonfly. The wings are used to mantle the thorax, thereby reducing convective cooling.

not to thermoregulate physiologically. But I speculate that perhaps the increased blood volumes required would increase the flight load and result in a decrease of flight maneuverability that could then impair territorial defense and mating success. Nevertheless, behavioral thermoregulation in some perchers is quite varied and complex and often quite precise (May, 1976a; Pezalla, 1979; Polcyn, 1988), while others are thermal conformers (Rowe and Winterbourn, 1981; Shelly, 1982).

Whereas flyers are often active before sunup, after sundown, and on overcast days, those perchers that are baskers are primarily active only at times and locations where sunshine is available. For example, in California in late June, *Libellula saturata* were observed only about an hour after the sun reached their local habitat, a sluggish stream. At air temperatures of 24–25°C they spent 100–200 seconds per basking bout, with the dorsum of the thorax and abdomen oriented nearly at right angles to the solar radiation (Fig. 3.10). At 27–31°C they perched (basked) for only 20–25 seconds. In the hot afternoon, at 33–37°C, the animals ceased to bask (Fig. 3.11), but instead spent long durations perched on plant stems in the shade with their abdomens inclined vertically instead of horizontally.

Flight activity in *L. saturata* is also strongly affected by air temperature. At 24–26°C mean flight durations at the California site were 67 seconds. Near midday, at 29–33°C, these "perchers" flew almost continuously. Uninterrupted flight durations could exceed 8 minutes, although others were as short as 10 seconds. At 36–37°C there was a precipitous decline in flight durations in all individuals, and mean flight time was dropped to 11 seconds. At air temperatures lower than 26°C, the ratio of flight to perching time was 0.6, but above 36°C the ratio declined nearly 9 times, to 0.07 (Heinrich and Casey, 1978).

May (1978) suggests that there is no convincing evidence that activity of dragonflies is prevented by high temperatures, because activity patterns of prey, humidity, and interspecific competition for space (May, 1977) could result in activity patterns correlated but not caused by variation in temperature. Nevertheless, the simultaneous changes of flight durations and perching times that are correlated with basking and heat-avoidance postures are indicative of behavioral modification to temperature, as such. If so, they indicate that these perching dragonflies, which have more massive thoraxes than the blood-cooled thoraxes of the flyers, are severely restricted in flight duration at air temperatures higher than 35°C because of their inability to physiologically cool their flight motor as the heat load from flight accumulates.

Dragonflies tethered in sunshine in wind-still conditions can heat to near 50°C (Heinrich and Casey, 1978), which is lethal to them (May, 1976a) in 2–3 minutes. Not surprisingly, perchers that are continually subjected to direct sunshine have mechanisms of avoiding overheating when perched. As already mentioned for *L. saturata*, and also seen in *Hagenius brevistylus* (Tracy, Tracy, and Dobkin, 1979), one mechanism is simply to minimize exposure to the sun, by perching in the shade, facing the sun, or orienting the body parallel to the sun's rays (May, 1976). Another common posture observed by Corbet (1963) and others (May, 1976a) is the "obelisk" position, in which the dragonfly appears almost to stand on its head and point the tip of the abdomen to the sun (Fig. 3.12). In this posture the animals appear to be using the

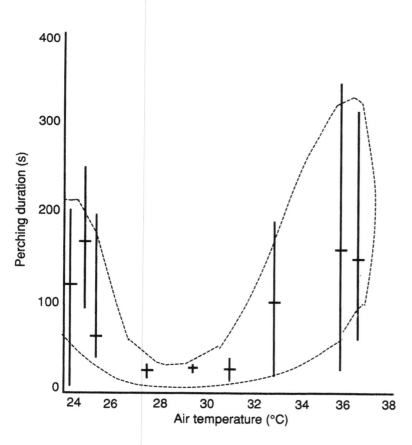

Fig. 3.11 Amount of time spent perching by *Libellula saturata* is a function of air temperature. At low temperatures the animals perch to bask, and at high temperatures they perch to avoid heating. (From Heinrich and Casey, 1978.)

Fig. 3.12 The obelisk posture of a percher dragonfly (with the sun directly overhead).

abdomen as a parasol to shade the thorax, somewhat as desert ground squirrels from North America (Chappell and Bartholomew, 1981) and South Africa (Bennett, et al., 1984) use their bushy tail to shield themselves from the direct sun.

Percher dragonflies usually have colorful abdomens used in sexual signaling. According to one hypothesis, at least in males of *Pachydiplax*, the dragonflies raise their abdomen not for thermoregulation but to display their colors. This posture is often assumed by females away from breeding sites, however. Furthermore, I've observed that if an individual in the obelisk position is shaded it will lower its abdomen, only to resume the position when the shade is removed. The animals also assume the obelisk posture in the laboratory when heated with a lamp, and it prevents or slows further rise in T_{thx}. The obelisk position is not seen in *Pachydiplax* until the animal is exposed to high air temperatures (May, 1976a). Similarly, it is not seen in *Erythemis collacata* in the forenoon at temperatures lower than 30°C, but it is routinely seen in the afternoon at high (>36°C) ambient temperatures (Heinrich and Casey, 1978). There seems little doubt therefore that the obelisk posture indeed serves a thermoregulatory function, but a simultaneous sexual-display function need not be excluded.

Although some of the larger species of Gomphidae shiver (May, 1976a), most other perchers probably warm up only by basking. At least in a sample of 9 perchers from Maine, none showed evidence of shivering warm-up, whereas all of a sample of 6 flyers did shiver (Vogt and Heinrich, 1983). When perchers were experimentally set out in the field at night so that they would either receive the morning sun or remain shaded, they remained grounded until T_{thx} increased (by rising air temperatures or by direct warming by the sun) to 4–12°C above the minimum for clumsy flight. Thus, those perchers receiving direct sunshine were able to take off several hours earlier than those perched in the shade. Presumably the animals waited to be warmed up until T_{thx} was sufficiently high for vigorous flight because as clumsy flyers they would be poor predators (they feed on other insects) as well as easy prey for birds. Nevertheless, some neotropical forest damselflies utilize shaded perches and maintain relatively low T_{thx} (Shelly, 1982), perhaps because shaded forest affords a large niche that is largely unexploited by heliothermic species.

Perchers typically bask, as butterflies do, by orienting their body at right angles to the solar radiation, thus maximizing the amount of sunshine intercepted. The degree of depression and pronation of the wings varies, but there is a clear tendency for perchers to elevate their wings at high air temperatures and to hold them down and forward around the head and thorax (Fig. 3.10) at low temperatures (May, 1976a). May (1976a) showed that when dead *Pachidiplax* with their wings positioned down were heated with a heat lamp, they tended to equilibrate to higher T_{thx} than specimens in the flat position. This mechanism for the elevation of T_b is presumably the same as that worked out by Wasserthal (1975) for dorsal-basking butterflies; holding the wings down and around the thorax causes the warm air to be trapped around the wing base, retarding convective cooling (Tracy et al., 1979).

Although most dragonflies are either heliotherms or endotherms, there are some species that combine both strategies. Gomphidae warm up by basking but during cloudy weather some of the larger species shiver during pre-flight warm-up (May, 1976a). In Panama, *Micrathyria atra* fly nearly continuously at air temperatures from 22 to 24°C, maintaining T_{thx} 10–15°C above air temperature (Fig. 3.13). At 29–30°C, however, they drastically reduce flight time to about 20 percent and prevent overheating by taking the obelisk posture (May, 1977). The transition from flight to perching occurs at T_{thx} of 33–38°C, which is, however, still several degrees below the maximum T_{thx} that is tolerated.

Few studies are available on the smaller, weak-flying damselflies (Zygoptera). All are "perchers" but, as expected because of their small size and weak flight (which together preclude appreciable endothermy in flight), shivering has not been observed in them, and they either bask or remain totally in shade, depending on the species (Shelley, 1982). The shade-seeking animals have T_{thx} nearly identical with ambient temperature, but sun-loving species in Panama have thoracic temperatures 4–8°C above air temperature.

Color Change

Bright colors in some dragonflies, as in butterflies, function in sexual signaling, and color may additionally be involved in thermal balance. Numerous species in five families of Zygoptera and Anisoptera, for example, undergo reversible temperature-dependent color changes (Jurzitza, 1967; Sternberg, 1987) from bright blues (or red) to dark browns by the migration of pigment in epidermal chromatophores. The distribution of these chromatophores varies among species, occurring in some on the dorsal and

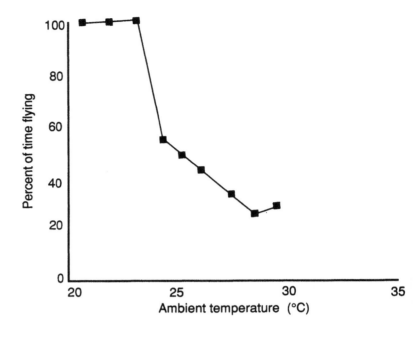

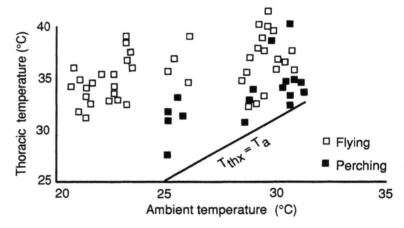

Fig. 3.13 Activity *(top graph)* and thoracic temperatures *(below)* of flying and perching *Micathyria atra* as a function of ambient temperature. (From May, 1977.)

THE HOT-BLOODED INSECTS

lateral surfaces of most of the body to only portions of one or two abdominal segments in others. All of the chromatophores, however, respond to temperature changes, and in general, those of species from warmer localities have higher thresholds for the dark-to-blue change than those from colder areas.

The dragonflies with ability to change color darken at night so that they are in the dark phase at dawn and throughout those days when it is overcast and they do not fly. Some percher dragonflies that show physiological color change (Veron, 1973, 1976) roost at night in locations where direct sunlight reaches them within a few minutes of sunrise (Veron, 1973, 1974). In general percher dragonflies do not begin flight until the sun strikes them, thus there can be several hours' difference in the initiation of activity between different individuals, depending on their perching site (Vogt and Heinrich, 1983). It is likely, however, that some weak flight may occur from shaded to nearby sunny areas where the animals can then bask and acquire optimal flight temperature (Veron, 1974).

Once in the sunshine, the dark-phase dragonflies orient within 1–3 seconds into a basking posture, where the body is at right angles to the sunshine (Veron, 1974). If the dragonfly lands in a shadowed spot it does not attempt to orient on landing, but if the shadow is removed then the insect reorients itself within 1–3 seconds. In the early morning before the insects have warmed up (and turned to the blue phase) the flights are short and generally oriented (as in stationary basking) at right angles to the sun's radiation, in contrast to later (after they have turned blue) when flights are longer and less accurately oriented to the sunshine. Veron (1974) released both blue- and dark-phase *Austrolestes annulosus* and *Ischnura heterosticta* in the field during the early morning, observing that both color phases orient, although "it was noticeable that the blue phase insects took 1–2 seconds longer to do so." When flying, the dark-phase insects remained oriented perpendicular to the solar radiation 95 percent of the flights observed, but blue-phase insects remained oriented in flight only if they happened to be launched at right angles to the incident sunlight (about 20 percent of the flights). Furthermore, when the eyes of 10 *A. annulosus* were coated with a rapid-drying, opaque black, water-based paint, they were still able to orient normally when perched.

It had been previously proposed (O'Farrell, 1963, 1964, 1968) that the dark-phase coloration at least in *Diphlebia* species enhances the absorption of solar radiation in the morning to permit early-morning activity. But no data show this. The difference in heat gain between blue- and dark-phase individuals of *Austrolestes annulosus* is relatively slight; the final thoracic-temperature difference (after 4 minutes of heating in sunshine) is only 0.23 °C higher in the dark phase than it is in the blue phase. This temperature difference is very small when compared with the change of up to 20 °C in body temperature that results from orienting the body in sunshine. Furthermore, most of the color change is local, often in the abdomen, and there is little or no reason to increase T_{abd}. How, then, can abdominal darkening aid in thermoregulation?

Veron (1976) proposes that the significance in the abdominal color change is related not to enhancing heating rate directly but to promoting rapid *orientation* to solar radiation, particularly in flight. He suggests that the dark-phase cuticle functions as a thermal receptor of incident solar radiation. It can be expected that the cuticle surface of the abdomen would heat very rapidly; it is very thin and has very little thermal inertia because it is insulated from the abdominal contents by a layer of insulating air sacs. Thermoreceptors underlying the cuticle could quickly detect changes of body-surface temperature that would in turn allow the insect to adjust its orientation. To my knowledge, however, there is presently no proof that such thermoreceptors exist and this hypothesis also remains untested.

Temperature Tolerance and Torpor

May (1976a) determined the "maximum voluntary tolerated" (MVT) temperatures of dragonflies by gradually heating animals with implanted thermocouples and noting T_{thx} when the insects avoided further heating by moving to the shade of the stick on which they were perched, assuming a special heat-avoidance posture, or flying away. The animals were then restrained and heated again until motor control was lost, to determine "heat torpor" (HT) temperature. The use of MVT is widespread in the insect literature, but I'm not always sure what it means.

The general assumption is that an *active* animal, such as a flying dragonfly, would act to avoid further heating when it encounters the MVT as determined (for a *resting* animal, as is always the protocol) in the laboratory. However, unlike vertebrates, resting insects commonly have mechanisms to keep *cool,* so that they may conserve energy (resting metabolic rates would be very high for any animal that tried to maintain its T_b at the high temperatures necessary for flight). In other words, I speculate that MVT, as commonly measured, may have little to do with the maximum voluntary heat tolerance during activity. The MVT temperature may greatly underestimate the T_b that may not only be tolerated during flight, but that the insects may even feel "comfortable" with during flight. Indeed, data show no correlation between T_{thx} in flight and MVT of several species of dragonflies, inasmuch as half, all, or none of the observed flight temperatures were *above* the lab-measured MVT in the flyers *Anax junius*, *Macromia taeniolata*, and *Myatheria marcella*, respectively (May, 1976a). Dragonflies spend at least half of their adult lives at rest, and there is no proof that MVT has any relevance to flight T_{thx} in flyers. In perchers, however, the laboratory situation could mimic the field situation more closely.

The temperatures at which heat torpor (HT) is reached, on the other hand, are clearly physiologically rather than behaviorally relevant. In most species examined, they are generally narrowly defined between 45 and 47°C (May, 1976a; 1978).

Both MVT and HT temperatures show seasonal and/or geographic changes. MVT tends to increase from cooler to warmer habitat, although HT tends to remain constant regardless of habitat (May, 1978). Minimum T_{thx} for takeoff are also lower for some dragonflies of northern than of southern distribution. Several species in Maine leave from their perches with T_{thx} of 17.5°C after their nighttime dormancy (Vogt and Heinrich, 1983), whereas similarly sized dragonflies from Panama and Florida are still unable to fly at thoracic temperatures as much as 2–3°C higher than that (May, 1976a).

As in other insects so far examined, body mass has a clear effect on thermoregulation in dragonflies; the MVT and HT are relatively independent of mass, but the larger the dragonfly, the higher the minimum T_{thx} for flight (Fig. 3.14), and the more independent is the T_{thx} from ambient temperature in flight (May, 1976a). Flyer dragonflies require a higher takeoff T_{thx} for any given body mass, as predicted from an evolutionary perspective, given their unavoidable heating in flight (Heinrich, 1977). Curiously, as in moths (see Chapter 1) and Diptera (Morgan, Shelly, and Kimsey, 1985), species with lower mass tend also to have lower wing loading (Fig. 3.14). Since lower wing loading allows insects to fly with lower T_{thx}, the inevitably lower T_{thx} due to small mass has

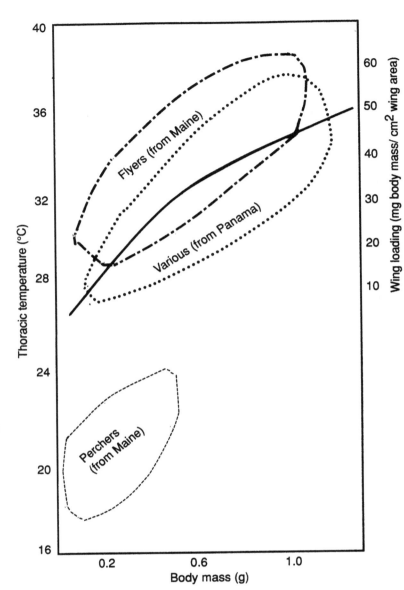

Fig. 3.14 Thoracic temperature at spontaneous initiation of flight in flyer and percher dragonflies in Maine (from Vogt and Heinrich, 1983) and in a variety of dragonflies from Panama (from May, 1976a). At least in the Panamanian dragonflies, higher T_{thx} at takeoff is associated not only with higher mass but also with higher wing loading. Wing loading of the various Panamanian species is depicted by the heavy line (adapted from May, 1978).

probably placed selective pressure on small insects to develop larger relative wing sizes.

In all insects the power of lift that the flight muscles can generate is a function of muscle temperature. However, quantitative descriptions of power generation (which includes force per wing beat and wing-beat frequency) vs. muscle temperature that are

THE HOT-BLOODED INSECTS

suitable for comparative studies of adaptation are rare. One set of data on the New England dragonfly *Libellula pulchella* by James H. Marden shows that power output during flight rises as T_{thx} increases steadily from 15°C to near 45°C, but then it declines precipitously at T_{thx} greater than 45°C (Fig. 3.15). Interestingly, in older males the power curve shifts to a higher and narrower T_{thx} range than is found for younger animals (Marden, personal communication).

The shift in thermal dependence of flight performance with age (Tsubaki and Ono, 1987) could presumably be caused either by changes in the muscles or by neurophysiological and behavioral changes. It is worth noting that immature males, who produce near maximum lift already at muscle temperatures of 28–30°C, are primarily sit-and-wait predators. They do little continuous flying. Mature males, on the other hand, spend a large portion of their time during their active period in flight (Pezalla, 1979), often engaging in strenuous maneuvers (contesting other males, chasing

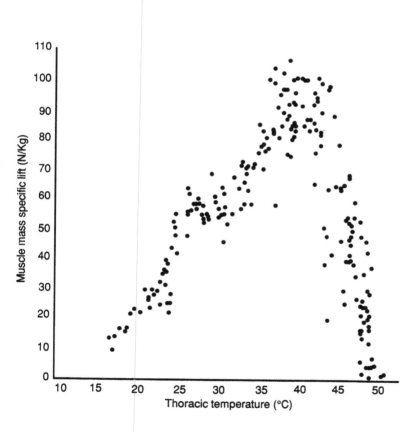

Fig. 3.15 The power or lift generated by the flight muscles of a nearly mature *Libellula pulchella* male (mass = 507 mg) as a function of T_{thx}. (From J. Marden, unpublished.)

females, guarding ovipositing females) that likely elevate their T_b. Thus, the changing thermal dependence of the flight motor (the thoracic muscles and neurons driving the wings) is reflected in changes of behavior in the field.

Protodonates, the World's Earliest Endotherms?

In all major taxonomic groups of insects so far investigated, T_{thx} in flight is largely a function of body mass. All medium- to large-bodied insects (0.2 g or more) that fly on their own power generate appreciable (15°C or more) temperature excess as a result of their flight metabolism (although insulation has a modifying effect on temperature excess, as does the vigor of flight).

Endothermy in flight, as such, need not imply thermoregulation (which, as noted before, we define as the independence of T_{thx} from ambient thermal conditions). Nevertheless, that correlation exists in insects showing a very large temperature excess—or *high endothermy*—as is usually the case for very large insects. The reason that conspicuous endotherms also thermoregulate is two-fold. First, the animal's biochemical machinery (enzymes, membranes, etc.) has to adapt to those (high) temperatures generated and encountered in flight if they are to continue to fly. Therefore, large animals will either have to remain continuously warm or have the means of warming up by some mechanism before beginning flight. Conversely, all small insects that do not heat up *in* flight do not warm up *prior* to flight, some even flying with T_{thx} near 0°C (see Chapters 1, 5, and 11).

In insects, shivering warm-up appears to have evolved independently in all of the major orders, as well as independently between some species (Kammer, 1970; May, 1976a). Shivering is physiologically very similar to flight itself, and there is therefore likely very little resistance to the evolution of shivering in those flying insects that require it. Thus, virtually all flying insects large enough to be endothermic also warm up prior to flight, and the Permian protodonate *Meganeura monyi*, an extreme in body size, would, over 300 million years ago, thus very likely have had pre-flight warm-up behavior and physiology.

Thermoregulation is also linked with large body mass because only large insects generate sufficient heat in flight that is not quickly dissipated by convection. Even moderately sized insects, such as medium-sized moths weighing less than a half-gram, generate a temperature excess of 25°C or more in flight. These relatively small insects could thus already be operating close to the maximum body temperature of about 45°C found in most terrestrial endotherms, at relatively modest air temperatures of near 20°C.

Any flying insect of mass greater than 1 g, flying at air temperatures higher than 25°C, faces a serious problem of overheating, and continued flight is not possible without some means of either reducing heat input or accelerating heat loss. In other words, if it flies at air temperatures above 25°C and weighs more than about 1 g, then it "must" thermoregulate. How do these generalities hold to Odonata, and what do we know of the ecology, flight behavior, and mass of the Protodonata?

The allometric relations of insects in general as they apply to thermoregulation also hold true for Odonata (May, 1981a,b), specifically; the larger the body mass of the species, the more precise the thermoregulation (May, 1978). Large species, like *Anax junius*, weighing near 1 g, generate thoracic-temperature excess of up to 17°C in flight (May, 1976a). Relative to percher dragonflies, continuous flyers like *A. junius* and *Aeshna multicolor* have increased blood volume and a larger heart used for dissipation of excess heat in flight (Heinrich and Casey, 1978), which allows them to remain in continuous flight at air temperatures up to at least 35°C (May, 1976a).

As already indicated, *Anax junius* has, unlike percher dragonflies, impressive capacities of pre-flight shivering and of heat loss by blood circulation. A dragonfly even larger than *A. junius* would need to have even better thermoregulatory capacities in order to fly for any but very brief intermittent flights.

At the present time there are few dragonflies much larger than *Anax junius*. But in the Upper Carboniferous and the Upper Permian epochs there were dragonfly-like insects that were of monstrous size in comparison with *Anax*. The wingspread of the protodonates *Meganeura monyi* and *M. americana* was 70 cm or more (Brogniart, 1894; Carpenter, 1943, 1947), whereas today's *A. junius* measures about 11 cm (Fig. 3.16). At an estimated body mass of 18 g (May, 1982), *M. monyi* was more than 18 times heavier than *A. junius*. The protodonates' large eyes, large mandibles, and heavily spined legs placed well forward of the wing bases (Carpenter, 1943) suggest that they took prey on the wing, as do present-day odonates (Wootten, 1981). If they were strong flyers, then they must have been hot-bodied.

Dimensional relationships of the wing to the body were also

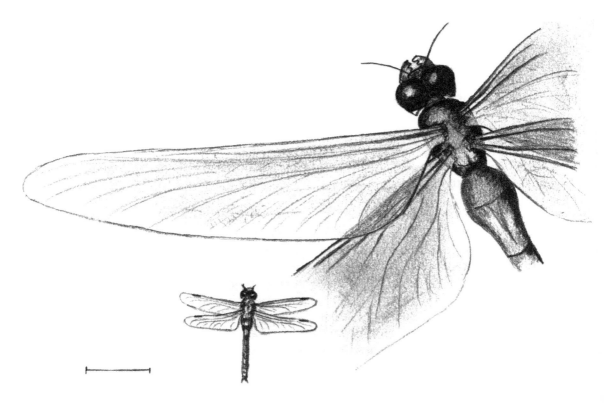

Fig. 3.16 The protodonate *Meganeura monyi (top)*, in comparison with the cruising dragonfly *Aeshna multicolor,* one of the largest species living today. Scale = 5 cm.

quite similar in Protodonata to those of Odonata (May, 1982). The more primitive Anisoptera (such as *A. junius*) are flyers, and they characteristically have longer wings and greater wing area per given body mass than perchers. Protodonata carried the wing morphology of the "flyer" dragonfly to an extreme; they were thus probably more like the flyer than the percher dragonflies of today. However, the low wing loading and the high lift-to-drag ratios associated with their long wings (Kukalova-Peck, 1978) could have allowed the protodonates relatively effortless flight, such as gliding flight. Nevertheless, maneuvers associated with prey capture must have required powered flight like that found in present-day dragonflies.

May (1982) has calculated various alternative estimates of metabolic heat gain and loss, using extrapolations from present-day odonates. His data show that, even if *M. monyi* was uninsulated (unlike present-day odonates, which have insulating air sacs; Church, 1960) it should still have generated a temperature excess of 62°C from flight metabolism alone at flight speeds of 2.5 m/sec, and 27°C at 10 m/sec. Both the Carboniferous and Permian fossil sites for the protodonates were tropical or subtropical (Wootton, 1981), and probably comparatively arid. Daytime temperatures were probably close to 30°C or higher, near those occurring in such areas today. Therefore, in the absence of thermoregulation, *M. monyi* would have generated, from its flight metabolism alone, thoracic temperatures from at least 57°C to 92°C, and possibly up to 104°C. In addition, the animal's large size strongly suggests that it flew not in but *above* the vegetation, where it would have experienced the direct effects of solar radiation as well.

Needless to say, *M. monyi* did not have a T_{thx} of 57°C to 92°C. Instead, it most likely defended an upper T_{thx} close to 45°C, near that of other endotherms (both vertebrate and invertebrate) in general and other odonates in particular (May, 1978). The remaining 12–47°C of the temperature excess would have had to be eliminated by thermoregulatory physiology or behavior, or both.

Given their large size, it seems likely that the protodonates would have had to employ both behavioral and physiological mechanisms to help reduce the potentially enormous thoracic-temperature excess. The animals very likely soared and increased blood circulation to the abdomen to dissipate heat from the flight muscles. Either mechanism is compatible with the responses of present-day dragonflies, and it seems almost certain that at least one was used. If so, following the general theoretical arguments (Heinrich, 1977) and specific suggestions (Heinrich, 1981; May, 1982), then the protodonates were thermoregulating endotherms, preceding the endothermic dinosaurs (Bakker, 1972) by at least 50 million years.

Summary

Dragonflies can be divided into two relatively distinct behavioral types having widely differing strategies of thermoregulating. One, the "flyers," remain in continuous flight (some over a range of at least 25°C) and thermoregulate physiologically by shunting excess heat to the abdomen, which serves as a convective heat radiator or by alternating between powered flight and soaring to reduce heat input at high air temperatures. Flyers shiver in pre-flight warm-up.

The second, more phylogenetically "advanced" dragonflies are the "perchers," most species of which thermoregulate almost exclusively by behavioral means. Perchers warm up by basking, using the wings as convection baffles much as dorsal-basking butterflies do. Their flight durations are short at high air temperatures to reduce endogenous heating, and they adopt at least two different kinds of postures while perched to reduce solar heating.

Some dragonflies have physiological color darkening on portions of the body (primarily the abdomen) in response to low temperature. The resulting "hot spots" may serve as temperature sensors for proper basking-posture orientation. Additionally, the color of some species could function in direct heating. As in other insects, elevated flight temperatures and the ability to thermoregulate are strongly a function of body mass. The larger the dragonfly, the higher T_{thx} it requires for flight and the more precisely T_{thx} is regulated in flight. At any given body size, however, "flyers" regulate higher T_{thx} than "perchers."

Much indirect evidence converges to suggest that the ancient protodonates were flyer dragonflies that regulated a high and stable T_{thx}.

Remaining Problems

1. Do color changes function as temperature sensors in some dragonflies?
2. Do dragonflies modulate flight effort to affect either heat production or rate of convective cooling?
3. Why do so many perchers show no shivering response?
4. Given that flyer dragonflies show immediate heat transfer to the abdomen long before lethal T_{thx} are reached, why do they not heat up the abdomen during shivering warm-up? Do females regulate T_{abd} for egg maturation?
5. Do the "breathing" movements of dragonflies augment blood circulation for thermoregulation?

Grasshoppers and Other Orthoptera

MORE is known about the biology of locusts than possibly any other insects. Our intense interest in the acrididine grasshoppers is due not to some peculiar trait of theirs but to their economic and ecological importance: they compete with mammalian grazers and with humans for grain crops in the world's temperate grasslands. As Daniel Otte (1984) states: "The impact of most North American species cannot even be roughly estimated, for they have not been studied, but perennially they are rated among the worst insect pests."

The orthopterans are a diverse group of some nine families. We have sufficient data on thermoregulation or temperature adaptation for analysis and review for species of less than half of these families. Most of the work done so far concerns the Acrididae (short-horned grasshoppers) and the Tettigoniidae (long-horned grasshoppers). Spotty and somewhat specialized but important insights can be gleaned from work with Gryllidae (crickets) and with Grylloblattidae and Blattidae (cockroaches).

Locusts

Although other acrididine species in aggregate likely consume far more vegetation, the most notorious grazers of all animals are about sixteen species of short-horned grasshoppers that have developed the habit of periodic mass migrations. Of these "locusts" the most well-known are the two African and Near Eastern species *Locusta migratoria* (the migratory locust) and *Schistocerca gregaria* (the desert locust) and the American species *Melanoplus spretus* (the Rocky Mountain locust). The latter species was once a very serious pest in the northwestern United States, but it has not been seen since the end of the last century, presumably because of changes of its environment due to agriculture.

Locust plagues are cyclical, generally following weather cycles. A sustained assault by desert locusts that began in 1948–49 in the Arabian Peninsula and ended in the mid-1960s affected many countries from West Africa across the African continent and to India. Dry years then prevented further serious outbreaks (Walsh, 1986), but as of this writing in 1991 they are again posing a serious threat in Africa.

A desert locust *S. gregaria* weighs about 3.5 g and eats its own weight of vegetation per day. Swarms may fly at 300 to 2,000 m altitude and appear as dark clouds in the sky; these "swarms" may consist of incredible numbers of locusts, some in the billions. A hundred such swarms may take flight during a plague, moving over 200 miles per day and devouring some 4,000 tons of vegetation per day along the way. Before the days of mass transportation and international aid, a locust plague was often devastating. For example, some 800,000 people died of starvation following a locust invasion in 125 B.C. in the Roman colonies of North Africa.

The Book of Exodus (written about 1500 B.C.) refers to plagues of locusts in Egypt: "The Lord brought an east wind upon the land all that day and all that night, and when it was morning the east wind brought the locusts. And the locusts went all over the land of Egypt . . . they covered the face of the whole earth . . . and there remained not a green thing, either tree or herb of the field, through the land of Egypt." Later it was noted that "an exceedingly strong west wind took up the locusts and drove them into the Red Sea."

Thanks to the leadership of Sir B. P. Uvarov, a White Russian emigré who became director of the London-based Anti-Locust Research Center, much of the mystery of the origin and movements of the giant locust swarms has been solved. For a long time no one could explain why or when a swarm would arise, probably because the behavior and morphology of the migrating locusts are distinctly different from those of the same species that are sedentary. Uvarov deciphered that the same grasshoppers could exist in two phases. For a long time prior to Uvarov's "phase theory," many of the migratory locusts were simply identified as separate species of unknown origin.

During years of recession individuals of various species of the "migratory" locusts spend their entire lives in the so-called statary phase at the breeding site. As local populations increase, however, the mutual mechanical stimulation from the crowding of the wingless nymphs causes the release of hormones that result in the production of adults with different coloration, body proportions, and behavior. These individuals in the migratory phase are "nervous" and flight-ready—they will take off and fly into the wind when provoked by disturbances overhead, such as other locusts swarming up above.

We now know that once they are in flight locust swarms are driven by the wind, as the biblical accounts recorded, from areas of high to low barometric pressure where rain has fallen or will likely fall. A locust swarm may start in Saudi Arabia, fly across the Red Sea on the trade winds to the Sudan, where it may encounter southwest winds. Rain would fall where the two wind-flows converge, and the locusts then lay their pods of eggs, about 100 per pod, into the moist earth. Six weeks later the ground is black with grasshoppers, and the offspring of the swarm later disperse, again flying on with the prevailing winds, perhaps now west to Nigeria, Algeria, and Mali. Some locusts have even reached England and central Europe.

The movements of the swarms are dependent not only on prevailing winds. Not surprisingly, they are also dependent on temperature, and thermoregulation is an important component of their and other acrididine biology. As noted above, locusts have been studied because of their great economic importance, and these studies have laid the foundation for our understanding of insect exercise physiology, which is relevant to thermoregulation physiology.

Much of the locust work originated with August Krogh, the dean of comparative physiology, at Copenhagen University in Denmark, along with Torkel Weis-Fogh (Krogh and Weis-Fogh, 1951; Weis-Fogh, 1952, 1964a,b, 1967), A. C. Neville (1963), and M. Jensen (1956) in the same laboratory, and with Peter L. Miller (1960) at the University of Cambridge and electrophysiologist Donald M. Wilson (1961) at Stanford University. The combined research of these researchers provides a broad framework reaching from the aerodynamics of flight to the complex integration of the neuromuscular and tracheal systems. Together the work is a triumph of experimental biology. It is beyond the scope of this book to elaborate in detail on the results, many of which now have little *direct* relevance to the mechanisms of insect thermoregulation. The enormous importance of their contribution cannot be overemphasized, however: by providing experimental protocols they freed the flying insects from the proverbial black box and made them available for the investigation of potential mechanisms; they helped frame and provide contexts for the studies of thermoregulatory mechanisms in other insects.

MECHANISMS OF BEHAVIORAL THERMOREGULATION

That locusts are highly dependent on temperature was first shown for the larvae. In a pioneering study (long before behavioral thermoregulation was explored even in vertebrate animals), Gottfried Fraenkel (1929) at the Zoological Laboratory at the Hebrew University in Jerusalem observed that the larvae of *Schistocerca gregaria* thermoregulate by postural adjustments and by aggregating into groups (to reduce convection?). At air temperatures of 20–27°C they remain on the ground and orient their bodies perpendicular to the sun's rays, thus maximizing the interception of solar radiation. At temperatures higher than 40°C, on the other hand, they crawl onto vegetation, thereby enhancing rather than reducing convective heat loss, and they then orient the long axis of the body parallel to the incident solar radiation. The precise T_b achieved, however, were not measured.

Fraenkel (1929) noted that the orientation response is executed with precision and quickness even in the presence of convective cooling. It therefore seemed to be inexplicable in terms of direct response to body temperature, and he supposed that the animals' behavior was a direct phototactic reaction rather than a thermotactic response.

Subsequently Fraenkel (1930) captured flying imagines of *S. gregaria* near Jericho, and these mature forms showed the same thermoregulatory responses as the larvae, even in captivity. He used the captive animals to study the mechanics of the thermoregulatory response and, contrary to his previous supposition, he discovered that the thermoregulatory orientations appeared to be in response to warmth, as such, and not to light.

In order to test whether light or temperature provided the stimulus for orientation, Fraenkel tried to separate the two responses first by blinding the animals with paint over one or both eyes and then by shading selected body parts. Both experiments indicated that body temperature, and not the light perceived by the eyes, is the primary stimulus for orientation. Totally blinded animals basked for the most part in exact perpendicular orientations to a heat lamp. Of those blinded only in one eye, 62 oriented with the

THE HOT-BLOODED INSECTS

seeing side to the light and 16 with the blinded side toward it. If the animals had oriented solely on the basis of warmth, 50 percent would be expected to line up in each direction: that is, 39 with blinded side toward the light and 39 with seeing side toward it. Therefore, the phototactic reaction plays a supplemental role, as a kind of insurance for correct orientation. Lubber grasshoppers *Taeniopoda eques* also exhibit correct basking responses (Fig. 4.1), even when their ommatidia were masked by black paint (Whitman, 1987).

Fraenkel (1930) observed (in 4 animals) that reorientation resulted when either head, thorax, or abdomen were shaded (by placing strips of cardboard onto the wire screen above perpendicularly orienting grasshoppers). When only the head was shaded the grasshoppers crawled into full sunshine on the average in 12 minutes. Abdominal shading resulted in reorientation within 17 minutes, while thoracic shading resulted in a response in only 4–6 minutes. Because of the very variable results in the small sample size, it is not possible to come to conclusions on possible differences with respect to thermal sensing abilities of different body parts. However, the results are consistent with the conclusion that the grasshoppers respond to temperature, as such, and that they have temperature sensors in or over their entire body.

In *Schistocerca americana* temperature sensitivity is associated with mechanosensory hairs located on the head, thorax, and tarsi.

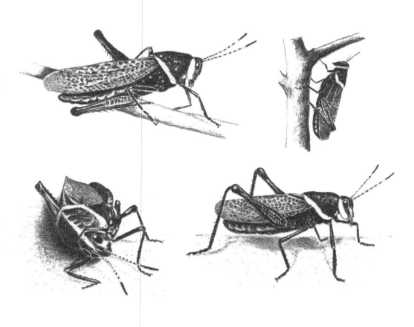

Fig. 4.1 Thermoregulatory postures of the lubber grasshopper *Taeniopoda eques*. Heating postures are flanking *(top left)* and ground flanking *(bottom left)*. Heat-avoidance postures are shade-seeking *(top right)* and stilting *(bottom right)*. (Drawings from photographs by Whitman 1987.)

Sensory neurons respond to mechanical displacement of associated hairs as well as to increasing temperature (Miles, 1985). Thoracic-hair neurons are more temperature-sensitive than the tarsal receptors and most head receptors, although the wind-sensitive head hairs (which are important for the initiation and maintenance of flight) show exceptional temperature sensitivity. Internal thermosensors are also likely present (see the section on cockroaches, below).

Carol I. Miles (1985) at the University of Washington has examined the regional differences in body temperature normally associated with the receptors. It appears that those hairs that are exposed to greater temperature variability are *less* sensitive to temperature than those exposed to less variability. Therefore, at least the tarsal hairs do not appear to be temperature sensors as such; instead, they seem to serve mainly other functions with neural temperature compensation added to reduce possible ambiguity of signaling. (But why do the others respond to temperature at all?)

Given the importance to society of being able to predict the movements of migratory locusts, much research has gone into predicting flight. Temperature is a primary variable for flight, and in general, locusts fly with a minimum T_{thx} near 30°C, which results from metabolic heat production and from solar radiation. Before sunrise, *S. gregaria* have body temperatures close to that of the air, but T_b rises 11–16°C above air temperature by basking after the sun rises (Gunn, 1942). In sunless conditions the locusts take off spontaneously only at air temperatures above 26°C (Waloff, 1963). However, during level flight at 3.5 m/s (on a flight mill) the temperature excess of the thorax is 6.5°C from metabolic heat production (Weis-Fogh, 1967), so the animals, once started, can presumably remain in flight even at about 20°C under overcast conditions. An additional 2–4°C temperature excess is possible in the field during flight as a result of radiative heating (Weis-Fogh, 1967).

Although takeoff for flight depends on sunshine and/or relatively high ambient temperatures, swarms may be lifted to 3,000 m or more by air turbulence. Temperatures at this altitude can be near 7°C (Rainey, 1958) and the insects then presumably have T_b too low for effective flapping flight (no direct measurements of T_b are available!). At this altitude, however, they often glide rather than flap their wings (Roffey, 1963). There has so far been no evidence showing that locusts regulate their T_{thx} while they are in flight.

FLIGHT AND FLIGHT MUSCLES

Weis-Fogh (1964b) measured wing-stroke parameters as well as "temperature excess" (the difference between thoracic and ambient temperatures) of tethered locusts as they flew into an artificial air stream against wind at the animal's preferred flight speed (at 30°C). He showed that temperature excess of the pterothorax in steady-state flight is a direct function of lift, and he concluded that temperature excess was a direct measure of power output. (Later work with other large insects, especially those in free flight, would show that temperature excess is not necessarily a measure of power output because heat loss can be regulated.) Weis-Fogh showed also that power output with increased lift was not necessarily a function of wing-beat frequency, and this work laid the foundations for understanding how power output was varied. But the answer to the mechanism of power variation at constant wing-beat frequency came from electrophysiological studies of the working muscles themselves.

A flying locust must maintain a minimum wing-beat frequency of 16–20 Hz, and thus the period of a wing stroke is then near 50–60 milliseconds (ms). However, the wing-stroke frequency depends on the duration of the muscle twitches, which must be much shorter, because both up- and downstroke muscles cannot contract at the same time and still move the wings. The mechanical change in a wing-muscle twitch must be almost completed within only a *half* of the wing-stroke cycle, or else the upstroke and downstroke muscles work against each other rather than on the wing. Increasing muscle temperature does little to affect the force per twitch, but it dramatically affects twitch duration (Fig. 4.2) and flight efficiency. At 22°C, for example (the minimum T_{thx} for takeoff in *S. gregaria*), the muscle-twitch duration is the same as the normal wing-stroke period, which means that some of the mechanical work in flight is spent against the antagonist muscles, rather than on the wings, and that much of the power is wasted. At 28°C the twitch durations are 43 ms, and much less of the work is wasted because there is less overlap in contraction. Finally, at 35°C, when twitch durations are only 30 ms, the contraction of the upstroke muscles is nearly completed before the downstroke muscles contract and work is then no longer

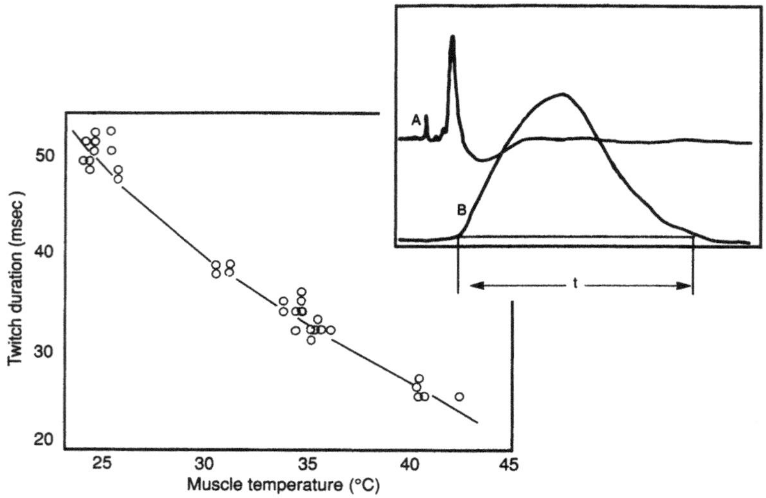

Fig. 4.2 The right panel shows a muscle action potential *(A)* and twitch force *(B)* of locust flight muscle. The graph at left shows total duration of single twitches decreasing with increasing muscle temperature. (From Neville and Weis-Fogh, 1963.)

wasted at the same (minimum) wing-beat frequency; in other words, almost all of the mechanical power is exerted on the wings and not on the antagonistic muscles except to help stretch them.

The situation changes somewhat when the muscles are activated doubly. During flight at medium intensity, the large dorsal longitudinal (DL) depressor muscles of the hindwings of migratory-phase *S. gregaria* are normally fired a single time per wing-stroke cycle, although during more powerful flight the locusts often double-fire these muscles per wing beat in closely spaced action potentials that result in stronger muscle contractions (Wilson and Weis-Fogh, 1962). During *closely spaced* (2–5 ms apart) double-firings, the resulting contraction resembles a smooth muscle twitch, but it is a stronger and longer twitch. Closely spaced double-firings can result in as much as 50–80 percent more force per twitch (fig. 6 in Neville and Weis-Fogh, 1963). The longer twitch durations can also result in more wasted work at low temperature, however, because the antagonist-muscle contractions overlap more. At 30–35 °C, on the other hand, there are no problems with work economy, even with double-firings, since both the upstroke and downstroke muscles are able to contract and relax in time for the next burst of activation.

The power-generating flight muscles of locusts are composed of only a few motor units—5 for the dorsal longitudinal muscle, the wing depressor (fig. 1 in Neville, 1963), 3 for posterior tergocoxal elevators (Mizisin and Ready, 1986), 2 for the metathoracic ter-gosternal elevator (Josephson and Stevenson, personal communication)—and the power produced by these muscles depends not only on whether they are activated singly or multiply but also on the number of motor units recruited during any one twitch or wing beat (Neville, 1963).

The basic model of flight control that emerged (Wilson, 1968) is that at any given temperature a neurogenic flyer such as a locust could potentially vary its power output (and/or heat production) in flight by recruiting varying numbers of muscle fibers and/or by activating them singly or multiply. Central nervous control directs the power output in response to flight demands, but performance is under severe constraints by muscle temperature. An oscillator in the thoracic ganglia controls the wing-muscle contractions, and the output from the oscillator is modulated by external sensory input, such as wind detectors on the head and stretch receptors on the wing hinges.

Neville and Weis-Fogh (1963) concluded that in the flying locust the muscles and not the nervous system present the primary limitation to flight over a wide range of temperatures. They assumed that wing-beat frequency, set by the central nervous system, remains relatively constant. Work efficiency, on the other hand, varies strongly as a function of temperature, because the twitch durations of the muscles are strongly temperature-dependent.

These studies on the migratory locust pointed to at least possible or potential mechanisms whereby some insects might conceivably vary their rates of power (and heat production). But none of the studies showed any hint that heat production or heat loss was varied for thermoregulation.

FLIGHT TEMPERATURE AND WATER LOSS

Low muscle temperature directly affects immediate takeoff and flight. But flight duration during migration is limited by water loss from the tracheal system, which is also a function of body temperature. The rates of both water production and loss can be calculated for a variety of temperatures and relative humidities, using the measured values for metabolic rate of 75.5 W/kg and a pterothoracic ventilation rate of 320 l/kg/hr (Weis-Fogh, 1967).

The rate of water production is derived from the metabolic rate, the knowledge that the locusts metabolize lipid in flight, and the knowledge that the combustion of 1.00 mg fat provides 1.07 mg water. But water is also lost as the animals ventilate while extracting oxygen from the air for burning the fat. The reasonable assumption can be made that the expired air is saturated with water vapor and that it is about the same temperature as that of the tissues from while it is expired. Tissue temperature is important, because the higher the temperature of the air in the tracheal system, the more water the expired air will release. A high muscle temperature can thus be a liability in terms of water balance, because it removes water from the body if the inspired air is not already saturated with water vapor.

Water-balance equations (Weis-Fogh, 1967) indicate that although locusts can remain hydrated in flight at 25°C in the absence of direct sunshine even in relatively dry air (35 percent relative humidity, RH) they require moist air (90% RH) in order to maintain a positive water balance during flight at 30°C in bright sunshine. Consequently, under hot or dry conditions they should ascend to high altitudes, where it is cooler, to minimize body temperature and the associated elevation of water loss. The dehydrating power of the air at any one temperature increases with altitude, however, so the one effect may cancel out the other. Biophysical calculations are required, given that empirical work is difficult to resolve the issue.

Water economy in the desert locust is likely facilitated by its breathing mechanism. Unlike some insects, such as bees (Bailey, 1954), the desert locust does not expire all of its inspired air from the hot thorax but instead shunts air from the thorax to the cooler abdomen before expiring it there (Weis-Fogh, 1967). Since the abdomen remains at a considerably lower temperature than the thorax, expiring air through the abdomen may be a mechanism (Heinrich, 1975) for condensing water out of the hot saturated air from the thorax, thereby conserving water despite the unavoidable heating of the thorax by the flight metabolism. Analogous mechanisms of water conservation have been deciphered for some desert rodents and the desert iguana (Murrish and Schmidt-Nielsen, 1970), which use the cool nasal passages to condense water from the saturated heated air coming from the lungs (Schmidt-Nielsen, 1972). We do not know if other orthopterans have a similar mechanism, because none have been examined.

EVAPORATIVE COOLING BY PANTING

Evaporation of water in locusts is apparently augmented by increasing the ventilatory rate to cause panting, which causes cooling under heat stress. These conclusions were derived from observations by Henry D. Prange from Indiana University both at his laboratory in Indiana and in Israel in the Negev Desert at the Blaustein Institute for Desert Research. Prange (1990) found that *Schistocerca nitens, Locusta migratoria* and *Tmethis pulchripennis* tolerated exposure to air temperatures higher than lethal (about 48°C) for more than a half hour, while maintaining T_b several degrees below T_a. The animals' rates of ventilation and evaporation remain relatively constant until about 45°C, above which they increase markedly.

I have also routinely observed "panting" (increased rates and amplitudes of abdominal ventilatory movements) in other insects when they are heated to near lethal temperatures (for example, in honeybees, bumblebees, and various beetles, moths, and dragonflies). Therefore, the panting could be a general phenomenon in insects. However, in the only other previous attempt to study the effect of panting (and to separate it from heat transfer by blood circulation) on possible thermoregulation (Heinrich, 1980), dead honeybees were forced by a mechanical device (using an electromechanical solenoid driven by electric pulses at frequencies from <80 to >600 times per minute) to "hyperventilate" while T_b was simultaneously measured. Both amplitude and frequency of abdominal "breathing" movements were varied, but no T_b depression beyond 1°C was noted. In one of the 8 bees examined however, the abdominal wall ruptured and T_{abd} then declined by 9°C as a result of evaporative cooling.

Other Short-Horned Grasshoppers

As the attractive color plates in Daniel Otte's book (1984) attest, the Acrididae are a diverse and beautiful group. Within the most numerous subfamily, the Oediponinae, there are species with blue, yellow, purple, orange, red, white, or multicolor hindwings. Many species have brightly colored hindlegs as well. The colors are thought to function in sexual signaling, and typically they are displayed with the accompaniment of sounds (loud crackling, snapping, or buzzing) produced with the wings during flight.

When not engaged in sexual displays, these grasshoppers are very cryptic, shy, and "in the heat of the day some species are virtually impossible to catch" (Otte, 1984). Therefore, they are likely hot-bodied. They typically inhabit dry, sunlit habitat with short vegetation where behavioral thermoregulation is a major feature of their lives.

Grasshoppers are well known to use postural adjustments that help maintain their T_b independent of ambient thermal conditions. Waloff (1963) describes the crouched and stilted positions of the desert locust for heating itself conductively from the substrate and for avoiding a hot substrate and cooling convectively. (The air higher above the substrate is cooler and has greater wind speed.) Anderson, Tracy, and Abramsky (1979) describe additional positions ("dropped," "straddle," and a combination of both) for heat dissipation in the grasshoppers *Psolaessa delicatula* from Colorado. As air temperature increased these grasshoppers first stilted, then dropped, then straddled, and finally simultaneously dropped and straddled. During straddling the animals spread their hindlegs laterally, presumably to enhance convective cooling. Animals in the "dropped" position are still stilted, but their abdomen is lowered to the ground.

The stilted posture (when the thorax is raised above the ground) appeared to be compromised when the abdomen is lowered to the ground. No explanation for the curious "dropped" posture has so far been given. But as a first guess I propose that perhaps the stance is assumed primarily when there is intense solar radiation from above but the substrate is not yet heated to lethal temperatures; the wings must get heated from above, and lowering the abdomen could act both to reduce contact with the hot wings (grasshoppers have no analogous space to some beetles' subelytral cavity) and to expose the top surface of the abdomen and increase convection over it.

Psolaessa delicatula is restricted primarily to bare ground, where surface temperatures sometimes approach 50°C. Another grasshoppers, *Eritettix simplex*, is also found in the short-grass prairie near Nunn, Colorado, but unlike *P. delicatula* it lives in vegetation and it has not been observed to exhibit any noticeable thermoregulatory posturing, even when it is on bare ground. Its body temperature is a strict function of air temperature.

Although posturing with respect to the substrate is a major thermoregulatory response in *P. delicatula*, this animal also thermoregulates in other ways. First, during early and midmorning hours the grasshoppers maintain positions perpendicular to the sun, thus maximizing the surface area of the body for solar heating. Later in the day, at air temperatures above 32°C, the insects avoid sunshine. When flushed at these high temperatures, individuals fly 1 m or less, and if they land in the sun they immediately move into the shade (Anderson, Tracy, and Abramsky, 1979).

In another study of two other grasshoppers coexisting in the mountains of central Colorado at near 3100 m, Gillis and Possai (1983) also showed thermal differences related to habitat partitioning. Both species, *Arphia conspersa* and *Trimerotropis suffusa*, appear to use the same food resources, although one, *A. conspersa*, tends to be found in more open habitat and it is active earlier in the morning and on cooler days than the other. The body temperatures of the two species are not conspicuously different from each other, showing almost perfect thermoconformity. Both maintain T_b approximately 6–7°C above ambient temperature over a broad range of ambient temperature. Nevertheless, it was concluded that "regression analysis of T_b on ambient temperature revealed significant differences in thermoregulatory performance between (them)." This is stretching a point, however; the slopes of the regression curves were 1.0 and 0.9, respectively, or nearly direct functions of ambient thermal conditions. It is curious that although both show almost total lack of thermoregulation in their T_b, they nevertheless are both reported to display perpendicular basking, shade seeking, and stilting.

It is not clear why individual animals may display thermoregulatory responses when the body temperatures of the population sampled in the field show no sign of being regulated. I suspect that, as in many insects, individuals thermoregulate only under certain conditions related to immediate activity. For example, there was a tendency for displaying males (those making the noisy flights) to have higher T_b than nondisplaying animals; at 18°C displaying males had body temperatures nearly 10°C higher than nondisplaying animals. As in other insects, the grasshoppers may not thermoregulate at all until they are ready to perform certain activities. Therefore, random sampling of body temperatures of the population in the field at large says little, if anything, about either the animal's thermoregulatory mechanisms or its capacity for thermoregulation.

In addition, as indicated by Kemp (1986) and Harrison (1988), linear regression of T_b on ambient temperature (T_a) in grasshoppers may be too coarse a comb to detect thermoregulation unless a very wide range of T_a is examined. Kemp and Harrison examined the T_b in the field of four species of range-land grasshoppers in western United States over a T_a range of 2–42°C. All species showed T_b close to T_a below 15°C, presumably reflecting their inactivity. At T_a of 15°C to 42°C they then showed good temperature regulation: temperature excesses ranged from up to 15° C above T_a (at 15°C) to 2°C below T_a (at 42°C).

A particularly instructive study of contrasting thermal constraints on grasshoppers was done by Mark A. Chappell from the University of California at Riverside on *Melanoplus sanguinipes* from a cold alpine tundra at 3,800 m and *Trimerotropis pallidipennis* from a hot desert environment near sea level (Fig. 4.3). Environmental air temperatures were 15–20°C higher throughout the day at Deep Canyon, the desert site, than in the White Mountains, the high-elevation site. However, the body temperatures did not reflect the same temperature differences; T_{thx} of the *T. pallidipennis* were on the average only 7.7°C (rather than 15–20°C) higher than those of *M. sanguinipes*.

The regression of body temperature on ambient temperature (from 37–48°C) is nearly linear for *T. pallidipennis*, apparently suggesting lack of thermoregulation. In this instance, however, the data could instead mean the opposite. Instead of not thermoregulation at all at these very high temperatures, these grasshoppers may continuously minimize T_b. Thermal conditions at the time of the study (early August) were almost always extreme: ground temperatures attained 60–70°C by 9:00 A.M., and the animals are then restricted to shade. When they were in open areas in sunshine for short durations they *always* stilted, and equilibrium temperatures during stilting were above the lethal limit near 50°C (Fig. 4.4).

In contrast, the main challenge of the alpine *M. sanguinipes* is to keep T_b high, and they remain in open areas in sunshine and crouch there onto the ground. They avoid shaded areas throughout the day. These grasshoppers are only rarely subjected to ground temperatures of 50°C. High T_b can be readily avoided since "stilted" grasshoppers are always 10–15°C cooler than the temperature of the substrate surface.

The desert species, *T. pallidipennis*, when flushed typically fly

Fig. 4.3 The thoracic temperatures *(squares)* of grasshoppers in two very different climates, the *Trimerotropis pallidipennis* in the hot desert *(bottom)* and the *Melanoplus sanguinipes* in a cool montane environment *(top)*, as a function of environmental temperatures throughout the day (from 8 A.M. to 6 P.M.). For comparison, temperatures are graphed for dead grasshoppers in deep shade *(solid line)*, in a crouched position in the open *(dashed lines)*, and in a stilted position in the open *(dotted lines)*. (From Chappell, 1983b.)

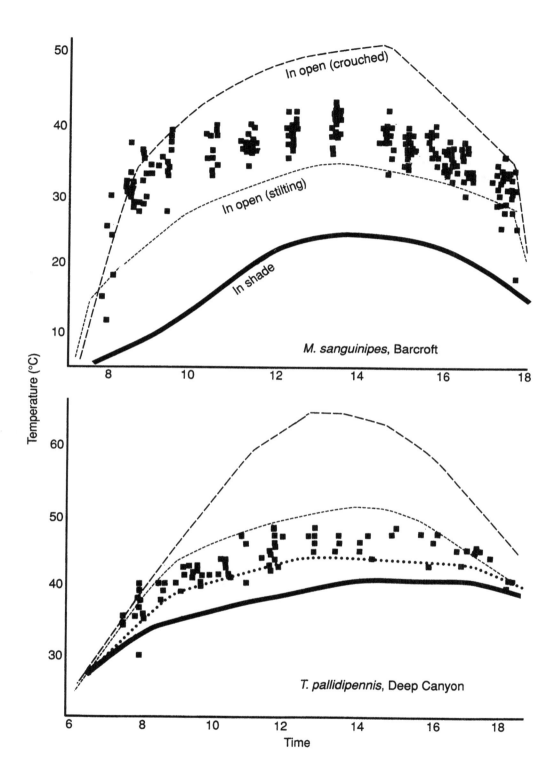

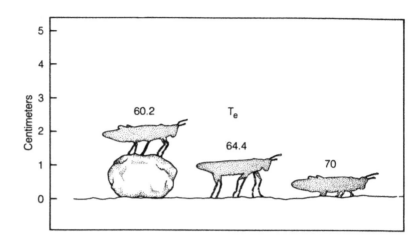

Fig. 4.4 Equilibrium, or operative, temperatures *(Te)* for dead *Trimerotropis pallidipennis* near ground level in Deep Canyon, California, indicate the importance of stilting and rock-climbing behavior for live grasshoppers. Note that the operative temperatures are far in excess of what live animals can tolerate. (From Chappell, 1983b.)

for 3–10 m and then freeze into immobility in a crouched position on bare, shadeless ground. Because they heat rapidly, however, they can perch there for only 1 minute or less. Do they have adaptations for tolerating the inevitably high T_b that they need to endure for predator avoidance?

The resting metabolic rate of *T. pallidipennis* is significantly lower at any one ambient temperature than that of the montane species. The significance of this difference is not clear. Since the resting metabolic rates of insects are by far insufficient to affect T_b, they cannot function as adaptive mechanisms either to increase or to decrease T_b. But could the lowered metabolism of the *T. pallidipennis* be a biochemical adaptation that aids survival at the extraordinarily high upper temperature limit near 50°C? (For example, reduced overall metabolism could be linked with lowered production of lethal heat-induced products.)

Additionally (or alternately to the above), different metabolic rates could also reflect adaptations to energy budget (Chappell, 1983b). Perhaps the high-temperature form conserves energy by reducing resting metabolic rate while the low-temperature species has a high metabolic rate that benefits "rapid growth and energy accumulation." Finally, the differences in the two species from two genera could reflect phylogenetic constraints.

As in Chappell's study, Hadley and Massion (1985) reported higher metabolic rates in high-altitude (Mount Evans) than in low-altitude (Boulder) Colorado populations of the grasshopper *Aeropedellus clavatus*. It was suggested that the high-altitude animals needed to metabolize foodstuffs more rapidly to develop in the shorter growing season. (One way the grasshoppers apparently adjust to the reduced growing season is to reduce the number of juvenile instars from 5 to 4; Alexander and Hilliard, 1969.) However, their data (fig. 3) refer only to relatively high ambient temperatures (20–40°C), which according to their environmental data the high-elevation hoppers would not normally encounter, especially at night. Instead, at the ecologically relevant temperatures (those between 5 and 20°C) the oxygen consumption rate of the two forms appears to extrapolate to nearly identical values. Given that an extrapolation of the metabolic rates of the high-altitude grasshoppers results in *negative* respiration at low temperatures (see their fig. 3), perhaps the simplest explanation is that the elevated metabolism of the high-altitude forms at high temperatures reflects increased activity in the respirometer, possibly related to discomfort since they normally do not experience these temperatures. (Similar patterns in respiratory rates as a function of altitude have been observed in three species of *Pogonomyrmex* harvester ants; Mac Kay, 1982). I conclude that it is not certain whether resting metabolic rates are adaptively altered in acrididine grasshoppers. Ninety-four species occur along the same altitudinal transect in the Front Range of the Rocky Mountains of Colorado near 40°N latitude (Alexander and Hilliard, 1969). It has not been determined if the same apparent or possible pattern of oxygen acclimation occurs in pairs selected at random.

WHY DO ACRIDIDINE GRASSHOPPERS THERMOREGULATE?

Acridids need to maintain relatively high T_b to be active, but thermoregulation serves different functions. Some functions (such as predator avoidance and courting) may be accomplished in a time frame of seconds or minutes, while others (feeding, digestion, and growth) may last for hours or days. Thermoregulatory mechanisms serving specifically the latter are difficult to detect, but they could be no less important.

Rate of Development. Developmental rates for acrididine grasshoppers, as in all insects, are highly temperature-dependent (Hilbert and Logan, 1983; Putnam, 1963). For example, in three species, *Camnula pellucida, Melanoplus bilituratus,* and *M. bivittatus,* total development time for the nymph phases under constant temperatures ranges from 53 days at 24°C to only 17 days at 38°C (Putnam, 1963). In many parts of their range these grasshoppers would be unable to attain sexual maturity if it were not for their ability to bask. In both field and laboratory conditions, the clearwing grasshoppers *Camnula pellucida* regulate T_{thx} at their optimal temperature for development of 38–40 °C by basking, if possible. Age-specific phenology data collected from field sites compared with simulation results (Carruthers et al., 1992) show that basking in sunshine is necessary for physiological maturation, as well as for preventing or reducing fungal infections. At some sites, such as near Alpine, Arizona, at elevations of 2,450 m, *C. pellucida* would reach the second instar only by the end of summer if development were dependent on air temperature alone. But the addition of solar radiation, which the animals use for basking to maintain T_b 10–15 °C above air temperature, allows them to complete all 5 instars within the season (Fig. 4.5). If thermoregulation is related to development, do grasshoppers stop regulating at a high T_b once they terminate development? (Or do the females then thermoregulate to increase development of the ovaries to boost egg production?)

Feeding and Digestion.   If the grasshoppers' choice of high temperatures is ultimately related to acceleration of development rate, then one might expect that their choice is more proximally related to food processing and/or molting, as has been shown repeatedly in vertebrate poikilotherms. Insects should also choose high temperatures when they are fed, to accelerate digestion, and prefer low temperatures when they are empty, to conserve the energy resources gained. Given the low-energy content of the food that grasshoppers eat, however, they must process food on a nearly *continuous* basis, possibly making short bouts of resting at low T_b of dubious value. In other words, they are normally seldom empty enough to test the hypothesis, and it has not been systematically tested. Changes of temperature do little to affect feeding in locust nymphs, except to depress feeding at T_a below 34 °C (Ellis, 1963).

Few data are presently available to evaluate the various possibilities of the relation of body temperature to feeding and its probable links to growth and development. Nevertheless, Chapman's (1965) detailed analysis of temperature preferences of migratory locust nymphs in a temperature gradient provides some possibly relevant data. In one experiment he allowed nymphs to become progressively more hungry over a 5-hour period and examined the distribution of the hoppers in a gradient every half hour. No change in preference was found. Well over 50 percent of the animals were within the range of 36–41 °C at all times. In a second experiment he starved nymphs overnight (for 21–26

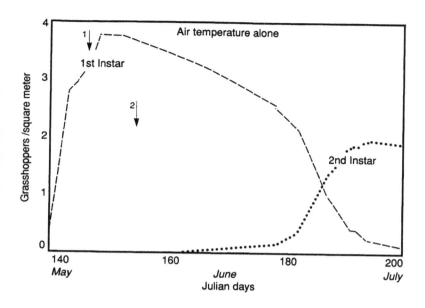

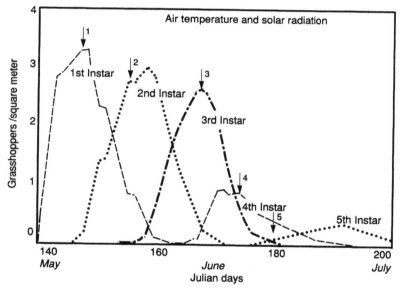

Fig. 4.5 Simulated growth rate *(lines)* and phenology of *Camnula pellucida* in Alpine, Arizona; temperature-dependent development rates were calculated using ambient temperature as an estimate of T_b *(top)* and using ambient temperature plus solar radiation for T_b *(lower graph)*. The arrows indicate when peak abundance of instars actually occurred in the field; calculations based on T_a plus solar radiation are obviously much closer to the field data. (From Carruthers et al., 1992.)

hours). Now most of the hoppers became aggregated at the low end (<22°C) of the temperature gradient. But Chapman (1965), who was not testing the hypothesis that the grasshoppers were manipulating their T_b for energy economy, assumed they were passive: "Following starvation there was a tendency to become inactive at low temperatures, the insects simply became trapped at the cold end of the gradient." Were they really "trapped" or did they regulate a lower body temperature?

In vertebrate poikilotherms, small body mass poses a severe problem for herbivores, because as mass-specific metabolic rates increase with decreasing body mass there is soon a limit to the energy returns to be had from a high-fiber diet such as leaves. Leaves may then become of marginal value in the diet, if not prohibitive. Some small reptiles get around this problem by feeding on fruits and flowers. But from a vertebrate perspective, it is a mystery how small insects can succeed by living on leaves. (Is their metabolism so low that food is not limiting?) Some reptiles increase speed of digestion by seeking higher temperatures after feeding. The thermoregulation by grass-feeding grasshoppers living at prevailing low ambient temperatures could at least in part also be explicable in terms of food assimilation.

Predator avoidance. Acrididine grasshoppers often live at very high temperatures. Buxton (1924) observed several species near Jericho in desert ground-surface temperatures of 60–62°C. As already indicated, very high thermal tolerance can function as a defense mechanism if it allows the organism to live where predators cannot be active. In many instances escape from the heat is possible. But Chappell (1983a) observed that the desert-dwelling grasshopper *Trimerotropis pallidipennis* escapes into the heat and tolerates body temperatures of almost 50°C, thereby escaping lizards and possibly also reducing predation by birds. Normally these grasshoppers avoid very high temperatures by resting under shrubs. However, attacks by predators elicit short leaps or flights into the open, where the insect then remains stationary, relying for protection on crypsis (Fig. 4.6). While maintaining crypsis in the open in sunshine on hot substrate, however, they cannot stilt and are then subject to severe heating. The higher the tolerable T_b, the longer the grasshopper can remain hidden. The two lizards in Chappell's study area (Boyd Deep Canyon Desert Research Station near Palm Desert, California) have much lower tolerable T_b (near 43–44°C) than their prey, the grasshoppers (near 50° C). The lizards do not hunt their prey in the open either because

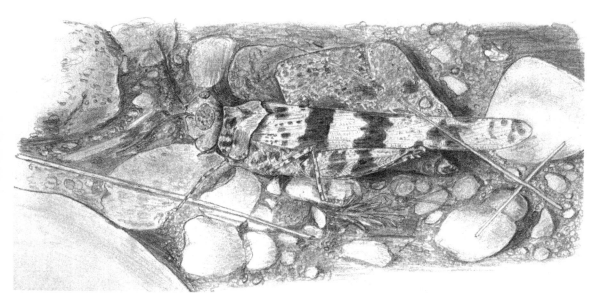

they cannot see their crouching prey at ground level or because ground temperatures are too high. (No data are provided to show if the lizards hunt in the cooler parts of the day or on cooler or overcast days). An important avian predator, the Say's phoebe *Sayronis sayii* hunts from perches above the ground, attacking those grasshoppers that reveal themselves by movement. Clearly, by tolerating the heat and not moving, grasshoppers should reduce their exposure to predation.

Lubbers—a case study of grasshopper thermoregulation.   The horse lubber grasshopper *Taeniopoda eques* (Fig. 4.1), a native of the hot Chihuahuan Desert of North America, provides an interesting contrast in thermoregulation to the other species of acridids so far described. Lubbers are one of the largest grasshoppers of North America (males weigh 1–3 g, females 2–10 g), and their size should impart high thermal inertia. Unlike *Trimerotropis pallidipennis*, this species does not need to have thermal responses for defense, because it is chemically defended. Lubbers rely on toxins advertised by bright coloration (black and yellow) that warn predators and deter attack. Furthermore, females of this grasshoppers are flightless.

Although my focus here is on *T. eques* because it has been studied in relatively good detail, I also mention the pioneering work of Mathew A. Parker of Cornell University, who studied a very similar animal, the barber pole grasshopper *Dactylotum bicolor* in New Mexico. This alarmingly colored (red, yellow, and blue), distasteful, flightless grasshopper of open grasslands from Montana to northern Mexico engages in an orderly sequence of diurnal movements within its host plant, *Baccharis wrightii*. It begins the day at the top of the plant and basks there in the morning and afternoon. At noon it retreats to the shaded basal portion of the plant (Parker, 1982). This animal chooses a body temperature generally less than 30°C, far below the maximum available in its environment and far below its lethal limit of about 47°C to 49°C (determined in a saturated atmosphere to preclude evaporative cooling).

In most other large insects, thermoregulation is associated with flight (and hence *all* of the activities related to flight), and as a consequence it can rarely be associated with any specific activity or function. As first suggested by John Alcock at the University of Arizona at Tempe (1972) and later shown by Parker (1982) and by Douglas W. Whitman at the University of California in Berkeley, the slow-moving, distasteful grasshoppers provide a particularly instructive example of the importance of activity patterns governing thermoregulation in a nonflying insect.

Like barberpole grasshoppers, lubber grasshoppers also regulate body temperature through a series of cyclical vertical movements (Alcock, 1972) between the low vegetation and soil (Whitman, 1987), as well as by a variety of body postures also found in other grasshoppers (Fig. 4.1). Thermoregulatory behaviors of these summer-active desert grasshoppers include flanking and crouching to heat up and stilting and shade-seeking to avoid overheating. However, these animals prefer considerably higher T_b (near 36°C) even though their lethal T_b appears to be lower.

Whitman studied the thermoregulatory responses of lubber grasshoppers near Portal, Arizona, during fall (September and October). Even at that time air temperatures still increased from near 15°C to 35°C throughout the day, and ground temperatures reached 65°C. The grasshoppers showed moderate, but not unusually high, thermal tolerances. When tested in a closed container at saturated relative humidity (to avoid evaporative cooling), they showed heat stress (rapid locomotion) at 42°C, impaired walking at 43–44°C, and the inability to right themselves at 45°C (Whitman, 1987). The primary food of these insects is on the ground, which is unreachable during part of the day because to descend to it is to risk increasing T_b of over 50°C and death. On the other hand, temperatures on the ground in the morning are, at near 15°C, so low that the animals either cannot move or can only barely move.

The lubbers negotiate their daily activity within these thermal extremes. At night they roost above ground in vegetation, where they are thought to be protected from ground predators (Alcock, 1972) but where they are also favorably situated to receive the morning sun. (They are there also favorably positioned to stay cool, to conserve energy resources.) They warm up by presenting their lateral (either right or left) surface to the sun. The hindleg of the shady side is raised, exposing its inner surface to the sun, and the hindleg on the other side is lowered, exposing the abdomen to the sun (Fig. 4.1). The abdomen is also lowered, apparently so as not to be shaded by the wings. This flanking position exposes 4.1 times more surface area to the sun than the animal would expose simply by facing it, and it results in at least a doubling of the temperature excess. Once warmed, the grasshoppers descend to the ground, where additional warming may occur as they simultaneously flank and crouch onto the substrate as it is warmed by the sun.

By midmorning the now very hot ground induces the lubbers to stilt, thereby allowing them to achieve T_b averaging 17–18°C below ground temperatures. Finally, by late morning the ground temperatures become intolerably hot and the animals seek thermal refuge by climbing back up into the vegetation (Fig. 4.7) and by perching in shade. As a consequence of these behaviors, body temperature remains relatively stable, near 36°C, despite the very large temperature fluctuations of the environment.

Whitman (1987) emphasizes that the lubbers have black coloration because it "most probably aids in temperature regulations," and he points out that although many species of grasshoppers thermoregulate, the lubbers do it best in the sense that their body temperatures are more independent of ambient temperature than are those of any others so far examined. The data are probably internally most consistent, however, when one stresses not the heating aspect of the black coloration but the heat *avoidance* that it could permit instead. Dark color may be important in determining grasshopper body temperatures (Buxton, 1924; Joern, 1982), or maybe not (Pepper and Hastings, 1952; Stower and Griffiths, 1966). In any case, the black color in conjunction with yellow markings makes the lubber grasshoppers very conspicuous

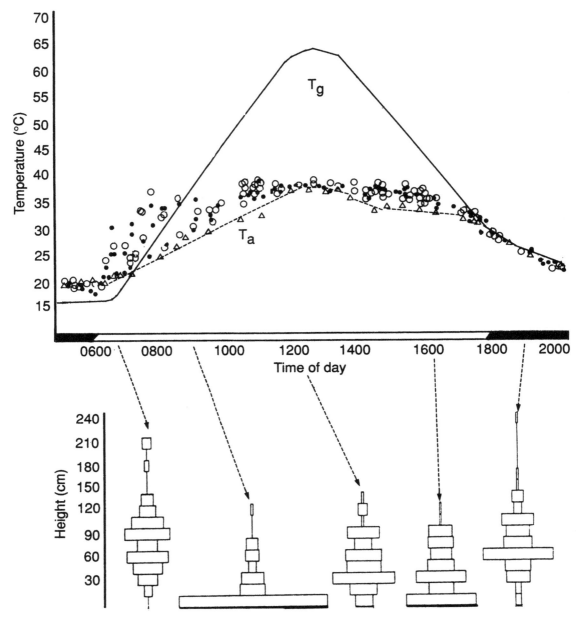

Fig. 4.7 *Top:* Ground *(solid line)*, air *(dashed line)*, and thoracic temperatures *(symbols)* of horse lubber grasshoppers *Taeniopoda eques* on a typical hot day in the Chihuahuan Desert in the fall. *Bottom:* Vertical distribution of the animals at various times of the day. (From Whitman, 1987.)

THE HOT-BLOODED INSECTS

on the light desert soil. It likely acts as an aposematic signal backed up by chemical protection (Whitman, Blum, and Jones, 1985). If the black color acts to advertise their noxiousness, then perhaps its primary benefit is to allow these animals the *freedom of movement* between substrate and bushes. (In contrast, the other desert acrididine grasshoppers discussed previously are edible, and they are forced to reduce movement and to remain immobile on the substrate, where they are camouflaged, in order to remain undetected by predators). The freedom of movement to escape lethal ground temperatures allows these animals to maintain a remarkably stable and relatively low T_{thx}. (A slight extra heating due to the color may be a minor cost at midday, and a minor benefit in the morning.) It is more likely, then, that the role of the color in thermoregulation is more explicable in terms of heat *avoidance* rather than of enhanced heating. Recall, also, that unlike the barberpole grasshoppers, these have *greater* danger of overheating (since they feed in a hotter environment, the desert soil as opposed to grassland vegetation, and in addition have lower lethal T_b). Furthermore, in contrast to the cryptic, ground-bound desert acridids, the lubbers with their freedom of movement may thus not have evolved high thermal tolerances; they can always escape. Since freedom of movement confers, by orders of magnitude, more indirect *cooling* than direct heating at very high temperatures, as well as potential heating at low temperatures, the black coloration of this desert organism (the melanin could also act as a sun screen) is probably not so enigmatic as has so far been presumed.

It has been argued (Whitman, 1988) that "thermoregulation" (temperature-elevating mechanisms?) in the lubbers is relevant for development (Fig. 4.8), because in the laboratory *T. eques* requires 850 DD (degree-days) for hatch-to-oviposition period (Whitman, 1986) whereas the number of degree-days available in the environment is only 700 DD. Two points need to be made, however. First, the DD of field-active hoppers will invariably be a large underestimate (as regards passive T_b) because the presence of sunshine means that the animals will increase their T_b whether they actively bask or not. Second, the DD comparison implies that selection has acted on the proximate mechanism of thermoregulation to counteract a low DD number. From the standpoint of evolution, however, faster or slower developmental times could have been selected (if necessary) for any of a large range of T_b, independent of proximate mechanisms of thermoregulation to alter T_b.

Curiously, for lubbers reared under constant laboratory condi-

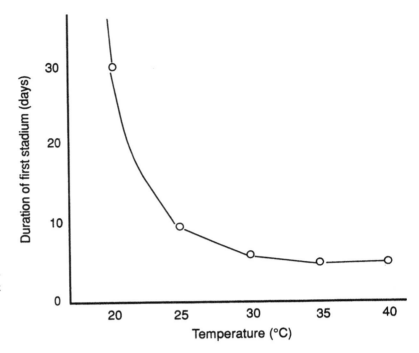

Fig. 4.8 Duration of first larval instars for lubber grasshoppers *T. eques* reared at different
ambient temperatures. (From
Whitman, 1988.)

tions the upper limit for long-term survival is 35 °C (Whitman,
1986). This limit is slightly *less* than the preferred T_b in the field
(Whitman, 1987), but at this temperature the duration of ecdysis
is shorter than at any lower temperature (Whitman and Orsak,
1985). The lubber's upper temperature limit for long-term survival
is also low in comparison with that of other grasshoppers; *Schistocerca gregaria* develop fastest at 38.3 °C and *Locusta migratoria* do
so at 42.2 °C (Hamilton, 1936, 1950). In general, most researchers
agree that grasshoppers of various species seek body temperatures
(Bodenheimer, 1929) close to but not above those which induce
highest developmental rates (Parker, 1930; Shotwell, 1941; Hardman and Mukerji, 1982; Kemp, 1986).

Thermometer Crickets

In 1897 a short paper by A. E. Dolbear appeared in the *American
Naturalist* in which it was claimed that a cricket (species not
specified) could serve as a thermometer.

Dolbear stated: "An individual cricket chirps with no great
regularity when by himself and the chirping is intermittent, es-

pecially in the daytime. At night when great numbers are chirping the regularity is astonishing, for one may hear all the crickets in a field chirping synchronously, keeping true as if led by the wand of a conductor."

He went on to state that it was not a conductor, but temperature, that did the synchronizing. At 60 °F the crickets chirped 80 times per minute, with a change of 4 chirps for each change of 1 Fahrenheit degree. If one knew N, the chirp rate, one could derive T, the temperature, from the formula $T = 50 + [(N - 40)/4]$.

Soon thereafter Bessey and Bessey (1898) confirmed these results by following individual male tree crickets, *Oecanthus niveus*, for up to three weeks at a time in their hometown of Lincoln, Nebraska. The crickets remained in the same tree and chirped at approximately the same rate in different parts of the city. However, chirping rate often varied from day to day, and on a cool evening a cricket caught and brought into a warm room "began to chirp nearly twice as rapidly as the out-of-doors crickets," at a rate corresponding to other (warm) evenings out of doors.

The calling songs of male crickets and katydids are achieved by scraping a series of pegs or teeth on one wing over a series of grooves on the other wing. In over 210 species tooth-strike rates and accompanying wing vibrations and sound frequencies during stridulation are set at species-specific rates that function in sexual communication. The sound frequencies commonly vary with ambient temperature (Walker, 1962, 1969a,b, 1975a,b). Therefore, these animals are presumably poikilotherms, at least while singing. Acoustic communication among conspecifics depends on the females recognizing the males' species-specific pulse and chirp rates. Temperature-dependent signal production would appear to present a problem.

One solution to the temperature-dependent alteration of the signal transmission is to have an equally temperature-dependent change in song reception—in other words, to have a temperature-dependent sound template in the female. The matching of signal generation and signal recognition at different temperatures has been termed "temperature coupling" (Gerhardt, 1978), and it has been demonstrated in a variety of poikilothermic crickets, grasshoppers, and in tree frogs (see Doherty, 1985, for references). However, temperature coupling is not the only solution to the problem of maintaining effective communication at different temperatures. Some species of acrididine grasshoppers, for example, may evaluate temperature-independent *ratios* between syllable and pause durations (Skovmand and Pedersen, 1983), while some species of katydids maintain their body temperature independent of ambient temperature (see below).

Maria Bauer and Otto von Helversen at the University of Erlangen-Nürnberg have made an attempt to understand underlying mechanisms of temperature coupling in two acrididine species, *Chorthippus parallelus* and *Ch. montanus*. In both of these species the females alter their innate sound template depending on their own body temperature, thereby distinguishing their song from that of the sibling species. When females are presented with songs of slower frequencies, they respond best if they themselved also have low body temperatures, and vice versa (Helversen and Helversen, 1981; Bauer and Helversen, 1987).

In experiments designed to test whether sound production and sound reception involved common neural mechanisms, grasshoppers were locally heated through small wires attached either to the thorax or to the head (Fig. 4.9). Temperatures were read with small thermocouples imbedded near the brain in the head and near the metathoracic ganglion in the sternum. Normally head and thoracic temperatures in these (presumably) poikilothermic grasshoppers are nearly equal, but the heating elements allowed for establishing a large temperature difference between the thoracic ganglia and the brain. The researchers then tested the responses of the grasshoppers using computer-generated songs that matched songs normally given at 16°C to 40°C, and even those expected (by extrapolation) but never heard at temperatures less than 16°C and above 50°C. (In nature the grasshoppers sang only between 16°C and 40°C.) Playback of the songs to females from a tape recorder by way of a loudspeaker showed that the females' temperature-dependent *choice* of song depends on their head temperature. On the other hand, the males' stridulatory movements of the hindlegs are a function of thoracic temperature. Thus, the recognition and production of song in these acridids is uncoupled, and the song generator cannot also be used as a template for song recognition.

Similar body temperatures and song frequencies can also be achieved by exploiting the regular diurnal changes of temperature. In general, insects wait to become active when environmental temperatures are appropriate. Thus, activity can shift from night to daytime, as temperatures becomes lower later in the season. For example, in the silkmoth *Antheraea pernyi* cold-treated females shift their release of sex pheromones to an earlier (warmer) part

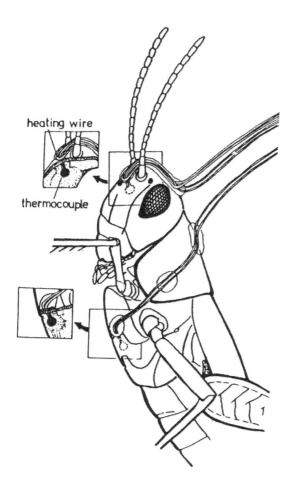

Fig. 4.9 Experimental setup for local heating of head or thorax (here head) to test temperature coupling in sound production and sound reception in acrididine grasshoppers. The thermocouples are placed in the head (near the supraesophageal ganglion) and in the mesosternum (near the metathoracic ganglion). The heating wire is shown on the top of the head. (From Bauer and Helversen, 1987.)

heating wire

thermocouple

of the night, and the males similarly advance their activity window (Truman, 1973). Similarly, in the field, males of the Australian field cricket *Teleogryllus commodus* call mainly during the night, when virgin females are also active. But calling and walking activities are under the control of circadian oscillators that are set by thermoperiodicities (Rence and Loher, 1975). Although calling frequencies would drop sharply during cold nights, in the laboratory the crickets do not drop calling frequencies when nighttime temperatures in the dark are reduced to 11–7°C; instead, they switch both female walking and male singing activities to the daytime, when it is normally warmer (Loher and Wiedenmann, 1981). Since both male singing and female walking are similarly switched, potential communication between the sexes is maintained and species-specific calling rates remained relatively constant despite wide temperature fluctuations of the environment.

Thermoregulating Katydids

Still a third mechanism that maintains species-specific calling rates is that observed by some species of katydids (Tettigoniidae, Fig. 4.10), in which similar T_b (and hence chirp rate) is maintained despite fluctuations of ambient temperature. The songs are very rapid series of chirps produced by overlapping the left forewing over the right and drawing a set of ridges or teeth, the file, on the underside of the left wing over the right scraper, the plectrum. (Crickets, family Gryllidae, do the same, except that the right elytrum overlaps the left, and only the *right* file and left scraper are functional.) Each chirp is produced by one movement of the wings across the set of teeth of the opposite wing. And in the katydid *Neoconocephalus robustus* a very loud, intense song is composed of chirps generated by wing movements driven by muscle contractions at the extraordinary contraction rate of 150–200 per second (Josephson and Elder, 1968; Josephson and Halverson,

Fig. 4.10 A katydid, or "long-horned grasshopper" (Tettigoniidae).

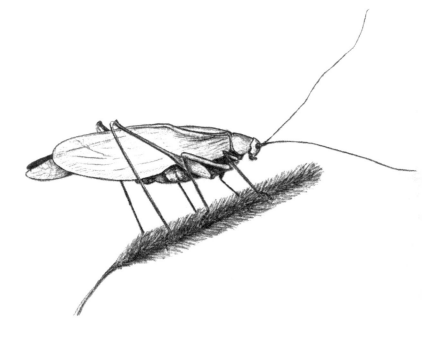

THE HOT-BLOODED INSECTS

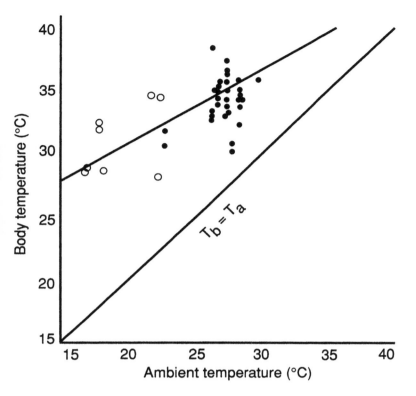

Fig. 4.11 Thoracic temperatures of singing katydids *Neoconocephalus robustus* as a function of ambient temperature. *Open circles:* field measurements; *filled circles:* laboratory measurements. (From Heath and Josephson, 1970.)

1971). To the human ear, the song is a continuous, loud (110 decibels at a distance of 10 cm) buzz.

A study by James E. Heath from the University of Illinois and Robert K. Josephson from the University of California at Irvine has shown that the unique singing capacity of *N. robustus* is associated with endothermic heat production and body-temperature regulation (Fig. 4.11). These katydids do not sing until shortly after sundown in the field, or after being darkened in the laboratory (Stevens and Josephson, 1977). Singing does not begin abruptly but is preceded by a period of shivering warm-up (Fig. 4.12). During warm-up the wings lie relatively motionless against the body (as opposed to raised during singing) but the wing muscles are active. Heath and Josephson (1970) recorded the electrical activity of the muscles both during warm-up and during singing. During warm-up all of the major wing muscles are activated to contract nearly synchronously (Heath and Josephson, 1970; Josephson and Halverson, 1971). As expected, during sing-

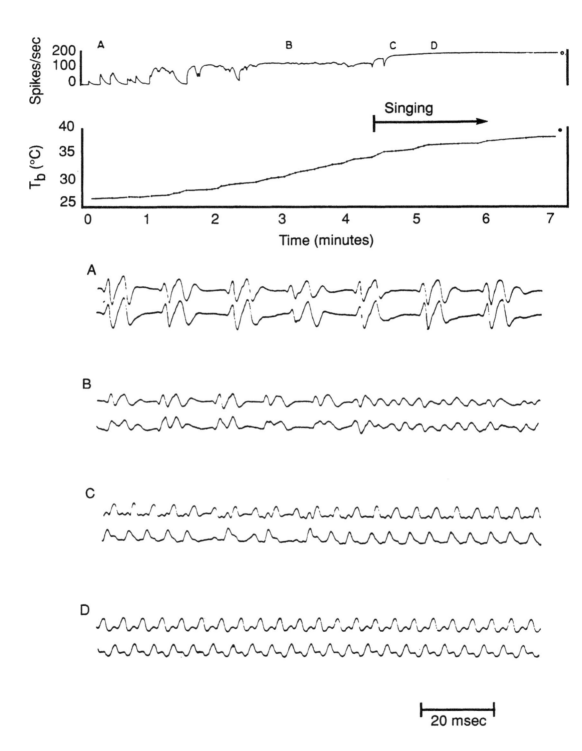

Fig. 4.12 The top two graphs show action-potential frequency (of the tergosternal muscle of the mesothorax) and T_{thx} during warm-up and singing
in the katydid *N. robustus*. The letters above the upper curve indicate the
portions where samples of action potentials are depicted below. Action potentials were recorded from the right subalar *(upper channel)* and right
tergosternal *(lower channel)* muscles of the mesothorax. Note that the synchrony between the two during warm-up changes to asynchrony during
singing. The small "bumps" in between the primary action potentials are
due to "cross-talk" from neighboring units; because of the small size of the
thorax, electrical isolation between recording channels is not complete.
(From Heath and Josephson, 1970.)

ing the antagonistic wing muscles are activated out of phase with
each other, causing the muscles to open and close the wings,
which may then be caused to rub together to produce sound by
friction (Fig. 4.13).

During warm-up (at 22–29°C) preparatory to singing, T_{thx} increases at 0.9 to 2.2°C/min and the activity results in about a
15-fold increase in energy expenditure (Stevens and Josephson,
1977). Singing does not begin until T_{thx} exceeds 30°C. Curiously,
the animals have, at 21.8°C, an extraordinarily high minimum
temperature required to return onto their legs after being turned
on their backs—which they must do to come out of torpor. (Was
this an artifact due to the animals' slow recovery from the cold
torpor to which they were subjected in the experiment?) Individuals sing in the field even at 17°C (Heath and Josephson, 1970).

The katydids show partial independence of thoracic from ambient temperature; those singing at 17°C, for example, have T_{thx}
near 30°C, while those singing near 28°C have T_{thx} averaging
35°C (Stevens and Josephson, 1977). Assuming a 15°C gradient
between T_{thx} and ambient temperature, Heath and Josephson
(1970) calculate that an 0.4 g male (thorax weight = 0.14 g)
must expend a minimum of 128 J/hr. This metabolic rate would
consume 3.2 mg of fat per hour (or about 0.8 percent of the

Fig. 4.13 Sound pulses
(fourth line) and action potentials *(top three lines)* from three
synergistic muscles (left first
tergocoxal, left second tergocoxal, and right second tergocoxal) during singing in the
katydid *N. robustus*. Note that
sound pulses are reduced or
absent on some action-potential cycles but are usually
coordinated with action potentials. Time *(bottom line)* is
marked in intervals of 5 milliseconds. (From Josephson and
Halverson, 1971.)

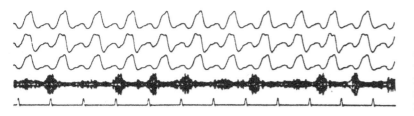

animal's wet weight). However, measured values of energy expenditure during singing are even higher (because of greater heat loss during singing, due to convection by wing movements and heat transfer to the abdomen), and the metabolic cost of singing is close to that of flight in many insects (Stevens and Josephson, 1977).

Why do these animals engage in such an energetically costly mode of communication? Heath and Josephson (1970) suggest that it could be due to signal competition, a case of reproductive character displacement. In the salt marshes in the vicinity of Woods Hole, Massachusetts, *N. robustus* occurs with a similar species, *N. ensiger*. (How wide is the geographic overlap?) Among tettigoniids, as in other orthopterans, song distinctiveness is presumably strongly selected, and *N. ensiger* (a presumably poikilothermic species) has a soft, intermittent song of only 10–15 chirps per second. Females are attracted to the loudest male calls they hear (Bailey, Cunningham, and Lebel, 1990), thus driving a sexual-selective pressure for increasing power output of male calling, which in turn can be proximally (and ultimately) increased by increasing muscle temperature.

Apparently, therefore, *N. robustus* may have, in the evolutionary sense, tried to "outshout" its congener by its more vigorous song. In doing so, endothermy has become first a consequence and then also a necessity. How the partial independence of thoracic from ambient temperature is achieved is so far not known. Possibly the lower difference between T_{thx} at higher than at low ambient temperatures is achieved by shunting excess heat to the abdomen, but this possibility has not yet been investigated.

In insects, wing movements at frequencies over 100 Hz during flight are generally achieved only in stretch-activated, asynchronous muscles, where numerous muscle contractions are associated with each action potential (this kind of activity is called "myogenic"). In *N. robustus*, however, the high contraction frequencies during singing are achieved by the synchronous (neurogenic) system (Josephson and Halverson, 1971) typical of all orthopterans. High frequencies in a neurogenic system are possible because of special muscle design.

Myogenic (also called fibrillar) muscles are characterized by almost complete absence of sarcoplasmic reticulum (SR). In contrast, fast neurogenic muscles have hypertrophied SR. In *Neoconocephalus* the SR is well developed, occupying about 19 percent of the total fiber volume (Elder, 1971). The metabolic requirements posed by continuous high-frequency activation are met by a large mitochondrial volume (44 percent of total fiber volume), and the fibers are of small diameter (10–25 μ), so the diffusion distances from the center of the myofibril to the nearest sarcoplasmic reticulum are minimal. Second, although the wing-stroke rate during stridulation is an order of magnitude higher than it is in flight, the metabolic rate during singing is the same or even *less* than it is in flight. The relatively low energetic cost per wing stroke during singing is due to the very low stroke amplitude and the little external work being done.

During flight, when the wing-beat frequency is low, the flight muscles during any one wing-beat cycle are activated by bursts of action potentials, whereas during singing each contraction is associated with a single action potential. During flight the bursts of typically 4 action potentials per wing beat increase the mechanical power of the wing stroke; singing muscles activated with bursts of stimuli in the normal flight pattern have at least 6 times the power output than flight muscles have when stimulated with a single stimulus (Josephson, 1985).

Robert K. Josephson has examined the requirements of a high temperature in the singing muscles in an elegant series of experiments on their contraction kinetics (Josephson, 1973, 1985). It is to be expected that most of the tension-generating phase of both the wing-opener and the wing-closer muscles would be confined to the time of half a wing-beat cycle, because otherwise muscles must do additional work to stretch their antagonists while they are still generating tension. Thus, given a wing-beat and muscle-contraction frequency of 200 cycles per second, the contraction rise of a singing muscle should be close to 2.5 ms for reasons of energy efficiency. As a model to examine contraction parameters of singing muscles, Josephson (1973) used a Southeast Asian katydid, *Euconocephalus nasutus* (which has been introduced to Hawaii). This species has similar song characteristics to the North American *N. robustus*, and it generates a sound-pulse frequency near 160 Hz. Like *N. robustus*, *E. nasutus* males are also endothermic while singing; they maintain a high (36°C) T_{thx} while singing, some 12°C above air temperature. Many of the muscles used for singing are used in different contexts, and hence at different frequencies of contraction also. For example, the tergocoxal muscles are used not only for warm-up, singing, and flight. They are also used for walking (Josephson and Halverson, 1971).

Temperature is of critical importance in the muscle-contraction kinetics, and there are sharp differences in temperature requirements between singing muscles and those used just for flight. Similarly, there are different temperature requirements for any *one* muscle when it is used for different functions, such as walking or singing. At 25°C the singing muscles produce a smooth tetanus when stimulated at the approximate singing frequency. When warmed to 35°C, however, the contractions become faster and the durations of rises in tension and relaxation are short enough to reveal individual twitches when the muscles are stimulated at 150 times per second (either by direct electrical shocks of the muscle or by stimulation of the nerves supplying it). When at 35°C, a singing muscle is able to relax about halfway between tension peaks when stimulated at 150/s.

The homologous nonsinging muscles are morphologically different, and they respond differently. For example, although female katydids do not sing, they have the same set of mesothoracic muscles as males. In females these muscles are smaller than in males, and they are white rather than pink, because they have fewer mitochondria. When the female mesothoracic muscles are stimulated at 150/s they show a smooth tetanus, as do metathoracic flight muscles of males (which are used for flight only). Twitch time in purely flight muscles (female muscle and male metathoracic muscle) is nearly twice as long as male muscle (mesothoracic) that is used for both singing and flight. Furthermore, singing muscle generates less tetanic tension than does flight muscle. The development of these sexual differences in muscle function has been examined by Ready (1986) and Novicki and Josephson (1987).

The performance of the katydids' singing muscles shows a clear example of the compromises encountered in muscle design. In neurogenic muscle, contractile activity is controlled by availability of calcium, which is released upon electrical stimulation from the sarcoplasmic reticulum and then re-sequestered to terminate the contraction. For rapid twitches there must then be an extensive sarcoplasmic reticulum within the muscle, and maintaining the high metabolism of continued rapid muscle contraction requires a large mitochondrial volume. The greater volume taken up by the sarcoplasmic reticulum and mitochondria in turn leave less space for contractile protein, thus lowering the tension that the muscle can develop. The cost of increased speed is reduced strength, and in those muscles requiring high speed the tradeoff has been made.

It is perhaps significant that increased muscle speed in the katydids' singing muscles is achieved not only by morphological adaptation but by elevations in temperature as well. It is intriguing to suppose that flight muscle in insects generally could therefore at least theoretically also be altered in a similar way to have faster contractions and hence to operate at lower temperatures. However, such adaptations would then obviously be at the cost of decreased flight vigor, if the design constraints seen in katydids (and possibly seen in geometrid winter moths discussed in Chapter 1) can serve as an example.

The Neurophysiology of Behavioral Thermoregulation in the Cockroaches

Cockroaches would appear to be the ultimate thermoconformers. First, they are largely nocturnal, and they avoid sunshine. They thereby avoid the major heat source used by most behavioral thermoregulators. Second, they seldom fly, and when they do (*Periplaneta americana*) their feeble flight generates elevations of T_b no more than a degree or two above air temperature (Farnsworth, 1972a,b; Janiszewski, 1984).

Cockroaches, presumably like other obligate ectotherms, nevertheless have temperature sensors (Herter, 1924) and they avoid temperature extremes. Some, like the desert cockroach *Arenivaga investigata*, even escape heat by living a subterranean life in the sand (Edney, Haynes, and Gibo, 1974). *Blatta orientalis*, acclimated at room temperature, for example, chooses temperatures of 27–28°C and avoids temperatures below 15°C and above 33°C (Gunn, 1934). I here focus on the physiological basis of temperature preferences and avoidance in the American cockroach *Periplaneta americana* (Fig. 4.14).

Bernard F. Murphy, Jr., and James E. Heath (1983) at the University of Illinois studied temperature preferences in this species in a shuttle box divided into five equal regions, each with a 10°C range of temperature within the overall range of 0–50°C. The temperature gradient was maintained by placing ice under one end of the shuttle box and hot water under the other end. The floor of the box was covered with wet paper toweling to minimize the humidity gradient. The time individual cockroaches

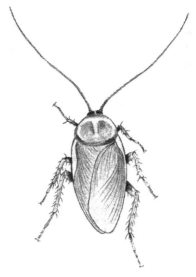

Fig. 4.14 The American cockroach *Periplaneta americana*.

spent in the five different areas of the box was scored in cold-acclimated, warm-acclimated, and room-temperature-acclimated cockroaches, and in antennaeless, tarsiless, antennaeless and tarsiless, and unaltered animals. Control runs were made in the box without the temperature gradient. Like desert locusts previously examined by Chapman (Fig. 4.15), the cockroaches showed clear temperature preferences (but near 20–30°C rather than near 35–40°C), and there was a tendency for more individuals to choose slightly lower or higher temperatures depending on prior acclimation (Murphy and Heath, 1983). It had previously been shown that cockroaches have temperature receptors on their labial palpi (Herter, 1924), antennae (Loftus, 1966, 1968; Alther, Sass, and Alther, 1977), and tarsi (Fig. 4.16).

In *P. americana* the thermal sensilla on the antennae are ring-shaped structures with a short hair emerging from the center. There are only about 20 thermoreceptors per antenna, located ventrally about one per segment on the distal half of the antenna. During electrophysiological recording from the receptor there is great variation in response to steady-state temperatures; hence, the receptor is a poor thermometer (Loftus, 1968), although it responds well to temperature *change* (Loftus, 1966). Electrical activity from the receptors increases greatly if temperatures are lowered below 13°C, although warming above 30°C also produces electrical activity.

Surprisingly, removal of either the antennae or the tarsi has very little effect on temperature preferences in the shuttle box (Murphy and Heath, 1983). With both of these exteroceptors missing, cockroaches still maintain their previous temperature preferences, although the preference is no longer quite as sharp.

Fig. 4.15 The distribution of ten desert locusts *Schistocerca gregaria* in a temperature gradient at hourly intervals over a 5-hour experiment. (From Chapman, 1965.)

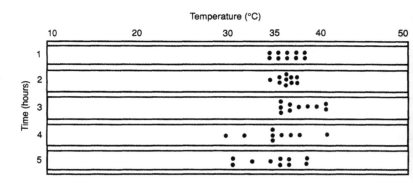

THE HOT-BLOODED INSECTS

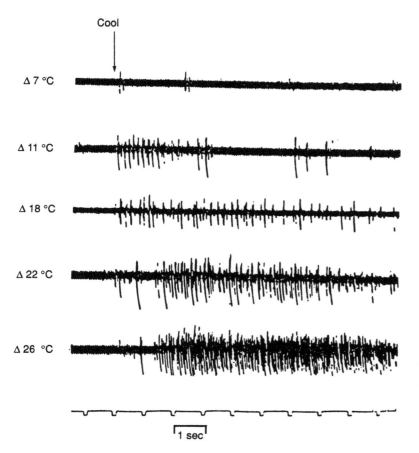

Cool

Δ 7 °C

Δ 11 °C

Δ 18 °C

Δ 22 °C

Δ 26 °C

1 sec

Fig. 4.16 The neuronal activity from the tarsus of the cockroach *Periplaneta americana* upon cooling by 7 to 26 ° C, as indicated. The tarsus was acclimated to 22 ° C. (From Kerkut and Taylor, 1957.)

(In the absence of a temperature gradient in the shuttle box the animals tend to choose the ends of the box, away from the region where the preferred temperatures were normally found.)

Behavioral and neurophysiological data suggest that the cockroaches prefer temperatures near their upper tolerable zone, and so a small rise in temperature is therefore more important to them than a small fall in temperature. They often become cold-torpid in the temperature gradient shuttle box, but they almost never become heat-torpid. However, antennaeless and tarsiless animals have a tendency to be found in the high-temperature end of the gradient (40–50°C) that normal animals almost always avoid. This lack of precision in thermoregulatory control suggests that the exteroceptors function primarily in responding to high temperatures (Murphy and Heath, 1983). I suspect, however, that the experiments so far may underestimate the role of these sensors. The temperature gradients in which the cockroaches were tested were smooth, continuous. Perhaps in nature the antennal thermosensors are important in detecting sharp, local temperature gradients, allowing the animal to decide whether or not, for example, to come out of a crack or from under a leaf into sunshine. In different species, they may also operate in different contexts. For example, in the blood-sucking bug *Rhodnius prolixus,* antennal thermoreceptors guide the insect's orientation to warm-blooded prey (Wigglesworth and Gillett, 1934).

Cockroaches (Kerkut and Taylor, 1956, 1958), locusts (Heitler, Goodman and Frazer-Rowell, 1977), and crickets (Janiszewski and Otto, 1988) additionally have temperature-sensitive neurons in their central nervous systems (Fig. 4.17), and Murphy and Heath (1983) examined the temperature-sensitive neurons within the prothoracic ganglion in the cockroach (Fig. 4.18). Some of the neurons responded by steeply increasing their firing rate as a thermode (heated or cooled by circulating water) placed underneath the ganglion was heated. Others, instead, responded by increased firing only upon cooling. These neurons were thus likely warm- and cold-receptors, respectively. (Other prothoracic neurons show little temperature response.) The implication of these results appears to be that *P. americana* regulate their T_b near 20–30°C primarily by use of their prothoracic temperature receptors, with relatively little involvement of the thermoreceptors on tarsi and antennae. Since the exteroceptors respond mainly to *changes* in temperature, however, it would be of interest to examine the speed at which the animals can choose their preferred T_b, relative to the heating and cooling rates of the body.

The role of the prothoracic temperature receptors also remains unclear. Janusz Janiszewski (1985) has shown that *P. americana* also has thermosensors in the head. Internal heating of the head results in lessened flight activity, but heating of the metathoracic ganglion does not. If flight and walking are regulated by different thermoreceptors, then the relative role of centrally located thermosensors in head, prothorax, and metathorax thus remains somewhat obscure, although superficially the results so far seem to suggest that the anterior ones regulate flight activity while the more posterior ones regulate walking activity.

Adding to the complexity are results from intracellular recordings of identified subesophageal neurons in the cricket *Gryllus*

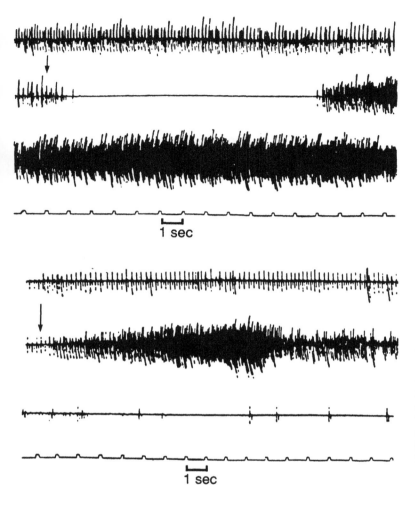

Fig. 4.17 The effect of sudden changes of temperature on isolated abdominal nerve cords in the cockroach. First trace in both sets shows activity before treatment. Second and third traces show activity during and after treatment. *Top:* At the arrow the temperature was increased from 13°C to 29°C. There is an immediate decrease in activity, followed by an increase. *Bottom:* At the arrow the temperature was decreased from 13°C to 4°C. There was an immediate increase in activity before activity fell to a lower level. (From Kerkut and Taylor, 1958.)

bimaculatus. Neurons which respond to changes in temperature (Janiszewski, Otto, and Kleindienst, 1987) also respond to auditory, visual and tactile cues and abdominal respiratory movements (Janiszewski and Otto, 1988; Janiszewski, Kosecka-Janiszewska, and Otto, 1988). One gets the impression so far that temperature is clearly of some considerable importance in these poikilothermic animals' integrative responses, but how the central integrative processes might be controlled by temperature is a mystery.

The correlation of activity with the response of some individual thermoresponsive neurons, on the other hand, is relatively good,

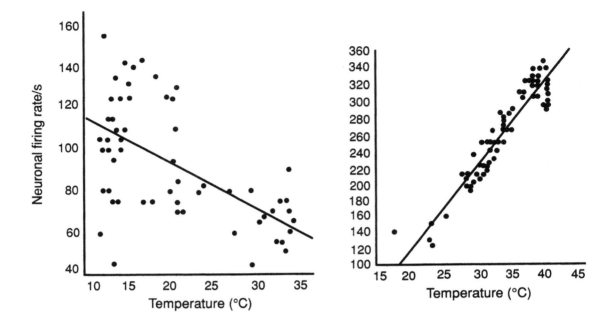

Fig. 4.18 Thermal sensitivity in firing rate of two different neurons in the prothoracic ganglion of the cockroach *P. americana. Left:* A high-sensitivity cold unit. *Right:* A high-sensitivity warm unit. (From Murphy and Heath, 1983.)

especially in locusts. In *Schistocerca* locusts, increases in temperature result in behavioral excitability, evidenced by increased incidence of jumping. The higher excitability is paralleled by decreased spike-threshold levels in identified neurons of the thorax after heating (Heitler, Goodman, and Frazer-Rowell, 1977; Abrams and Pearson, 1982). Nevertheless, there is considerable variability in the physiological response to heat, even in identified neurons. Indeed, as observed by Goodman and Heitler (1977), strains of locusts may be selected in the laboratory with very little of the normal increase in "jumpiness" with increase in temperature. These individuals show an abnormal increase in current threshold, an abnormal invariance in the voltage threshold, and an abnormal decrease in membrane resistance in the motoneurons of the fast extensor tibiae muscle (Goodman and Heitler, 1977). This abnormality does not affect the entire nervous system; certain other identified neurons appear normal in their response to temperature.

ACCLIMATION AND ADAPTATION

Cockroaches respond to temperature mosaics in their environment and then choose specific body temperatures, generally near 20–

30°C. Immediate behavioral options are often limited, however, and the animals are then unable to make up the difference by physiological temperature regulation. Instead, they acclimate. Acclimation has long been known for chill-coma (Mellanby, 1939). For example, *Blatella germanica* acclimated to 35, 25, 15, and 10 °C had chill-coma temperatures of 7.0, 5.5, 4.4, and 4.3°C, respectively (Calhoun, 1960). Acclimation is seen in the oxygen consumption of *Periplaneta americana* (Dehnel and Segal, 1956) and in enzyme activities, such as muscle apyrase (Muchmor and Richards, 1961). Optimum temperature for apyrase activity decreases with the lowering of the chill-coma temperature during cold-acclimation. Similarly, in several species of cockroaches the electrical activity of the nervous system also changes with acclimation (Kerkut and Taylor, 1958; Anderson and Muchmor, 1968). In isolated ganglia from the cockroach *P. americana*, steady activity rate is highest near 22°C in animals kept for 4 weeks at 22°C before the experiment, and it is highest at near 31°C for animals kept at 31°C (Fig. 4.19) for 4 weeks (Kerkut and Taylor, 1958).

On a still longer time frame, it is likely that different temperature optima have been selected through evolution (Bullock, 1955). Although most orthopteroids inhabit warm climates or are active in warm microhabitats, one group, the alpine *Grylloblatta* species, is a notable exception that illustrates evolution for activity at very low temperatures.

Grylloblattids, primitive grasshopper-like animals, are nocturnal scavengers on snowfields at 2,700 m in Mount Rainier National Park, Washington, where they can run at 2 cm/sec and where they have been observed to travel at least 120 m over snow in one night at temperatures of 1–3°C (Edwards and Nutting, 1950; Morrisey and Edwards, 1979). These insects have temperature preferences near 0°C, and their giant interneurons continue to elicit action potential down to at least −3.5°C; in contrast, the house cricket *Acheta domesticus* shows no sign of central nervous activity at near 4°C (Morrisey and Edwards, 1979). The specialization to low temperature in the grylloblattids is at the price of sensitivity to higher temperatures; heat coma occurs at 15–20°C.

Onset of Thermogenesis

In mammals and birds, the ability to thermoregulate by heat production typically develops gradually after birth or hatching.

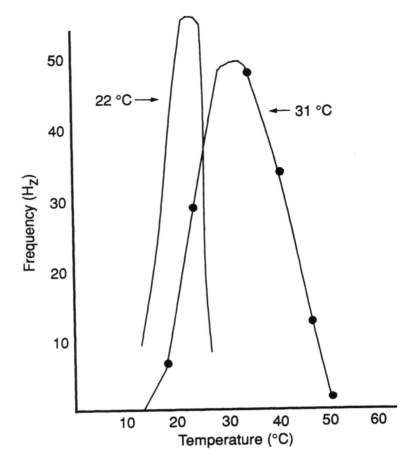

Fig. 4.19 Steady rates of electrical activity were recorded from isolated ganglia from the cockroach *Periplaneta americana* at two acclimation temperatures. Cockroaches were kept at 22 °C or 31 °C for 4 weeks prior to the experiment. (From Kerkut and Taylor, 1958.)

The full heat-generating response by the muscles is usually correlated with the ability to achieve sustained high levels of exercise, which are limited developmentally. To some extent this limitation is circumvented by nonshivering thermogenesis using brown fat. Larval and pupal insects have surface areas that are too high and metabolic rates too low to affect body temperature. However, since flight generally occurs within hours of eclosion, so does the very high metabolic rate and with it the concomitant heat production that elevates T_b.

Although the transition from near inactivity to flight is abrupt, there is nevertheless developmental preparation for it. During insect metamorphosis there is a dramatic reorganization involving both the degeneration of old muscle and the growth of new. Both

THE HOT-BLOODED INSECTS

of these developmental changes are a direct result of hormones acting on the muscles (Schwartz and Truman, 1984). There are also some suggestions of behavioral preparation, or "exercise," but whether this exercise enhances muscular function and increases T_b has so far not been shown.

In flightless nymphs of both locusts and crickets there is already neural stimulation (firing) of the muscles that will be used for flight in the adult stage. When nymphs of the field cricket *Teleogryllus commodus* are mounted in a wind tunnel and subjected to flight stimuli, they often assume a flight posture and firing of the flight muscles then occurs (Bentley and Hoy, 1970). In the desert locust *S. gregaria* (Kutsch, 1971) the Australian plague locust and *Chortoicetes terminifera* (Altman, 1975), the flight motor neurons also fire in the nymphs, and this firing is rhythmic and approximately simultaneous, reminiscent of the pre-flight warm-up patterns of many insects. Although the muscles that will ultimately be used for flight fire in response to flight stimuli in at least the last four larval instars, the coordinated motor pattern necessary for flight is absent. That pattern appears only gradually, around the time of the last molt to the adult stage.

In *Chortoicetes* the adult motor pattern becomes established during a period of about two days before, and three days after, the last molt. Fixing the wings after the final molt to eliminate practice wing beating does not prevent the establishment of the normal, alternating motor pattern between the antagonistic flight muscles (Altman, 1975). In crickets the adult motor pattern (for song) is fully developed already in the last larval instar but is normally not expressed. The connections are apparently complete and then switched on, perhaps hormonally, at or before the final molt (Bentley and Hoy, 1970). Meanwhile, in crickets neural input is required for muscle growth (Novicki, 1989b) and muscle development continues even after the adult molt (Novicki, 1989a).

The flight pattern of the adult is under the influence of peripheral inputs and central neural programs (see references in Wilson, 1968). However, after the adult ecdysis, the wing-beat frequency in at least two species of locusts continues to rise (nearly tripling) during the first week and up to the third week of adult life (Kutsch, 1971, 1973). In Orthoptera there are also changes in the morphology of the muscles (Mizisin and Ready, 1986; Novicki, 1989b), in their enzyme patterns (Brosemer, Vogell, and Bücher, 1963), and the contraction kinetics of the muscles (Novicki and Josephson, 1987) during the first week of adult life. Thus, at least in orthopterans, the development of the flight motor is both morphologically and neuronally a gradual process that starts before the adult stage and continues after it. A gradual increase in metabolism, and the capacity for endothermy, presumably accompanies this development.

In a holometabolous insect, however, the transition from an inactive, poikilothermic stage to a flying, endothermic stage is abrupt. For example, a moth or butterfly emerges from its pupa slowly inflates and dries its wings, and then flies off. Thus, over the course of a minute or less it raises its metabolic rate at least a hundred-fold above what it has been for many days previously, to a level near the most intense physical performance in the animal kingdom. However, the transition could be less abrupt than it appears.

In a study of the development of the flight-motor patterns in the sphinx moth *Manduca sexta,* Ann E. Kammer and Sue C. Kinnamon (1979) recorded for up to 8 days the activity of the thoracic muscles of moths developing within the pupae. Curiously, pupae that showed no movement at all nevertheless showed apparently spontaneous bouts of activity of the flight muscles. The bouts of muscle activity (recorded from fine wire electrodes) lasted from about 30 minutes to 2 hours, alternating with periods of rest of similar duration. Activity was recorded from the ninth day after pupation and on all successive days until typically the nineteenth day. During the 3 days just before emergence from the pupa, the rhythmic patterns typical of pre-flight warm-up and flight appeared. Kammer and Rheuben (1976) were able to elicit muscle contractions from immature animals by electrically stimulating the appropriate motor neurons. The muscle activation recorded from the pupae is therefore likely also associated with muscle contraction, or "exercise." However, these muscle contractions are very weak, because the pupae that showed the muscle patterns similar to those of shivering or flight did not show corresponding increases in body temperature.

Summary

The Orthoptera, in general, are not known for their endothermy, presumably because few (except migrating locusts) are continuous flyers. However, because of their huge economic importance (particularly migratory locusts and other acridids) and their large size, and ease of maintenance in the laboratory (locusts, crickets, cockroaches), they have been ideal and popular subjects for the study of basic physiological mechanisms. Much information from orthopterans provides a background for the thermal responses of insects in general.

The development of the flight motor in the hemimetabolous orthopterans is a gradual process that starts before the adult stage and continues after it. The capacity for limited endothermy presumably accompanies it. However, no orthopterans have so far been observed to warm up by shivering before flight (although tettigoniid grasshoppers are capable of shivering warm-up). Some may cool through evaporative cooling by panting at near-lethal temperatures.

Some of the largest flying orthopterans, like migratory locusts, achieve modest endothermy in flight, although there is so far no evidence that they regulate their T_b in flight. On long migrations, a high T_b may be a liability in terms of water loss, but this is in part counterbalanced by specialized respiratory patterns that make use of regional body heterothermy to conserve water.

A number of species of acrididine grasshoppers inhabiting variable thermal environments, especially environments with very low and very high ambient temperatures, show strong independence of T_b from T_a. Thermoregulation is accomplished by microhabitat selection and by a complex repertoire of postural adjustments. On the other hand, many species (especially cockroaches) show relatively strict conformity to ambient temperature.

Individuals of poikilothermic species (cockroaches) acclimate their thermal tolerances following exposure to either lower or higher temperatures. Acclimation affects temperature preferences, the response of central neurons, and rates of oxygen consumption.

Within the Orthoptera there is an extraordinary range of thermal tolerances. Some species have adapted to forage on glaciers, at near 0°C; others survive in deserts and tolerate T_b close to 50°C. Extraordinarily high thermal tolerances allow some desert acrididine grasshoppers to escape from vertebrate predators more sensitive to heat than they are.

Even the most thermoconformers (relative to the temperature of the environment where they come to rest) have temperature sensors. Both cold- and warm-receptor neurons have been identified in the central nervous system. Cold- and warm-exteroceptors have also been identified (on tarsi and antennae in cockroaches), but these sensors apparently play a secondary role in adjusting thermal preferences.

Some katydid males are highly endothermic. They regulate their thoracic temperature to produce high and specific chirp rates independent of air temperature, as required to attract females. The males of most crickets and katydids, however, produce chirp rates at strictly temperature-dependent rates. Such variable rates are nevertheless recognized by the females of the appropriate species by "temperature coupling"—a matching between sender and receiver by way of the same temperature dependence, which nevertheless involves different neural elements in males and females.

The ultimate evolutionary significance of maintaining an elevated T_b in different orthopterans has been traced, in different species, primarily to predator avoidance, sexual signaling, and food acquisition and processing.

Elevated T_b by basking in many species of acrididine grasshoppers accelerates growth to allow the completion of larval instars within a temporally limited season. No reproduction is possible without the completion of the life cycle, and empirical studies supported by modeling show that the environmental ranges that some species can occupy depend on basking behavior.

Remaining Problems

1. What are the central neural mechanisms whereby the thermal responses are initiated and integrated?
2. What are the biochemical changes that result in temperature adaptation?
3. Is thermoregulation in temperate-zone grasshoppers a prerequisite for food processing?
4. What is the role of temperature exteroceptors in poikilothermic orthopterans?

Beetles Large and Small

TO the ancient Egyptians, scarab beetles were of great religious significance. The pharaohs were placed in their tombs with replicas of the scarab, precisely and uniformly carved according to instructions in the Book of the Dead. These beetles—which build deep underground chambers where their eggs are laid and develop through the larval, white grub stage, until as adults they emerge from the earth as beautiful creatures capable of flight—were symbols of resurrection for the Egyptians, symbols of the continuity of life. For those who understand their habits as dung scavengers, they are also a symbol of, and the central participant in, the continuous cycles of the ecosystem in which they live.

Scarabs may be among the most familiar of beetles but, like other beetle groups, they are by no means uniform in appearance, ecology, and thermal biology. At some 300,000 species, there is more variety in Coleoptera worldwide than in any other order of insects. Beetles are noteworthy not only for their diversity but also for their wide range of body size. They include members of the Ptiliidae measuring 0.25 mm to 1 mm and probably weighing less than 0.1 mg to the heaviest modern insects known, the 11–35 g tropical elephant beetle *Megasoma elephas*, which weighs up to 10 times more than the smallest birds, bats, and shrews. Beetles are acknowledged to be a predominant life form and they are known from the fossil record to have existed 250 million years ago. Indeed, when the great British biologist J. B. S. Haldane was asked what the Lord might have to show us through His creations, he responded unhesitatingly that He must have had "an inordinate fondness of beetles."

Are they good thermoregulators? Most beetles move slowly and fly rarely. Many are also nocturnal or remain hidden in the daytime. It is not likely that this majority thermoregulates under these circumstances. However, those few species that do thermoregulate (and sometimes impressively) provide instructive examples into the evolution of thermoregulation.

Desert Tenebrionids

The tenebrionid beetles are perhaps one of the most diverse of any of the beetle groups. Certainly the strangest and for comparative physiologists and thermobiologists the most interesting members of this family are the ground-dwelling and mostly flightless species that are a conspicuous part of the fauna in many deserts (Koch, 1961, 1962). Thanks in large part to the Gobabeb field research station in the Namib Desert of southwestern Africa, we now have considerable knowledge of thermoregulation in this group of animals, and most of the following account refers to beetles studied near there.

Possibly the first person to examine the thermal adaptations of these beetles was P. A. Buxton (1924), who observed the tiny black tenebrionid *Zophosis punctatus* near Jerusalem running on bare soil heated to 60–62°C in the sunshine. In a large black species, *Adesmia ulcerosa*, of Algeria and Palestine he observed that live animals had lower body temperatures than dead ones. Niels Bolwig (1957) later also became interested in the problem of thermal balance in these beetles. On a trip to the Kalahari Gemsbok Preserve he recorded that "in a relatively cool period in February" the soil temperature was nevertheless 57°C, while the air temperature at 12 cm above the ground was 46°C. The "cool" weather notwithstanding, he was surprised to see very black tenebrionid beetles active "even in the hottest part of the day when the sand was so hot it was most uncomfortable for me to walk about barefooted." Buxton's and Bolwig's observations were pivotal in drawing attention to the remarkable behavior and physiology of these animals that has since attracted many other researchers.

As one adaptation to the hot desert environment, the beetles have very high upper lethal temperatures. Beetles from the sand-dune environment tolerate T_b near 50°C for half an hour, while those from mesic woodland have tolerances more than 5°C lower (Edney, 1971; Holm and Edney, 1973). Second, the animals have morphological adaptations that help reduce body heating. These

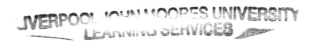

include the long, stilt-like legs of *Stenocara phalangium* (Fig. 5.1) that raise the body above the hot substrate (Koch, 1962; Henwood, 1975b). Alternately, some have raised elytra that act as a sunshield in tenebrionids (Fig. 5.2) and possibly in other beetles (Fig. 5.3). No studies are available that systematically examine the thermal effects of the raised elytra on body temperature, al-

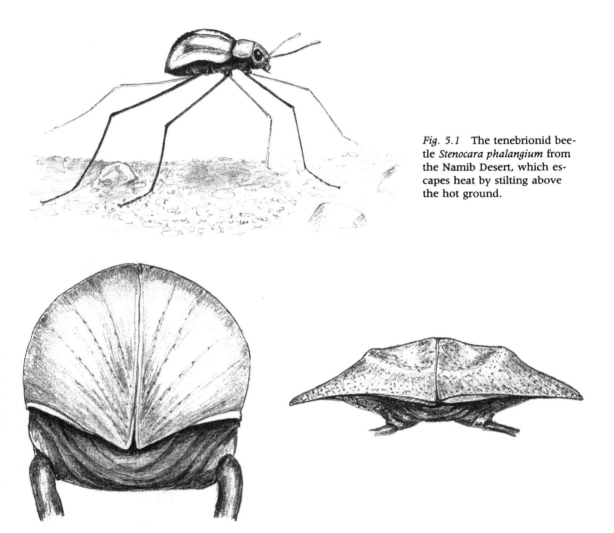

Fig. 5.1 The tenebrionid beetle *Stenocara phalangium* from the Namib Desert, which escapes heat by stilting above the hot ground.

Fig. 5.2 Elytral morphology of two tenebrionids. At left is a rear view of *Onymacris unguicalaris*, a diurnal, sunloving beetle with highly convex elytra and consequently a large subelytral cavity. To the right is a nocturnal tenebrionid, *Stips stali*, whose elytra are flatter and whose subelytral cavity is therefore smaller. (From Koch, 1961.)

Fig. 5.3 The blister beetle *Cysteodemus armatus* (Meloidae), active in sunshine on the ground in the Mojave Desert of California, has raised elytra and a large subelytral space.

though in *Adesmia ulceros*, a large black tenebrionid of Algeria and Palestine, the temperature of the subelytral space is reportedly 1 °C lower than body temperature itself (Buxton, 1924). Undoubtedly this is a large underestimate of the maximum shielding effect possible, which relates also to the temperature of the elytra themselves. (It is difficult to measure the temperature of an elytron since one cannot insert a thermometer into it, and to my knowledge adequate measurements have not yet been reported.) I see no reason to suppose why the black elytra under direct sunshine would not heat to near ground temperature, as high as 60 °C. Undoubtedly, T_b is at least 10 °C lower.

Elytral-surface temperature is also affected by color. Some species have white elytra that reflect sunlight and produce T_b from 1–3 °C lower when convection is minimized (Bolwig, 1957; Hadley, 1971; Edney, 1971; Hamilton, 1973; Turner and Lombard, 1990). In addition, in the Namib Desert many beetles have reversible colors due to wax blooms that are not normally found in other arthropods. These wax blooms are yellow, white, red, or pink, and they cover an otherwise black cuticle. Elizabeth McClain at the University of Namibia has extensively studied the physiology and ecology of wax secretion in Namib tenebrionid beetles. Using equipment originally designed to measure colorimetric properties of diamonds, she and co-workers (1991) showed that the waxes reflect light relatively uniformly over a large portion of the visible spectrum (380–750 nm). In *Zophosis mniszechi* the total amount of light reflected by the surface wax represents about half of the radiation of the visible spectrum. The wax blooms secreted onto either part or all of the body act not only in reducing water loss but also in reducing temperature excess (McClain et al., 1984a, 1984b, 1991). The waxes retard water loss directly. Reductions in body temperature reduce the saturation deficit, and that should secondarily reduce additional water loss as well. Some frogs also have waxy secretions that minimize water loss, but these waxes break down just below the critical temperature of the frogs, allowing for evaporative cooling in emergencies.

In *Onymacris plana*, variations of the percentage of the cuticular surface covered with a wax bloom varies along a climatic gradient. Beetles from the cool, coastal fog belt have little or no wax bloom while those from the hot, dry, inland desert are completely covered with wax (McClain et al., 1985). And in *Stenocara phalangium* (Fig. 5.1) wax bloom is regenerated within a day in response to high temperatures and low humidity (McClain et al., 1984b).

Taken together, the combined data suggest that elytral color and wax blooms are both adaptive responses that permit activity at high temperatures under desiccating conditions. The white and wax-bloomed species are thought to have the ability to stay out longer in the heat of the day and to reemerge earlier in the afternoon, when air temperatures are still high (McClain, Kok, and Monard, 1991). Nevertheless, Koch (1961) has pointed out that not all white, reflective beetles are diurnal, and Turner and Lombard (1990) indicate that color affects T_b only when convection is less than that normally encountered in the field. There is no reason to suppose, however, that selection occurs under "normal" conditions. More likely it exerts its effects primarily under extremes.

Behavioral adaptations of the desert tenebrionid beetles range from the long-term to the immediate. For example, when Bolwig (1957) heated *Onymacris bicolor* under a heat lamp, he observed that as T_b approached 40°C there were great fluctuations in the air temperature under the elytra (Fig. 5.4), caused by strong ventilation of the normally tightly sealed subelytral cavity. Heat-stressed beetles also showed rhythmic protractions and retractions of the head, as well as exposure of the genital apparatus. These emergency measures may allow beetles to tolerate high temperatures for short durations but they have not been documented in the field. Instead, in the field, heat-stressed *Onymacris plana* escape lethal temperatures by burrowing into the sand (Koch, 1961; Henwood, 1975b). This is remarkable behavior, since in order to reach cool sand layers the animals must first subject themselves

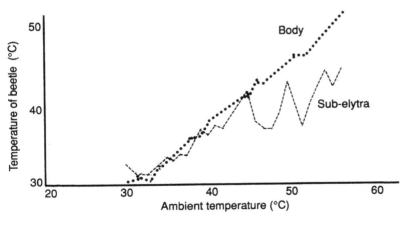

Fig. 5.4 Body *(dotted line)* and subelytral *(dashed line)* temperatures of *Onymacris bicolor* as a function of ambient temperature under experimental conditions. (From Bolwig, 1957.)

Fig. 5.5 A male of the Namib Desert tenebrionid *Onymacris plana*, a very swift runner.

to contact with the very hot surface layers, which may exceed 60°C (Buxton, 1924).

Tenebrionid beetles of various species bring their body temperature within a favorable range by shuttling between sunshine and shade when available (Edney, 1971; Wharton, 1980) and, as *O. plana* do, by burying themselves and/or avoiding activity in the hottest part of the day or the season (Hamilton, 1971; Holm and Edney, 1973; Henwood, 1975b; Wharton, 1980; Kenagy and Stevenson, 1982). Diurnal activity shifts occur in the Sahara carabid beetle, *Thermophilium sexmaculatum;* the southern subspecies is nocturnal throughout the year, whereas the northern subspecies is diurnal in winter and nocturnal in summer (Erbeling and Paarmann, 1985). Tenebrionids in eastern Washington, in the United States, also have day-night shifts throughout the season (Kenagy and Stevenson, 1982), but no such shifts from day to night have been shown in African tenebrionids.

Two Namib tenebrionids occupying thermal extremes may provide an instructive example of the role of ecology in the evolution of thermoregulatory strategy. One species, *Onymacris plana* (Fig. 5.5), is a large (1.0–1.5 g), shiny black, and dorso-ventrally flattened beetle. (The males have a broader, flatter abdomen than the females.) This species is a strictly diurnal and predominantly summer-active species. It is omnivorous, feeding on green plants, dead animals, and on plant detritus. It is active throughout the sand-dune system except for the crests. Body temperatures for *O. plana* in the field range from near 30°C in the morning up to near 43°C at noon, before further activity ceases (Henwood, 1975b; Nicolson, Bartholomew, and Seely, 1984), and the beetles take refuge by burrowing. Upper lethal temperatures are near 50°C (Edney, 1971).

Onymacris plana is faced with the high end of the temperature spectrum in the desert. Not surprisingly, therefore, beetles emerging in the early morning are obliged to warm themselves in direct sunshine, principally on the eastern slopes of the sand dunes, where they often squat "belly down" to heat themselves by contact with the sand (W. J. Hamilton III, personal communication). By late evening the beetles that remain active move to the westerly slopes, where they catch the last rays of the sun (Henwood, 1975b). Their elytra have elevated transmittance (as opposed to reflection) to short-wave infrared radiation, which is thought (Henwood, 1975a) to be used to increase heat gain both at the beginning and at the end of the day. However, since the infrared radiation is even *more* intense at midday (although its total percentage is less), when the beetles must retreat from the heat, enhanced transmittance can be of benefit in the way suggested only if the time the animals lose during the day is more than compensated for by increased time gained in the morning and evening and on cooler days. (But if heating is a primary adaptation of the transmittance to the near infrared when the sun's rays are at a low angle, why then are the *visible* rays simultaneously reflected and not absorbed like the infrared? It should make no difference to the beetles whether they are getting heat from the infrared or the visible portion of the spectrum.) The hypothesis on the role of reflectance and transmittance of *particular* wavelengths on the thermal balance may be debated, but the fact remains that the relatively flat *O. plana* beetles are subjected to an enormous heat load in sunshine in the middle of the day. Might they be able to compensate for the large midday heat load by cooling convectively by running?

Onymacris plana is the swiftest runner of all the Namib beetles, possibly the fastest running arthropod ever recorded. The beetles can run at speeds over 1 m/s, and on the average they cover near 90 cm/s or the equivalent to 50 body lengths per second (Nicolson, Bartholomew, and Seely, 1984). When a beetle is running "it appears to float along like a low flying bumblebee." (Bartholomew, Lighton, and Louw, 1985). There is no evidence for endothermic elevation of T_b during this activity (Nicolson, Bartholomew, and Seely, 1984), and in male *O. plana* the metabolic cost of running on a treadmill is the same at running speeds of 13 to 22 cm/s (Bartholomew, Lighton, and Louw, 1985). Since while they run the beetles are 1.0–1.5 cm above the hot substrate, where air temperatures are about 15°C lower than at the surface itself (Holm and Edney, 1973), it follows that they dissipate some of their heat load from radiation by convective cooling. Whether they run *in order to* cool has not been examined. They have at least two speeds at the same temperature (W. J. Hamilton III, personal communication), however, and on-off running activity is a potential option for thermoregulation.

The hypothesis that running results in cooling has been critically examined by Carole Roberts, at the National Museum in Namibia, and her associates (Roberts et al., 1991), who showed that although the rate of body-temperature increase from the metabolic heat can be close to 1°C when a beetle runs at full speed for a minute, the cooling during running while solar heated for the same period of time can result in T_b near 2°C lower than that without running. The critical variables for effective convective cooling by running are high radiation, low air temperatures, low wind speed, and/or movement in a direction other than that of the ambient wind. Before we can conclude that running is adaptively varied for cooling, behaviors relative to these variables must be examined.

Stenocara phalangium (Fig. 5.1) present an interesting contrast to *O. plana*. These beetles inhabit a relatively level, gravelly environment. Although they apparently engage in basking and ground hugging on cool mornings, they generally do not appear on the gravel plains (from where?) until air temperatures 1 cm above the ground reach 34°C, which allows them to maintain a predicted 36°C T_b (Henwood, 1975b). At much higher air temperatures than 34°C they encounter potential problems. First, unlike *O. plana*, these beetles have no dune slopes to which they can retreat to decrease solar heating near midday. Furthermore, they are not morphologically suited for digging, so they cannot readily bury themselves to escape the heat, either. However, like *O. plana*, they have a unique morphology and thermoregulatory behavior associated with it.

These beetles have the longest legs relative to their body of any of the tenebrionids of the Namib, and at very high ambient temperatures, when their legs are insufficiently long to escape the hot surface of the ground, they minimize further heat gain by elevating themselves even more above the searing gound by mounting small rocks and stilting upon them. They may then also point the white posterior of their abdomen to the sun, which reduces heat gain even more by lowering both the absorptivity (McClain et al., 1984b) and the body-surface area receiving direct radiation (Henwood, 1975b). Using a model beetle to predict body temperatures, Henwood (1975b) concluded that for one particular hot day (May 6, 1973) a *S. phalangium* would lower its predicted T_b of 50°C to 44°C by standing on a 5 cm rock, and then reduce T_b another 2°C by stilting on the rock. (The larger *O. plana* model under the same conditions that day would heat to over 55°C on the same rock.) Henwood noted that rock climbing started at about 9:30 A.M., and by 10:30 most beetles were each defending a particular rock. Beetles pushed off their rock made vigorous attempts to climb back on or to mount another, and those denied the opportunity to climb a rock became comatose after 15–20 minutes and died of heat prostration. Such episodes of high thermal load may not be common. But they could be a critical bottleneck that strongly selects for beetles capable of remaining active at a high T_b because it would reduce the chance of death directly if stilting were not used. A high thermal tolerance would also reduce the necessity for time-consuming stilting behavior on rocks.

For the most part, the variables and mechanisms that affect T_b in terrestrial beetles are no longer great mysteries. More problematical are questions of why specific T_b are maintained. Why, for example, do many desert tenebrionid beetles regulate such high body temperatures?

W. J. Hamilton III at the University of California at Davis has espoused the "maxithermy" theory (Hamilton, 1971). He proposes the animals try to achieve the highest T_b possible to achieve higher metabolic rates and thereby enhance competitive ability and reproductive output (Henwood, 1975b). That in any one animal these functions are indeed enhanced nearer the upper as opposed to the lower end of its normal temperature range is not disputed. But that is a proximate or physiological explanation. It does not explain why particular temperature *ranges* have been chosen by evolution. For example, contrary to the Namib tenebrionid beetles, the surface-active tenebrionids from eastern Washington State (Kenagy and Stevenson, 1982) operate at a body-temperature range some 20°C lower than that of most Namib beetles. This proves, therefore, that tenebrionid beetles are not unique animals that as a group *require* a high T_b to be active; they can also evolve to be active at low T_b. Presumably the Namib and Washington beetles have both evolved to be active in a temperature range near that to which they have been exposed through their evolutionary history. Maxithermy does not explain why many desert tenebrionids in Namibia subject themselves to sunshine in diurnal (rather than nocturnal) activity patterns. The answer of why they do likely resides in ecological adaptation, not an extrapolated proximate physiology.

Wharton (1980) suggests that in the Namib Desert the major predators of the beetles are nocturnal or crepuscular so that the beetles, forced to be diurnal to reduce their exposure to predation, face the thermal challenges of the daytime instead. If competition or predation were the evolutionary driving force for the occupation of a thermally stressful habitat, physiological and behavioral adaptation could have then followed until the previous thermal stress is no longer normally stressful. Indeed, at the present time neither temperature nor desiccation as such appear to explain the beetles' immediate behavior underlying burying and emergence patterns (Seely and Mitchell, 1987).

A second idea is that the beetles compete for resources and partition them diurnally (Hamilton, 1971). Since the tenebrionid beetles eat vegetation and organic detritus, however, food availability should not be a function of time of day. It is therefore not immediately obvious why it should be advantageous for the beetles to search for food at a certain time of day. Still a third idea is that the various sympatric species only incidentally "divide the environment" among themselves by functioning at different activity times, which have evolved primarily to facilitate mutual encounters for mating and to avoid the disadvantages of interspecific courting (Holm and Edney, 1973). If this hypothesis is correct for these visually orienting beetles, then it suggests further that the bright, white elytra found in some beetles could have an additional purpose besides the afore-mentioned thermoregulatory and aposematic effects.

I speculate that in a windy, open environment vision may play a larger role than scent in mate recognition from a distance. Would mistakes be made if all species were uniformly black in color? Can white beetles be more easily detected and differentiated by conspecific mates so that time on the ground surface can be minimized? This idea has not been examined; we know very little about mate recognition in the context of potentially competing signals from other species. At least in the black Namib tenebrionid beetle *Physadesmia globosa*, however, males that find females spend much of their time following them (Marden, 1987). The longer they are able to follow, the more likely they will copulate. If beetles of another species were present, would *P. globosa* follow females of its own species longer if they were of a unique color? Selective pressures such as food availability, meeting of mates, predator avoidance, and possibly still others are probably all contributing factors to different degrees to the evolution of any one trait in different species at different times. It seems, however, that it is probably more parsimonious to suppose that (for whatever reason) the beetles, rather than becoming "maxithermic" to achieve maximal activity rates, evolved to occupy a particular ecological niche and then concurrently adapted to deal with the temperatures there encountered.

The temperatures encountered because of ecological adaptation may be benign or severe. For example, some myrmecophilic beetles have "broken the code" of chemical communication with ants and now live within ant brood chambers, nest chambers, and garbage dumps (Hölldobler, 1972). Obviously, to be able to reproduce there they need to also adapt to the temperatures in the nests that these social hymenopterans maintain. On the other hand, temperate carabid beetles living in moist environments (Thiele, 1977; Erbeling and Paarmann, 1986) have adapted to relatively low temperatures. Most carabids require fairly moist and cool conditions, but the larvae of several species in very arid regions are found in ant nests, possibly surviving in an otherwise unhospitable hot and dry desert environment because of this refuge (Paarmann, Erbeling, and Spinnler, 1986).

Tiger Beetles

Tiger beetles (Cicindelidae) are swift-running, thin-legged, ground beetles inhabiting open sandy areas (Fig. 5.6). Like tenebrionids, they are often subjected to hot substrates and high solar radiation. Unlike tenebrionids, however, they are predators and they not only run but also routinely fly. Tiger beetles are almost exclusively diurnal heliotherms (Dreisig, 1980; Morgan 1985) lacking any appreciable endothermy even in flight (Morgan, 1985). Metabolic rates during rest and activity are like those of other beetles (May, Pearson, and Casey, 1986).

Several species often occur in the same area at the same time. Food is often a limiting resource to them (Pearson and Knisley, 1985), however, and the pattern of species' co-occurrence may be influenced by food availability and competition for that food. (Pearson and Mury, 1979; Pearson and Stemberger, 1980; Ganeshaiah and Belavadi, 1986).

Cicindelids must hunt when their specific kinds of prey are available. In contrast to the tenebrionids foraging from detritus or plants, tiger beetles do not have food available throughout the 24-hour period. Instead, they must often hunt when it is hot. For example, Hans Dreisig from the University of Copenhagen noted that near Mols Laboratory in Denmark the tiger beetle *Cicindela hybrida* had available only 0.16 prey/m^2 at a sand surface temperature of 20°C. Prey density increased to a maximum of 0.40/ m^2 at 36°C, and then declined to 0.08 at 44°C (Dreisig, 1980, 1981). The beetles hunt at the moderately high temperatures, when the prey is the densest.

Unlike tenebrionid beetles, most tiger beetles probably show only small physiological differences in thermal tolerances of co-

Fig. 5.6 A tiger beetle (*Cicindela* species) is subjected to full solar radiation on open, sandy substrate.

Fig. 5.7 The relation of body temperature to ambient temperature (at beetle height) for *Cicindela tranquebarica* that were basking *(open triangles)*, searching for prey or mates *(filled triangles)*, or stilting *(squares)*. (From Morgan, 1985.)

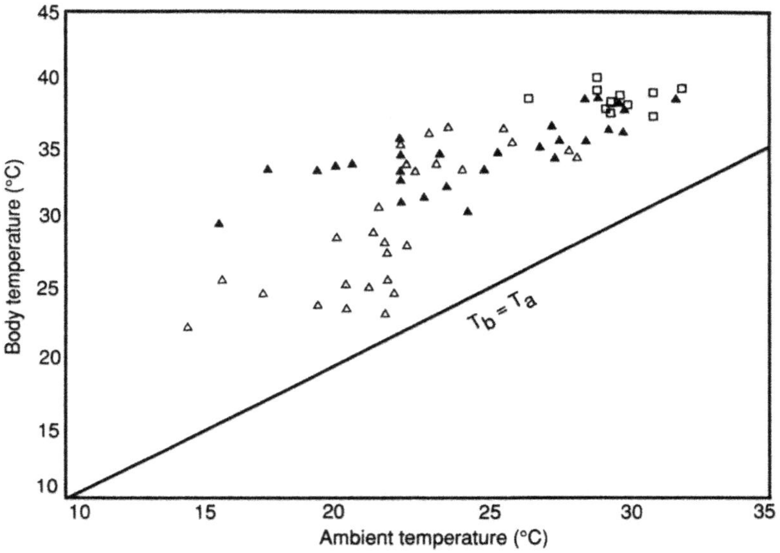

$T_b = T_a$

Body temperature (°C)

Ambient temperature (°C)

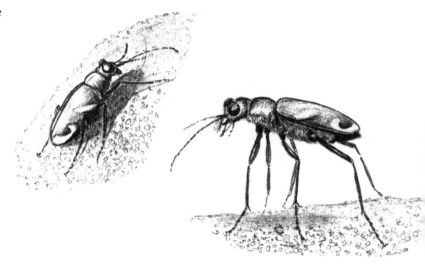

Fig. 5.8 Body postures of the tiger beetle *Cicindela hybrida* during basking at sand temperatures near 19 °C *(left)* and during stilting after soil temperature exceeds 40 °C. (Drawn from Dreisig, 1980.)

occurring species (Schultz and Hadley, 1987a). Tiger beetles maintain T_b near 30–35°C (Fig. 5.7) by a variety of behavior patterns (Dreisig, 1980; Guppy, Guppy, and Hebrard, 1983; Morgan, 1985). After emerging from their burrows in response to soil warming to near 19°C, they seek a sunny spot and press themselves onto the warm substrate (Fig. 5.8), where they will be heated by conduction and where they also minimize convective heat loss because of the boundary layer of air near the ground. The beetles also align at right angles to the sun, thereby maximizing solar heating. After T_b exceed at least 30°C they begin to search for prey by running over the substrate and stopping at frequent intervals, apparently to scan visually. As air temperature and solar radiation increase throughout the morning they spend more and more time moving around searching for prey and correspondingly less time basking. Finally, at high heat loads they begin to stilt (Fig. 5.8).

At sand-surface temperatures below 35°C, the resting beetles allow their undersides to touch the ground, but when sand temperatures exceed 36°C they keep the body raised above the sand, while they are both running and resting. At this time the legs are still kept sideways and bent at an angle, so that the body is raised about 3 mm above the sand. At high sand temperatures the body is raised higher, and at 40°C the legs are extended straight beneath the body so that the underside of the body is up to about 1 cm above the substrate. A few individuals exhibit this stilting behavior already at a soil temperature of 40°C, and at 44°C all are stilting (Dreisig, 1980), with the long axis of the body pointed at the sun, thus exposing the minimum body-surface area to solar radiation. Depending on wind speed, the difference in temperature between the ground-surface and air temperature at a beetle's body height (Fig. 5.9) ranges from at least 1.5 to 8.7°C, so that a stilting beetle usually has T_b several degrees lower than air temperature near the ground surface (Dreisig, 1990; Morgan, 1985). Stilting is not always adequate as a cooling mechanism, and during the hottest part of the day the beetles take frequent short flights of a meter or more above the hot ground that result in convective cooling to reduce T_b by 2°C to 4°C (Fig. 5.10). The stilting behavior results in longer daily activity periods and a T_b closer to the optimum for foraging (Dreisig, 1990).

Many tiger beetles have metallic green or blue iridescent cuticles, although some species are nearly black or white. Do the metallic colors serve to reflect heat? Apparently not (Schultz and

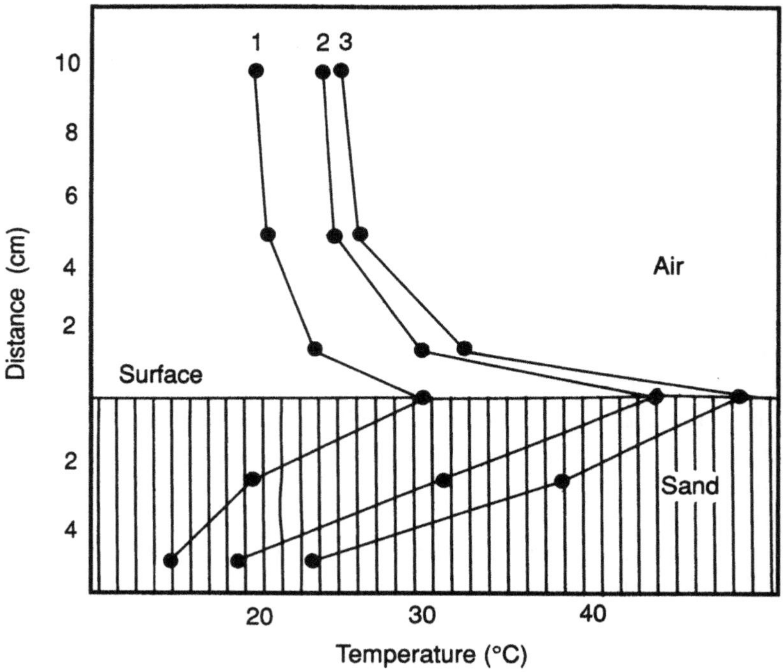

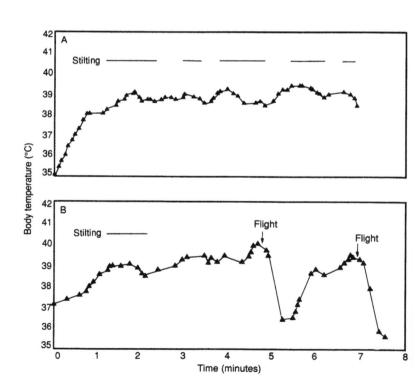

Fig. 5.9 Temperature profiles above, at, and below the sand surface illustrating conditions during three different activities by the tiger beetles *Cicindela hybrida*: (1) at time of emergence, when beetles bask; (2) when all animals stilt; (3) when all animals have disappeared from the sand surface. (From Dreisig, 1980.)

Fig. 5.10 Nearly continuous records from *Cicindela tranquebarica* showing T_b during stilting (the parts of the graphs under the solid horizontal lines at top) and during flight. Only a small decrease in T_b follows stilting, but a large decrease follows flight. (From Morgan, 1985.)

Hadley, 1987b). The metallic colors are "structural" colors and are not effective in reducing heat gain from solar radiation. But beetles with white elytra experience significantly lower body temperatures than either black or iridescent beetles under given environmental conditions (Schultz and Hadley, 1987b). In the tiger beetle *Neocicindela perhispida,* from the coastal beaches of the North Island of New Zealand, one subspecies occurs on black sand beaches, another on yellowish-brown beaches, and one on white quartz sands. The reflectance of the beetles' elytra in turn matches that of the beaches where they live, and it is concluded that the beetle that lives on the dark sands sustains considerably higher rates of solar heating from short-wave radiation than the other two (Hadley, Schultz, and Savill, 1988). Apparently, darker coloration is a thermal cost that is compensated for by greater predator protection from the cryptic resemblance of the beetle to the substrate.

Proximally, the elevation of body temperature in tiger beetles increases both time available for searching (Morgan, 1985) and searching speed (Fig. 5.11). Running speed is a direct function of body temperature (Morgan, 1985), as it is in other insects. The tiger beetles *Cicindela tranquebarica* run up to 40 cm/s when their T_{thx} is 28°C. At $T_b = 35$°C (the average T_b during searching), the predicted running speed is 74 cm/s. (Searching speed is less than running speed because searching beetles pause at frequent intervals.)

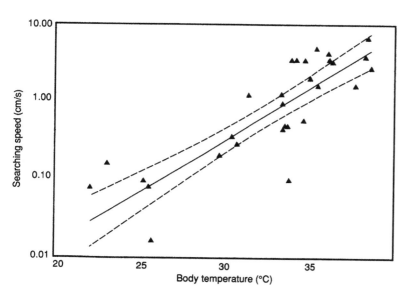

Fig. 5.11 The relation of searching speed (includes stop times) to T_b in the field in *C. tranquebarica.* The dashed lines are the 95 percent confidence limits for the regression line. (From Morgan, 1985.)

African Dung Beetles

There are about 7,000 species of beetles that feed on animal dung, having evolved from ancestors feeding on plant litter and humus. Dung beetles consist of a small number of families in the super-family Scarabaeidea, and the "true" dung beetles, subfamily Scarabeinae, are perhaps most remarkable for the sheer range and perfection of their parental investment and nesting habits. The beetles locate fresh animal droppings by following odor plumes encountered in rapid cruising flight or by perching with out-stretched antennae. Flying upwind on the odor plumes, they land on the droppings, where sexual pairs are formed and contests with rivals may also occur. Freshly formed pairs may cooperate through the breeding cycle to construct underground nests, pro-visioning the nests, copulating, and then (in some species) re-maining to care for the brood. Removal of the dung to underground nests allows for escape from competition, predation, parasitism, and often adverse climatic conditions above ground. In turn, the large amount of parental investment has selected for low fecundity. In at least one species, each pair rears only a single offspring per breeding cycle (one or two per year), and the fe-male's ovaries remain physiologically suppressed as long as she is caring for her young.

In most (but not all) species both adults and larvae feed on dung. However, the adults feed on the high-quality liquid com-ponent of the dung, whereas the larvae feed on the solid, low-quality component. Larvae have fermentation chambers in the gut with cellulose-digesting bacteria, and by repeatedly re-eating their own feces they take advantage of the steady dietary improvement caused by the symbiotic microorganisms.

Dung beetles have an interesting biology that evolved as a consequence of their competition for an extremely rich resource (see Hanski and Cambefort, 1991, for review). However, they are perhaps better known for their crucial role in nutrient cycling and enriching the soil; they forge a link between large herbivores and their ecosystem. It is estimated that in African savannas dung beetles may bury one metric ton of herbivore dung per hectare per year. In the arid Sahel, dung beetles play the role that earth-worms do in more humid biomes. In many developing countries they are highly beneficial for their removal and burial of human wastes. Dozens of species have been imported to Australia to handle the dung of cattle introduced from abroad that the native species were not equipped to handle. In addition to directly making millions of more acres of pasturage available in Australia, the introduced beetles were a huge success in controlling the bothersome flies that otherwise would breed in cow dung.

There is likely nothing special about those dung beetles from Africa, with regard to thermoregulation, but they are common and so several studies have been done on them. I therefore slant my coverage toward these particular animals, although they are found the world over (Hanski and Cambefort, 1991).

Africa boasts an extraordinary assemblage of large mammalian herbivores, and more than 2,000 species of beetles feed on their dung. These beetles have evolved into an amazing assortment of shapes and sizes. The largest weigh more than 20 g and they are 7 or 8 times larger than the smallest birds, bats, and shrews. The smallest weigh only milligrams.

Foraging strategies are diverse, but three main patterns are found. In one, adopted by the largest beetles, such as the elephant dung beetle *Heliocopris dilloni,* the animal works like a giant earth-moving machine (Fig. 5.12). Backward-directed spines and processes on its thick, short legs give traction in the soil while the beetle tunnels by lowering the blade-like front of its head into the earth and making powerful upward thrusts. These beetles, the tunnelers, burrow directly underneath a fresh dung pile, pulling

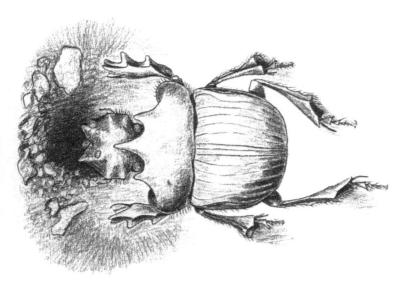

Fig. 5.12 The elephant dung beetle *Helicopris dilloni* excavates by using its blade-shaped head as a shovel and its short, stubby legs as traction.

Fig. 5.13 A pair of *Kheper platynotus* rolling away a ball freshly made out of elephant dung. The male rolls while the female, who has met the male at the dung pile, rides on his ball.

down legfuls of dung to feed from later or to fashion into brood balls that serve as food for their larvae.

Another strategy, adopted by some more gracile, medium- to small-sized beetles, is to take the round dung pellets, such as those of antelope, and to roll them away and bury them at some place where the soil is suitable for digging. Some of these beetles, the rollers (including the famous *Sisiphus*), have also evolved to utilize the dung of large herbivores (buffalo, rhinocerus, and elephant) by tearing bits off the larger dung boluses and fashioning them into balls suitable for transport to distant burial sites (Fig. 5.13). The famous Egyptian *Scarabaeus sacer* is an example of this type.

Still a third example are the endocoprids or dung dwellers. These generally tiny, little-studied beetles feed directly in the dung pile where it is deposited.

For ball rollers, possession of a dung ball is in many species a prerequisite for mating. Balls made by males may function as nuptial offerings in that they attract females to the male (Sato and Imamori, 1988), and these nuptial balls may be eaten or serve as brood balls. In those species that bury dung deep in the soil under the pile where it is dropped, sexual pairs are formed at the fresh dung pile and then work as a team. Generally the female makes the initial burrow and the male then brings dung down to her, grasping it in his front legs. He backs down her burrow and drops his load on her as soon as the tip of his abdomen strikes the top of her head. She then takes the dung down the rest of the way and fashions it into pear-shaped balls and, after mating in the burrow, deposits an egg into each brood ball. The female sometimes remains to care for the brood balls.

Competition for dung can sometimes be intense. In Tsavo Park, in Kenya, more than 4,000 beetles of less than a dozen species came to one single half-liter sample of elephant dung within the first 15 minutes after it was deposited. Thirty-liter samples were (at dusk) transformed into spreading, heaving mats within half an hour, as clouds of beetles swept in, landed on the dung, tunneled in, or fashioned it into balls to roll away (Heinrich and Bartholomew, 1979). The awesome spectacle of thousands of beetles converging on the dung commences at dusk and proceeds throughout most of the night (in the rainy seasons), whenever fresh dung is put out anew.

There is relatively little competition for dung in the daytime; the hordes of beetles first appear at dusk. It is at dusk, too, that *Scarabaeus laevistriatus* first appears. *S. laevistriatus* is usually extremely rapid in its movements. These beetles run rather than walk, and their front legs move in a blur while patting on the balls to compress them. Speed is clearly of the essence in the scramble competition for the rapidly diminishing dung in the evening, because a dung pile may be consumed in less time than is required to tear off a piece and compress it into a ball ready for rolling. The higher the T_{thx}, however, the faster the beetle can make a ball. (From casual observations I did not detect differences in the size of dung balls made as a function of T_b, but this possibility is not ruled out. Colder beetles may have to make smaller balls to escape the competition sooner.) Additionally, the higher the T_{thx} of the ball roller, the faster it escapes with its prize once made (Fig. 5.14). At a metathoracic temperature of 28°C, the ball-rolling velocity of *S. laevistriatus* is 5 cm/s, but the velocity increases by 400 percent as T_{thx} increases to 43°C (Bartholomew and Heinrich, 1978). As indicated below, the running causes very little measurable increase in T_{thx}. Therefore, the beetles are heated *to* run, rather than being heated by the running.

A dung pile can be rendered useless to ballmakers in 15 minutes, but some large, well-formed balls take over half an hour to make (when the dung is experimentally screened to exclude the competitors), although most are completed in 10 minutes or less (Heinrich and Bartholomew, 1979). Balls that are not completed and quickly rolled away are invaded by endocoprids and simply disintegrate within minutes.

In addition to scramble competition for dung, there is also contest competition there direct combat (Fig. 5.15). At the dung piles where thousands of beetles converge at dusk, the *S. laevis-*

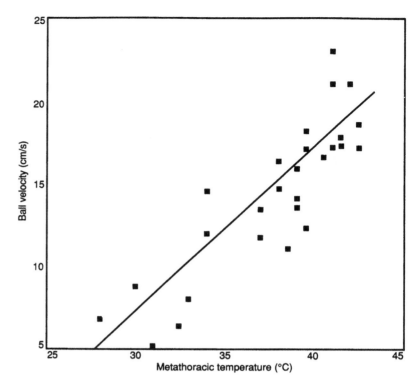

Fig. 5.14 Ball-rolling velocity as a function of metathoracic temperature in the nocturnal *Scarabaeus laevistriatus.* Mean body mass is 3.3 g; mean mass of dung balls, 35 g. Ambient temperature = 23–27 °C. (From Bartholomew and Heinrich, 1978.)

triatus arriving there often first run all over the dung pile, attempting to steal wholly or partially made balls from conspecifics already on the job. Immediately after flight they are still hot due to their flight metabolism, while those that have been rolling balls for a while may either have cooled passively or may have kept themselves heated by shivering. In the combat over already-made dung balls, the winner (the one who retains the ball) is almost invariably the hotter of the two contestants, often despite a large size disadvantage (Fig. 5.16). Both the mean mass of the winners and their T_{thx} are significantly greater than those of the losers, but the more than 1,000-fold difference in the level of significance shows temperature to be the more important variable than mass, although mass probably would be a very important variable, if its range were larger. Similarly, in the flightless tenebrionid beetle *Physadesmia globosa,* males challenging other males in possession of females tend to be about 1 °C warmer than their opponents (Marden, 1987). But in these beetles the males "in possession"

THE HOT-BLOODED INSECTS

Fig. 5.15 Combat over an already-made dung ball by the nocturnal roller *Scarabaeus laevistriatus*. Here the victor has flicked the loser off the ball with its powerful front pair of legs.

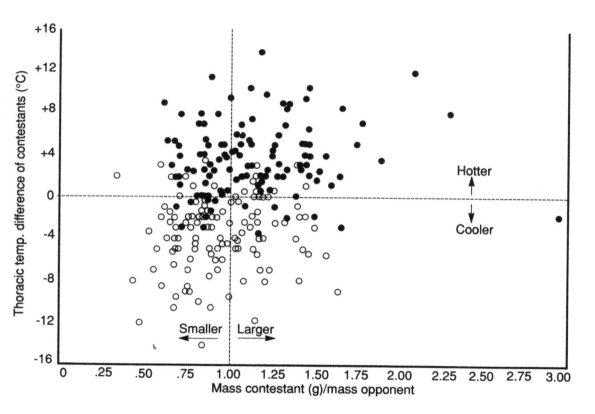

Fig. 5.16 Size and thoracic temperature of winners *(filled circles)* and losers *(open circles)* of *Scarabaeus laevistriatus* in contests over dung balls. Ambient temperatures = 23–26 °C. (From Heinrich and Bartholomew, 1979.)

are following females (usually in shade) and the challenging males have the option to get heated first in the sunshine. Whether unpaired males move into sunshine before initiating contests is not known.

In dung beetles fights are brief (2–3 seconds) and since they seldom result in injuries, the beetles have little to lose by risking a fight, especially if they start it immediately when they have the advantage of being hot after flying in and before they cool down. Whether they cool or remain heated by shivering after they land might depend on the risk of being challenged, and on the need for speed, which is correlated with the extent of inter- and intra-species competition at the site. The fact that the large dung beetles active in the daytime, when there are few competitors for fresh dung, are only slightly endothermic supports this idea, but it so far remains untested.

Why do some beetles save energy and cool down while proceeding to make a dung ball, while others remain heated up? One possibility is that the behavior of keeping warmed up (which permits rapid ball construction as well as ball pirating) is a conditional strategy that depends on the amount of competition already at the dung pile. If many beetles are present, then the energy-economical strategy of cooling down may not be possible, because cooled beetles will never be able to finish making a ball quickly enough before it is tunneled through or eaten by the tens or even hundreds of thousands of other beetles. The only chance a beetle has then is to bear the added cost of sustained endothermy or else gain nothing. Second, perhaps beetles with lower energy reserves are less willing to "gamble" on expending a lot of energy when they cannot be certain they will end up with a ball at the end. Such energy conservation is observed in bumblebees (Heinrich and Heinrich, 1983).

Rather than a conditional strategy, or a strategy of avoiding risks, another possibility is that there is involved here a genetic mix of two strategies, along the lines of Maynard Smith's hawk-dove analogy. In this case the same ideas might apply to a "pirate-producer" analogy. The pirates (ball snatchers) would be at a large disadvantage when they are very common and producers (poikilothermic ball makers) are rare. On the other hand, they would be at a great advantage when they are in the minority, when producers are common. Perhaps a balanced genetic polymorphism (in the same species, *S. laevistriatus*) could result. So far no data are available to distinguish among the alternatives.

Little is known about the thermoregulatory strategies of the beetles *before* they contact dung. Some species search for food by flying close to the ground. Others remain hidden, where they are presumably poikilothermic, while waiting for an odor plume to reach them. (Remaining warm and ready for takeoff by endothermy is unlikely, given the high energy cost of shivering.) At least one neotropical species, *Canthon spetemmaculatus*, perches on leaves that are located in sunflecks, thereby remaining flight-ready through basking (Young, 1984). However, these beetles could potentially perch in sunflecks for reasons other than flight-readiness.

Cetoniine Flower Scarabs

In one group of scarabs, the cetoniids, the adults have specialized to feed on nectar, pollen, and even the petals of flowers and ripe fruit. These beetles are often agricultural pests (Donaldson, 1981). (The larvae feed on decaying plant material, as did the ancestors of the dung beetles.) These very colorful and very fast-flying beetles vary greatly in size and they have a nearly cosmopolitan, primarily tropical distribution. Very few species occur in the north-temperate region.

Mechanisms of thermoregulation have been examined for one species, *Pachnoda sinuata* (Fig. 5.17), while the beetles were foraging from rose petals in a large rose garden in Johannesburg, South Africa (Heinrich and McClain, 1986). Like other cetoniine

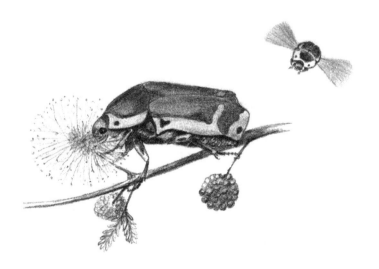

Fig. 5.17 The cetoniid flower scarab *Pachnoda sinuata*, feeding on pollen and nectar of an acacia blossom.

beetles (Nicolson and Louw, 1980; Chappell, 1984), *P. sinuata* warm up prior to flight. In the laboratory *P. sinuata* are either unwilling or incapable of warming up at air temperatures below 20°C, although when they attempt warm-up at 20°C their T_{thx} rises quickly (at 3–4°C/min until takeoff), until a T_{thx} near 35–38°C is achieved (Heinrich and McClain, 1986). The latter temperatures are near the optimum for phosphorylase and actomyosin ATPase activity in their flight muscles (Barnett, Heffron, and Hepburn, 1975).

In the field these beetles prefer to warm up by basking rather than shivering, always crawling onto the top of flowers to bask in direct sunlight for about one minute prior to takeoff. Recently killed beetles placed into sunshine on top of flowers in the "basking" posture heat up at 4.5°C/min (close to the rate of shivering, endothermic warm-up in shade at 20–23°C). Therefore, most of the T_{thx} increase during pre-flight warm-up in the field in sunshine can be accounted for by solar heating alone.

Dead beetles placed into the sunshine on flowers heat up to near 50°C, but live beetles maintain T_b near 32–34°C by quickly (within 13 seconds) crawling into the shade of the flower petals upon landing on a flower. In contrast, on a cool day (20–23°C) they remain perched on top of the flowers on which they feed (Fig. 5.18).

The perched beetles attempt to maintain T_{thx} considerably below their optimum for flight. Even though they could easily keep heated up by the "free" energy of the available sunshine or by shivering, they instead *actively* reduce T_b within seconds after landing by seeking shade. Possibly their "regulated" hypothermia is related to the absence of competition for food. Since resting metabolism increases sharply (from 0.3 to 1.8 ml O_2/g/hr at 15 to 40°C) as a function of T_b, there are thus, as there are for bumblebees (Heinrich and Heinrich, 1983), considerable energy savings by staying cool if the food resources harvested offer a low energy payback. In contrast, another cetoniid, *Cotinus texana*, of about the same mass (near 1–2 g) that feeds on fruit usually maintains its T_{thx} above 35°C (Chappell, 1984). Fruit, having a higher sugar content than flower petals, may make it energetically less costly for *C. texana* to remain warm-bodied than for *P. sinuata*.

Both *C. texana* and *P. sinuata* regulate T_b behaviorally when they forage, but neither has been shown to regulate T_{thx} in flight. At T_a below 15°C none of the *Pachnoda* fly for a full 2 minutes without stopping (a criterion for "continuous" flight). At such air

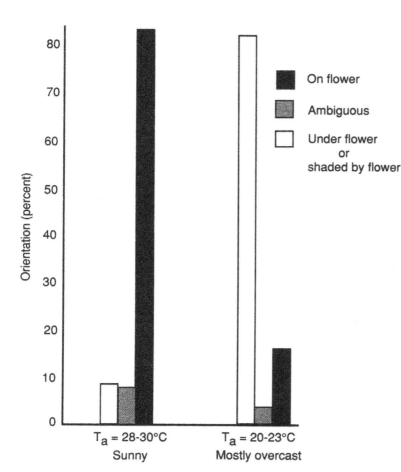

Fig. 5.18 Orientations assumed by flower beetles *Pachnoda sinuata* on rose flowers at high (28–30 °C) and low (20–23 °C) ambient temperatures. (From Heinrich and McClain, 1986.)

temperatures their T_{thx} declines to less than 30 °C and they then lose the power to remain airborne. Their T_{thx} averages 10–13 °C above air temperature from 20 to 32 °C (Fig. 5.19), suggesting temperature conformity rather than temperature independence. In contrast, *C. texana* are reported to regulate T_{thx} in flight, since T_{thx} of beetles caught in flight in the field averages about 12 °C above air temperature at 20 °C and only 4 °C above air temperature at 35 °C (Chappell, 1984). However, these data refer to beetles that were flying only *when caught*. (It was not differentiated whether a "flying" beetle was one that had just taken off after a pre-flight warm-up, or whether it had stabilized its T_{thx} after 2 minutes or more of continuous flight.) Thus, the field data cannot confirm whether the animals thermoregulate in flight, or if they

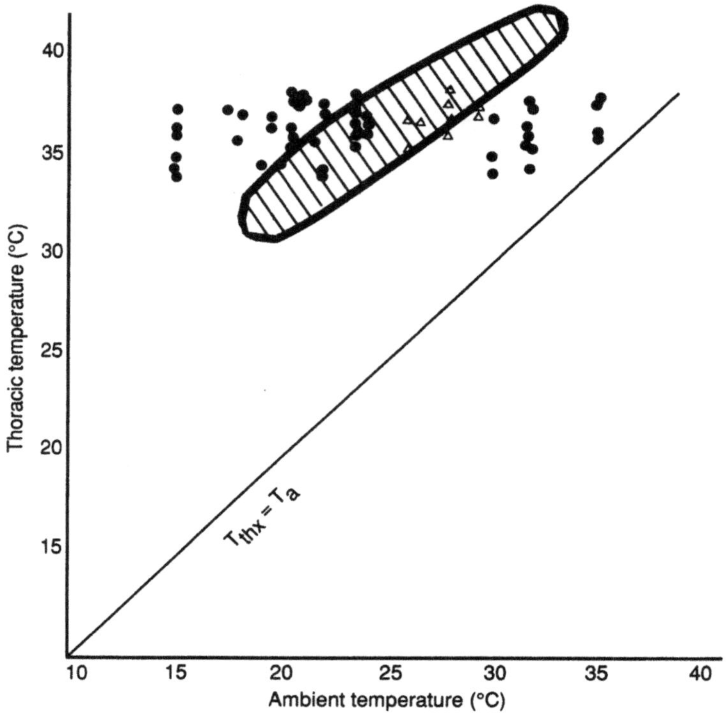

Fig. 5.19 Thoracic temperatures of *Pachnoda sinuata* at takeoff as a function of ambient temperature in the laboratory *(circles)*, when perched on rose flowers in the field *(triangles)*, and when in continuous free flight *(enclosed line)* both in the field and in the laboratory. (From Heinrich and McClain, 1986.)

do, they do not distinguish whether they have a physiological mechanism of thermoregulating *in* flight or whether they have a behavioral mechanism of regulating T_{thx} *with* flight.

Rain Beetles

The *Plecoma* species (Scarabaeidae), or rain beetles (Fig. 5.20), are particularly interesting because they are, unlike all the other large (near 1 g) beetles so far studied, a cold-weather animal. In the mountains of California, Oregon, and Washington the males fly in search of flightless females at night in the winter even at $-1°$ C. Kenneth R. Morgan (1987) shows that they sometimes fly with a T_{thx} as high as $38.6°C$ at air temperatures of $2.9°C$ (the females, which outweigh the males by about 8 times, are incapable of flight and they do not elevate their T_b).

Rain beetles show impressive endothermy. They also exhibit pronounced independence of T_{thx} from air temperature when

THE HOT-BLOODED INSECTS

captured in flight in the field. Thoracic temperatures are near 38–40°C at air temperatures from 3 to 15°C (Fig. 5.21). As with the data for *C. texana*, however, it is not known whether the individual data points represent T_{thx} of beetles that had flown long enough to stabilize T_{thx}. Until such data are available, it remains unproven whether beetles regulate T_{thx} *in* flight or whether they maintain stable T_{thx} through different durations of on-off episodes of shivering and/or flight. Unlike almost all other scarabs, *Plecoma* have

Fig. 5.20 A male rain beetle, *Plecoma australis*. (From photograph by K. Morgan.)

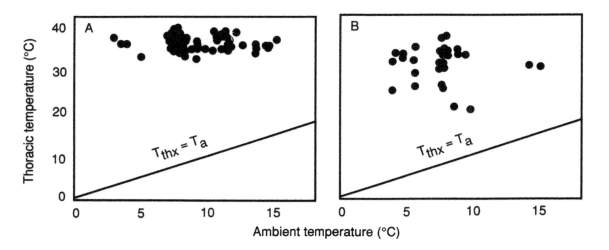

Fig. 5.21 The relationship between T_{thx} and ambient temperature for *(A)* 83 males of five different species of rain beetles during mating flights and *(B)* 34 males of four different species during ground searching for females. (From Morgan, 1987.)

Fig. 5.22 Male rain beetles in
flight; note the thoracic "fur."
(From specimens provided by
K. Morgan.)

a heavy layer of thoracic pile (Fig. 5.22) that undoubtedly acts to
retard convective cooling in flight.

All of the endothermy in rain beetles is fueled by stored energy.
(The adults possess only vestigial mouthparts and are thus unable
to feed.) Given their energy limitation as adults, the rain beetles
(so named because they typically fly for several days following
rains) are thus undoubtedly the most impressive example of endothermy in any beetle. It is not known how many days the
beetles live and how long they sustain their endothermy, but in
the laboratory males maintained elevated T_b for as long as 4.2
hours (by shivering) while walking in a respirometer chamber at
9°C. As in other beetles, walking speed increased exponentially
with T_b (Morgan, 1987). The males must dig vigorously for the
females at their emergence burrows, and numerous males may
simultaneously be attempting to secure copulation with one female (Ellertson, 1958). The maintenance of a high T_b is therefore
undoubtedly a great proximal advantage for the males during the
scramble and contest competition that ensues when they land
near a virgin female.

A possible scenario for the evolution of the extraordinary endothermy in *Plecoma* may be related to the contrast between the

great predator pressure facing the adults (Ellertson, 1958) and the little predator pressure on the subterranean larvae. Originally, the adults probably fed on vegetable matter in the daytime, as other scarabs do. They could then have been partially heliothermic, as many scarabs still are. Predator pressure could have increased the premium for becoming winter-active (avoiding hibernating mammalian predators) and nocturnal (avoiding avian predators). Such pressure could have placed more and ultimately full reliance on endothermy for activity.

Curiously, the rate of pre-flight warm-up in the northern, winter-active *Plecoma* is close to that predicted (extrapolated) from the "lazy," equatorial *Pachnoda* flower beetles that are unwilling to warm up even at air temperatures of 20°C. When the flower beetles *do* warm up they show a very rapid rise of T_{thx}, and extrapolations suggest that if a hypothetical tropical *Pachnoda* did warm up from 7°C it should do so at 1°C/min. This rate is very close to the 1–3°C/min observed in the rain beetles at that air (and body) temperature. Has a large part of the evolutionary adaptation of *Plecoma* been the behavioral willingness to attempt warm-up and flight at ever-lower temperatures, or do these beetles also have specific biochemical adaptations even though when in flight they operate at the same and even higher T_{thx} than the tropical beetles?

The great similarities in body temperature maintained between two kinds of scarab beetles in flight, even though one is active under the equatorial sun while the other is nocturnally active in a cool season in the northern hemisphere, prompts a more general examination of beetle endothermy in flight.

Endothermy in Flight

Beetles, like other insects, inevitably generate much heat when they fly, and this heat elevates T_{thx} if the insect is large. As expected, the T_{thx} of dung beetles in continuous free flight (arriving at fresh dung) is related to their mass. At air temperatures of 25–26°C the T_{thx} of the small beetles is close to air temperature, then increases steeply to near 40°C at body weights near 2.5 g (Fig. 5.22). Although the larger beetles are endothermic in flight, they may still be strict thermoconformers (with a constant temperature excess) at all air temperatures. As already indicated, flower scarabs *Pachnoda sinuata* (mean mass = 0.93 g) generate T_{thx} approximately 12°C above air temperature while in continuous free flight

(Fig. 5.19), and that T_{thx} excess is relatively independent of air temperature. Similarly, the T_{thx} of the highly endothermic cock-chafer ("june bug") *Melolontha melolontha* (1–2 g) at different air temperatures (Schneider, 1980) also shows a typical thermocon-forming response.

Thermoconformity may be impossible as beetles exceed a certain body size. Every gram increase of body mass yields approximately 8°C more excess of T_b. Thus, at 25°C, a 1 g beetle in flight heats to 33°C, at 2 g animal heats to 41°C. Without cooling (or stopping flight), a 4 g beetle should heat to 57°C, and the extrapolated T_{thx} of a 20 g beetle exceeds 100°C! No such high T_{thx} are of course ever attained, and 14 g beetles fly with T_{thx} near 46°C, the same as 4 g beetles (Fig. 5.22). Like most large insects, beetles defend an upper limit of T_{thx} below 45°C. However, whether large beetles (such as those weighing near 20 g) thermoregulate in continuous flight has so far not been determined. (The reason for the lack of data is undoubtedly related to the great reluctance of most large beetles to fly continuously. Most large beetles with wings spend the overwhelming majority of their time "grounded." Does their potential overheating in flight have something to do with this behavior?)

The data so far indicate that morphological constraints have been the main determinants of body temperature in flight. Except possibly for the large beetles, biophysical considerations mainly dictate T_{thx}, and evolution must then largely accommodate. There are three main considerations: generating sufficient wing-beat frequency at given body temperatures; pre-flight warm-up to generate sufficient T_{thx} to take off, which is largely dictated by body size; and preventing overheating when T_{thx} threatens to be too high.

Depending on the species, a specific minimum wing-beat frequency (and T_{thx}) is required for flight, because, as is true of other muscles, a high flight-muscle temperature (proximally) increases the frequencies over which the muscles can oscillate (Machin, Pringle, and Tamasige, 1962). As shown in *M. melolontha*, wing-beat frequency in flight is closely related the T_{thx} (Schneider, 1980). Small beetles that fly with much lower T_{thx} than larger ones, however, do not fly with lower wing-beat frequencies. Instead, they fly with much higher wing-beat frequencies despite their low T_{thx}. Wing-beat frequency has apparently evolved to accommodate to temperature, and there are great differences among species of beetles in the sensitivity of their wing-beat frequency to T_{thx} (Oertli, 1989).

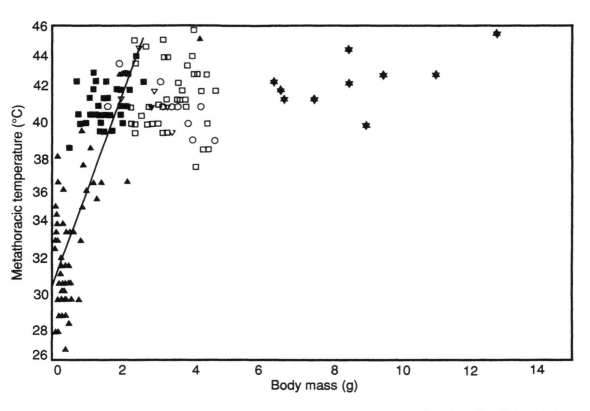

In the dung beetles mentioned above, flight initiation occurs after a sometimes prolonged pre-flight warm-up (Fig. 5.23). Thus, only the large beetles are obliged to warm up; the small ones can take off and fly almost "instantly" without warm-up at the air temperatures normally encountered in their environment (20–25°C). Pine beetles *Dendroctonus frontalis*, which weigh about 2.1 mg, are almost certainly not endothermic in flight, and they have been reported to be able to fly at air (and hence thoracic temperatures) as low as 7°C (Moser and Thompson, 1986).

Fig. 5.23 The relationship between metathoracic temperature during flight to body mass of dung beetles. The least-squares regression line is fitted to data for beetles with a mass less than 2.5 g. The different symbols denote different species. (From Bartholomew and Heinrich, 1978.)

Endothermy during Terrestrial Activity

August Krogh and Eric Zeuthen's (1941) classic work on the dorbeetle *Geotrupes stercorarius* established much of what we know about the mechanisms of endothermy in nonflying but flight-capable beetles. It had earlier been observed that beetles "pump" with the abdomen prior to takeoff for flight. These abdominal respiratory movements were assumed to serve the function of filling the tracheal system with new air, to raise the concentration of oxygen. However, Heinz Dotterweich (1928) had already shown that they had something to do with heating up the body. Krogh and Zeuthen (1941) confirmed this, making continuous recordings of body temperature and showing that the abdominal pumping movements were indeed associated with shivering and a rise in thoracic temperature, but they did not cause T_{thx} to rise. The abdominal pumping merely served as a bellow does to aerate the thoracic muscles while they were shivering. The beetles attempted to fly when T_{thx} reached 30–35°C, and those of this species that attempted to fly at the lower T_{thx} did not succeed. Obviously, *G. stercorarius* is a relatively large beetle. The rise in T_{thx} was restricted to the mesothorax, which contains the powerful flight muscles. The temperature of the pterothorax, which contains the muscles used for walking and digging, increases only slightly.

The abdominal pumping movements of *G. stercorarius* increase from 20–25/min (when at rest at 22°C) to 180–240/min in pre-flight warm-up. But Krogh and Zeuthen (1941) saw no visible movement of the wings during warm-up, even while making close observations with a hand lens after removal of the elytra. Nevertheless, they recorded action potentials from the flight muscles. Direct proof that the flight muscles are mechanically active during warm-up was later provided by Leston, Pringle, and White (1965), who recorded sound produced during pre-flight activity in the water beetle *Acilius sulcatus*. The flight muscles are therefore generating mechanical oscillations, although the presence of a click system keeps the wings effectively decoupled from the muscles when in the folded position, presumably by disarticulation of certain ball-and-socket joints at the wing base. As in all other insects during pre-flight warm-up, the pulse (muscle-contraction) frequency rises as T_{thx} rises. Similarly, Bartholomew and Casey (1977a) observed vibrations of thermocouple leads implanted in the scarab beetle *Strategus aloeus* and in the cerambycid beetle *Stenodontes molarium*, thereby also demonstrating that the flight muscles are mechanically active during warm-up even though the wings and elytra remain motionless.

Almost all endothermy in active beetles can be attributed to the production of heat by the flight muscles. For example, beetles may sometimes be strikingly endothermic while walking, but even the largest dung beetles (up to 20 g), such as *Heliocopris dilloni*, can walk vigorously without measurable elevation of T_{thx} (Fig. 5.24). Perhaps the best proof that increases in the body temperature are

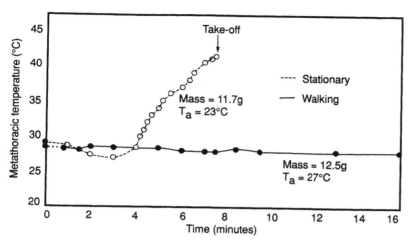

Fig. 5.24 Metathoracic temperature during pre-flight warm-up and during continuous walking in *Heliocopris dilloni*, showing the independence of body temperature from locomotor activity. Note the length of time this animal, which is obligatorily endothermic in flight, must invest to get ready to fly. (From Bartholomew and Heinrich, 1978.)

associated with the activity of the wing muscles, in contrast to the walking muscles, can be seen in the large (3.7–11 g), flightless, dung beetle *Circellium bacchus* from southern Africa. This beetle has vestigial wings and its large prothorax is filled with muscles associated with its powerful legs. Although other, much smaller dung beetles that fly may at times show appreciable endothermy (by shivering) when walking or dung-ball making and rolling (Bartholomew and Heinrich, 1978; Heinrich and Bartholomew, 1979), *C. bacchus* shows no endothermy during any of these activities (Nicolson, 1987). It remains strictly ectothermic despite its large size. Another flightless dung beetle, *Scarabaeus rodriguesi* from the Namib Desert dunes, also shows no evidence of endothermy. All of the many species of flightless tenebrionid and carabid ground beetles are also never endothermic during terrestrial locomotion. As already mentioned, even the fairly large (0.6–1.3 g), flightless, tenebrionid beetle *Onymacris plana*, which runs at the extraordinarily high speed of up to 1 m/s, could theoretically generate an endothermic temperature excess only of about 1 °C (Roberts et al., 1991). In other words, when flight is lost, then endothermy is lost as well.

Only large beetles with flight muscles have the capacity for warming up endothermically. Since they must achieve a minimum high T_{thx} before they can fly and escape, it is not surprising that many of them warm up when they become startled. Several large beetles maintain modest elevated T_{thx} (Bartholomew and Casey, 1977a,b) under conditions that could be interpreted as being stressful, such as when being confined in a small space or placed on their back so that they have to struggle to right themselves. The larger the beetle, the greater the elevation of T_{thx} in such possibly alarmed animals.

The tropical elephant beetle *Megasoma elephas* (11–35 g) elevates its metabolism in response to cooling to 20°C (Morgan and Bartholomew, 1982), displaying a homeothermic response otherwise found only in birds, mammals, and incubating bumblebees (Heinrich, 1974). The beetle can fly, but is reluctant to do so. It defends a lower setpoint temperature (close to minimum T_{thx} for flight?) by initiating heat production when T_{thx} falls below it. Thus, there are more frequent warming cycles by shivering when heat-loss rates are higher. Little is known of the ecological relevance of this shivering. It is not yet known if the endothermy associated with experimental conditions noted above, in captive beetles, is a natural occurrence in the wild (as it is in dung beetles competing for food) or if it is an alarm response of animals preparing to escape.

Summary

Most beetles are relatively slow-moving animals that fly very little, and so they are only infrequently or not at all subjected to high heat loads. However, a number of species inhabit environments of very high ambient temperatures and high loads of solar radiation. These beetles show impressive behavioral thermoregulation. The most well-known of these are certain tenebrionid beetles active in summer at midday in African deserts. Predictably, these beetles have evolved very high thermal tolerances, and they function at T_b near their upper limits.

Close relatives of the cool-loving carabid beetles, the tiger beetles, have also evolved to occupy hot, open, sandy areas where they rely on their vision to prey on small arthropods. In order to maximize encounter rates with prey, these beetles must hunt at high air temperatures, and they are also subjected to radiant heat loads. Tiger beetles have evolved behavioral thermoregulatory mechanisms similar to those of the desert tenebrionids. In addition, they take short flights above the hot substrate that augment convective cooling.

Both Namib Desert tenebrionids and tiger beetles operate at high T_b and they must heat up prior to activity in the morning. Both use similar basking strategies, but the tiger beetles, subjected to somewhat less intense heat loads than many Namib tenebrionids, operate at somewhat lower T_b. In turn, tenebrionids from a cool climates operate at even lower T_b. Different beetles are adapted to their different respective thermal environments, not to any specific T_b per se.

Beetles do not show appreciable endothermy in continuous flight until they reach a mass of near 1.0 g. Small beetles take off in flight with a low T_{thx}, without pre-flight warm-up. Large beetles are also adapted to the highest temperature to which they are subjected (during flight); they must warm up by pre-flight shivering. During warm-up the wing muscles contract but they are mechanically decoupled from the wings.

Sustained endothermy during terrestrial activity is possible only in very large beetles with flight muscles. In some dung-ball-rolling beetles, the muscles associated with the legs are secondarily heated when the flight muscles are exercised, and hot-bodied beetles have a pronounced edge in scramble and contest competitions against beetles of their own kind and other species that cool down. In the absence of intense competition for food (as in diurnal dung beetles and in flower scarabs), the animals cool down almost immediately after flight.

The most impressive beetle endotherms so far examined are the nocturnal males of rain beetles (*Plecoma* species), which are active in winter on mountains in the northwestern United States. (They maintain nearly the same T_{thx} in flight as do tropical beetles of their size in the sun.) It is likely that these beetles evolved from diurnal progenitors, being driven to a low-temperature niche possibly to escape predation or parasitism on adults. The specific body-temperature ranges over which particular species operate are largely driven by biophysical constraints, and evolution then acts to allow activity at the temperatures generated.

Remaining Problems

1. Do some of the larger beetles thermoregulate in flight, and if so, how?
2. Do dung beetles competing for dung adjust their body temperature so as to gain a competitive edge? If so, what are the stimuli to which they respond?
3. Why do some dung beetles of the same species cool down while making a dung ball while others maintain prolonged endothermy?
4. Given that the *capacity* for heat production by the cold-adapted

rain beetles comes close to that of a tropical flower beetle, what accounts for one warming up at very low air temperatures while the other does not?

5. What has driven members of some beetle groups to occupy hot (or cold) environments where they are obliged to thermoregulate?

6. What are the neural mechanisms of thermoregulation in beetles that regulate T_{thx} by on-off heat production?

7. To what extent are the colors of ground-dwelling tenebrionids adaptations for thermal balance, as opposed to responses to other selective pressures?

8. Is running speed in swift-running Namib beetles varied to affect body temperature?

Bumblebees out in the Cold

IN HIS now-classic book on bumblebees of 1934, the Harvard entomologist Otto Plath (father of poet Sylvia Plath) wrote: "Like all cold-blooded animals, honeybees and bumblebees have no means of regulating their body temperature, and this exposure to cold invariably results in lethargy, and often death." As I hope to show here, we have come a long way in understanding bumblebees (and honeybees) since Plath repeated that almost-universal assumption. To give just one example: the bumblebees' distribution in northern or cool climates would be almost unthinkable without their phenomenal ability to regulate body temperature.

Bumblebees live throughout the temperate regions and on cool mountaintops in the tropics. They are also permanent residents above the Arctic Circle. Wherever they live, they are important pollinators of native vegetation (Fig. 6.1), thus acting as keystone organisms of their environment (Heinrich, 1979). Two species live further north than any humans. Yet, like us, the bumblebees maintain a body temperature above 37°C when they are active, even though their rate of heat loss is necessarily (because of their small size) hundreds of times greater than ours. If we were to cool as rapidly as a bumblebee, then we would perceive an air temperature of 10°C to be at least 200°C below zero. Despite our highly vaunted evolutionary advancement and our massive size, it is still not easy for us to regulate our body temperature near 37°C even at air temperatures near 10°C—at least not without the aid of extensive artificial insulation. Bumblebees do it easily, sometimes even to near 0°C. Our own puny thermoregulatory abilities make the feats of bumblebees all the more impressive. If

Fig. 6.1 The bumblebee *Bombus terricola* collecting pollen from wild roses in a northern New England bog.

any animal on earth has mastered temperature, it is surely bumblebees.

Other bees (such as carpenter bees, *Xylocopa* species) have mastered living in very hot climates. Still others, like our familiar "honeybees" *Apis mellifera* (most of the thousands of bee species are "honey" bees in the sense that most make and eat honey), can live almost anywhere. Thermoregulatory mechanisms of bees are diverse, and much research has been done to elucidate them. I here break the bees up into three chapters to reflect that research and to highlight the contrasting thermoregulatory strategies in different thermal environments that to some extent also respect taxonomic affinities. The taxonomic division is somewhat artificial, however, because to some degree each of the capacities of bees exhibited in one group are also found (to a greater or lesser degree) in the other. The distinctions among various bees' thermal adaptations are in the sum total of their behavior and physiology—as would be expected, because animals adapt to any one

THE HOT-BLOODED INSECTS

contingency generally with any and all means they have at their disposal.

Heat Production by Bumblebees

Although in Plath's time there had been no study showing thermoregulation in individual bees, their capability to produce heat had been known for a hundred years. Already in 1837 George Newport, a member of the Entomological Society of London and the Royal College of Surgeons, took temperature measurements of various species of bumblebees, using tiny mercury thermometers pressed against the abdomen. In the introduction of his paper entitled "On the Temperature of Insects, and its Connexion with the Functions of Respiration and Circulation in this Class of Invertebrate Animals," he wrote: "Every naturalist is aware that many species of insects, particular hymenopterous insects, which live in society, maintain a degree of heat in their dwellings . . . but no one, I believe, has hitherto demonstrated the interesting fact that every individual insect when in a state of activity maintains a separate body temperature considerably above that of the surrounding atmosphere . . . and that the amount of temperature varies in different species of insects, and in different states of those species."

It soon became apparent, however, that bumblebees do not automatically regulate "a" body temperature. First, M. Girard (1869), in France, using thermocouples, showed that there are differences between thoracic and abdominal temperatures, and A. Himmer (1925), in Germany, measured thoracic temperatures averaging 8.7°C and 10.1°C above air temperature in queens and drones, respectively. Danish researchers August Krogh and Eric Zeuthen (1941) briefly examined *Bombus horti* and found that a bee tethered on a thermojunction shows a rise of T_{thx} from 25°C to 30°C in 2 minutes of flight while the angular excursions of its wings increased from 30° (more buzzing than "flight") at 25°C to 120° at 30°C. Krogh and Zeuthen's observations established that heat production takes place in the flight muscles, therefore accounting for the very low T_{abd} relative to T_{thx}.

The above studies laid the foundations for much of the later work on the mechanisms of bumblebee thermoregulation. They also led to misunderstandings. First, Krogh and Zeuthen's (1941) observations suggested indirectly that bees warm up by buzzing, and Torsten B. Hasselrot (1960), in his extensive and detailed studies on nest temperature of Swedish bumblebees, therefore concluded that "worker bumblebees increase their metabolism through vibration of their wing muscles or through running energetically to and fro in the nests, buzzing with their wings to maintain nest temperature above certain critical levels."

We now know that buzzing is normally an alarm response. Even though bees have a capacity to warm up while they are walking or running, running in even the largest insects does not generate sufficient heat for endothermy (see previous chapter). Buzzing (while running or stationary) necessarily results in heat production, but in all bees most endothermy, we now know, is achieved without any sound whatsoever. While warming up, the bees invariably make rapid pumping motions with the abdomen (Kammer and Heinrich, 1972; Heinrich and Kammer, 1973), but they are silent and they may either walk or remain stationary. As in beetles and other insects, in bees abdominal pumping is correlated with warm-up but it serves mainly to aerate the thoracic muscles that are the site of heat production.

PRE-FLIGHT WARM-UP MECHANISM

A bumblebee perched quietly, as though asleep, may suddenly show a dramatic rise in T_{thx} without showing any wing vibrations. Within a minute, for example, T_{thx} can shoot up from 24°C to 37°C (Fig. 6.2). (Abdominal temperature, however, tends to remain low during warm-up, although it may rise immediately after the thoracic temperature has stabilized.) Much research has been done to examine the mechanism of warm-up, and all agree that the temperature increase during warm-up is accomplished exclusively by heat production by the thoracic muscles. Furthermore, when action potentials from the major fibrillar flight muscles—the dorsal longitudinal muscles (DL), the wing depressors, and the dorso-ventral muscles (DV), the wing elevators—and temperature are measured simultaneously, there is never any increase in T_b without accompanying action potentials (Fig. 6.3), and vice versa (Heinrich and Kammer, 1973). In other words, there is no heat production by the fibrillar muscles unless they are activated, and heat production is a direct function of action-potential frequency (Figs. 6.4, 6.5). In order to understand more fully what happens during pre-flight warm-up, we must examine the physiology of the fibrillar muscles that have the "myogenic" mode of action.

The power-producing muscles of the wings in Hymenoptera (and in Diptera, Coleoptera, and Hemiptera) are of the fibrillar

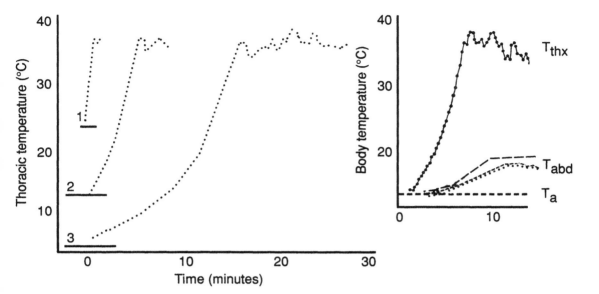

Fig. 6.2 Body-temperature changes during warm-up of a *Bombus vosnesenskii* queen. The graph at left shows T_{thx} during warm-up from ambient (and thoracic) temperatures of *(1)* 24°C, *(2)* 13°C, and *(3)* 6.2°C, and subsequent regulation of T_{thx} above 35°C. At right, along with T_{thx}, T_{abd} is recorded at 3 locations simultaneously: long-dashed line = ventral, anterior; short-dashed line = ventral, posterior; and dotted line = dorsal, anterior. (From Heinrich, 1975.)

type. Fibrillar muscles have the unique capacity to be stimulated to contract upon being stretched as well as upon receiving an impulse from a nerve (which is the "conventional" or neurogenic mode). During flight, contraction of the downstroke muscles not only depresses the wings, it also stretches the upstroke muscles, causing them, in turn, to contract; similarly, contraction of the upstroke muscles stretches the downstroke muscles, and so on. As a result, oscillations of upstroke and downstroke muscle contractions are produced at much higher frequencies than neuronal stimulations can generate, and the muscle operates "myogenically." However, *isolated* fibrillar muscles, such as the DL and DV of bumblebees (Boettiger and Furshpan, 1954; Ikeda and Boettiger, 1965) and of other myogenic flyers (Pringle, 1954; Machin and Pringle, 1959; Boettiger, 1960), behave as typical slow or neurogenic muscle fibers; they show a small mechanical response (rise in tension) to a single neural stimulus, followed by a slow rate of relaxation. The rate at which the tension rises in experimentally isolated fibrillar muscles under multiple stimulation depends on the frequency of stimulation (Fig. 6.6). During warm-up there can be no oscillation of muscle contractions, as there is in flight, because the opposing muscles are not stretched since there are no wing beats. Instead, the muscles can contract only as they do when isolated on the lab bench, as "slow" muscles.

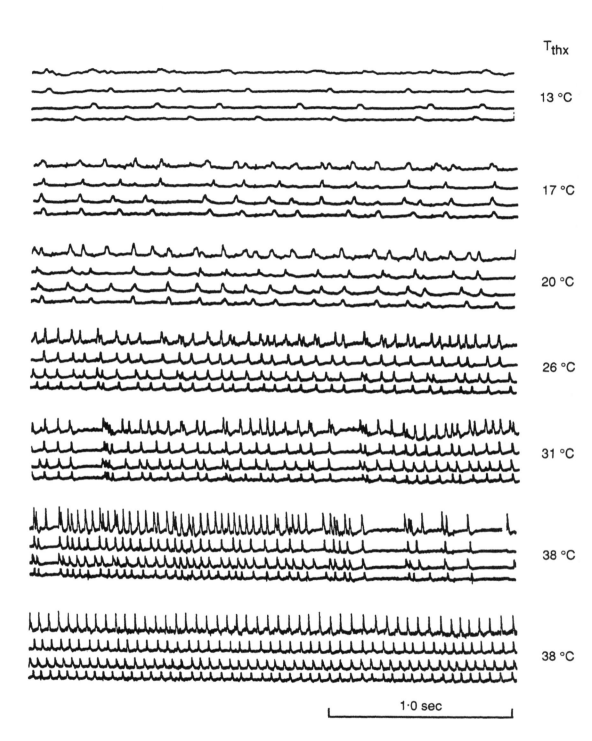

T_{thx}

13 °C

17 °C

20 °C

26 °C

31 °C

38 °C

38 °C

1·0 sec

THE HOT-BLOODED INSECTS

Fig. 6.3 Electrical activity (action potentials) recorded from 4 different motor units of the fibrillar flight muscles of a *B. vosnesenskii* queen during warm-up (first 6 sets of recordings) and during flight (bottom set). Note that the amplitude and frequency of action potentials increase while duration decreases with increasing T_{thx} (indicated at right). Action-potential frequency increases from about 3–4 per second at $T_{thx} = 13\,°C$ to 22–28 per second at $T_{thx} = 38\,°C$. This bee started to stabilize T_{thx} at 38 °C, hence the reduction in action-potential frequency at the end of the warm-up. (From Heinrich and Kammer, 1973.)

Given that bumblebees during shivering warm-up activate the DL at up to at least 40 times per second (Mulloney, 1970; Kammer and Heinrich, 1972; Heinrich and Kammer, 1973), and that stimulation frequencies of only 15 per second are enough to cause complete mechanical tetanus (Ikeda and Boettiger, 1965), it is clear that a shivering bumblebee need not exhibit much thoracic vibration because tetanic contractions produce much tension and

Fig. 6.4 Experimental setup used to get data shown in Fig. 6.5.

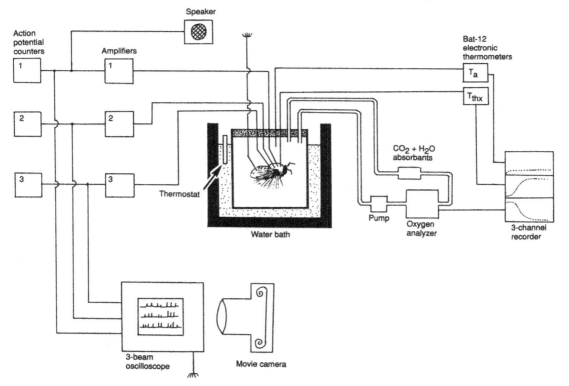

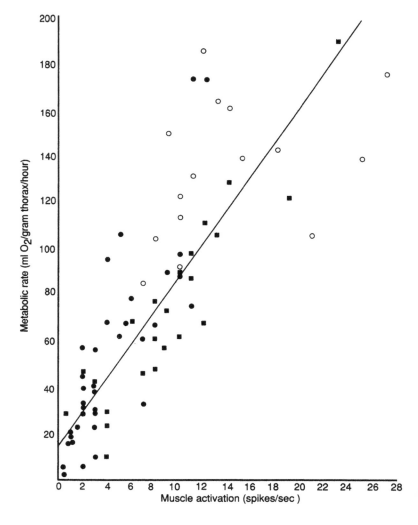

Fig. 6.5 Metabolic rates during warm-up (squares = workers, filled circles = queens) and flight (open circles = queens) as a function of action-potential frequency recorded from one of the motor units of the fibrillar flight muscles. (From Kammer and Heinrich, 1972.) For comparison with data from whole animals, see Fig. 6.18.

little motion. Nevertheless, there could be minor motion, but addition mechanisms may be involved to dampen it. (The advantage of dampening motion could be many-fold: eliminating noise that might hinder communication within the nest, eliminating vibrations that might add wear to the wings, and eliminating shake-off or loss of pollen that is collected in the body hairs.)

I shall now examine several more mechanisms of how vibrationless warm-up may be accomplished. First, the amount of tension (and energy and heat production) and motion can be modified by the central neural patterns of muscle activation. In

THE HOT-BLOODED INSECTS

bumblebees, as in other myogenic insect endotherms, the extra-cellular action potentials recorded from the fibrillar flight muscles are identical during warm-up and during flight (Mulloney, 1970; Kammer and Heinrich, 1972; Heinrich and Kammer, 1973). The only difference is in their timing. In general, during *flight* the different motor neurons (five) innervating the DL do not fire in any fixed temporal pattern one relative to the other. But during warm-up the motor units of both the DL and DV fire relatively synchronously, in bursts (Fig. 6.3). The synchronous bursts of action potentials of antagonistic muscles during warm-up should reduce mechanical motion of the muscles, because the opposing muscles are contracting even more synchronously against each other. (There is an already-existing partial tetanus, as previously indicated, because the muscles are operating in the slow or neurogenic mode.) The synchronous activation of the fibrillar muscles

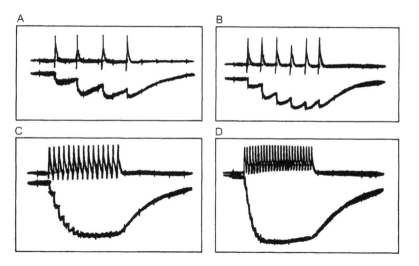

Fig. 6.6 The response of bumblebee fibrillar flight muscle (dorsal lateral) varies with the frequency of impulses carried by the nerve innervating the muscle. In each set of traces, the lower line records the mechanical activity of experimentally isolated muscle, and the upper line records electrical stimulation of the nerve at *(A)* 3.2 stimuli per second, *(B)* 5.0 stimuli/s, *(C)* 15 stimuli/s, and *(D)* 25 stimuli/s. Temperature = 23–28°C. Note the smooth-tetanus contraction of muscle stimulated 15 times per second. The frequency of natural stimulation in intact animals (see Fig. 6.3) equals or exceeds this frequency. Therefore, at least at these temperatures, shivering flight muscles of intact bumblebees are predictably in tetanus. (From Ikeda and Boettiger, 1965.)

in bumblebees during warm-up is comparable to the synchronous activation of the same muscles during warm-up in neurogenic flyers (Kammer, 1968).

Comparative studies indicate, however, that the neurophysiology of warm-up differs from that of flight in other subtle (but probably very significant) ways besides a strong tendency for synchronous activation. For example, honeybees (Bastian and Esch, 1970) and syrphid flies (Heinrich and Pantle, 1975) can warm up and change from flight to buzzing to vibrationless heat production with little or no noticeable difference in the activation pattern of the heat-producing flight muscles, except for the ratio of DL/DV activation. In honeybees (Esch and Goller, 1991) and in bumblebees (Esch, Goller, and Heinrich, 1991) the DL is, curiously, activated at a greater frequency during warm-up than in flight (the DL/DV activation ratio is intermediate during buzzing). What possible function can be served if one set of wing muscles is activated at a slightly different frequency than another? First, since they are operating neurogenically during warm-up, it would mean that one contracts slightly more than another, even though both are in near tetanus.

Harald Esch at the University of Notre Dame has determined, further, that a consequence of increased DL activation is the stretching of the DV until the scuttelar arm, an externally visible projection in the back of the thorax, is pulled up against the dorsal shell of the thorax, the notum, so that a cleft between them is closed. He interprets this to mean that closing the cleft (mesonotal suture) creates a mechanical stop preventing further movement of the muscle (and mechanical vibrations) and preventing, as well, stretch activation of the antagonist muscles, the DL. Significantly, during flight-initiation there is a momentary *alternate* activation of the DV and DL muscles (rather than simultaneous activation as before), so that the opposite muscles can be fully stretched and the myogenic contraction cycle is initiated.

The work of Ikeda and Boettiger (1965) suggests still a fourth theoretical possibility (that has not been previously considered) of how a physiological damping of antagonistic muscles could conceivably occur so that vibrations are eliminated. Ikeda and Boettiger (1965) distinguish four axons (A_1, A_2, B_1, and C_2) innervating the *Bombus* indirect flight muscles (Fig. 6.7). The A_1 and A_2 may carry branches of the same axon, and stimulation of the A-type axons gives large action potentials in all muscle fibers of both the DL and DV. The action potentials, given single stim-

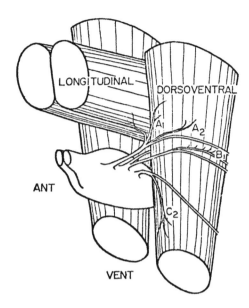

LONGITUDINAL

DORSOVENTRAL

A_1

A_2

B_1

ANT

C_2

VENT

Fig. 6.7 Nerves leading to the second thoracic ganglia of the *Bombus* dorsal longitudinal and dorso-ventral flight muscles. (From Ikeda and Boettiger, 1965.)

ulation, are followed by a small rise in tension and there is, as expected, a marked summation or tetanus following multiple stimulation. The B_1 axon runs posteriorly and laterally around the DV and branches on its anterior surface send action potenials that also cause appreciable contraction of the muscle. (B_2 and C_1 send no branches to either fibrillar muscle.) The C_2 innervates the anterior surface of the DV, but activity from this nerve does not cause muscle contraction, and Ikeda and Boettiger (1965) suggest instead that this nerve serves to hyperpolarize or *erase* the depolarization caused by the other, excitatory, nerve fibers. No function for such organization was suggested. Could it serve to cause the DV to contract less than the DL, in which case would it cause the same effect as the increased DL/DV activation just described and if so, is it related to flight control or reduction of thoracic vibrations?

Aside from the above proven and potential mechanisms of reducing wing movement despite strong contractions of the wing muscles, there is also a coupling mechanism analogous to a clutch in an automobile. In bees (both honeybees and bumblebees) a small muscle, the third axillary, folds the wings and uncouples them from the power-producing muscles (J. W. S. Pringle, personal communication). In many other myogenic insects, especially flies, another muscle, the pleurosternal, serves the same function (it otherwise acts as a servomechanomuscle to control wing pitch and power in flight). To further understand the shivering response of bumblebees, we must first evaluate the claims for non-shivering thermogenesis in bees.

NON-SHIVERING THERMOGENESIS?

To most people, sick pigs hold little resemblance to bumblebees. But to biochemists, truth and beauty are sometimes more than skin deep. Swine suffer from a malignant syndrome that results in a sharp rise in body temperature associated with muscle rigor. The muscle rigor is not due to shivering. To the contrary, it is due to the same reason that pigs (and other animals) get stiff after they die: their muscles stop working because of the depletion of ATP from those muscles, a condition commonly referred to as rigor mortis.

As described some years ago by Michael G. Clark and co-workers (1973b) from the Institute of Enzyme Research at the University of Wisconsin, the *live* pigs are suffering from heat prostration and ATP depletion, with premature "rigor mortis," because of uncontrolled cycling of fructose-6-phosphate. Their thesis is that high activities of the enzyme phosphofructokinase (which metabolizes the ATP-requiring reaction of fructose-6-phosphate to fructose-1,6-diphosphate) occur simultaneously with very high activity of fructose diphosphate (catalyzing the reverse reaction). In a sense, the reaction is thought to rage on uncontrollably, with the net result of burning off all the ATP.

I have not brought attention to this peculiarity of a mammalian disease because I am particularly interested in pigs, nor because I'm a biochemist. However, I've wrestled with this problem for some 20 years now, because it has been claimed in some dozens of papers from Clark's group and several others that this is the *normal* physiology of bumblebees when they produce heat. In my various correspondences with people working on the biochemistry of non-shivering thermogenesis (NST), I've got letters saying from "the data for it [the cycle] is overwhelming" to "such a cycle is simply outrageous." I'm not prejudiced against pigs, but I nevertheless tend to the latter view with regard to bumblebees. (The fate of the pigs I'll leave to the biochemists.)

To begin discussing NST, we've got to clear up definitions first. Shivering concerns quivering due to muscle contractions. *Non-shivering thermogenesis, by definition, implies the absence of muscle contractions, either twitches or tetanus resulting from multiple electrical stimulation of the muscles. Furthermore, as regards bumblebees, NST implies heat production in the absence of muscle contractions, not just the experimenters' inability to detect thoracic vibrations.

The necessary proof of NST is simple: demonstrating heat production to elevate T_{thx} (in *any* amount) in the absence of muscular contraction. Such proofs are legion for mammals, but no such proof exists for any bumblebees. Nevertheless, since the thesis of NST for bumblebees is accepted as gospel truth in the general literature (Hochachka and Somero, 1973) and since numerous biochemically oriented publications almost to the present day (Greive and Surholt, 1990; Surholt et al., 1990) claim that bumblebees can produce heat to elevate T_{thx} without shivering, it is unfortunately necessary to discuss the topic even though it is not relevant to any insect, as far as we know.

High activities of phosphofructokinase are found in bumblebee flight muscle simultaneously with very high activity of fructose diphosphatase. How can this be explained? The thesis of numerous papers that bumblebees have a "futile cycle" to use up ATP for heat production for non-shivering warm-up (Newsholme et al., 1972; Clark et al., 1973a; Surholt and Newsholme, 1981) was questioned early on (Kammer and Heinrich, 1978) because the heat generated from the amount of substrate cycling proposed can only be a tiny fraction of that actually needed to raise body temperature. Furthermore, the biochemical measurements were derived from bumblebees cooled to 5°C and *presumed* to be endothermically maintaining a high T_b by NST. No attempts were made to determine if the bees were indeed endothermic or if their flight muscles were inactive when they *were* endothermic. (Given what we know about bumblebees, I presume they were *not* endothermic under the experimental conditions described.)

There are other gross inconsistencies. For example, non-shivering thermogenesis has been reported for some species of bees and not for others, to be active in small bees but not in large of the *same* species, and to be strictly turned off in flight (Newsholme et al., 1972). Nevertheless, in subsequent measurements the precise cycling rates were calculated at between 0.3–1.0 μ moles/minute/g of fresh weight for male bumblebees at "rest" (no attempts were made to determine whether the bees' muscles were indeed at rest), with values of 1.93 and even up to 76.9 for bees in flight (Surholt et al., 1986). Apparently the cycle was *not* turned off in flight, as had previously been supposed as a precondition for NST by this mechanism. It is highly unlikely that bees sacrifice power in flight by futile cycling, when by not cycling they would *still* produce nearly the same amount of heat while still enjoying the benefit of power. Curiously, one report (Prŷs-Jones, 1986) suggests the enzymes associated with the presumed non-shivering thermogenesis are found primarily in workers of *Bombus* foraging on *massed* as opposed to single flowers. And it is precisely on massed flowers that bumblebees are often not endothermic but become torpid instead (see figs. 4 and 5 in Heinrich, 1972c; Heinrich and Heinrich, 1983a,b).

In other experiments living bees were injected with [2-^{3}H, 2-^{14}C]-glucose and then freeze-clamped in liquid nitrogen after either "rest" (again, non-shivering was assumed but not examined) or flight. The glucose and hexose monophosphate were then isolated from the tissues and ^{3}H and ^{14}C radioactivities were measured (Surholt and Newsholme, 1983). Rates of *in vivo* cycling could then be calculated on the basis of ^{3}H/^{14}C ratios and measured rates of glycolysis. In marked contrast to Clark et al. (1973a), who report a shutdown of cycling during flight in bumblebees, Surholt and Newsholme (1983) report a marked increase of cycling in flight in the large (2.3 g) death's-head hawkmoth *Acherontia atropos*. As in bumblebees, it is highly unlikely that an insect (especially a very large one at moderate air temperatures) would relinquish power in flight for boosting heat production; sphinx moths that are considerably smaller than *A. atropos* (2–3 g) generally face the *opposite* problem of getting rid of excess heat in flight (see Chapter 1) in order to be able to continue flight.

Surholt et al. (1990) argue that NST exists in *B. terrestris* queens because (1) they produce more heat than do *Xylocopa* bees, (2) no thoracic vibrations are detected in *B. terrestris* whereas thoracic vibrations are detected in a moth and in *Xylocopa*, and (3) high biochemical cycling rates have been found in bumblebee muscle. However, as previously indicated, the inability to detect thoracic vibrations does not preclude muscle contractions. Surholt et al. (1990) confirm that, as in all previous *Bombus* studies, there is heat production only in the presence of action potentials from the muscles, yet they discount the fact that action potentials are always associated with muscle contractions that produce a smooth tetanus at high activation frequencies (Ikeda and Boettiger, 1965). Higher rates of heat production in bumblebees than in *Xylocopa* is simply the expected consequence of a general pattern of bee adaptation to thermal environment (Stone and Willmer, 1989). Thoracic vibrations in an endothermic moth are, of course, obvious; one can clearly see the wings move with the naked eye.

Any demonstration of a biochemical NST mechanism must do more than show that vibrations are more obvious in one species than in another. One must answer how the process might be controlled. Greive and Surholt (1990) show that free calcium ions strongly inhibit the fructose diphosphatase (= fructose-bis-phosphate) reaction that recycles fructose-1,6-diphosphate back to fructose-6-phosphate. They propose that action potentials in bumblebee flight muscles during warm-up are at low frequencies and that such low action-potential frequencies "obviously do not lead to muscle shivering but they probably cause a gradual release of calcium ions." (If true, this requires a new theory of excitation-contraction coupling.) They then propose that the activity of fructose diphosphatase is increased and the rate of substrate cycling "is markedly intensified until a pCa between 7 and 6 is reached, and heat is generated by this non-shivering thermogenesis mechanism." Unfortunately this calcium-dependent model has problems: (1) muscle contraction is assuredly not abolished at low action-potential frequency (see next section), (2) action-potential frequencies at any one temperature are not lower during warm-up than in flight, and (3) there is no evidence yet of any "gradual release of calcium."

A DIRECT TEST OF SHIVERING THERMOGENESIS

One would predict from the anatomy of a bee that contractions of the power-producing flight muscles (DL and DV) should result in slight movement of the scutellum, even though the thorax itself remains stable. In order to detect possible scutellar movement associated with thoracic muscular contractions, Esch, Goller, and Heinrich (1991) mounted bees by suspending them from a small wooden rod glued onto the notum of the thorax with tacky wax (Fig. 6.8). A small plastic mirror was glued onto the middle of the scutellum and a horizontal bar of direct-current light was reflected off this mirror onto a photovoltaic cell, which was partially shielded so only a downward pointing triangle of the cell was exposed. In this arrangement an upward deflection of the light beam on the triangle resulted in increased voltage, and a downward deflection to the narrow portion of the exposed photocell resulted in decreased voltage. Electrodes in the DL and DV muscles were used to record action potentials, and thermocouples

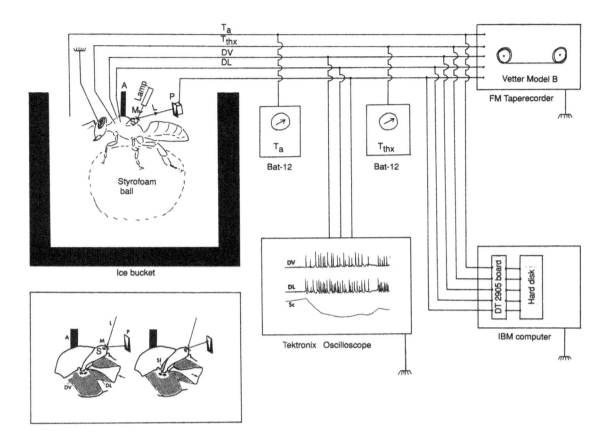

Fig. 6.8 The experimental setup for getting data on shivering. At lower left is a detailed view of the thorax (which is attached to rod *A*) in two positions. When the DL muscles (wing depressors) contract *(left)*, the scuttellum *(S)* rotates downward. Attached to the scutellum is a mirror *(M)*, which also moves. The light beam *(L)* striking the mirror is deflected to the point at the bottom of the triangle on the photocell *(P)*, decreasing current output. When the DV muscles contract *(right)*, the scutellum rotates upward. As it moves, the scutellar fissure or cleft *(Sf)* opens and the light beam is deflected up to the wide top of the triangle, which increases current from the photocell. (From Esch and Goller, 1991; Esch, Goller, and Heinrich, 1991.)

recorded T_{thx} and air temperature. All of the signals were observed on oscilloscope screens and simultaneously recorded on an instrumental tape recorder, stored, and analyzed in a computer.

This setup was used in a comparative study of honeybees *(Apis mellifera)*, carpenter bees *(Xylocopa virginica)*, and bumblebees *(Bombus impatiens)*. The bees alternated between bouts of rest (defined as no action potentials and no endothermy), numerous bouts of warm-up, and occasional buzzing and fligl. . The start of action potentials in all three species was always associated with scutellar movement and with increases in T_{thx}. There was never any increase in T_{thx} without action potentials, and no action potentials without associated mechanical muscle activity. Indeed, in all three types of bees, even single action potentials to the DV and DL muscles could be correlated with individual muscle twitches (scutellar movements). In all three kinds of bees, DL activation during warm-up in the absence of detectable DV activation was associated with downward rotation of the scutellum, as in flight. Conversely, DV activation in the absence of DL activation was associated with upward scutellar movement (Fig. 6.9). Normally, however, during warm-up all of the bees rapidly activated both their DV and DL muscles simultaneously, although the DL muscles were activated more frequently during heating than in flight, especially in *Apis* and *Bombus*.

The effect of the increased DL/DV activation ratio is to stretch the opposing DV muscles until the scutellar fissure (see Fig. 6.8) is closed (verified by microscopic examination). That is, the DV is stretched until a mechanical stop prevents further shortening. Given the highly contracted state of the DL and the "permanent," only partially stretched state of the DV during shivering in *Apis* and *Bombus*, there should, a priori, be very little scutellar movement unless there is simultaneously a *stronger* activation of one vs. the other muscle. Such is indeed the case. There is a smooth tetanus (Fig. 6.10) in the expected upward or downward direction, depending on which set of wing muscles is stimulated relative to the other. A smooth tetanus is ensured since the muscle twitches are considerably longer than the frequency at which they are activated, and (in *Bombus*) tetanic contraction is facilitated even at high T_{thx} because both the DV and DL muscles are activated, as mentioned earlier, by synchronous bursts of impulses.

These results on the *in situ* muscle contractions prove that heat production is, in *Bombus*, as well as in *Apis* and *Xylocopa*, based on shivering thermogenesis. Surprisingly, the mechanism of endothermic shivering is very similar to the mechanism of pre-jump preparation in poikilothermic leaping insects (see Chapter 15). In both mechanisms, two sets of opposing muscles contract isometrically to produce little or no external movement. In both, the contractions are stabilized by having one set of muscles (the stronger) pull against the other and/or against a mechanical stop, usually while at a mechanical advantage by way of a lever (in the jumpers). And in both there occurs a relaxation of one muscle, or an alternate activation of the two, at the initiation of activity.

Although there is now ample evidence for shivering thermo-

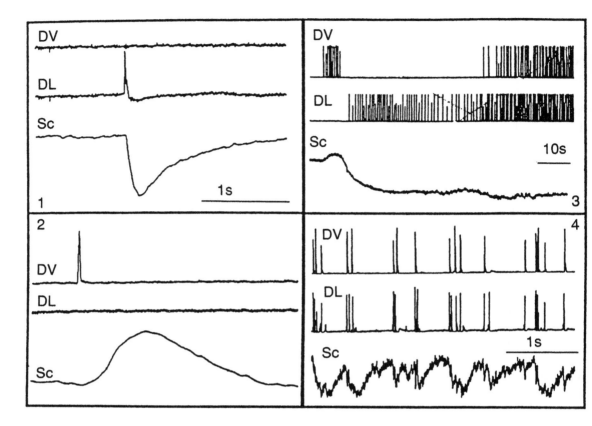

Fig. 6.9   In situ bumblebee muscle contraction *(Sc)*, as determined by scutellar movement (see Fig. 6.8), in response to single muscle potentials in the DL *(top left panel)* and the DV muscle *(bottom left)* at $T_{thx} = 13°C$, and in response to multiple stimulation at higher T_{thx} *(two right panels)*. The scutellar traces on the left show the long durations of single muscle contractions and subsequent relaxation at low muscle temperature (single action potentials are rare and were chosen here only to show the effect on muscle contraction). The graphs at top right (at $T_{thx} = 14°C$) and bottom right (at $T_{thx} = 25°C$) show the masking effect of antagonists and faster twitches: the muscles contract against each other and so summation of the contractions occurs. Both left panels were drawn to same time scale. (From Esch, Goller, and Heinrich, 1991.)

genesis in bumblebees, there is still not one shred of experimental evidence for NST. Nor is there any known function for it in bumblebees (as opposed to shivering). Furthermore, almost all of the circumstantial evidence for NST in bumblebees is based on either unwarranted or on wrong assumptions. For example, those situations when the high cycling rate was demonstrated (and it was presumed that the bees were endothermic) were precisely those when the bees were predictably poikilothermic—namely,

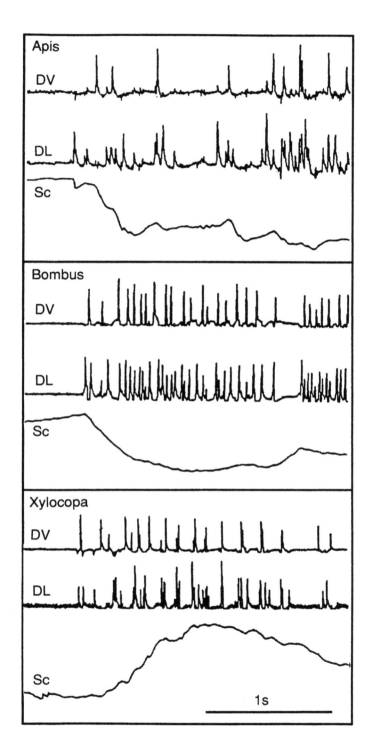

Fig. 6.10 *In situ* muscle response to the typically observed simultaneous activation of DL flight muscles during shivering in *Apis* (at T_{thx} = 29°C), *Bombus* (at T_{thx} = 27°C), and *Xylocopa* (at T_{thx} = 25°C). The scutellar trace *(Sc)* shows summation and tetanus. Note that in *Xylocopa* the activation frequencies of the DV and DL are more equal to each other during shivering than they are in the bumblebee, and the net effect is that the DV contraction is greater than the DL contraction (resulting in a net upward scutellar movement).

small bees at low temperatures, and on panicled flowers (Heinrich, 1972c; Heinrich and Heinrich, 1983a,b).

The biochemical data are not in question, and an explanation must be found for them. Since no other hypothesis has been offered, much less tested, I can only offer a speculation: It is difficult to envision evolution bringing about a *non*-shivering thermogenesis when the shivering already works so superbly. There can be no energetic advantage, because the goal is mechanical inefficiency. If there is need for more heat, then the mechanism that already exists should be perfected. It should not be dumped in favor of a new mechanism. On the other hand, it makes very good evolutionary sense that a bee enters torpor to conserve energy (and the bumblebees with the presumed NST *were* most likely torpid). If the quick switch to torpor is advantageous, it should go one step further than torpor, by *reducing* metabolic rate even less than normal resting metabolism. Could the high rates of enzyme activity in the *backward* direction (away from the "forward," energy-yielding Krebs-cycle activity, which needs no explanation) be part of a mechanism of "backpedaling" to conserve energy (by shutting off substrate to the Krebs cycle)? If so, the presumed "NST" is a mechanism that is precisely the opposite of what has been uncritically presumed. It would act instead to *reduce* heat production.

The Ecology of Heat Production

By whatever combination of mechanisms bumblebees use to warm up, they are nevertheless in part dependent on a warm shelter—the nest—to be active in the field, as they commonly are, at air temperatures below 6–7°C. After they are heated up they can (depending on body size and availability of fuel) produce prodigious amounts of heat and even forage and fly at near 0°C.

Bumblebees that come to rest outside the nest (or before they have a nest) often cool to near ambient temperature. They become torpid and are incapable of flight until they warm up through muscular thermogenesis. Large (0.25–0.60 g) queens of *Bombus vosnesenskii* and *B. edwardsis* can warm up from temperatures as low as 6°C (Heinrich, 1975). Small (0.10–0.13 g) workers have not been seen to warm up from such low temperatures after they have been fully cooled to ambient temperature, and unlike the large queens, they do not forage at temperatures below 10°C (Heinrich and Heinrich, 1983a), presumably because their surface-volume ration is too large to permit enough heat retention. Nevertheless, like queens, they are physiologically capable of *beginning* to shiver at near 6–7°C, but not below (Goller and Esch, 1990).

The increase in T_{thx} during pre-flight warm-up from 6 to 7°C starts out slow, near 0.7°C/min, and then the rate increases to near 11°C/min at 25°C (Fig. 6.2). The rate of heat production depends on T_{thx}, however, and is independent of ambient temperature (Heinrich, 1975). For example, the bees produce about 50 J/g thorax/min when T_{thx} = 30°C during warm-up at air temperatures of 6°C as well as at 25°C. As in other insects, the rate of heat production increases steadily throughout warm-up as the rate of pre-flight warm-up is near the maximum that T_{thx} will allow.

Warm-up is energetically costly, but because the bee *maximizes* energy expenditure during a warm-up to expedite warm-up rate, the energetic cost of any one warm-up is minimized. On the basis of calculations using heat-transfer measurements and body-temperature changes (Heinrich, 1975), I estimate that a *B. vosnesenskii* queen spends about 12.1 J (in 1.0 minute) during warm-up from 24°C, while it costs her 65.7 J (and 17 minutes) for a warm-up from 6.5°C. These values underestimate total energy cost at low temperatures because they do not take into account some inevitable heat loss to the abdomen.

Regulation of Body Temperatures

Bumblebees at times stabilize T_{thx} independent from ambient thermal conditions—while perched and stationary, while foraging from flowers, during continuous flight, and while incubating their brood. If we are to understand the mechanisms of thermoregulation, we must examine each behavior in strict isolation, because the mechanisms could potentially vary depending on the activity.

Stationary (or walking) bumblebees often allow T_{thx} to fall to near ambient temperature. If they regulate T_{thx}, however, they do so by varying the intensity of shivering. For example, as the temperature excess (and heat-loss rate) doubles as air temperature is lowered and the bees attempt to maintain the same T_{thx}, the action-potential frequency that can be recorded from the flight muscles is then also doubled (Heinrich and Kammer, 1973), and this doubling is reflected in a doubling of oxygen-consumption rate, or heat production (Fig. 6.5). Generally, however, bumblebees do not indiscriminately engage in this typically homeo-

Fig. 6.11 Bumblebee foraging from fireweed *(Epilobium angustifolium)*.

thermic response well known for vertebrate endotherms. It is usually too energetically costly. But shivering activity is used to heat the nest (Chapter 16), when the animals are disturbed and not at rest, and when energy supplies are ample and flight activity is frequent.

During foraging the bees engage in intermittent flight, and since the flight metabolism itself generates heat, thermoregulation could potentially vary depending on ambient conditions and the kinds of flowers visited. Some bumblebees may forage at near 0°C when there is frost on the ground, as well as on the hottest days. Sometimes they spend more than 60 percent of their time in flight while foraging, and sometimes they perch for long durations to extract nectar, spending less than 1 percent of their time in flight. Do they stabilize T_{thx} as a function of ambient temperature?

The first attempts to answer this physiological question were performed under natural conditions in the field, where efforts were made to eliminate or reduce the effects of direct sunshine and wind. To determine the scale of the problem, natural cooling rates needed to be determined. Bumblebee workers have, in the absence of heat production (when dead), a very rapid cooling rate. For example, a *Bombus vagans* worker (weight 85 g) with a T_{thx} of 32°C at an air temperature of 24°C cools at a rate of about 17°C/min (Heinrich and Heinrich, 1983a). Newton's law of cooling dictates that the passive cooling rate is a direct function of temperature excess. Thus, at an air temperature of 8°C the same bee maintaining the same T_{thx} (triple the temperature excess) would now cool at the phenomenal rate of 51°C/min. Body temperature measurements of *B. vagans* in the field foraging from fireweed (Fig. 6.11) show, however, that T_{thx} is nevertheless stabilized independently of ambient temperature (Fig. 6.12), despite the at least three times greater differences in cooling rates at the temperature extremes.

During foraging (Heinrich, 1976b, 1979), bumblebees usually stop and start at intervals measured in seconds or fractions of seconds. One mechanism of thermoregulation might be behavioral modulation of flight activity for heat production to keep warm. Another mechanism could be to shiver or not shiver during the inter-flight intervals, when the animals are perched on flowers while extracting nectar or pollen (this option is not available to sphinx moths, because they hover while sipping nectar.)

The hypothesis that the bees produce heat to keep warm during the inter-flight intervals while working the flowers was tested by

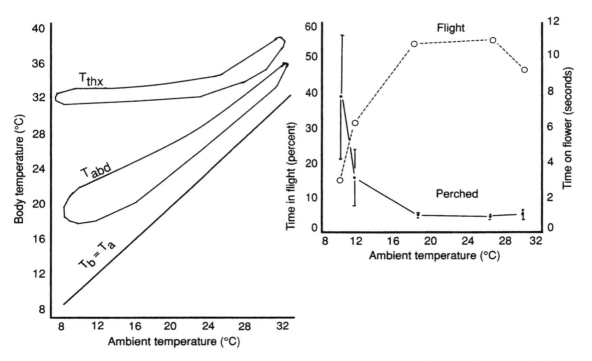

artificially increasing the amount and viscosity of "nectar" in some flowers, which caused bees that happened to visit those flowers to stay for up to a minute or more. Normally they required only a second or two to take up the minute amounts of nectar or pollen available per flower. If the foraging bees stay heated up from the flight metabolism alone, then they should have started to cool immediately after stopping, as dead bees do, and T_{thx} would then be a strict function of perching duration after landing. However, the data showed conclusively that they did not cool after landing. Indeed, stationary bees imbibing nectar or sugar water (in shade) were generally *warmer* than flying bees. (The more concentrated the "nectar," the hotter they got.) Their rate of heat production to counteract passive cooling could then be calculated. Given the expected cooling in the absence of heat production (derived from dead bees), the rate of heat production was calculated (Heinrich, 1972a) at near 2.1 J/min, which is as high as that resulting from flight metabolism. Foraging bees thus keep warm not only from the heat produced as a by-product of their flight metabolism, they may also work their flight muscles as vigorously as they do in flight at low air temperatures to counteract cooling.

Fig. 6.12 Thoracic and abdominal temperatures of *B. vagans* workers foraging from fireweed in relation to ambient temperature (in shade or overcast). The graph at right shows the bee's activity (percent of time in flight and time perched on flowers) during this foraging. Note that the lower levels of T_{abd} and T_{thx} are regulated independently of flight or perching activity. (From Heinrich, 1972a.)

The above results indicate that foraging costs include not only those of time and energy to fly between flowers, but also thermoregulatory or shivering costs while on flowers. Is the energetic cost of shivering justified? Comparing the energy costs of flight and thermoregulation with the sugar contents of flowers (and the rates at which flowers are visited) shows that the bees adjust their foraging behavior in accordance with an energy budget (Heinrich, 1972a,c, 1973, 1974). For example, one flower of fireweed *(Epilobium angustifolium)*, that has been screened from nectar foragers for 24 hours contains on the average of 1.85 mg of sugar, sufficient fuel for a *B. vagans* worker to allow it to remain heated up and flight-ready at air temperatures near 10°C for 14 minutes. [Note: 1 mole or 180 g of glucose yields 686 kcal (or 180 mg yields 686 cal = 2,870 J). Thus the 1.85 mg sugar yields 30 J (2,8760 J/ 180 mg × 1.85 mg = 30 J). At an energy expenditure of 2.1 J/ min, these 30 J would last 14.3 minutes.] Since *B. vagans* workers visited 10–20 flowers of fireweed per minute, the calories expended are easily replaced if the flowers contain their full complement of nectar. (In addition the bee gets pollen, too.)

There is usually intense competition for the nectar and pollen supplies offered by the flowers. Unscreened flowers to which there was free access by all foragers contained (in the above study) a mean of only 0.086 mg sugar. (Exact amounts would of course vary from one site to the next, depending on the local pollinator populations.) But a bee should make an energetic profit even at these already-visited flowers at low air temperatures if it extracted the sugar from about 2 flowers per minute. According to these approximations, the bees (normally visiting 10–20 flowers per minute) are energetically justified to expend the large amounts of energy for heat production, since when they do not do so they are unable to fly at all or their rate of flower visitation becomes drastically reduced.

One of the more intriguing observations about bumblebee thermoregulation is that T_{thx} is not regulated at any one setpoint but is adjusted with regard to food supply. If nectar supplies and air temperature are both very low and the bees forage on inflorescences with hundreds of florets that can be reached by walking, then they allow T_{thx} to decline well below the minimum for flight while continuing to forage (Heinrich, 1972c, 1973; Heinrich and Heinrich, 1983b). They may then warm up again by endothermic warm-up or by basking and fly to the next inflorescence to resume foraging. Since bees without fuel *necessarily* become hypothermic,

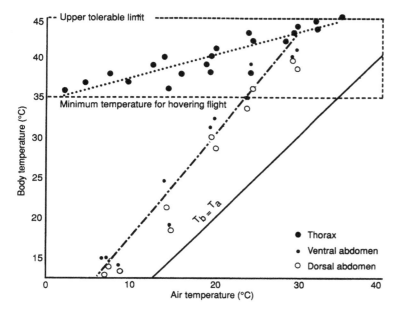

Fig. 6.13 Thoracic and abdominal temperatures of a number of *B. vosnesenskii* queens in continuous flight as a function of ambient temperature. Whereas the thoracic-temperature excess decreases greatly at higher ambient temperatures, that of the abdomen increases instead (From Heinrich, 1975.)

it might appear that the hypothermic foraging bees are also simply out of fuel. Such is not the case; torpid foraging (and stationary) workers and drones may be bloated with large amounts of nectar (Heinrich and Heinrich, 1983b). Therefore, their hypothermia is not merely a physiological response to low food reserves; it is a response to external factors and it functions in energy economy.

During continuous flight a bee has the option neither of shutting off heat production nor of shivering to produce additional heat. Nevertheless, large-bodied bumblebees forced to remain in continuous flight maintain some independence of T_{thx} from air temperature (Fig. 6.13). Queens of *B. vosnesenskii* and *B. edwardsis* maintained their T_{thx} at 36 to 45°C during flight over a 33°C range of air temperature, from 3 to 36°C (Heinrich, 1975). The difference between thoracic and air temperature was thus only 9°C at an air temperature of 36°C, while it was more than three times greater (33°C) at 3°C.

The mechanism of thermoregulation in flight in queens does not involve regulation of heat production, because although their temperature excess varies more than 3-fold, there is no 3-fold increase in heat production at low air temperatures. Indeed, heat production is nearly fully independent of air temperature, although it is highly variable; the metabolic rate of bumblebees flying in the confines of a 4-liter respirometer ranged from 50 to 100 ml O_2/g/hr, and the more than doubling in metabolic rate of both flying bumblebees (Heinrich, 1975) and honeybees (Wolf et al., 1989) can be accounted for by increased load (up to doubling of body weight) carried in the bee's abdomen. (Load can be experimentally manipulated by allowing bees to tank up on variable amounts of nectar or honey.) Subsequent measurements in a 7-liter respirometer, using simultaneous O_2 and CO_2 gas analysis (Bertsch, 1984), have confirmed a value of 56 ml O_2/g/hr for flying drones who were relatively unloaded and flying at 20°C. Finally, direct measurements of sugar utilization over prolonged periods of flight in *B. terrestris* drones (which were relatively lightly loaded) in their normal patrolling behavior in captivity showed a fuel consumption of 7.6 μ mol sugar/min/g body mass, which is equivalent to 61 ml O_2/g/h (Surholt et al. 1988). Thus, different independent methods of measurement all converge to a metabolic rate of near 60 ml O_2/g/h for lightly loaded bumblebees in free flight. Ambient temperature does not affect this rate, nor does flight speed (Ellington, Machin, and Casey, 1990).

Elimination of heat production as a thermoregulatory mechanism leaves heat loss as the alternative, and since the passive rate of cooling is a direct function of the difference between body and ambient temperature, the bees flying at 3°C were passively losing heat from the thorax at rates more than 3 times greater than when they were flying at 36°C. They were, of course, losing the same amount of heat also at 36°C, but the balance must therefore have been accomplished by active dissipation. That is, thermoregulation was accomplished by varying heat loss exclusively and the rate of active heat dissipation was therefore varied by some 300 percent, presumably by preventing overheating when T_{thx} becomes dangerously high (see below).

Although bumblebee queens demonstrate relatively good independence of T_{thx} from air temperature in flight, their thermoregulation is not achieved by regulating T_{thx} about a specific temperature setpoint. Bumblebees flying with lower power output (supported in flight from a tether) may quickly cool even while still in continuous flight, and they then continue to "fly" at T_{thx} much lower than they maintain or can fly at in the field (Heinrich, 1972b).

Although overheating and heat dissipation are largely functions of ambient temperature, the load carried in flight is also a variable that affects the T_{thx} maintained. First, metabolic rate or work output in flight is a direct function of load carried, and the load that can be carried depends on muscle temperature. Heavily loaded bees often fail to lift off at their first attempts, but they then shiver to increase T_{thx} to 38–44°C before trying it again. Thus, by regulating flight effort for liftoff or to stay aloft, they automatically maintain T_{thx} at least above the minimum for flight (Heinrich, 1975).

In contrast to the above data and conclusions on flight metabolism and thermoregulation in flight, Unwin and Corbet (1984) suggest that bumblebee workers regulate T_{thx} in flight by varying wing-beat frequency (and presumably heat production.) However, they measured wing-beat frequency (from recordings of flight tone or sound) of only two individual B. pratorum and one B. pascuorum while foraging on borage. The B. pascuorum showed no change of wing-beat frequency at air temperatures from 14 to 23°C, while the two B. pratorum showed decreases of wing-beat frequency of 8 and 4 percent, respectively, over the same range of air temperatures. Considerably more extensive data by Joos, Young, and Casey (1990), on the other hand, showed wing-beat frequency of a number of bumblebee species was *not* correlated with ambient temperature. Since bumblebees routinely have a high flight tone (and rapid flight and high T_{thx}) when they are foraging from very highly rewarding flowers, and since they have a low flight tone (personal observations) and barely fly or do not fly at all when they are foraging from low-reward flowers (Heinrich, 1972c), a more conservative explanation for Unwin and Corbet's results is that there was more nectar and pollen available from the borage in the morning (when it was cooler) than in the afternoon, so they had higher T_{thx} then. In any case, a presumed 4–8 percent increase in heat production would have had a negligible effect in regulating T_{thx}, inasmuch as the passive heat loss over the doubling of T_{thx} excess (assuming a T_{thx} near 34°C) would have required at least a full 100 percent increase in metabolic rate.

PUBESCENCE

The pile or "fur" covering the head, thorax, and to a lesser extent the abdomen of bumblebees has a considerable insulating effect, and hence it allows the bees to conserve energy and extend their activity range to low temperatures. Norman Stanley Church (1960) was one of the first to examine the insulating effect of this pile by heating bees (and other insects) with an internal resistor.

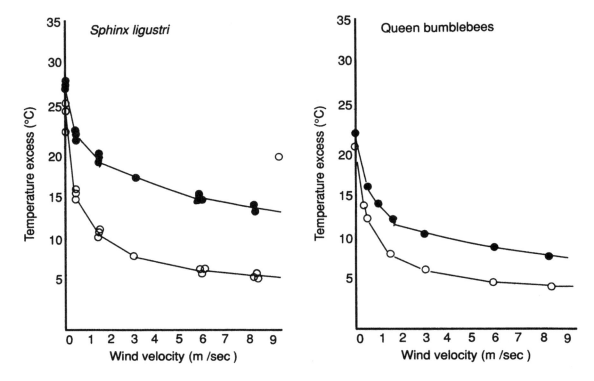

Fig. 6.14 The temperature excess of artificially heated dead insects with insulating pile intact *(filled circles)* and with insulating pile removed *(open circles)*, as a function of wind velocity. The comparison between a sphinx moth and bumblebees shows that the sphinx moth is better insulated than the queen bees. Nevertheless, in both the presence of insulation nearly doubles their temperature excess. (From Church, 1960.)

He mounted the animals facing the air stream in a wind tunnel and then measured the temperature excess with the hair or pile intact and with all the insulation removed (with a pipe cleaner and with paraffin wax). The hairy legs (which when drawn up as they are in flight undoubtedly also insulate the body) were cut off and the cut ends were sealed. In six species of *Bombus*, the difference in temperature excess between fully furred and depilated bees was 50–73 percent, averaging 60 percent, at a wind velocity of 300 cm/s. These values in temperature excess between intact and depilated animals, though impressive, were nevertheless only about half those for several moths (Fig. 6.14). Comparisons between different species of *Bombus* having either long or short fur, as well as between individuals of one species with hair clipped to about one-third its original length, showed that hair density rather than length is most important for insulation.

As expected, there is little difference in cooling rate between insulated and uninsulated insects in still air when the ambient air "layer" affords insulation. However, the effect of wind speed is strong at low wind speeds. Cooling rate almost triples until about 3 m/s, then quickly levels off at higher wind speeds. For example, a tripling of wind speed at wind speeds above 3 m/s results in only a one-third further greater cooling rate (Fig. 6.14). Since bumblebees fly at several m/s, the data show that these heavily insulated animals have limited options for regulating heat loss by varying flight speed.

What explains the independence of cooling rate from wind speed at high wind speed? A solid object should have a cooling rate linearly related to wind speed. To explain the curvi-linear function, I speculate that the fur may become even less leaky to heat at higher wind speeds because it becomes matted down and denser, more effectively trapping the insulating air layer. If the fur did not bend down and remained totally stiff, then air resistance, turbulence, and heat loss should continually increase with increasing flight speed.

The pile, which plays such an obvious role in bumblebee thermoregulation and hence in their ability to be active at low temperatures, may be short and densely packed or it can be relatively long, loose, and shaggy. Colors usually range from black, yellow, white, and orange or red, and the patterns of these colors may differ strikingly on different body parts. It is reasonable to ask whether or not the pubescence or pile has evolved physical characteristics and color to fit specific thermoregulatory needs in different species and castes having different geographical ranges and activity regimens.

Aside from its role in thermoregulation, the pile functions also in pollen gathering. But if pollen collecting were the sole function of pile, then many of the other kinds of bees should be furry as well. Since tropical-adapted bees (various species of honeybees, carpenter bees, and stingless bees) are relatively glabrous or have only short hair, it is probable that the conspicuous furriness of bumblebees is indeed an adaptation for living under cold conditions, or else the glabrousness of other bees is an adaptation for living in a hot climate. Does bumblebee pile show fine-tuning to specific thermal regimes, or can fur variation in these animals be explained by other contingencies, such as protection from predators?

Bumblebees are distasteful to predators because of their stings, and predators recognize them by their characteristic pubescence (Brower, Brower, and Westcott, 1960). Numerous convincing lines of evidence (Plowright and Owen, 1980) converge to point at the powerful role of mimicry in determining bumblebee-pile characteristics. Flies as a rule are relatively glabrous. Yet even among the Diptera, species from several families have converged to become furry and to produce striking resemblances to sympatric bumblebees in terms of body size, behavior, furriness, and details of the fur's color patterns (Gabritchevsky, 1926). However, as Plowright and Owen (1980) point out, the numerous Batesian fly mimics (edible animals that mimic noxious ones) as well as the stingless bumblebee males themselves, must exert a "Batesian load" upon the stinging female bumblebee population. The more the effect is diluted by "mimics" of their own (the males) and other species (flies), the less effectively the females can be protected by their conspicuous color patterns that warn predators.

One way to reduce the effect of the Batesian load is through Müllerian mimicry (noxious species resembling each other)—where different species of bumblebees in any one area mimic each other. Plowright and Owen (1980) provide numerous convincing examples of Müllerian mimicry for a set of 47 species of bumblebees from the United Kingdom and North America. The most remarkable examples of color resemblance are those of the various parasitic *Psithyrus* species, which closely mimic their *Bombus* hosts. At least the *color* of bumblebees' coats is therefore not likely the result of only thermoregulatory considerations.

The amazing Batesian mimicry between flies and bumblebees and the Müllerian mimicry complexes among various species of bumblebees of very different phylogenetic origin indicate that coat color is a very plastic characteristic. It is therefore curious that male bumblebees sometimes differ from their females in fur characteristics. Drones should be palatable to avian predators because they do not sting, and one might expect that they would very closely mimic the stinging females in their appearance. In general, however, males have slightly longer (and less dense) and lighter-colored pile than females (Stiles, 1979). The sexual dimorphism in pubescence increases at higher latitudes. What can account or this sexual dimorphism? (Are the differences sufficient for predators to differentiate males from females?)

Stiles (1979), who has examined the sexual dimorphism of bumblebees from the western hemisphere, suggest that both color patterns and pubescence characteristics of male bees can in part be explained by advantages gained in thermoregulation. He presumed (no data provided) that males, when they are searching for or attracting mates, engage in longer periods of sustained flight than females do, and that they might then have heat-dissipation problems (no evidence given). Seventy-six percent of the males of 41 species studied from 30°N, the center of their latitudinal range, had longer hair than females, whereas only 11 percent of those south of 30°N had longer hair. Similar patterns are found with both the density and color of the pile. It was presumed that the lighter and less dense fur reduces problems of overheating both through reflection and convection. (Are the differences sufficient to affect biologically significant cooling rates?)

In Scandinavia both the males and the females of 12 of the 26 species are melanic or have melanic forms in some populations, again suggesting that the color variation may have a thermoregulatory significance (Pekkarinen, 1979). However, so far there are no data to support the assumption that the darker forms gain more heat. There are also no studies that show systematic differences of the effect of either coat color or coat length on temperature excess in Arctic vs. temperate bumblebees. Church (1960) provides data from only seven *Bombus* and one *Psithyrus* species from Britain, but no consistent pattern emerges with regard to coat density among them. In general, Arctic species are thought to be bigger, fuzzier, and darker (Pekkarinen, 1979) than those farther south because of thermoregulatory advantages. However, the queens of *Bombus polaris* from above the Arctic circle were no heavier (or darker) than most of the queens from near Burlington, Vermont (Heinrich and Vogt, 1991). Furthermore, the Arctic bees' cooling rates in a wind tunnel also do not show differences in insulating capacity of the pile from those of temperate bumblebees. The presumed color effects on external heating would be small for endotherms that usually forage in shade of dense vegetation or on cloudy days, and I conclude that the data available at the present time show that bumblebee pile is of critical importance for insulation, but the idea that the pile varies geographically for thermoregulatory considerations remains a speculation.

BODY MASS

Although naturally occurring coat color and pile characteristics likely affect heat transfer, their efforts are nevertheless small and possibly negligible. Instead, the effects of body mass on cooling and thermoregulation are massively obvious and easily demonstrable by even the crudest direct measurement. The female bees in any one colony may vary in mass by an order of magnitude, and over a size range as large, say, as from 60 mg to 600 mg, cooling rates vary 4-fold (Fig. 6.15). Thoracic temperature, however, is not determined by mass and surface-volume ratios but is set instead by the flight machinery as such of that species (Stone and Willmer, 1989).

Since small bees cool convectively several times faster than the largest bees, the ability to thermoregulate in continuous flight is very much related to body size. Indeed, continuously flying workers of *B. edwardsis* are unable to maintain a minimum T_{thx} of 30 °C at air temperatures less than 10°C, and at temperatures from 10 to 20°C the difference between T_{thx} and air temperature in workers is relatively constant, averaging near 20°C (Heinrich, 1975). At high T_a the difference between T_{thx} and air temperature declines several degrees. Therefore, there is active regulation only at higher temperatures, when T_{thx} of both queens and workers converges near 45°C, the upper tolerable limit of T_{thx}. In the laboratory only the large bees, the queens, show conspicuous thermoregulatory ability and continuous uninterrupted flight at low ambient temperatures (Fig. 6.13). As predicted, it is probably also thermoregulatory constraints that prevent small-bodied bumblebees from foraging at very low temperatures, when large-bodied bees can still be active (Heinrich and Heinrich, 1983a).

During foraging, however, the bees stop at flowers, and they can then shiver and rewarm after having cooled convectively in flight. This behavior makes a thermoregulatory compensation for small size possible, and both large and small bees maintain similar T_{thx} (Fig. 6.15) despite differences in cooling rates of up to 400 percent. These data therefore provide another impressive documentation of the bee's thermoregulatory response. The physiological compensation (shivering) is not without cost; small bees must shiver whereas large bees can often dispense with it.

The obvious energy constraint that small-size bees face when foraging at low temperatures should provide selective pressure for larger body size, but to my knowledge geographical size variations have not been demonstrated. In general, worker size increases gradually throughout the season in healthy colonies, although colonies disrupted or otherwise in stress also produce smaller workers (Knee and Medler, 1965). The availability of resources or their redistribution by the attendant workers to the larvae proximally determines size, and workers could presumably make few large workers or many small ones. But so far there is no evidence that resource allocation is adaptively related to variations in size advantageous for thermoregulation and foraging in differ-

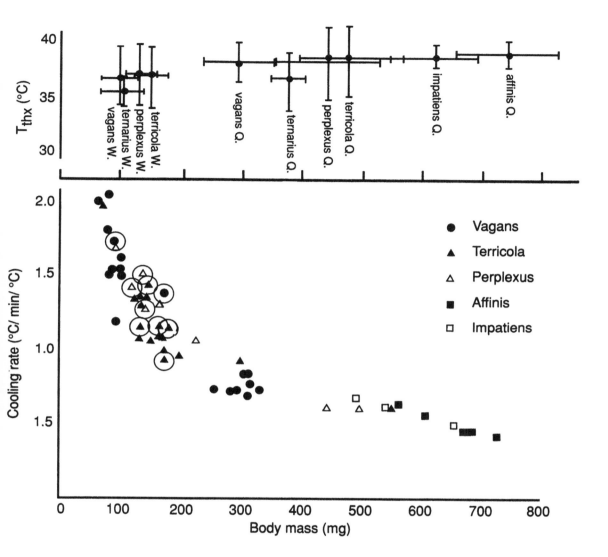

Fig. 6.15 At top, the T_{thx} of various species of bumblebees are plotted in relation to mass. Q = queens foraging from raspberry, apple, self-heal, leatherleaf, and lupine. W = workers foraging from raspberry. Ambient temperature varied from 2.5°C to 22°C. All the bees weighing more than 200 mg are queens. Note that T_{thx} are very similar despite large differences in mass. The lower graph shows the cooling rate of these bees as a function of mass. (From Heinrich and Heinrich, 1983a.)

ent climates. There are large bee species in almost all tropical areas, and both small and large bee species in cool alpine meadows at high latitudes.

At least some of the workers of a bumblebee colony can be small even at low ambient temperatures at no great loss to the colony; rather than acting as foragers, the small bees can function as nest helpers. Again, presumably various complex selective pressures other than thermoregulation may be exerted on shaping body size, and these selective pressures presumably take precedence over temperature regulation. For example, given *small* nectar and pollen rewards of flowers, it would be energetically prohibitive to harvest them with transport vehicles (workers) of a large mass, even though thermoregulation of a large-sized body is physically more feasible than one of small size. Thus, although small bees have a thermoregulatory disadvantage in foraging at low temperatures, they nevertheless may have an energetic advantage in foraging over large bees. This advantage applies over all temperatures; body size could be more related to the size of food rewards to be harvested than to the energetic cost of thermoregulation.

BROOD INCUBATION

Long before data on bumblebee thermoregulation became available, it was already suspected that bumblebees might incubate their brood. As early as 1837 George Newport established, by direct measurement with small mercury thermometers, that bumblebees when pressed upon their brood have a warm abdomen. Although Newport claimed the bees were incubating their larvae, Girard (1869) established that bumblebees often have a relatively *cool* abdomen with respect to the thorax (the brooding and non-brooding bees were assumed identical), and Buttel-Reepen (1903) then concluded that the bumblebees described by Newport (1837) were not incubating at all. Instead, he thought they perched upon the brood (Fig. 6.16) in order to heat themselves by it. (Compare with Fig. 6.17.) At that time, there appeared to be great concern with precision-measuring T_b to a tenth of a degree but almost no concern that the biological context of the measurement may make a difference of tens of degrees.

Others have also described the brooding behavior (Sladen, 1912; Himmer, 1933; Plath, 1934; Free and Butler, 1959; Hasselrot, 1960), but no proof of the incubation hypothesis was available until both the T_{thx} and T_{abd} of incubating queens (Heinrich,

Fig. 6.16 The founding queen of *Bombus vosnesenskii* incubating her initial brood clump with pupae and new egg clumps. She is facing her honeypot, which she refills during intermittent foraging trips. Her elongated abdomen is pressed tightly upon the brood. (From photograph by B. Heinrich.)

Fig. 6.17 A worker *B. edwardsii* carrying pollen loads and regurgitating honey into a honeypot. Note its relatively compact abdomen. (From photograph by B. Heinrich.)

1972d, 1974), as well as the brood temperature with and without incubating bees (Richards, 1973), were measured.

Temperature measurements of single queens with their initial brood clumps show that the bees indeed incubate, and the process of incubation is even more intricate, behaviorally and physiologically, than at first presumed. First, the increase and decrease in temperature of brood clumps (experiments were performed without nests to eliminate the possibility of heating the brood secondarily by warmed air in the nest) is directly correlated with the presence or absence of the incubating bee. Brood clumps (with larvae and pupae) begin to cool within seconds when the attending bee steps off, and their temperature rises immediately when the bee again perches on them. Temperatures of brood clumps without bees soon approach those of the environment. Incubating queens (all castes incubate in the laboratory) straddle the brood clump or pupae by spreading the legs laterally, extending the abdomen, and closely pressing the abdomen onto the substrate being incubated.

The T_{abd} of incubating bees is usually within 2–3°C of T_{thx}, and the high T_{abd} functions to heat the brood. The significance of the high T_{abd} was tested by implanting an electrical resistor into the abdomen of a dead bee fixed onto the brood clump and then simultaneously measuring T_{thx}, T_{abd}, and brood-clump temperature (Heinrich, 1972d). Electrical heating of the thorax resulted in a negligible temperature increase in the brood. Abdominal heating, however, resulted in the typically observed rise in brood temperature. Thus, even though the heat-generating mechanism is located in the thorax, the abdomen serves as the organ of heat transfer to the brood. (The thorax has legs and coxae protruding from it and does not allow for a smooth contact, but the abdomen is ventrally smooth and uninsulated, except for some long hairs that can bend down. Birds, in contrast, have a bare brood patch which can be covered by feathers from the sides, when not in use.)

If the primary heat flow out of the bee into the brood is through the abdomen, then after the abdomen is heated and applied to the brood it should quickly cool. Quite the contrary occurs: the T_{abd} of incubating queens remains high and close to T_{thx} (Heinrich, 1974). Both T_{thx} and T_{abd} of incubating queens are therefore regulated, the thorax at 35–37°C and the abdomen at 30–36°C, over an ambient temperature range of 3–33°C (Heinrich, 1974). These results are in marked contrast to the body temperatures observed during flight and foraging, when T_{thx} is regulated (but not so precisely as in incubation) and when T_{abd} is *un*regulated (an exception will be discussed later) and usually only a few degrees above air temperature during free flight (Heinrich, 1972a,c).

Brood incubation through the abdomen involves heat transfer from the thorax to the abdomen. Stationary incubating bees regulate T_{thx} by regulating heat production, presumably by shivering—the same mechanism used for warm-up. The metabolic rate of an individual stationary incubating queen on an uninsulated brood clump is, as in a typical vertebrate homeotherm during thermoregulation at low ambient temperature, inversely related to air temperature. Metabolic rates of incubating bees at 5°C rise to as high as those of flying bees (Fig. 6.18), increasing linearly some 160-fold from near "resting" rates at high ambient temperature to 160 ml O_2/g thorax/h at the low ambient temperature of 5°C (Heinrich, 1974). This relative increase in metabolic rate in response to low temperature is impressively greater than that achieved by small winter-adapted finches capable of withstanding −50°C. Finches can achieve increases of only 5–6 times basal metabolic rates under maximal metabolic challenge, and their maximal shivering metabolism is lower than their metabolic ex-

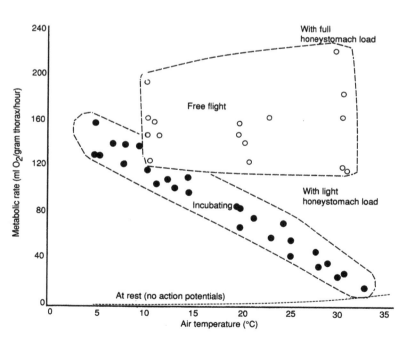

Fig. 6.18 The rates of oxygen consumption of *B. vosnesenskii* queens as a function of air temperature while incubating the brood, during continuous free flight, and while at rest (no flight-muscle activity). (From Heinrich, 1975; Kammer and Heinrich, 1974.)

penditure during flying (Marsh and Dawson, 1988). Thus, bumblebees mount a considerably greater metabolic challenge for thermoregulation than do some of the most impressive of vertebrate endotherms.

In the experiments with bumblebees the nest insulation, as mentioned, was removed in order to control air temperature near the bee. In the field, however, bumblebee nests are normally highly insulated. We may conclude, then, that much lower metabolic rates are required in the nests than those measured, or perhaps bees would be able to heat broods at environmental temperatures much lower than the 0°C limit imposed in the laboratory.

OVARY "INCUBATION"

Except during brood incubation, as indicated above, T_{abd} of insects is generally low and unregulated. For example, some winter-flying endothermic moths are capable of maintaining an impressive T_{thx} excess of some 35°C in part by a counter-current mechanism that greatly retards heat leakage into the abdomen at all temperatures (Heinrich, 1987). It therefore seemed reasonable to suppose that Arctic bumblebees might be designed in a similar way—that is, able to retain heat in the thorax to help maintain an elevated T_{thx} even at very low air temperature. Contrary to expectations, however, queens of high Arctic bumblebees, principally *B. polaris* (Fig. 6.19), maintain high (near 30°C) and quite well-regulated T_{abd} (even at air temperatures near 5°C) while they are foraging shortly after emerging from hibernation in the spring (Fig. 6.20). In contrast, though, the T_{abd} of bumblebee queens from New England under similar conditions are several degrees lower, even though their T_{thx} are identical. Since both the T_{thx} and T_{abd} of Arctic drones and workers are indistinguishable from those of bees in temperate regions (Fig. 6.20), this difference indicates that the high T_{abd} of the Arctic queens is a unique adaptation.

Why might Arctic queens, specifically, find it advantageous to maintain a higher T_{abd} than their more southerly congeners and Arctic drones or workers? One plausible hypothesis is that they incubate their ovaries. The very short growing season in the high Arctic places unusual demands on speeding up the colony cycle. A non-social insect such as the caterpillar *Gynophora* (see Chapter 13) may spend 14 years to complete one life cycle, whereas the social bees that are constrained (presumably by evolutionary inertia) by the need to avoid freezing in the immature stages must

Fig. 6.19 A *Bombus polaris*
queen alongside Arctic willow.
(From photographs by B.
Heinrich.)

complete *several* life cycles (generations) to build up the colony in
just a single season. The buildup of the colony within a single
season is a race against time, and the outcome of the race depends
on temperature. While they are in the nest, the eggs, larvae, and
pupae are maintained at a nearly tropical climate (near 30°C),
even when prevailing outside temperatures are commonly below
5°C. However, the eggs mature not only in the nest. They also
mature within the body of the queen, who emerges from hibernation with undeveloped ovaries. If before a nest is built the
queen's abdomen remains close to air temperatures of near 5–
10°C, then weeks or possibly the whole season could presumably
elapse before the ovaries develop and before the eggs grow to
maturity in the ovaries. The bees must produce several generations
within only 4–6 weeks, however, so temperature elevation is
critical. An elevated temperature is critical in the thorax for foraging, in the abdomen for egg production, and in the nest for
brood rearing (and possibly for initiation of foraging flights at
temperatures below the minimum for shivering warm-up).

Abdominal-temperature regulation to speed up egg maturation
may be a more general phenomenon. For example, thermopre-
ferences of pregnant female houseflies, *Musca domestica*, are correlated with the optimal developmental temperatures of the young

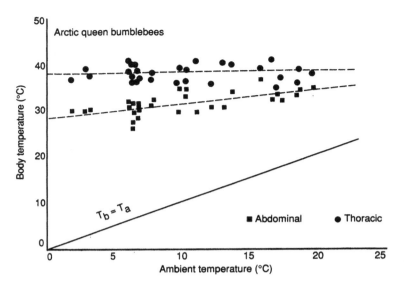

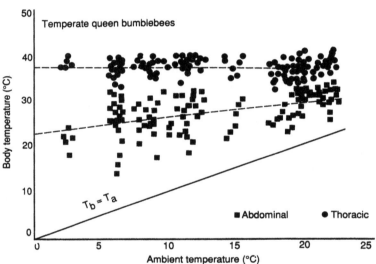

Fig. 6.20 Thoracic and abdominal temperatures of High Arctic and temperate queen bumblebees foraging in the field as a function of air temperature. Analysis of a linear model reveals that Arctic queens forage at significantly higher abdominal temperature than temperate queens ($p < 0.001$). (From Heinrich and Vogt, 1992.)

larvae that hatch, develop, and are carried within the ovaries (E. Thompsen, 1937, cited in Heran, 1952).

Circulatory Anatomy

As in other insects, the main features of the circulatory system in bumblebees are the dorsal vessel or heart (H) and a ventral diaphragm (VD). The heart continues into the thorax as the aorta

THE HOT-BLOODED INSECTS

after passing through a narrow "waist," the petiole. The heart generally pumps blood anteriorly, while the ventral diaphragm propels it posteriorly by wave-like undulatory movements (Fig. 6.21).

As shown in *B. vosnesenskii* and *B. edwardsii* queens (Heinrich, 1976a), the heart is attached closely to the dorsal surface of the abdominal wall. It bends down sharply along the air sacs at the anterior end of the abdomen and then forms a ventral loop that is loosely attached to the same air sacs as it droops down loosely

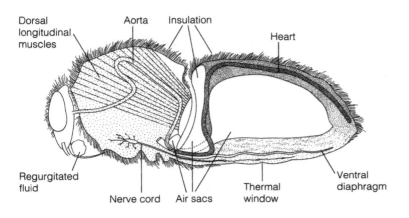

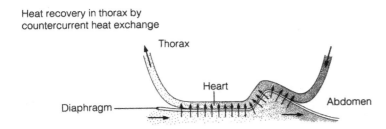

Fig. 6.21 Sagittal section of a bumblebee illustrating the features that act to retain or transfer heat from the thorax. The thorax and dorsum of the abdomen are heavily insulated with pile, but the ventrum of the abdomen is relatively free of pile and acts as a "thermal window." The narrow petiole between thorax and abdomen and the air sacs at the anterior portion of the abdomen act to retard heat flow to the abdomen. Cool blood *(dark grey)* is pumped anteriorly from the heart to the thorax, where it is warmed *(stippling becomes lighter)* in the aorta when it passes between the right and the left dorsal longitudinal muscles. Undulations of the ventral diaphragm propel warm blood posteriorly. (From Heinrich, 1976.)

near the petiole onto the VD. After traversing the narrow petiole and entering the thorax, the heart vessel (now called the aorta) curves dorsally, passing between the right and left dorsal longitudinal muscles, where it makes a sharp loop before entering the head.

The VD originates in the petiole. It is a thin transparent sheet of primarily transverse muscle that overlies the ventral surface of the abdomen above a blood-filled space enclosing the ventral nerve cord. In the vicinity of the petiole, the VD forms a valve which when in the "up" position can release blood into the abdomen, and when it is "down" it closes the aperture through which blood passes into the abdomen (see Fig. 6.21). The VD can thus release blood into the abdomen in pulses.

In the petiole area, cool blood entering the thorax is separated from warm blood entering the abdomen by only several cell layers. The close physical association between the heart loop and the ventral diaphragm in the petiole and anterior portion of the abdomen conforms to a counter-current heat exchanger. Heat in the blood flowing out of the thorax should follow the large temperature gradient (potentially over 30°C), and some of this heat could be returned to the thorax.

Relatively little heat reaches the abdomen during shivering preflight warm-up, when there are then often long periods without a heart beat (Heinrich, 1976). Circulation cannot be cut off indefinitely, however, because the working muscles must be supplied with sugar from the honey-stomach in the abdomen when the glycogen reserves of the thoracic muscles become exhausted. When blood flow becomes essential to supply this sugar, perhaps the counter-current mechanism can then retard heat loss.

HEAT TRANSFER BY THE CIRCULATORY SYSTEM

Despite the counter-current anatomy as described above, bumblebees nevertheless (as previously indicated) heat up their abdomen. Experiments clearly implicate blood as the vehicle for the transfer of this heat from its source, the thoracic muscles. For example, when heat is focused exclusively onto the thorax in tethered bees, T_{abd} may increase sharply after T_{thx} has stabilized near 42–44°C (Fig. 6.22). However, in both dead bees and in live bees with the heart in the abdomen tied off (with human hair, using a surgeon's eye needle) and made inoperative, the same heat input causes T_{thx} of live bees to soar to lethal temperatures without causing an appreciable increase in T_{abd}. The elaborate anatomical arrange-

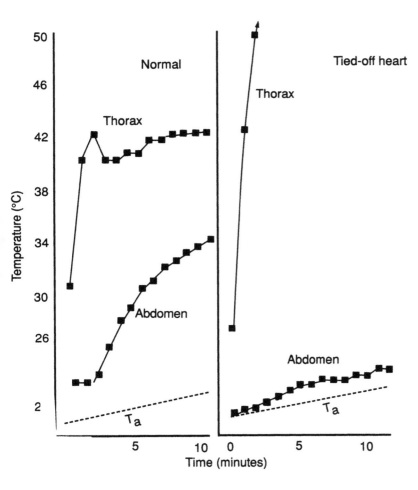

Fig. 6.22 Thoracic and abdominal temperature during two heating experiments of a tethered bumblebee. Heat was applied only to the thorax. The normal bee *(left)* prevented T_{thx} from exceeding 42 °C by shunting excess heat to the abdomen. When the heart was made inoperative *(right)*, the same input of heat to the thorax killed the bee because it no longer dumped excess heat into the abdomen. (From Heinrich, 1976a.)

ments that should act to retain heat in the thorax (Fig. 6.21), and the simultaneous demonstration that the abdomen can nevertheless greatly increase in temperature despite them, suggest that a remarkable physiology exists that can be activated to circumvent the anatomical heat-retention arrangements. This physiology is used when bees stabilize T_{thx} during continuous flight by shunting excess heat into the abdomen. The same mechanism is presumably also used to heat the abdomen when incubating brood. How, then, does the flow of blood during flight at low air temperatures result in relatively small increases of T_{abd}, whereas blood flow between thorax and abdomen in incubating bees at all temperatures results in high T_{abd}? In other words, is there a mechanism

Diaphragm beats

Temperature pulses in ventral abdomen

Seconds

Fig. 6.23 The concurrently recorded ventral-diaphragm beats and associated micro-changes in T_{abd} in a bumble-bee with elevated T_{thx}. Each beat is followed by a minute (<0.1 °C) but sharp temperature pulse in the ventrum of the abdomen as a pulse of warm blood enters. Successive diaphragm beats result in a step-wise increase of T_{abd}. (From Heinrich, 1976a).

of transferring heat, by way of the blood, that is independent of or in addition to volume regulation of blood flow?

I here present a model and data supporting an *alternating-*current flow, which would partially neutralize the *counter-*current heat exchange and which could account for heat transfer to the abdomen. According to this model, the anterior portion of the ventral diaphragm operates like an alternating switch. During the abdominal-expansion phase of the breathing cycle, the ventral diaphragm is in the up position and warm blood then rushes into the ventrum of the abdomen (Fig. 6.23) from the thorax while air simultaneously enters the open spiracles into the abdominal air sacs. During the abdominal-contraction phase, the abdominal spiracles close and the air from the abdominal air sacs is pushed into the thorax. The ventral diaphragm is down at that instant in the petiole area, shutting off blood flow to the abdomen but leaving space above it for the heart to expand and deliver a bolus of cool blood into the thorax. In this way the warm and cool blood are *temporally* separated within the anatomical arrangement of the counter-current heat exchanger, because they pass through it alternately, at different times. (The system is, however, not "perfect": some heat exchange is inevitable because of the heat storage of the tissues.)

THE HOT-BLOODED INSECTS

If the bee tries to conserve heat in the thorax it must minimize T_{abd} increase. Aside from shutting off blood flow entirely, this reduction of heat exchange to the abdomen can be accomplished by maximizing counter-current heat exchange, and the latter requires that the blood *flows* simultaneously in both directions rather than being forced through in alternate pulses. In other words, chopping the blood flow into discrete pulses should help to transfer heat, while a slow continuous stream should facilitate counter-current heat exchange to aid in the retention of heat in the thorax.

This model was examined by simultaneously recording the abdominal pumping (ventilation), heart activity, ventral-diaphragm activity, and the temperatures in thorax and abdomen (Heinrich, 1976a). The interrelationships of all of these variables showed striking differences between stressfully heated (on thorax only) vs. non-overheated bees. In resting bees or bees not heated to over 40°C, the mechanical activity of the ventral diaphragm is very variable. Heart beats are irregular as well, but they are usually very rapid and feeble, showing fibrillations but usually no distinct beats (Fig. 6.24). Heart, breathing, and ventral-diaphragm beats are all independent of one another and at different frequencies. Since the heart pulsations are very rapid and shallow, there is likely a continuous stream of blood, rather than flow by discrete pulses.

Large-amplitude pulsations of the heart emerge only at very high T_{thx}, and the first significant heating of T_{abd} then begins. The beating of the ventral diaphragm then also converges with abdominal breathing movements (Fig. 6.24); there is thus one heart beat and one VD beat for each in-out breathing movement associated with heat transfer to the abdomen. The ventral-diaphragm activity in the petiole clearly acts as a switch damming and releasing warm blood into the ventrum of the abdomen with each beat, because when the thorax is heated, each beat of the ventral diaphragm is associated with a small step increase of T_{abd} (Fig. 6.23).

There is no direct proof that the pulses of warm blood *out* of the thorax do not coincide (rather than alternate) with pulses of cold blood coming *in*, as the model demands. The space limitations in the petiole, however, and the pressure changes associated with abdominal ventilatory movements combine to argue against this possibility. It is physically unreasonable to suppose that blood enters the abdomen at the same instant when the diaphragm is down (damming the passage for blood) and while the abdomen is expanding in volume, and that blood enters the thorax when

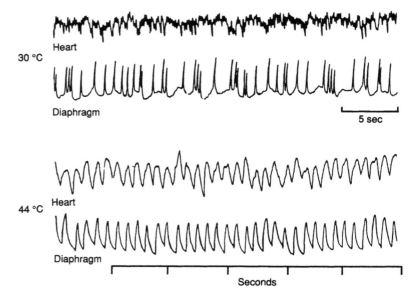

Fig. 6.24 Concurrently recorded mechanical activities of the heart and the ventral diaphragm at thoracic temperatures of 30 °C and 44 °C during heating of the thorax in a tethered bee. At low thoracic temperatures the heart beat is rapid but shallow while the diaphragm beats at irregular intervals. The heart-beat pattern results in a thin, continuous stream of blood entering the thorax, which allows counter-current heat exchange to occur. At high thoracic temperatures the pulsations of the heart are of large amplitude, and the frequencies and inter-pulse intervals of both the heart and the diaphragm become regular and identical with each other; blood then flows in pulses both in and out of the thorax, creating an alternating current of warm and cool blood. (From Heinrich, 1976a.)

there is a suction created in the *opposite* direction (as the abdomen expands during air intake). Spiracular openings relative to abdominal movements were not observed in bumblebees, but the pattern is presumably like that described for honeybees (Bailey, 1954) and as is presumed for the above model (Fig. 6.25).

Calculations suggest that the maximum observed heat exchange in a queen bumblebee could be accomplished with about 60 μl of blood per minute, or about 0.2 μl of blood per beat at 300 heart pulsations per minute (Heinrich, 1976a). Blood volume in a 220 mg bumblebee is about 27 μl, or about 12.3 percent of body weight (Surholt et al., 1988). Therefore, a 600 mg queen would have nearly 74 μl of blood, and with 0.2 μl/heart pulse and 300 pulses per minute, the bee would need to circulate most

(6/7) of its blood once per minute between thorax and abdomen to account for the maximum observed heat transfer.

EVAPORATIVE COOLING

Bumblebees heated to 44–45 °C sometimes struggle violently and regurgitate fluid from their honey-stomach or crop. In-out flexing of the proboscis keeps the regurgitated fluid droplet in motion. This behavior can also be observed in resting bees when they are gorged and concentrating their nectar. However, the behavior also

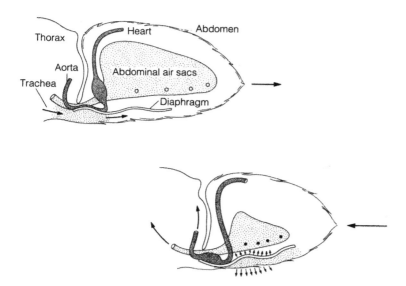

Fig. 6.25 A highly schematic diagram showing the probable sequence of heart and diaphragm pulsations relative to abdominal pumping movements, whereby the counter-current heat exchanger is reduced. The abdomen expands *(top)* and air is drawn into the abdominal sacs from outside the animal through the spiracles *(four small circles)*. At the same time, the diaphragm is raised and the suction created by the expanding abdomen also allows blood to enter from the thorax. When the abdomen contracts *(bottom)*, the abdominal air sacs deflate by forcing air into the thorax. At this time the diaphragm is lowered in the petiole, and in this position it simultaneously enlarges the air passage into the thorax while reducing the passage for blood out of the thorax. However, blood can enter the thorax by way of the heart. Small arrows indicate passage of heat from the warm blood (in the lightly stippled area below the diaphragm) that has entered the abdomen by bypassing the cool blood (in the dark grey heart) entering the thorax. In this manner, the bee is able to shunt heat to the abdomen to prevent overheating of the thorax or for incubating the brood. (From Heinrich, 1976a.)

functions in temperature regulation. None out of 20 *B. vosnesenskii* flown at 25–31°C showed the behavior either during or immediately after flight, but 8 out of 11 showed it after 3 minutes of continuous flight at 42°C. Furthermore, placing a droplet of fluid on the proboscis causes reductions of head temperature of nearly 2°C, but only a slight (about 0.4°C) decrease of T_{thx} (Heinrich, 1976a). The mechanism is thus apparently an emergency response to affect head temperature only. Although the crop-regurgitation response is normally a relatively minor one in thorax thermoregulation in bumblebees, it assumes major importance in honeybees (see Chapter 8).

Summary

Few other organisms on earth show such a high degree and sophistication of thermoregulation as bumblebees. These insects can regulate a T_{thx} of above 35°C at air temperatures down to near 0°C while remaining stationary, while foraging from flowers, while in continuous flight, and while incubating brood. Head, thorax, and abdominal temperatures can be regulated independently of each other. But the degree and occurrence of thermoregulation is adaptively varied with respect to body mass and energy supplies as well as on ecology. The pile covering the bees assumes major importance in thermoregulation, and without it bumblebees would be unable to forage even at modest ambient temperatures.

Large bumblebees can warm up from temperatures near 7°C or less and fly with a T_{thx} near 35°C. They regulate T_{thx} over relatively wide ranges of ambient temperature. Small individuals can warm up and fly only at air temperatures over 10°C. They are poor thermoregulators in flight but they can fly at ambient temperatures higher than 30°C, when large bumblebees suffer heat prostration. Thermoregulation in flight is accomplished by shunting excess heat out of the thorax and into the abdomen.

Foraging bees stop at flowers at frequent intervals and at low air temperatures they generate heat to offset cooling, provided the energetic costs of thermoregulation and foraging do not exceed the potential energy gains. Foraging bees are better able to regulate T_{thx} than those in continuous flight. Although heat production is regulated to stabilize T_{thx} when the bee is foraging on high-reward flowers, the T_{thx} is allowed to decline while it is foraging on low-reward massed flowers, where it can forage by walking rather than flying for short intervals.

Queens, workers, and drones incubate their brood in the nest via the abdomen, and Arctic queens incubate their own ovaries in the abdomen as well. The brood is heated primarily via the abdomen, even though the heat is produced in the thorax by the power-producing flight muscles. The transfer of heat from the thorax to the abdomen is accomplished by blood circulation. Normally a *counter*-current heat exchanger in the ventrum of the abdomen greatly reduces heat leakage from the thorax. But during heat transfer this heat exchanger is partially obliterated by a unique *alternating*-current mechanism that shunts hot and cold blood through this structure in alternate pulses.

During warm-up (for preparing for flight, for regulating T_{thx} while foraging, and for incubating brood), the flight muscles are activated by the central nervous system as in flight, but the temporal pattern of this activation of antagonistic muscles is more synchronous. During warm-up there are no mechanical movements of the wings or thorax, but experiments with isolated muscles as well as *in vivo* measurements show that the flight muscles engage in tetanic contractions. Heat production in the absence of mechanical contraction of the flight muscles has never been observed, the numerous claims for "non-shivering thermogenesis" to the contrary. Several neurophysiological mechanisms have been identified that allow bees to shiver without showing mechanical vibrations.

The ability of bumblebees to be active at low temperatures resides in the capacity to activate the flight muscles at low temperature, the presence of insulating pile, the use of physiological heat exchangers to retard heat loss, and efficient foraging methods that compensate for the energetic costs of thermoregulation, which are very high because of the bees' small size.

Remaining Problems

1. How are the different body temperatures in bumblebees "set" for different kinds of activity?
2. To what extent are heating and cooling of the abdomen in active bees related to the control of volumes of blood flow or the control of a counter-current heat exchange?
3. When heat flow to the abdomen is modulated, does the ventral diaphragm admit different volumes of blood per beat into the abdomen at different temperatures?
4. Do bumblebees regulate head temperature?
5. Do bumblebees use the head to get rid of excess heat?

6. Is blood flow to the abdomen shut off during pre-flight warm-up?
7. Is bumblebee fur uniquely designed for thermoregulation?
8. Given that the smallest bees of some species have the same mass-specific rates of heat production but 4 times greater rates of heat loss, do they "know" the lower ambient temperatures they can warm up from? If so, how?
9. What is the function of the very high activities of the enzymes phosphofructokinase and fructose-1,6-diphosphatase that have frequently been observed in bumblebees?

Tropical Bees

B E E S are well known for their endothermic heat generation, and they are subjects of numerous comparative studies. For example, one study of 55 species of bees from 6 families showed that both body mass and thermal environment are important factors in endothermy (Stone and Willmer, 1989). Smaller species generally have a relatively greater rate of heat production (and lower T_{thx}) per unit mass than larger bees, but rates of heat production are relatively constant per unit mass within any one species, regardless of body mass. This general relation is often blurred by the effects of the thermal regime in the field, however. In general, those species that encounter lower minimum air temperatures have a greater capacity for endothermy at the minimum temperatures to which their thermogenic system is adapted (Stone and Willmer, 1989).

Euglossine Bees

Bumblebees occur even above the Arctic circle, but the majority of the world's bees live in the tropics, often in the equatorial lowlands, and these insects are adapted to high temperatures. One of the major and conspicuous groups of lowland tropical bees are undoubtedly the euglossine (or orchid) bees. Euglossine bees, a diverse group, are well known for their importance as pollinators in the neotropics (Janzen, 1971; Dressler, 1968). They are highly constrained by temperature when they fly, and they generally avoid air temperatures below 25°C and above 31°C (Armbruster and McCormick, 1990). The euglossines range in mass from 60 mg to 1.1 g (Casey, May and Morgan, 1985). Some of the larger species, such as *Eulaema*, are densely pubescent and resemble bumblebees. Others are glabrous and colored in metallic blues and greens (for example, *Exaecrete* and *Euglossa*). Most of the smaller species, such as *Euglossa*, lack pubescence. Males are unique in that they collect non-nutritive scents from certain orchid flowers, and they presumably use these scents in some way as sex perfumes (Kimsey, 1980). Males can be conveniently collected in large numbers by attracting them to scents, including cineole, skatole, and methyl salicylate (Casey, May, and Morgan, 1985).

A great deal is known about the flight energetics of euglossine bees in relation to morphology and wing-stroke parameters (Casey, May, and Morgan, 1985), but under natural conditions these bees are not subjected to temperature extremes or to a great range of thermal conditions and their thermal relationships are not well studied. On Barro Colorado Island in Panama, thermal conditions (for an insect) are very benign, with temperatures fluctuating only from 20 to 30°C, and over this temperature range the small *Euglossa* bees (typically < 100 mg and non-pubescent) maintain thorax, abdomen, and head temperatures almost parallel with air temperature (Fig. 7.1), with T_{thx} reaching 40°C at near 30°C. Pubescent euglossines maintained T_{thx} primarily above 35°C; similar and slightly higher T_{thx} have been observed previously (Inouye, 1975). Regression equations of T_b (May and Casey, 1983) on air temperature in both the glabrous and pubescent bees reveal a slight deviation from absolute thermal conformity. It was concluded that "regulation of T_{thx} is shown by the fact that the slope of air temperature is significantly less than 1.0." However, since the slope is closer to 1.0 than it is to 0.0 in both groups (but more so in the glabrous *Euglossa*), it seems to me that thermal *conformity* is thus more apparent than thermoregulation. Thermal conformity reflecting little or no thermoregulation would be expected in small insects under environmental temperatures where there is little danger of overheating.

The larger, pubescent euglossine bees, in contrast, fly with considerably higher thoracic-temperature excess, and when flying at air temperatures near 27°C it appears that the bees try to keep T_{thx} from rising much above 40°C (May and Casey, 1983). So far these data over this narrow temperature range allow us to say very little more about possible thermoregulation. The T_b are as expected, and the physiological limits of these bees and how they deal with them remain to be explored. It is intriguing, however, that the smaller bees, the *Euglossa*, have 5–6 loops of the aorta

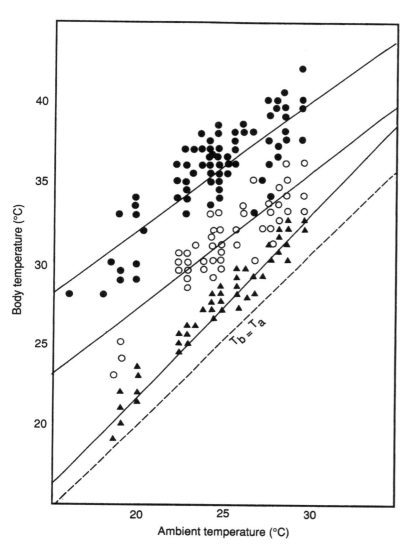

Fig. 7.1 Thoracic *(filled circles)*, head *(open circles)*, and abdominal temperatures *(triangles)* of small *Euglossa* bees of numerous species from Panama. (From May and Casey, 1983.)

(Wille, 1958), an arrangement that apparently acts as a countercurrent heat exchanger for heat retention in honeybees (Heinrich, 1980).

Although most of the tropical bees live under relatively benign thermal conditions, the larger species still need to warm up by shivering prior to flight (Stone and Willmer, 1989). As other large insects they inevitably heat in flight, and the high T_{thx} normally generated must than dictate the evolution of biochemical adaptations that then necessitate warm-up. Nevertheless, air temperature varies little, and these bees are not usually subject to great thermal stresses.

Desert Bees

In contrast, many other bees live in hot deserts, where they face not only high air temperatures but also intense solar radiation. In large species T_{thx} could easily exceed 50 °C in flight, barring active counter-measures. How do large desert bees cope with these potential thermal stresses? What counter-measures to overheating do they possess?

Body mass obviously makes a big difference in keeping a thermal balance, as does the diurnal and seasonal time of activity. But plants constrain bees' activity times. The bees are active when their host flowers are in bloom, but in a co-evolved system it is likely that the blooming times of a habitat shift to accommodate the permissible flight times of the pollinators (Heinrich, 1976). Therefore, over evolutionary time, the bees' activity likely determines the flowering time (Heinrich and Raven, 1972), whereas moisture availability probably determines vegetative growth. But in deserts, where there is less opportunity for continuous vegetative growth in all but those species with very deep root systems, flowering times must be close to times of vegetative growth—namely, in summer after rains.

As might be expected because of potential heat stress in summer, the most numerous bees of deserts are small enough to be only marginally endothermic and to cool rapidly by convection. These include *Perdida* species and small megachilids, panurgids, halictids, anthophorids, colletids, melittids, andrenids, and others. All of these weak-flying insects are active in the hot portions of the day (Linsley, 1958, 1960a,b). In many cases small size also permits them to crawl into the flowers and work in the shade.

Many of the larger desert bees that could overheat at midday are active instead at dawn or even before daylight (Linsley, 1960a,b). In the desert of California, Arizona, and New Mexico these include *Colletes stephani*, *Martinapis luteicornis*, various species of *Centris*, *Anthophora*, and *Tetralonia*, and the giant colletids *Caupolicana yarrowi* and *Ptiloglossa* species (Linsley, 1958). These bees may be active again late into the evening (MacSwain, 1957). Similar temporal shifts of activity of solitary bees of differently sizes are also observed in the Negev Desert of Israel (Avi Shmida,

Fig. 7.2 A hovering male anthrophorid bee, *Centris pallida*.

personal communication). Concentrated work on a few select species of endothermic desert bees highlights the required behavioral and physiological adjustments.

CENTRIS PALLIDA

This anthophorid bee species (Fig. 7.2) of the southwestern United States takes the brunt of the heat the desert has to offer. And to do so it must operate with a thoracic temperature (Fig. 7.3) close to the lethal limit. It regulates T_{thx} quite well (Chappell, 1984), and always above 40°C, but males and females differ in their behavior and in their T_{thx}.

Females spend most of their time foraging: flying from flower to flower, landing, and flying to the next. Males, in contrast, not only forage but also hover for long durations searching for females at the food sources. Aggregations of hundreds of males may gather around both at good nesting areas and at food sources, such as palo verde bushes. Chases and aggressive encounters are almost

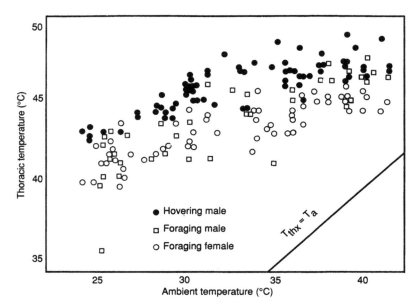

Fig. 7.3 Thoracic temperatures of *Centris pallida* in the field as a function of ambient temperature. (From Chappell, 1984.)

continuous among males at such sites, even in the hottest part of the day. The patrolling, chasing, and particularly the hovering at the hottest part of the day should be thermally stressful, because the bees' flight metabolism is high and is not (and probably cannot be) reduced to lessen the heat load at high ambient thermal conditions. Indeed, at high air temperatures males maintain T_{thx} of 46–47 °C for hours; sometimes T_{thx} even reach 48–49 °C (Fig. 7.3), which is close to their already unusually high lethal tolerance of 50–51 °C. Foraging females, despite their larger size (197 vs. 134 mg), have T_{thx} about 5 °C lower than hovering males.

The bees' T_{thx} excess of about 17 °C at 25 °C is not unusual for their size. But it declines to only 5 °C at 40 °C, which indicates impressive thermoregulation. The mechanism of thoracic-temperature stabilization isn't clear, but the constant abdominal-temperature excess of 4.5 °C at all air temperatures (Fig. 7.4) gives no evidence of physiologically facilitated heat loss to the abdomen (Chappell, 1984). It is difficult to account for the temperature regulation, since neither heat production nor heat loss seems to be physiologically controlled. The key could reside in behavior.

Although no quantitative data are published, Chappell (1984) reports that in the morning when air temperatures are low (near 25 °C) "bouts of hovering are of noticeable longer duration than

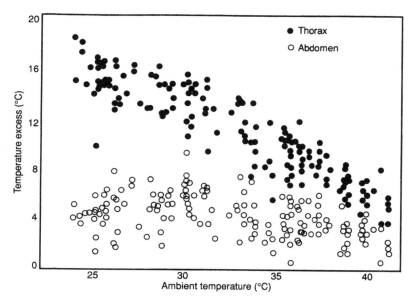

Fig. 7.4 Temperature excess in the thorax and abdomen as a function of ambient temperature in *Centris pallida*. (From Chappell, 1984.)

during the hotter portions of the day." By foraging instead of hovering, bees intersperse frequent stops (at flowers). These stops must result in thoracic cooling. Chappell (1984) also saw the bees shifting activity from the open, sunny areas to more shaded areas within the foliage during foraging. Both behaviors would help to stabilize T_{thx}, but no laboratory data are so far available to eliminate other alternatives.

Carpenter Bees

Carpenter bees are primarily tropical bees and are often found in deserts. Many weigh up to 10 times more than *Centris* bees, which are already large enough to be operating close to the lethal temperature limit. Carpenter bees are probably the largest bees known, some species weighing 2 g, or about 4 times more than Arctic bumblebee queens. Carpenter bees have high wing loading, and thus they cannot reduce their power (and heat) output in flight. At all air temperatures their metabolic rate in flight remains high and independent of air temperature, being nearly identical to that of the hottest insect endotherms known, the bumblebees and the sphinx moth, and to that of *Centris pallida* (Nicolson and Louw, 1982; Chappell, 1982).

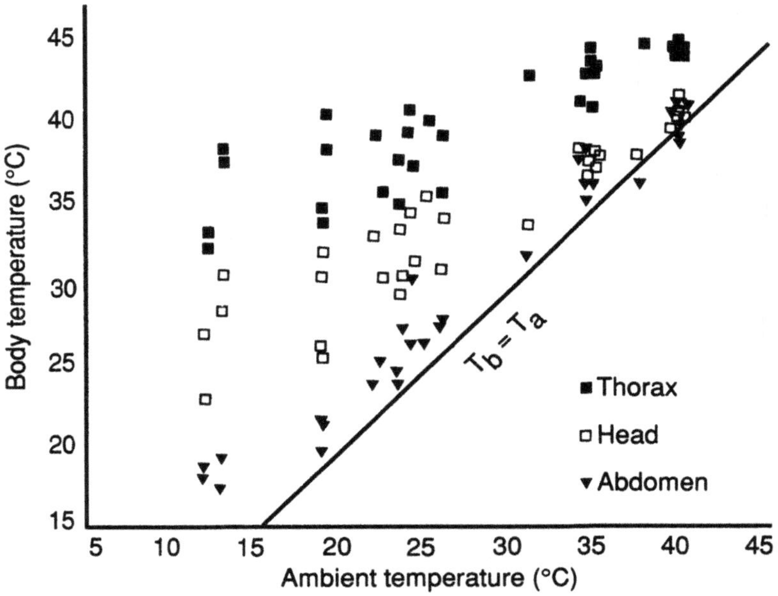

Fig. 7.5 Thoracic, head, and abdominal temperatures during continuous free flight as a function of ambient temperature in the carpenter bee *Xylocopa varipuncta*. (From Heinrich and Buchmann, 1986.)

Despite the apparent thermal disadvantage resulting from both very large size and very high rates of heat production, carpenter bees are common in hot deserts both in the New and the Old World. Considering extrapolations from data from other bees, one might expect them to be active in the cooler parts of the year, and at night. Amazingly, however, they fly in the hottest summers on the hottest days, at air temperatures of at least 40°C, and in bright sunshine at noon (Nicolson and Louw, 1982; Louw and Nicolson, 1983; Chappell, 1982; Baird, 1986; Willmer, 1988).

Carpenter bees regulate T_{thx} fairly well. In continuous free flight at 12 to 40°C *Xylocopa varipuncta* maintains T_{thx} from 33.0 to 46.5°C (Fig. 7.5). The T_{thx} of *Xylocopa* in the field are generally in the high thirties (Nicolson and Louw, 1982), even at 16°C (Willmer, 1988), but they can reach up to 48.5°C at air temperatures above 40°C (Chappell, 1982; Baird, 1986).

Like *Centris pallida*, carpenter bees are able to withstand exposure to T_{thx} even higher than 48.5°C. Bees heated for 2–5 minutes to a T_{thx} of 52°C were able to fly normally after T_{thx} had again cooled to 42–45°C (Chappell, 1982). For the most part, however, the T_{thx} of *Xylocopa* during free flight are "only" between 38 and 45°C, similar to the T_{thx} of large moths and large beetles. How these extremely large and energetic bees normally prevent themselves from overheating in flight is still an intriguing biological puzzle, although insights into the process are now available.

One potential heat-loss mechanism at very high temperatures is evaporative cooling. However, neither Chappell (1982) nor Heinrich and Buchmann (1986) observed any extrusion of stomach contents in overheated bees. Susan W. Nicolson and Gideon N. Louw (1982) from the University of Cape Town, South Africa, were able to simultaneously measure water loss, oxygen consumption, and thoracic temperature during flight in *X. capitata*. As expected, the bees showed high rates of water loss in flight, but this loss was closely parallel with the very high metabolic rate (which was temperature-independent), and it was significantly correlated with air temperature and water-vapor deficit. Water was therefore lost by evaporation in amounts as predicted (Heinrich, 1975) for *passive* loss on the basis of the ventilatory rate and assuming that the excurrent air is fully saturated. Thus, although water loss undoubtedly cools the bees, this cooling is not physiologically varied; it cannot account for temperature regulation. Furthermore, since it (being passive) is similar to that of other flying insects, it does not account for the ability of *Xylocopa* to fly at high air temperatures.

Another obvious possibility for accelerating heat loss is, to shunt the excess heat from the thorax to the abdomen, as several other large endothermic insects do. The abdomen of *Xylocopa* are glabrous, large, and flattened, and thus ideally constructed as heat radiators. Tethered *X. californica* (Chappell, 1982) and *X. varipuncta* (Fig. 7.6) physiologically transfer large amounts of heat into the abdomen; dead control animals show negligible abdominal heating but head temperature still closely tracks T_{thx}. In both species, live animals showed impressive abdominal-temperature excesses of about 80 percent of thoracic-temperature excess. During flight, on the other hand, T_{abd} was (in *X. varipuncta*) either almost identical or close to air temperature at all air temperatures (Heinrich and Buchmann, 1976), or considerably elevated (5–10 °C) above air temperature (in *X. virginica*) with the greater temperature excess occurring at *low* rather than at high air temperatures (Baird, 1986). Except possibly for Chappell's (1982) suggestive results from *X. californica* at the Deep Canyon Desert Research Station showing higher abdominal-temperature excess at high rather than at low air temperatures, these data by themselves offer no proof of facilitated heat transfer to the abdomen in flight as the mechanism for achieving independence of T_{thx} from

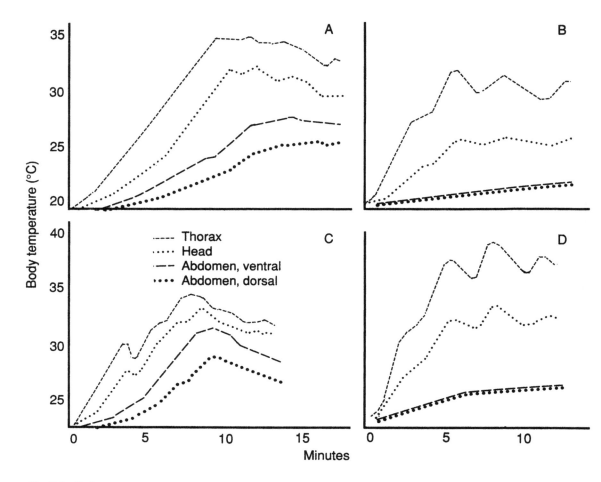

Fig. 7.6 Body temperatures of tethered *X. varipuncta* during pre-flight warm-up at ambient temperature of 19 °C *(A)* and at 23 °C *(C)* compared with T_b in the animals when killed and with the thorax artificially heated (*B* and *D*). Note that head temperature closely tracks T_{thx} in both live animals and dead controls. (From Heinrich and Buchmann, 1986.)

air temperature. The main problem with the T_{abd} data in "flight" in these three studies is that they are not strictly comparable. They refer to field data. We do not know how long or how fast the bees had been flying, so we cannot evaluate the effect of convective cooling. Taken together, however, they do suggest that physiologically facilitated heat transfer to the abdomen is not the primary or the only mechanism.

The most promising possibility so far is that these bees accelerate convection in flight at high temperatures, so that large temperature gradients in the abdomen are not maintained despite active heat transfer to the abdomen. For example, *X. varipuncta* flies much faster at high than at low air temperatures. Because the conductance of the glabrous abdomen (Chappell, 1982; Heinrich

THE HOT-BLOODED INSECTS

and Buchmann, 1986) is greatly increased with a small speedup in air flow over it, the heat loss from the abdomen would have been accelerated even as the same temperature gradient was maintained.

Additional large quantities of heat are lost by convection from the head (Heinrich and Buchmann, 1986). Unlike most solitary bees, which are ground nesters, carpenter bees are notable for chewing nest galleries into solid wood (Gerling, Hurd, and Herfetz, 1983), and for this they require massive mandibles, associated muscles, and heads to mount them in. The large heads (Fig. 7.7), in turn, are pre-adapted for ease of convective heat dissipation.

The head is greatly flattened and fits with a solid contact like a cap onto the front of the thorax. During forward flight, the front of the head faces the wind stream and convective heat loss is inevitable, provided the head heats up above air temperature (Fig. 7.6). Since any temperature increase of the thorax is immediately reflected in a rise in head temperature as heat passively follows the temperature gradient, T_{thx} is ultimately "controlled" by head temperature.

Heat loss from the head is also about 2–3 times more sensitive to variations in wind speed than heat loss from the isolated thorax and abdomen (Fig. 7.8). For example, the cooling constants (conductance) of the heads of *X. varipuncta* increase from 0.5 to 3.0°

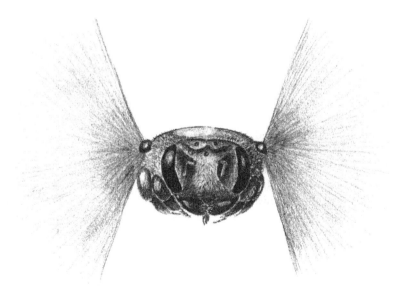

Fig. 7.7 A carpenter bee in flight. The thorax is naked or thinly insulated, and the very large, naked head facing the airstream during forward flight provides a large area for rapid convective heat loss.

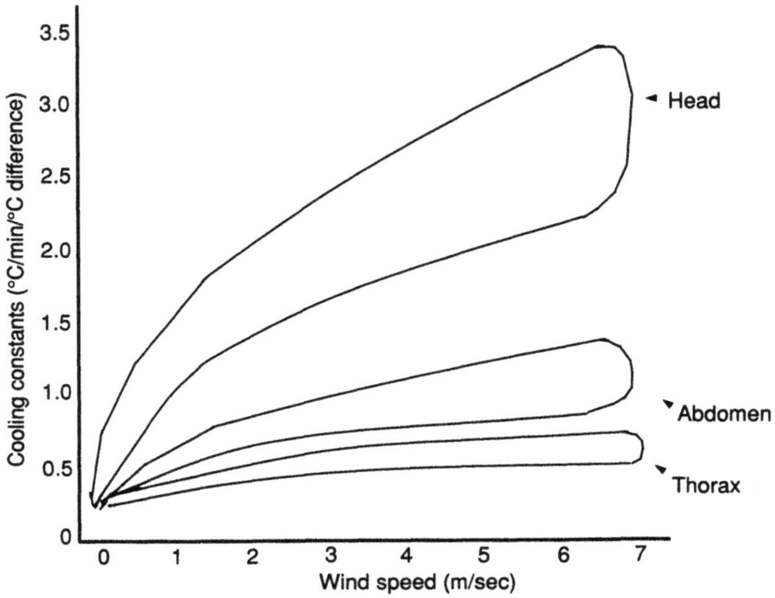

Fig. 7.8 Cooling constants for the head, thorax, and abdomen of dead *X. varipuncta* as a function of wind speed. The body parts were severed from each other but held in position near each other (but without touching) by a slender pile of wax. (From Heinrich and Buchmann, 1986.)

C/min/°C difference of T_{hd} from T_a, at wind speeds from 0 to 6 m/s, whereas conductance of the abdomen only increases from 0.5 to 1.0°C/min°C body-temperature difference over the same range of wind speed (Heinrich and Buchmann, 1986). Since carpenter bees regularly fly at speeds up to 12 m/s (Chappell, 1982) and since metabolic rate and heat production are *independent* of flight speed from at least 0 (hovering) to 4 m/s in bees (Ellington, Machin, and Casey, 1990), it is obvious that carpenter bees can cool themselves merely by increasing their flight speed. The extent of this mechanism has so far not been documented in the field, but we know it exists; carpenter bees in a temperature-controlled room dramatically increase their flight speed at higher air temperatures (Heinrich and Buchmann, 1986).

Xylocopa therefore have two major windows for "regulated" convective heat loss: the head and the abdomen. (The thorax is glabrous or it has a light layer of fuzz.) There is predominantly passive heat flow from the thorax to the head, because of the close physical coupling between these two body parts. In addition, there is physiologically facilitated heat transfer both to the head and to the abdomen, and this transfer is via the circulatory system. Increased heating of the thorax results in increased amplitude and frequency of the abdominal heart beat (Heinrich and Buchmann, 1986) and large increases of abdominal temperature. As in bumblebees (Heinrich, 1976), abdominal pumping ("breathing") movements in *Xylocopa* are synchronous with the heart pulsations when heat transfer to the abdomen is maximized. Curiously, the aorta has seven loops in the petiole (Wille, 1958; and Fig. 7.9), which could act as a physiological counter-current heat exchanger. (Do the bees stretch these out during the abdominal breathing movements when the abdomen is elongated with each "breath," thereby eliminating counter-current heat exchange?)

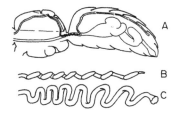

Fig. 7.9 Circulatory anatomy of a *X. varipuncta (A)*, with enlarged lateral *(B)* and dorsal *(C)* views of the aorta in the petiole. (From Heinrich and Buchmann, 1986).

The implications of thermal balance in foraging, given the convective cooling mechanism, are discussed by Chappell (1982). *Xylocopa* foraging behavior can be divided into three major activities: fast forward flight, hovering, and perching or walking on flowers. Hovering in sunlight would result in rapid heat buildup, and in full sunlight above 35°C sustained flight should be possible only if it occurs at high speeds; hovering should be limited to short bursts during which T_{thx} should increase. Chappell (1982) calculates that at about 38°C a carpenter bee can maintain a T_{thx} of 45.5°C at flight speeds of 5–10 m/s, but if it begins to hover T_{thx} will exceed 48°C within 95 seconds, and if the bee lands T_{thx} will fall to 43°C in 3.3 minutes (at wind speeds of 1 m/s). As air temperature increases, bees will cool more slowly when landing. At very high air temperatures, bouts of hovering must be short, to be followed by very long bouts of immobility. Perhaps this helps to explain why *Xylocopa* can feed on desert agaves *(Agave deserti)* at very high temperatures, when foraging from palo verde *(Cercidium floridum)* becomes difficult. On the agave inflorescences the bees can easily visit dozens of individual flowers without flying, which means that intervals between flights may last several minutes and cooling can occur. While foraging from palo verde, on the other hand, they need to fly every 10–15 seconds (Chappell, 1982), and potential heat stress could occur. Flower choice relating to thermoregulation is also a prime concern in bumblebees (Heinrich, 1972).

Some carpenter bee species are almost totally glabrous whereas others have a short layer of dense, yellow fuzz on the thorax, and in still others the males are fuzzy and the females are glabrous. So far the meaning of these differences is not clear, but it is possible that it has at least some thermal significance, since carpenter bees fly not only at noon in the hottest deserts. In the Negev Desert of southern Israel, a glabrous bee *Xylocopa sulcatipes*, for example, exists alongside *X. pubescens*, which has slightly greater mass and yellow thoracic pubescence. As predicted, the *X. pubescens* begin to forage earlier in the morning while it is still cool, and they generate T_{thx} at least 1 °C higher than *X. sulcatipes* in flight (Willmer, 1988). Similarly, the high-temperature *Xylocopa californica* from the southwestern United States is almost completely glabrous, while the *X. virginica*, from much cooler environments in the eastern United States, has thoracic pile.

Summary

Except for possibly some of the large euglossines, bees in the moist lowland tropics live in a relatively benign temperature environment and do not face overheating. They only need to warm up prior to flight. Desert bees may face much higher air temperatures and intense solar radiation as well. Some of the larger species avoid the potential high-temperature stresses through seasonal or diurnal shifts of activity time. But there are exceptions. Some *Centris* and *Xylocopa* bees are very active flyers in the hottest part of the season on hot days. Both have very high thermal tolerances, flying with T_{thx} of up to 47–48 °C. Both groups regulate T_{thx} in flight by preventing further increases in T_{thx} near their lethal limit. The mechanisms of thermoregulation in these two groups are not fully elucidated, but a large part of the regulation involves variations of flight behavior to affect convective cooling. The male *Centris* bees at high air temperatures hover less and switch from patrolling to thermally less stressful activities such as foraging. Heat loss in *Xylocopa* is very rapid, especially from the head, and very sensitive to changes in wind speed. Since heat flow from the thorax to the head is also rapid and since heat production is independent of flight speed (and air temperature), these bees need only fly faster at high temperatures to cause head (and hence thoracic) cooling. Additionally, *Xylocopa* also shunt heat to the abdomen. The abdomen is uninsulated and very sensitive to wind speed in convective cooling.

Reliance on forced convection in *Xylocopa* is possible because they lack insulation and their heads are large and touch the thorax. Large head size for thermoregulation is a preadaptation from a large chewing apparatus used in boring nesting cavities in wood.

Remaining Problems

1. Data are needed on the flight durations and flight speeds of individual large, desert-adapted bees *(Centris, Xylocopa)* to determine the extent of thermoregulation occurring through variations of flight behavior.
2. The circulatory anatomy of no large, hot-weather bee has yet been examined with respect to possible thermoregulatory function.
3. *Xylocopa* transfer large amounts of heat into the abdomen despite a coiled aorta in the petiole. How is this potential countercurrent heat exchanger circumvented?

EIGHT

Hot-Headed Honeybees

T H E honeybee, *Apis mellifera*, is a highly atypical flying insect. It is the only one that has refuge year-round, winter and summer, in a warm environment stocked with food. Honeybees can always seek the warmth of their companions in the nest—or, to put it another way, they are unavoidably subjected to heating by them.

Honeybee biology is probably also shaped by the constant access the individuals have to food because of the social nature of the species. The degree of socialization in honeybees (and its effects on thermoregulation and its intimate associate, energetics) is readily apparent from a simple experiment. If you confine a fly, a moth, a beetle, or any other of a vast array of insects in a jar, it will (unless you disturb it) soon become torpid and remain in torpor for many days and sometimes weeks. Honeybees are different. They normally have energy supplies nearly constantly within reach, and so they do relatively little to conserve them. As if there were no tomorrow, a captured honeybee remains active and endothermic, and it dies within hours.

Olavi Sotavalta (1954) described honeybees flying non-stop for 10–15 minutes on a roundabout flight mill until "exhaustion." Afterward, the bees "usually died in 5–10 minutes unless food was given." Bees that were in death struggle when their fuel ran out could easily be saved by offering a drop of sugar solution to the tip of their tongue.

As I am writing this I similarly removed 10 bees from my hive and kept them at 23 °C in a jar in front of me after first offering them sugar syrup. In the first 2 hours all were lively, which for honeybees also means that they were endothermic. In 5 hours half of them were dead, and the others only crawling. In 7 hours all were dead. All 10 bees began the experiment bloated and ended it with empty honey-stomachs.

Honeybees generally remain endothermic as long as they have sugar in their honey-stomach. When the food is gone and they are only able to crawl, they also soon exhaust their tissue reserves and then die. No solitary insect comes even close to such a lethal endothermic response. To put the bees' performance in perspective, I also put 4 Japanese beetles (*Popillia japonica*, Scarabaeidae) that I collected off grape leaves into a jar at the same temperature, near 23°C. (The beetles are of similar mass to the bees.) After 2 days without food, a pair was copulating. At 9 days 2 out of 4 were still alive.

Fig. 8.1 Honeybee worker, *Apis mellifera*, in flight. This bee is carrying pollen loads, and suspended from its mouth is a regurgitated droplet of fluid from its honey-stomach.

Had I made my observations at a 10°C lower temperature, the bees—in trying to maintain the same T_{thx} (therefore double the temperature excess)—would have had twice the heat-loss (and heat-production) rates and should have lasted half as long, about 2–3 hours. The beetles, on the other hand, because they are thermoconformers, would in contrast have *slowed* their metabolism by half because of the Q_{10} effect, and they should have lived twice as long, or to about 18 days. The difference between death in 2–3 hours vs. 18 days demonstrates the potential maladaptive nature of hot-bloodedness in a small animal, unless that animal has continuous access to very *high-energy* food, such as nectar or honey (i.e., sugar). The Japanese beetles could not maintain continuous endothermy even if they had continuous access to their food (foliage) because they cannot process enough of it and extract energy quickly enough from it.

A honeybee worker is solitary only during foraging, and then only for a very short time. A honeybee isolated from its hive is, in some ways, like the proverbial fish out of water. Except in *physiological* studies of thermoregulation where the functioning of one organ system has meaning when simultaneously compared with others in order to decipher how the systems are integrated, the only biologically meaningful thermoregulation in honeybees is that related to the social context and the hive economy. Ultimately honeybees may have evolved the requirement of a relatively high body temperature because they unavoidably experience one in the hive.

As an objective means of measuring the minimum body temperatures from which honeybees can warm up by shivering to become active, Harald Esch (1988) of the University of Notre

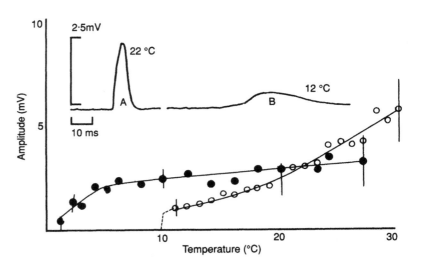

Fig. 8.2 The effect of temperature on the amplitude and duration of muscle potentials in honeybees *(open circles)* and cuculiinid (Noctuidae) winter moths *(filled circles)* during preflight warm-up (shivering). The inset plot at the top shows representative traces of muscle potentials in bees at 22 °C and 12 °C. (From Esch, 1988.)

Dame measured muscle potentials as a function of muscle temperature. The amplitude of muscle potentials in honeybees is strongly temperature-dependent; they are extinguished totally at 10°C (Fig. 8.2). Similarly, the durations of honeybee muscle potentials rise asymptotically already near 10°C, whereas they do not do so until near 0°C in winter moths (Esch, 1988). These data suggest that whereas winter moths can shiver vigorously over a wide range of T_{thx} until a suitable flight temperature similar to that of honeybees is achieved, honeybees are physiologically unable to shiver at 10°C or less and they require a relatively high T_{thx} before they can begin to shiver vigorously. Acclimation (essentially complete in 24 hrs) may modify the temperature response of the muscle potentials (Esch and Goller, 1990) and the chill-coma temperatures (Free and Spencer-Both, 1960) by about 1–2°C. The above considerations on thermal requirements are useful for trying to evaluate the vast amounts of data on honeybee body temperature and thermoregulation—one of the most complete and beautiful pictures of the thermoregulatory behavior and physiological mechanisms of any insect.

Heat Production during Warm-Up

Bees must generate heat to warm up prior to flight, maintain an elevated T_{thx} while foraging at low air temperatures, and also help regulate nest temperature (see Chapter 16). As in all other insects, heat is generated by the flight muscles during shivering.

Beyond descriptive studies of body temperature (Pirsch, 1923; Himmer, 1925) and the energy expenditure (Kosmin, Alpatov, and Resnitschenko, 1932; Heusner and Roth, 1963) of individual bees, most of our basic knowledge of the physiology of heat production in honeybees is due to an elegant series of technically difficult studies by Professor Harald Esch and his associates. Their work now spans some 30 years and it is still in progress.

Esch was the first person to make systematic studies of thermoregulation in individual honeybees. He measured their T_b during various activities within the social context of the hive (such as dancing, dance following, and clustering) and while feeding outside the hive (Esch, 1960). He found that the great variability in T_{thx} is, as in other myogenic flyers, most directly due to shivering by the flight muscles. Increases in muscle temperature are always associated with neural activation (detected by extracellularly recorded action potentials), but as in other myogenic flyers, wing movements and other thoracic vibrations are not always detectable in honeybees during shivering (Esch, 1964).

The oxygen consumption per action potential (Bastian and Esch, 1970) is constant during shivering (1.16 μl/g/min) and during flight (1.14 μl/g/min), even though there are about 10 muscle contractions for every action potential during flight whereas in warm-up there is only one muscle contraction per action potential. Power output and T_{thx} are obviously under direct neural control. The higher the T_{thx}, however, the greater the action-potential frequency and heat-production rate that is possible. Bees that are not in flight can also activate their flight muscles at any of a wide *range* of frequencies to generate heat at varying rates, provided their T_{thx} are high.

In both honeybees (Esch and Bastian, 1968) and bumblebees (Kammer and Heinrich, 1974) the shape and duration of the action potentials is identical during warm-up and during flight. Similarly, in both isolated, indirect flight muscles from the bumblebee (Ikeda and Boettiger, 1965) and in the dorso-ventral muscle of intact honeybees (Esch and Bastian, 1968), there is a direct relationship between action-potential frequency and the amount of contraction of the muscle being activated (Fig. 8.3). There therefore seems little doubt that muscle contractions (shivering) occur during warm-up in bees, but curiously there are seldom vibrations of the wings or thorax despite the high metabolic rate of the muscles. But the mystery of the shivering mode of action of the two pairs of fibrillar muscles that nearly fill the bees' thorax

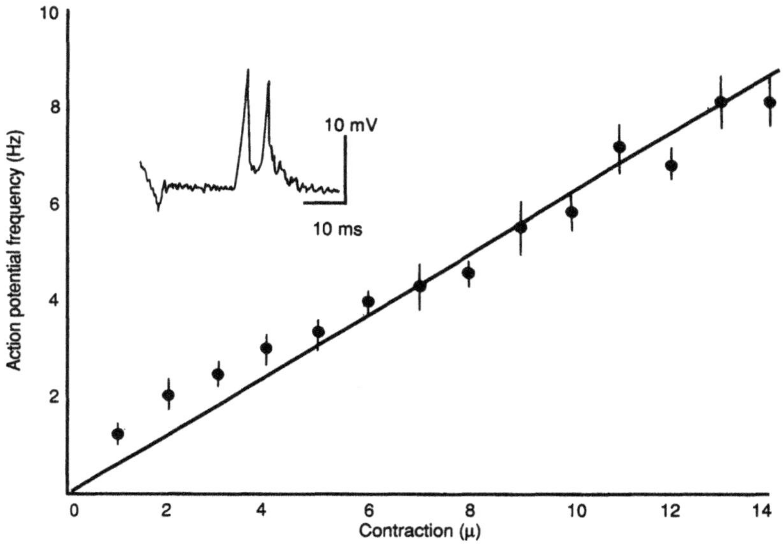

Fig. 8.3 The relationship between action-potential frequency and contraction of the dorso-ventral muscles in an intact honeybee. The inset shows two action potentials from the muscle. (From Esch and Bastian, 1968.)

and that cause most of the heating has recently been largely solved by Esch and Goller (1991).

Since fibrillar muscles do not exhibit a one-to-one relationship between muscle potentials and contractions during flight, they are called "asynchronous." Asynchronous muscles have a stretch-activating mechanism; quick stretches during oscillations initiate contraction. Antagonistic action of the two big muscle pairs, the wing elevators and wing depressors, supported by the resonance properties (the frequency that requires the least energy to maintain, as in a tuning fork) of the thoracic capsule, keep the wing oscillations going.

During shivering, as previously indicated, the muscles act differently. They then form conventional twitches with a one-to-one relationship between muscle potentials and contractions, like typical "synchronous" muscle. To examine the shivering mechanics in more detail, Esch and Goller (1991), using methods already described in Chapter 6 (see Fig. 6.8), simultaneously recorded from up to 6 different electrodes from the dorso-longitudinal and dorso-ventral flight muscles while T_{thx} and air temperature were also monitored. Simultaneously a very light plastic mirror was glued to the middle of the scutellum. A beam of light was focused onto the center of the mirror and reflected onto the surface of a photovoltaic cell. During DL and DV contractions, angular move-

THE HOT-BLOODED INSECTS

ments of the scutellum occurred, and these movements were detected by movements of the reflected light, which were quantified by the voltage output of the photocell. The whole system was computerized and sensitive enough to record muscle contractions caused even by single action potentials.

The results from the experiment show, surprisingly, that during shivering (unlike in flight) the action-potential frequencies in the DL muscles are *higher* than those in the DV muscles (Fig. 8.4). (Intermediate DV/DL frequency ratios occur during buzzing.) The action-potential ratios are correlated with the amount of tipping of the mirror (the angular displacement of the scutellum). In general, the activation of the DL and DV muscles increases in parallel, and the increase in action-potential frequency causes increasingly stronger contraction (Fig. 8.3), which results in closing of the fissure (Fig. 6.8) between the scutellum and the thorax. Finally, at higher DL activation, the scutellar fissure is closed as the mesoscutellar arm has moved as far as it can. (The mesoscutellar arm is moved by contracting the DL muscles.)

An interpretation of these results is that the DL muscles contract more than their antagonists do during shivering warm-up, when a number of contractions in a row result in summing or tetanus. At high activation frequencies (and heating rates) of the DL, these muscles are not only pulling against the antagonistic DV muscles, they are also pulling in a tetanic contraction against a mechanical strut. The shortening of the DL until it feels the resistance of the

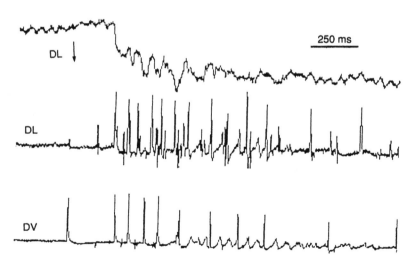

Fig. 8.4 Shortening (top trace) of honeybee DL-muscle at beginning of thoracic heating by shivering. Note that the ratio of action potentials favors the DL over the DV, and the net movement of the scutellum (see Fig. 6.8) is downward. Some of the muscle twitches are due to the firing of action potentials in distant muscle units, some of which are detected as *small* spikes. (From Esch and Goller, 1991.)

strut prevents the stretching of the antagonists, which might, if they were stretched even further, be stimulated to cause the contraction oscillations observed in buzzing and in flight.

What causes the switch from the isotonic tetanus before flight to the *oscillation* of contractions during flight? As explained in detail elsewhere (Esch and Bastian, 1968), a command to start the oscillations for wing movements seems to be a momentary high burst of action potentials in the DV muscles that shortens the dorso-ventral (DV) or wing-elevator muscles 20–40 μ. In these myogenic or stretch-activated muscles, this shortening is likely just sufficient to stretch the opposing dorso-longitudinal (DL) or wing-depressor muscles, and vice versa, and set the myogenic contraction cycle in motion (Boettiger, 1957) that is then continually sparked by another action potential at approximately every 10 contractions. The direct flight muscles that affect the articulations of the wings would presumably also be involved in releasing the wings from their locked positions so that they can oscillate.

Heat Production in Flight

Is heat production varied during flight itself? During continuous free flight at air temperatures over 15°C T_{thx} is necessarily above 30°C, the minimum for flight, and the power output is by definition continuous. As shown by measuring the rate of oxygen consumption of honeybees during free flight over a 22°C range of air temperature, the power output and hence the rate of heat production is constant (Heinrich, 1980b). Thus, although the bees often regulate T_{thx} by adjusting heat production while they are grounded and the muscles are responding in the neurogenic mode, they cannot use the same mechanism of regulating heat production while they are in flight and the muscles are operating myogenically, when heat-production rates are necessarily always very high. Nevertheless, like bumblebees (Heinrich, 1975), honeybees are *capable* of nearly tripling their power output while in flight in direct proportion to the mass of the abdomen—in other words, the honey-stomach load (Wolf et al., 1989). However, airborne bees apparently vary power output only in response to flight parameters, and not in response to their T_{thx}.

To examine the physiological basis of variations in power output during flight, Esch (1976) and co-workers (Esch, Nachtigall, and Kogge, 1975) attached animals in an aerodynamic balance where

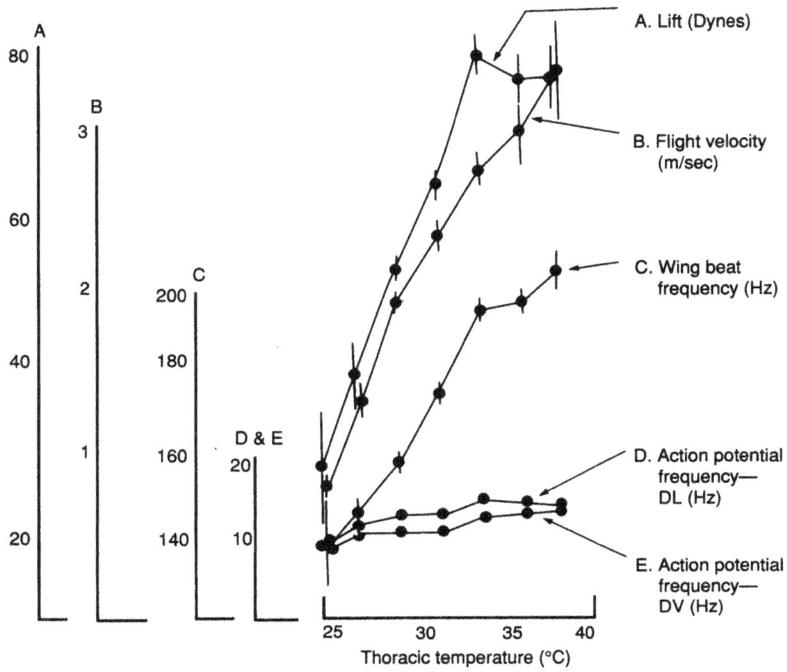

A
80
60
40
20
B
3
2
1
C
200
180
160
140
D & E
20
10
25
30
35
40
Thoracic temperature (°C)
A. Lift (Dynes)
B. Flight velocity (m/sec)
C. Wing beat frequency (Hz)
D. Action potential frequency— DL (Hz)
E. Action potential frequency— DV (Hz)

Fig. 8.5 Five flight parameters over a range of T_{thx}. Honeybees were measured in fixed flight on an aerodynamic balance where the animals selected their own power output. The scales *(A–E)* correspond to the similarly labeled graphs shown at right. (From Esch, 1976.)

the bees controlled their own flight velocity. Through a servomechanism on this apparatus the bees held the wind velocity meeting them to exactly the opposite of their flight speed; thus they selected lift and thrust freely. (The bees were induced to change their lift and thrust during tethered flight by experimentally altering the optical field to which they were exposed by means of a rotating striped drum.) The parameters of lift, "flight" velocity, T_{thx}, and the occurrence of action potentials in the DV and DL muscles were then simultaneously measured and their correlations with respect to the flight performance were analyzed.

When a bee is induced to start flight (by taking away a piece of paper it is holding), there is at first a burst of action potentials of 20–60 Hz in the DV and DL muscles, and then during steady flight (of up to several hours) the action-potential frequency stabilizes at 10–30 Hz (Fig. 8.5). The absolute number of action potentials depends on the specific motor unit (of 5 per muscle) probed, the lift produced, and the T_{thx}. (It is possible to get flight parameters at T_{thx} considerably lower from tethered bees than with free-flying animals.)

As expected, T_{thx} is very important for flight performance. Wing-beat frequency, lift, and flight velocity all vary in parallel as a function of T_{thx} (Fig. 8.5). A T_{thx} of near 33 °C is required to generate sufficient lift to support the empty body weight. In free flight, however, the animals can sacrifice flight velocity for additional lift (Esch, Nachtigall, and Kogge, 1975). Sudden increases in lift and thrust are accompanied by a distinct rise in action-potential frequency (Esch, Nachtigall, and Kogge, 1975). However, when lift and flight velocity change together with increasing T_{thx} (25 °C to 38 °C), there is little change in action-potential frequency (Fig. 8.5). That is, the DV and DL muscles are not activated more frequently, yet power increases.

The above data appear to be not in strict agreement with other data for honeybees (Bastian and Esch, 1970) and bumblebees (Kammer and Heinrich, 1974). In these studies *oxygen consumption* was found to be a direct function of action-potential frequency over a limited range of T_b. The apparent discrepancy—that *mechanical-power* output by the bee increases with temperature at a *given* action-potential frequency even though energy expenditure per action potential remains constant—is likely due to a physico-chemical phenomenon affecting the muscles's viscosity (possibly less friction during contraction?) that is temperature-dependent. The tension-length of the muscle is virtually unaffected by temperature. But a rise in temperature increases the range of frequency over which the muscles can do oscillatory work, and also increases the amount of work per work cycle (Machin, Pringle, and Tamasige, 1962). Perhaps more work can be obtained when the muscle becomes less viscous at higher temperatures.

The alternate hypothesis, that there is loss of mechanical energy (as measured on the flight mill) at lower temperatures because of overlapping contractions, is probably not plausible for myogenic flyers. In neurogenic flyers the nervous system could at least potentially activate antagonistic muscles before their antagonists have relaxed, which means that the contractions may overlap and internal power may be wasted. In myogenic flyers, however, for which stretch-activation rather than neural innervation determines the exact timing of the contractions of antagonistic flight muscles, the simultaneous contraction of antagonistic muscles is prevented since it is the tension in one that creates the lengthening in the other.

A recent series of four publications by Werner Nachtigall and colleagues (Jungmann, Rothe, and Nachtigall, 1989; Feller and

Nachtigall, 1989; Nachtigall et al., 1989; and Rothe and Nachtigall, 1989) at the Universität des Saarlandes in Germany have examined in great detail the T_{thx} and energetics of honeybees in flight on a round-about flight mill and in a wind tunnel. They have largely confirmed previous notions of the interrelationships of flight energetics and body temperature.

Heat Conservation

The relatively small mass of honeybees (near 90 mg empty weight for the European honeybee *A. mellifera* and near 61 mg for the African honeybee *A. m. adansonii*) means that passive-convective heat loss is very rapid. On the other hand, having evolved a highly social system with tens of thousands of individuals massed into a nest, honeybees are unavoidably subjected to convective heating, and they have probably adapted to fly near the relatively high T_b that they experience. Their small size, however, has probably been dictated by foraging constraints (see Chapter 16). The need, due to evolutionary history, for a relatively high T_{thx} in flight despite relatively small body size, and the physical difficulty of maintaining a high T_{thx} in flight because of that small size, have undoubtedly put a premium on foragers for heat retention in the thorax when they leave the hive.

Honeybees have a thoracic pelt of short pile that acts as insulation (Southwick, 1985). In a small insect an increase in thoracic insulation for heat retention soon reaches a point of diminishing returns, because it increases the air resistance in flight and therefore the energetic cost of flight. Not surprisingly, most small insects are uninsulated or only very poorly insulated.

The problem honeybees have in retaining sufficient heat during flight is evident from the T_{thx} of animals leaving and returning to the hive. Bees leaving the hive have a T_{thx} near 37°C over a relatively wide range of air temperatures (Fig. 8.6), and these T_{thx} are presumably the T_{thx} of choice and not of necessity. Once flight is initiated, rapid convection occurs and thoracic temperature cannot be maintained more than about 15°C above air temperature (Heinrich, 1979b). Thus, bees starting to fly at an air temperature of 10°C rapidly cool until T_{thx} reaches about 25°C, some 5°C below the minimum for vigorous flight. However, bees do return to the hive even at air temperatures of 10°C or less, and these bees still have T_{thx} near 30°C (Fig. 8.6).

The above observations suggest that the lower-than-preferred

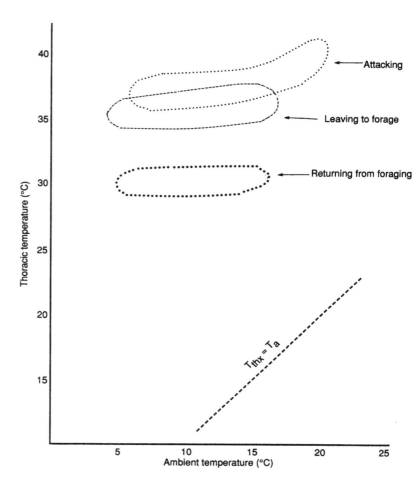

Fig. 8.6 Thoracic temperatures of *A. mellifera mellifera* when leaving the hive to forage, when attacking, and when returning to the hive. (From Heinrich, 1976b.)

T_{thx} of bees returning to the hive are the minimum that they try to achieve. They keep heated up at low air temperatures by periodic *stopping,* to reduce convective heat loss, and then shivering to regain a suitable flight temperature. Bees may occasionally even leave the hive to forage at 4.5°C, although they then stop flight approximately every 10 seconds to warm up (H. Esch, personal communication). Previous to our knowledge of thermoregulation in bees, it was supposed that honeybees needed more time to *return* from a feeder than to fly to it because with their heavy burden they "stop to rest" (Frisch, 1967). They do not stop to rest. Instead, they stop to work, to produce heat by shivering, and the greater their load, the higher the T_{thx} that are needed (Hein-

THE HOT-BLOODED INSECTS

rich, 1975; Wolf et al. 1989), and thus the longer their "rest" stops at low air temperatures.

Harald Esch (personal communication) has exploited the honeybee's very large dependence on energy expenditure for thermoregulation to test a long-standing theory (Frisch, 1967) that honeybees gauge distance to a food source by the energy expended to get there. Esch had the bees get food for the hive by walking through a tube at high temperatures (when energy expenditure was low) and at low temperatures (when energy expenditure was very high). (In either case, the distance indicated by the dance was identical. Therefore, the bees' perception of distance to a food source, as given by their dances, is not based on energy expenditure.)

The durations of warm-up stops when the bees must travel long distances by flight have likely evolved to be reduced by the development of an anatomical adaptation. The large body of recent work on the thermoregulatory role of insect blood circulation in general, and a set of experiments on honeybees specifically, support the hypothesis that the nine aortic loops in the honeybees' petiole (Fig. 8.7) function as counter-current heat exchangers that help prevent heat flow from the thorax to the abdomen despite presumed blood flow between these two body parts. Blood heated in the thorax necessarily flows over and around the loops before returning to the abdomen. The coils would then function in three ways to promote counter-current heat exchange. First, they would obliterate discrete pulses of blood that could otherwise be quickly shuttled through the petiole, as occurs in bumblebees when they dump heat into the abdomen (Heinrich, 1976). Second, the loops would create a large surface area for heat exchange. And third, by providing resistance to slow down the blood they would increase the time for heat exchange to occur, allowing the otherwise cool blood from the abdomen to pick up heat leaving in the blood

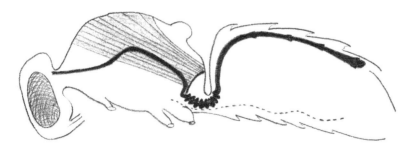

Fig. 8.7 Sagittal section of a honeybee showing the abdominal heart, the thoracic aorta, and the convolutions of the circulatory tube in the petiole area. The dotted line depicts the ventral diaphragm. (From Freudenstein, 1928.)

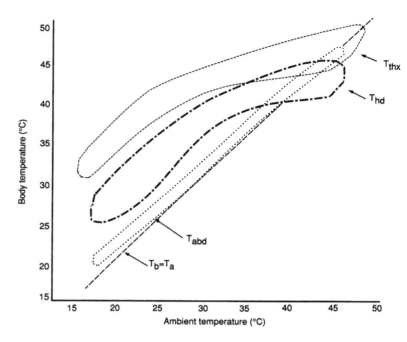

Fig. 8.8 Head, thoracic, and abdominal temperatures of honeybees in continuous free flight as a function of ambient temperature. These temperatures were taken from bees in the laboratory rather than from bees returning to the hive, when flight duration and the possible effects of solar radiation would obscure the physiological variables of interest. (From Heinrich, 1979b, 1980a,b.)

from the thorax. As predicted by this hypothesis, honeybees retain their T_{abd} very close to air temperature (Fig. 8.8), and they do not transfer heat into the abdomen even when the thorax is experimentally heated to near lethal temperatures (Heinrich, 1980b; Coelho, 1991).

The curious loops in the honeybees' petiole have long been known. They were first noted by W. J. Pissarew in 1898 and then confirmed by Freudenstein (1928). They are not found in wasps, hornets, leafcutter bees, or bumblebees (Snodgrass, 1956). Pissarew (1898) presumed that the loops in the honeybees' petiole helped to amplify the pumping of the heart. Others suggested instead that the loops function to prevent backflow of blood from the thorax (Arnhart, 1906) or that they provide slack so that the heart attached to the thoracic phragma would not tear during the rapid vibrations of the thoracic muscles during flight (Zander, 1911). Freudenstein (1928) suggested that the convolutions acted as lungs to help aerate the blood. However, none of these alternative hypotheses are substantiated either by supporting experimental evidence or by comparative data.

Thermoregulation and Heat Loss in Flight

Since heat production during flight is necessarily continuous and high (Withers, 1981; Harrison, 1986) and since its rate is not varied as a function of the air temperature at which the bees fly (Heinrich, 1980b), one might therefore expect a constant thoracic-temperature excess in flying bees at all temperatures. Indeed, bees in free flight generate a constant temperature excess of about 15 °C over a relatively broad range of air temperature from 17°C to 25°C (Fig. 8.8). However, temperature excess dramatically declines at air temperatures over 30°C. And unlike any other highly endothermic insects of its size so far investigated, honeybees have the unique capacity to fly even at the extraordinary high air temperature of 46°C, at which they astonishingly have an average T_{thx} of 1 °C *less than* air temperature (Heinrich, 1980b).

In most large insects, heat loss is augmented by increasing the *area* for convection, by shunting heat into the abdomen, which then acts as a heat radiator. But this mechanism is not feasible for honeybees because the circulatory adaptation just described prevents heat from being dumped into the abdomen. Furthermore, convective heat loss is possible only as long as body temperature exceeds ambient temperature. Only evaporative water loss can drive T_b below air temperature. And honeybees dissipate enough heat to depress T_{thx} some 16°C below that which normally occurs by convection at lower air temperatures. A major portion of this heat dissipation is by evaporative cooling from the head.

In honeybees, the primary site of evaporative cooling is the head, and the head is physiologically (by blood circulation) and morphologically (by close physical contact) coupled to the thorax, so that heat loss from the head is practically equivalent to heat loss from the source where it is produced, the thoracic muscles. The isolated head of a honeybee cools three times as fast as the thorax. In intact, live bees as well as in dead bees, however, head temperature closely tracks T_{thx} (Fig. 8.8), primarily because of passive-conductive heat flow.

When tethered bees are heated on the head with a focused beam of light, they regurgitate nectar from their honey-crop when T_{hd} reaches 44–46°C (Fig. 8.9). Heat is then transferred from the thorax to the head and from the head to the fluid droplet (by sucking it in and out) and then to the air. Evaporative cooling creates a steep temperature gradient for heat loss from the head.

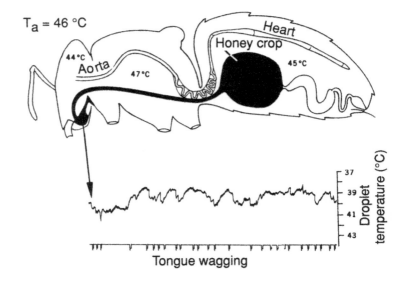

Fig. 8.9 Anatomical features of a honeybee related to thermoregulation, including typical body temperatures maintained during continuous free flight at an ambient temperature of 46 °C. The graph shows continuously recorded droplet temperature of a tethered bee heated on the head. Tongue-wagging events are shown as downward-projecting ticks on the horizontal axis. Note the low droplet temperature despite the high T_{hd}. (From Heinrich, 1979a,b.)

Highly amplified signals from thermocouples implanted in the head indicate that the aortic activity is often matched, beat by beat, with temperature changes in the head. This suggests that in addition to primarily passive heat conduction along the thoracic-head temperature gradient (Heinrich, 1979a), heat is also actively pumped in the blood from the thorax to the head (Fig. 8.10), the site of evaporative cooling. Thus, the thorax is cooled (Fig. 8.11).

Regurgitation of nectar and the re-ingestion of the cooled droplet has now also been observed in the giant honeybee *Apis dorsata* (Marden and Kevan, 1989). In *A. dorsata* the number of bees engaged in the "gobbetting" behavior increases as air temperature rises and, as in *A. mellifera* (Cooper, Schaffer, and Buchmann, 1985), gobbetting giant honeybees have lower T_{thx} than non-gobbetting bees.

The high temperature setpoint that initiates the cooling response in honeybees probably resides in the head, because *thoracic* heating to lethal temperatures does not result in the activation of the cooling response unless the head is heated simultaneously (Heinrich, 1980b). Head temperature is regulated, but therefore apparently only by defending an upper setpoint. Head-temperature regulation in honeybees could be functionally related not only to regulation of T_{thx} but perhaps also to temperature receptivity of sensory receptors of the eyes (Duruz and Baumann, 1968). However, the *lower* setpoint (near 36 °C) below which the bee shivers

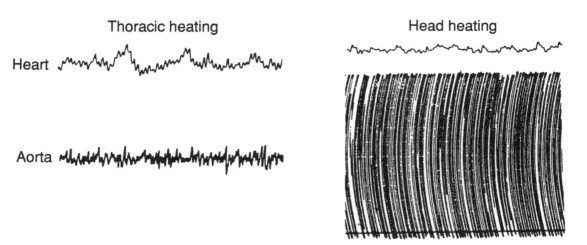

Fig. 8.10 Concurrently recorded mechanical activity of the heart and the aorta while the thorax and then the head of the same bee were heated to near 45 °C. The records each span 13.5 seconds. (From Heinrich, 1979b.)

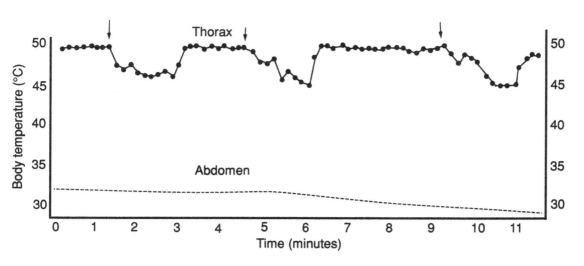

Fig. 8.11 Thoracic and abdominal temperatures of a bee that initiated three bouts of droplet extrusion (at arrows) during experimental overheating of the thorax (at an ambient temperature of 28.5 °C). Note that the bee did not shunt excess heat into the abdomen. (From Heinrich, 1980b.)

appears to be controlled in the thorax, as would be expected since the neuro-muscular coordination of the thoracic muscles resides there.

Lensky (1964a) had previously noted that honeybees have relatively low water consumption (3.5–4.5 mg/bee/hr) within the range of air temperatures from 28 to 45°C, but water consumption rapidly increases to 13.3 mg/bee/hr (equals about 150 mg/g/hr) starting at 45°C. Apparently because of this high rate of water consumption bees are able to stay alive at 50°C for durations up to 45 minutes (Lensky, 1964b). In a more detailed accounting of the water budget of honeybees, Louw and Hadley (1985) compared the rate of metabolic water production to the rate of water lost through evaporation in simulated desert conditions (30°C, <5 percent relative humidity). Under these conditions the extremely high metabolic demands while ferrying 52 mg of water (65 percent of body mass) to the hive produces almost sufficient water (74 mg/g/hr) to balance simultaneous evaporative water loss (80 mg/g/hr).

Although the above data show that flight metabolism can produce almost enough water to balance evaporative water loss in loaded bees at 30°C, that water is produced in the thoracic muscles, where it then may not be available for thoracic cooling. Thus, dilute honey or water in the honey-crop may still be a prerequisite for evaporative cooling from the mouth, despite high rates of metabolic water production. Nevertheless, some of the sharp rise in water loss in heated bees is apparently from sites other than the mouth. Water loss rises sharply when a bee extrudes water from its mouth, but at T_{thx} of 45°C bees with their mouth parts sealed with a wax-resin mixture still show a water-loss rate of 61 mg/g/hr, whereas unaltered bees have water-loss rates of 139 mg/g/hr (Louw and Hadley, 1985).

Thermoregulation by evaporation of water at high air temperatures could have implications for foraging behavior. For example, in the Sonoran Desert, the T_{thx} of pollen-collecting honeybees (who carry on the average only 1.3 μl honey-crop contents, in comparison with 7.9 μl in other bees) returning to the hive at air temperatures near 40°C have T_{thx} over 2°C higher (46.1 vs. 44.0°C) than those of nectar gatherers (Cooper, Schaffer, and Buchmann, 1985). At 20°C no bees returning to the hive carry a fluid droplet on the tongue, but at 40°C many (40 percent) do so (Fig. 8.12). In general, at air temperatures above 40°C, pollen collecting appears to be curtailed. Some bees that do not curtail

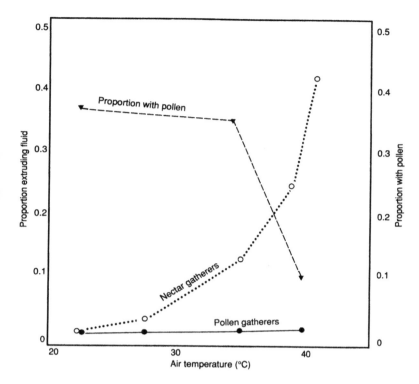

Fig. 8.12 Proportion of bees returning to the hive and extruding fluid from the tongue as a function of ambient temperature. (From Cooper, Schaffer, and Buchmann, 1985.)

pollen foraging at high ambient temperatures suffer T_{thx} increases to the near-lethal temperature of 50°C. The drop in pollen collecting at high air temperatures could thus be due both to less pollen production by flowers and to greater thermoregulatory constraints in pollen vs. nectar foraging.

Foraging

The physiological limits that affect T_b as detailed above provide a rough framework from which to examine other behavior. Foraging involves four separate responses that have unique thermoregulatory requirements and constraints: warm-up prior to takeoff, intermittent flight between flowers, perching or walking on flowers, and continuous flight to and from the hive. Each of these responses may be associated with a different T_{thx}, depending on air tempeature. In a study of the African honeybee *A. mellifera adansoni* at about 2,700 m elevation in Kenya, there was nearly maximal traffic into and out of hives in sunshine at about 8:00

in the morning, even at air temperatures as low as 8–10°C. The T_{thx} of bees leaving the hive was near 37°C, on the average 5°C higher than those of bees in the brood nest (Heinrich, 1979b). Thus, the bees had warmed up (shivered) in the hive prior to takeoff. Even so, the warm environment of the hive allows them to get a head start over non-social bees, since bees without a hive would have been physiologically incapable of initiating flight because the muscles are incapable of shivering when they are cooled to below 15°C (Esch, 1988). Similarly, as a result of the overnight refuge of their hive, bumblebees fly and forage at air temperatures near 0°C (see Chapter 6) even though they are incapable of warm-up if muscle temperature cools to 6°C (Goller and Esch, 1990a).

At air temperatures less than 25°C the T_{thx} of honeybees during continuous flight equilibrates at about 15°C above air temperature but upon cessation of flight (once reaching a feeding place, when perching begins), convection is immediately reduced and solar heating may then occur (Cena and Clark, 1972). Stationary bees also have the option to cool passively or to shiver and produce heat at variable rates up to those in flight. Not surprisingly, unlike during flight at low air temperatures, the T_{thx} of bees while foraging from flowers (or while feeding from an artificial feeder) is adjusted by several thermoregulatory options that are available—being heated by the sun, shivering, and/or cooling convectively. Thoracic temperature is almost always kept above 30°C, the minimum for free flight to the next flower. But there is no single, unvarying temperature setpoint.

As in bumblebees (Heinrich, 1979c), the T_{thx} of foraging honeybees varies markedly in direct response to the richness of food rewards. Using a thermovision camera in which the color of the continuously recorded image codes for surface temperature, Sigurd Schmaranzer and Anton Stabentheiner from the Karl-Franzens University in Graz, Austria, directly observed the immediate rise in T_{thx} of bees given sucrose solutions of different concentrations (Stabentheiner and Schmaranzer, 1986, 1987, 1988; Schmaranzer and Stabentheiner, 1988). Keith D. Waddington (1990), at the University of Miami, also measured the T_{thx} of honeybees foraging at different sucrose concentrations and confirmed that T_{thx} varies as a function of immediate food rewards. Bees maintain T_{thx} on the average at 36°C while imbibing 40–60 percent sucrose, but when imbibing 10–30 percent sucrose their T_{thx} declines to 33°C (Fig. 8.13) before shivering is initiated and

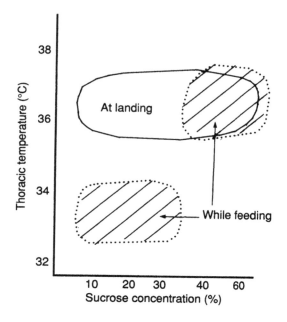

Fig. 8.13 Thoracic temperatures of foraging honeybees in relation to the concentration of sucrose from which they fed; T_{thx} were recorded in bees after landing and while feeding. Takeoff T_{thx} is near 36°C. (From Waddington, 1990.)

a T_{thx} of 33°C is maintained. However, all bees leave the feeder only after warming back up again to 36°C.

The above results on T_{thx} cannot be without implications for energetics, but as worked out by Waddington, the results are not intuitively obvious. Why do bees that feed on dilute sugar solutions allow their T_{thx} to decline? The reduction in T_{thx} saves them considerable energy expenditure (20 percent) but the reduced T_b also greatly reduces foraging pace. (Pace depends in part on sucking time, and although the highly concentrated syrup is viscous and should have required more time to imbibe, the bees sucked up the concentrated sucrose very quickly and took 3 times longer to suck up the diluted solution; see Fig. 8.14.)

Waddington (1990) calculates that the bees *could* have had an immediately improved net gain if they maintained a higher T_{thx} even at the lower food rewards, and he concludes that the payoff for operating at a lower than immediately optimal metabolic rate (with a T_{thx} of 33 rather than 36°C) might be a long-term one. This logic incorporates the idea that the bee's life span may be reduced by increasing metabolic rate (Pearl, 1928; Neukirch, 1982; Calder, 1984; Schmidt-Hempel and Wolf, 1988; Wolf and Schmidt-Hempel, 1989). When profits are potentially very high, it pays to invest energy to harvest resources and maximize short-

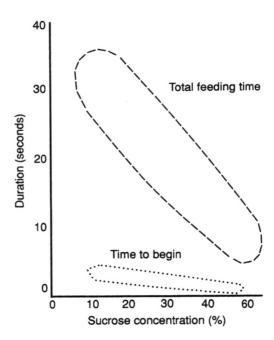

Fig. 8.14 Duration of feeding and time to begin feeding in relation to the concentration of sucrose in the food source. (From Waddington, 1990.)

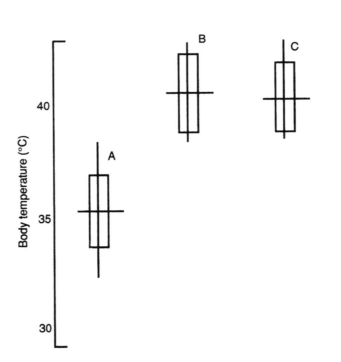

Fig. 8.15 Thoracic temperatures of freely moving dancing bees, determined by a thermovision system: (A) "Zitter" dances (unrelated to food), (B) round dances (indicating food at <100 m), (C) waggle dances (food at 100–160 m), (D) waggle dances (food at 600–1,800 m). (From Schmaranzer, 1984.)

term rate of gain, because the long-term gain is not likely ever to be any better. But if profits are low at any one time, then a bee can reduce energy costs (and thereby sacrifice short-term but lower net profits) in favor of long-term gain to the colony through improved potential contributions in the future (see also Chapter 9, on ants).

The relation of food rewards and T_{thx} extends to dancing bees and those following them within the nest (Fig. 8.15). By continuously recording T_b using a thermovision camera, Schmaranzer (1983, 1984) determined that in the hive, with combs at 30.5°C, the T_{thx} of round dancers indicating nearby food was near 40°C. Waggle dancers indicating food at a distance of only 110 m also had T_{thx} near 40°C, whereas bees indicating food at 600–1,800 m during the waggle dance had T_{thx} averaging only 37.5°C. When a dancer feeds a follower, then her T_{thx} drops by about 1°C, whereas the T_{thx} of the follower imbibing the food offering rises (Esch, 1960) to near the same T_{thx} as that of the dancer.

Age and Acclimation

Not all honeybees have identical thermal responses. Bees exposed to a new temperature for relatively short periods (about 24 hours) will acclimate to new chill-coma temperatures (Free and Spencer-Booth, 1960; Goller and Esch, 1990a) and to new temperature limits for muscle potentials, including their amplitude and extinction (Goller and Esch, 1990c). Thus, winter bees understandably have lower chill-coma temperatures than summer bees.

Thermal response is correlated not only with season but also with age. Young bees only gradually (over a few days) develop the capacity for endothermic heat production (Himmer, 1932; Allen, 1955; Harrison, 1986; Stabentheiner and Schmaranzer, 1987). Before they have developed the capacity to shiver, new workers tend to stay in the warm brood nest (Free, 1961, as quoted in Free and Spencer-Booth, 1960), where temperatures are maintained at near 35°C in summer. That is, freshly eclosed or young bees are strictly thermo-conformers, but like many poikilotherms they thermoregulate behaviorally in a temperature gradient (Heran, 1952).

The division of labor of the hive economy as a function of the age of honeybees (Lindauer, 1952; Rösch, 1925; Free, 1965) is correlated with thermoregulation. Young bees initially perform hive duties, and after developing the capacity to shiver they become foragers. Foragers tend to stay away from the brood nest and so are then subjected to lower temperatures than the younger house bees encounter, particularly at night and when foraging in the daytime. The chill-coma temperatures of these foragers then decline accordingly (Free and Spencer-Booth, 1960).

The task-related physiological change that occur as a hive bee develops flight (and shivering) capacity have been investigated by Jon M. Harrison (1986) at the University of Colorado in Boulder. Peak foraging activity occurs between 15 and 32 days of age, depending on the season. Maximal thorax-specific rates of oxygen consumption increase dramatically with age for about a month. However, pyruvate kinase (PK) and citrate synthase (CS) activities increase (tenfold) up to only 4 days of age, and then they gradually decline in older bees. In the first few days the increase in thorax-specific maximal metabolic rate (in other words, the ability to fly and/or shiver) closely corresponds to the increase in enzyme activities. Although the increase in thorax-specific metabolic rate (with the constancy of thoracic mass) strongly suggests that the flight muscles are developing their oxidative capacity, the muscles must also be maturing in other ways besides increases of PK and CS. What this maturation involves is not known. Possibly either exercise or some other thoracic-muscle enzymes could be involved, since older bees (foragers) and 9–10-day-old lab bees have different enzyme profiles (Hersch et al., 1978). Furthermore, lab-reared workers, presumably little-exercised, show flight-muscle mitochondrial volume and cytochrome concentrations increasing up to 20 days of age (Herold, 1965; Herold and Borei, 1963), suggesting that exercise could stimulate more rapid development of oxidative capacity. These changes are presumably mediated through hormones, specifically juvenile hormone, which has a potent effect on muscle growth in insects.

The final large jump in total-body, mass-specific increase in oxygen consumption (and the capacity to haul nectar) is due to a single event: defecation. Young bees eat much pollen, and that results in the accumulation in the hindgut of a large mass of feces. But bees do not defecate in the hive and the feces in the young bees necessarily accumulate until they have developed flight capacity. Possibly the need to defecate induces the "orientation" flight (Free, 1965) that then helps cause house bees to become foragers outside the hive.

The decrease in rectal mass due to defecation on the orientation flight is responsible for 65 percent of the total weight loss that bees experience before they become foragers (Harrison, 1986). That weight loss then reduces the energy requirements for flight, and greater foraging loads can be carried as a result of the greater lifting capacity as well as the greater space then available for nectar in the abdomen.

Caste

Workers but not drones isolated from the hive continue to maintain a high T_{thx} by shivering thermogenesis, and thus the lower the air temperature (to about 12°C), the higher their metabolic rate (Allen, 1955; Esch, 1960; Cahill and Lustick, 1976). As in bumblebees (Heinrich, 1975) and other insects, in honeybees elevated metabolism at low air temperatures (Fig. 8.16) is not "resting" metabolism even though the bees may sit quietly without moving their wings; the shivering wing muscles of these "resting" bees may be working as hard as in flight.

Drones do not forage and do not need to be flight-ready at all times, and their thermal biology is much different from that of workers even when they do fly. Drones have twice the body mass of workers (Fig. 8.17) and largely because of their greater mass they generate T_{thx} several degrees higher than those of workers in flight (Coelho, 1989; Goller and Esch, 1990b; Coelho, 1991).

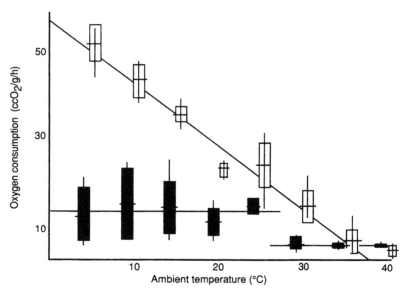

Fig. 8.16 Metabolic rates of non-flying (shivering) honeybees—10 workers *(unshaded)* and 7 drones *(shaded)*—as a function of ambient temperature during the first hour following placement into a respirometer. Note that the drones, who do not maintain an elevated T_{thx}, do not elevate their rate of heat production at low temperatures. (From Cahill and Lustick, 1976.)

Fig. 8.17 A worker and a drone honeybee. Drones have twice the body mass of workers and much larger eyes.

Like workers, however, drones also have the aortic coils in the petiole area that should retard or prevent heat dumping into the abdomen. Drones can be heated on the thorax until they die, and they still fail to dump heat into the abdomen; nor do they regurgitate nectar to cool themselves evaporatively (Coelho, 1991). But these results are inconclusive in deciding whether or not they can cool themselves, because honeybee workers act similarly, although they have superb mechanisms of cooling; they must be heated on the *head* (Heinrich, 1979a, 1980a) to initiate their cooling responses. Whether they regulate T_{thx} in continuous flight also is not known.

A heat-budget model for drones predicts that their T_{thx} would increase dramatically with decreased flight speed (Fig. 8.18). It is possible therefore that drones returning to the hive (when they necessarily slow down) may often have higher and closer-to-lethal T_{thx} than they normally experience during their prolonged flights after virgin queens in the field (Witherell, 1971). Nevertheless, their maximum muscular force is produced at slightly higher muscle temperatures than in the case for workers (Coelho, 1990).

Drones have a much greater tendency, compared with workers, to become and to remain torpid (Goller and Esch, 1990b) or to warm up only sporadically (Cahill and Lustick, 1976; Fahrenholz, 1986). Drones also acclimate to temperatures about 2 °C higher than workers do (Goller and Esch, 1990c), although they do not have enhanced tolerances for higher T_{thx} (Coelho, 1991). In short, in comparison with workers, drones are unable to warm up from temperatures as low, they are less willing to expend energy to maintain a high T_b by shivering, and they more easily attain near-lethal temperature excess in flight.

Species

The temperature excess of insects is largely a function of body mass, and deviations from this generalization therefore point out potentially interesting phenomena. For example, two races of the common honeybee, *A. m. adansonii* and *A. m. mellifera*, have the same average T_{thx} (Heinrich, 1979) even though the former (the African honeybee) is 30 percent smaller than the European race. Since the smaller bee has a faster rate of cooling, it follows that it must have a higher rate of energy expenditure, or "pace." Presumably some ecological factor has led to the evolution of the difference in pace and/or the reduction in body size with no change in T_b.

THE HOT-BLOODED INSECTS

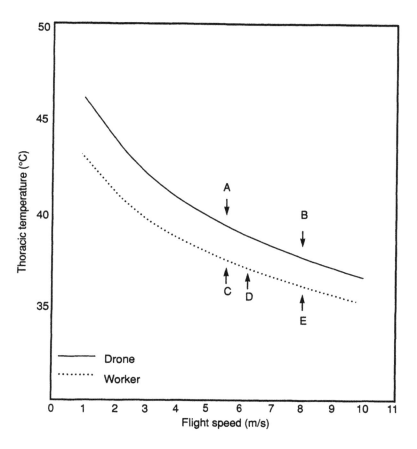

Fig. 8.18 The effect of flight speed on predicted T_{thx} in drones and workers at an ambient temperature of 30°C. Arrows indicate actual flight speeds: *(A)* drone mean; *(B)* drone maximum; *(C)* worker mean returning to hive; *(D)* worker mean departing hive; *(E)* worker maximum. (From Coelho, 1991.)

The reason for and possible implications of pace and endothermy were raised by Fred C. Dyer and Thomas D. Seeley in a comparison of the T_{thx} of the 3 Asian honeybee species (genus *Apis*) and the "European" species, which together span a five-fold range in body mass. As expected, the rate of convective heat loss from the thorax of the 4 species is a direct function of body size. However, the T_{thx} maintained by the animals does not scale with body size. The intermediate-sized *A. cerana* and *A. mellifera* consistently have higher T_{thx} than the largest species, *A. dorsata*, and these higher T_{thx} are associated with elevated mass-specific metabolic rates in flight, as suggested by their greater wing loading (hence wing-beat frequency) and flight speed. The fast-flying, high-powered bees both nest in enclosed cavities, whereas both species of the slow-flying, low-powered bees have exposed combs that are protected by curtains of interlinked worker bees.

Dyer and Seeley (1987) propose that the open-nesters are protected from microclimatic change as well as from predators only by the curtains of bees surrounding their combs, whereas the hole-nesters have physical protection. They therefore speculate that the open-nesters have nearly 5 times as many workers (relative to numbers of comb cells) than cavity-nesters have for added protection. If cell numbers in the combs limit the number of young that can be raised, then it follows that larger numbers of workers per cell must be explained either by a shortening of egg-to-adult development time or by extended longevity. Dyer and Seeley (1987) indicate that development time for all 4 species is similar—about 18–22 days. Therefore, workers of the open-nesting species are probably longer-lived than cavity-nesting workers.

As already indicated, there seems to be a causal link between the physiological tempo of workers and their longevity. The need for increased longevity (which ensures there will be enough workers available to protect open-nesting honeybees, such as *A. dorsata* and *A. florea*) may have constrained workers of these species to a low-tempo existence. On the other hand, hole-nesters, such as *A. cerana* and *A. mellifera*, have been released from this constraint and were then able to evolve a high worker tempo that enhances immediate foraging profits at high-reward food sources.

Size, as such, has presumably been a very malleable worker characteristic in *Apis* honeybees. Presumably worker size is related to the energetics of food ferrying to the hive. For example, large loads harvested from a distance would favor large body size in the same way as large trucks are favored for shipping large loads across the country whereas small trucks are better suited for carrying a few parcels across town—fuel economy makes the choice obvious. Worker size in honeybees thus probably evolved independently from major constraints related to pace and/or thermoregulation.

Bees in the Nest

The high temperature generated by thousands of bees crowded into one aggregation greatly facilitates the regulation of T_{thx} of individual bees at low temperatures, but it can create problems of overheating at high temperatures. The problems and mechanisms associated with regulation of body temperature in this environment are discussed in more detail in Chapter 16; here I give only an overview of the effect of colony thermoregulation on the individuals.

Individuals do not know nor "care" whether the temperature they experience is a result of the effects of others or of the effects of the weather. However, there are in the nest large temperature gradients, from near 35°C at the core to near environmental temperatures at the periphery. By exploiting these gradients to regulate their own T_b, bees necessarily affect the thermal environment of others as well as their own. Although bees are specifically attracted to brood, they are otherwise relatively free to exploit the temperature gradients that they find within the colony to regulate their own T_b and hence colony temperature.

Most nest temperatures can be explained by the following broad responses: (1) attempts by individuals to regulate their own T_b by following temperature gradients when possible, (2) shivering and active cooling to regulate individual T_b when other means do not suffice, (3) acclimation, which affects temperature preferences and the distribution of individuals within the hive, and (4) attraction to brood, and attempts to maintain a regulated T_b near brood in the same way that an electric blanket remains near a set temperature independent of what lies under it.

Individual bees have only a limited capacity to stabilize their T_{thx}, and individual T_{thx} generally fluctuate (Himmer, 1925, 1926; Esch, 1960; Heinrich, 1981b), but when bees are gathered together in ever-larger groups (Free and Spencer-Booth, 1958) body-temperature stabilization becomes ever more precise because of the reduced thermal inertia of the larger mass. Thus, for example, a classic study by John B. Free and Yvette Spencer-Booth at the Rothamsted Bee Research Station in England showed that whereas individuals in even small groups of 10 bees could maintain T_{thx} near 30°C, at 15°C it required groups of 200 before individuals could maintain the same T_{thx}. The maintenance of the high T_{thx} was at the cost of greater sugar consumption (Free and Spencer-Booth, 1958) or, in other words, at the cost of greater metabolic rate (Allen, 1955; Cahill and Lustick, 1976). Thus, whereas groups of 50 bees each individually consumed 58 mm^3 of a 66 percent sugar solution daily at 15°C, sugar consumption declined to 6 mm^3 at 40°C. Presumably, sugar consumption would have been even higher if the bees had not clustered (one assumes they were attracted to each other's warmth), thus effectively reducing their collective body-surface area for heat loss and hence conserving colony resources. The percentage of bees clustering increases markedly in bees of all group sizes (10 to 200 bees) as air temperatures decrease to 10°C.

The responses of very small groups of bees such as those studied by Free and Spencer-Booth (1958) can be extrapolated to estimate the behavior of bees that find themselves in a colony with 50,000 or so inhabitants. Bees in a large group (such as within a swarm cluster) no longer individually experience the full extent of external air temperatures. Instead, they experience primarily the heat (and insulation) generated by their colony mates.

In a large cluster of some 30,000 bees, for example, the individual bees shielded within the core have a passive cooling rate of only $0.001315°C/min/°C$ (Heinrich, 1981b). (Individuals external to the cluster instead have a cooling rate over 608 times greater, $0.80°C/min/°C$; Heinrich, 1980b). Thus, if the swarm were to maintain a 30°C difference from air temperature (say 35°C at 5°C), then the individuals within it would have a cooling rate of $30 \times 0.001315 = 0.0395°C/min$ if it maintained a core temperature of 35°C. As calculated elsewhere (Heinrich, 1981b), a bee situated at the swarm's core would require a metabolic rate of about 0.39 ml $O_2/g/hr$ to produce sufficient heat to oppose this cooling and stabilize its T_{thx} at 35°C. However, this calculated metabolic rate is nearly 11 times *lower* than the actual resting (that is, non-shivering) metabolic rate of a bee with a T_{thx} of 35°C. Since bees are unable to reduce their metabolic rate below the resting rate, it follows that in such a large swarm cluster at an air temperature of 5°C there is a considerable excess of heat produced, enough to reduce the individual comfort of the bees in the center. Core bees must either tolerate higher T_b and allow core temperature to rise, or else they must crawl to the swarm periphery where it is cooler. Both responses occur.

At very low air temperatures, bees on the swarm or colony mantle attempt to warm themselves by crawling inward, thereby effectively blocking air channels for heat loss from the core. Core temperatures may (generally temporarily) then sometimes rise to 46°C even at an air temperature of 1°C (Heinrich, 1981a). The higher core temperature should then increase the heat production in the core still further, because of the Q_{10} effect on resting metabolic rate. However, overheated core bees vigorously escaping to the exterior will (inadvertently) loosen the swarm to allow heat to escape and a thermal equilibrium to become established.

Young, as yet poikilothermic bees seek the core, where they are passively warmed. Older bees can shiver and prefer lower temperatures, and these bees seek the cooler peripheral areas of the swarm or nest. Thus, as a result of different thermal preferences, an equilibrium in temperature gradients is established and maintained in the colony, which reduces constant reshuffling of bees each trying to maintain its own optimal body temperature.

Summary

Honeybee workers are unlike most other insects (which normally do not have a constant food supply available) in that they attempt to maintain an elevated T_{thx}, even if it means death by starvation in an hour or less when they are artificially moved from food (the hive). Young bees are poikilothermic and maintain an elevated T_b by behavioral means, by exploiting temperature gradients within the hive. Within a few days, however, hive bees develop the capacity to shiver and to fly. Thoracic temperatures of individual bees that are capable of shivering are quite variable at low ambient temperatures, but collectively, in the hive context, bees produce enough heat so that the ambient temperature near each bee is sufficiently high for these individuals to maintain their T_{thx} close to 35°C. Before takeoff to leave the hive, bees shiver and achieve T_{thx} near 37°C.

During shivering, wing and thoracic vibrations are generally not detectable and bees may appear to be quiet and "at rest." The indirect, power-producing flight muscles are activated during shivering, however, to generate work outputs at any of a wide range of rates up to near those of free flight. The thoracic muscles contract isometrically during shivering, and energy expenditure is, as in flight, a direct function of action-potential frequency.

During flight the animals must continually generate enough power at least to support their own weight. The rate of heat output is high and continuous enough at most air temperatures to generate a temperature excess of some 15°C. Since the minimum T_{thx} for free flight is near 30°C, bees cannot fly continuously at air temperatures lower than 15°C. At low temperatures they stop flight occasionally to shiver and warm up again to continue flight.

Flight endurance at low temperatures is likely enhanced in honeybees by a series or aortic loops in the petiole. The series of loops acts as a counter-current heat exchanger that prevents leakage of heat to the abdomen. Heat retention in the thorax reduces energy costs by reducing frequencies and durations of individual warm-ups during foraging at low air temperatures.

There is little or no evidence for regulation of T_{thx} during flight until air temperatures exceed 30°C. However, honeybee workers have the ability, unusual for a highly endothermic insect, to fly continuously at air temperatures up to 46°C and to regulate T_{thx} at near or even slightly below air temperature. The remarkable thermoregulation of individual worker bees at high temperatures is due to evaporative water loss from regurgitated honey-crop contents. Droplets of regurgitated liquid cool quickly and are sucked back in to pick up heat, and regurgitated again, and so on, in quick, successive cycles that also involve spreading of the droplet with the tongue and sometimes the forefeet. The heat-dissipation response is apparently activated by a sensor in the head and not the thorax. However, head and thoracic temperatures are tightly coupled, because the two body parts touch each other and because there is blood flow between them. The bees regulate head temperature, and T_{thx} stabilization follows.

The specific body temperature regulated by honeybees affects pace and the optimization of short- vs. long-term gains. These energy gains to the colony are related to longevity and availability of immediate profits. Both vary among different species and according to different food rewards.

Remaining Problems

1. Is the increased thermoregulatory ability of honeybee workers with age a function of exercise experience, maturation, and/or specifically of shivering experience?
2. If longevity is correlated with body temperature and pace, do bees that maintain higher T_b die sooner because they somehow "burn out"? Or do they die because of energetic accidents in the field (running out of energy before reaching flowers or the hive)? Or are they genetically programmed to have a specific physiological life span?
3. How (proximally) and why (ultimately) do African honeybees maintain nest and body temperatures similar to European bees despite their smaller size?
4. Nothing is known about the neurobiology of the presumed temperature setpoints in the central nervous system. How do the sensors "work"? How are they set and read?
5. Are drones able to fly continuously and regulate T_{thx} at high air temperatures, as workers are?

The Tolerance of Ants

IN some climates, such as the rainforest, ants may account for nearly a third of the entire animal biomass (Hölldobler and Wilson, 1990). This huge mass is fragmented into millions of tiny bodies, however, and since tiny bodies tend to lose heat rapidly, this chapter on the ability of individual ants to regulate body temperature is a short one. In the nest the individuals are affected by temperature more than they affect it. And outside the nest they quickly assume a T_b dictated by the physical environment.

The generally exclusively terrestrial locomotion of ants (except in the winged reproductives) also precludes endothermy and physiological regulation of T_b, but some species at times operate at T_b higher than that of any other insects on earth. I discuss the ants and, in the next chapter, their close relatives the wasps because they provide an interesting perspective on strategies of thermal adaptation where endothermy is lacking. In addition, some ants, like honeybees, have a social thermal response, which is discussed elsewhere (Chapter 16).

Almost complete dependence of ants' activity on temperature was shown by the famous astronomer Harlow Shapley (1920, 1924). Concerned principally with the study of globular star clusters, Shapley determined that our sun is located not in the center of the Milky Way galaxy but at its outer edge. In so doing, he earned a reputation for doing for the Milky Way system what Copernicus had done for the solar system. But his hobby was ants, and his results on ant "thermokinetics" were published in the prestigious *Proceedings* of the National Academy of Sciences.

Shapley (1920) observed the trails of *Liometopum apiculatum* at the Mount Wilson Observatory in Pasadena, California, within a few yards of a special station maintained there by the United States Weather Bureau. Some of the ant trails did not change over at least two years, and Shapley set up some "speed traps," 30 cm long, in the path of the ants. Using a stopwatch, he recorded the time required for individuals to traverse the trap. Twenty individuals were timed for any one temperature. The ants ran along the trails when the relative humidity was as low as 5 percent and as high as 100 percent. They ran at all hours of the day or night (presumably noticed by Shapley when the stars were not visible), and at temperatures from 8 to 38°C. They ran at the same speed whether going to or from the nest. Nothing affected running velocity except temperature.

Curiously, Shapley plotted temperature as a function of running velocity, as if he were trying to use the ants as a thermometer, possibly to compete with the weather station. Indeed, from a single point of running speed on his graph, temperature can be predicted within 1°C. I have here reversed the axes (Fig. 9.1) to make running speed the dependent variable, even though ambient temperature in the field is probably much less reliably measured with a thermometer than an ant's speed may be measured with a stopwatch. As temperature rose 30°C, running speed changed 15-fold, increasing uniformly from 0.44 to 6.60 cm/s. Shapley (1924), who subsequently moved to the Harvard Observatory, then extended his results to *L. occidentale*, *L. luctuosum*, *Dorymyrmex pyramicus*, *Iridomyrmex analis*, *I. humilis* and *Tapinoma sessile*. He observed the latter species running in columns through a room in which the temperature could be changed with the aid of a gas-burning furnace. Shapley found that when room temperature reached 40°C (106°F) the ants no longer passed through.

The effect of temperature on running speed is likely even more precise than Shapley supposed if individual variation in weight is factored out. Increasing the burden that an ant carries can slow down its pace (Rissing, 1982; Lighton, Bartholomew, and Feener, 1987). And in the army ant *Eciton hamatum*, body mass varies and also greatly affects pace. Minor workers weighing 2 mg run at 7 cm/s while 12 mg workers run at 10 cm/s, whereas 25 mg soldiers speed along at about 12 cm/sec (Bartholomew, Lighton, and Feener, 1988).

The pace of individual ants is relatively precisely pre-determined; they are poikilotherms but they are nevertheless capable

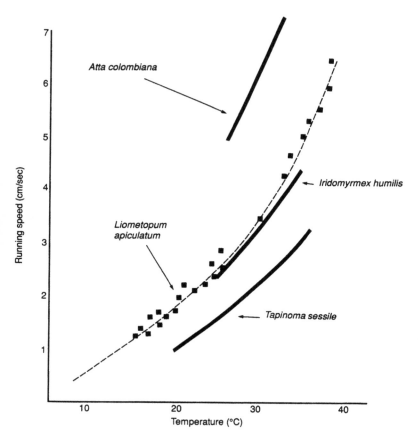

Fig. 9.1 Ant running speed as a function of temperature. Points (given only for *Liometo-pum apiculatum;* from Shapley, 1920) are mean times derived from 20 individuals that traversed a 30-cm speed trap. (Unlike in Shapley's original publication, running speed is here plotted as the dependent variable). Data for *Iridomyrmex humili* and *Tapinoma sessile* from Shapley (1924); data for *Atta colombiana* are adapted from Lighton, Bartholomew, and Feener (1987). Note the different pace of the different species at any one temperature.

of being active over a wide range of temperature. Although ants apparently have no choice in how fast they run, they can and do "decide" *when* to run, and timing profoundly affects pace. Proximally, activity depends on temperature, on food availability (Bernstein, 1974), and on relative humidity or moisture. For example, soil wetting causes intense activity in the harvester ant *Pogonomyrmex rugosus* (Whitford and Ettershank, 1975). The above variables can produce diurnal as well as seasonal activity shifts. Thus, during cooler parts of the year peak foraging may be at midday, with nocturnal activity occurring in the hottest part of the season, or activity peaks may be bimodal—in the fore- and afternoon (Marsh, 1987, 1988; Moser, 1967; Sanders, 1972; Gamboa, 1976; Kay and Whitford, 1978). However, the thermal preferences of different species varies enormously. For example, in the Chihuahuan Desert, the *Pogonomyrmex* harvester ants have maximum foraging activity at 45°C whereas another harvester ant, *Novomessor cockerelli*, peaks at 20°C (Whitford and Ettershank, 1975). Hölldobler and Wilson (1990, pp. 380–381) present a table listing the temperature at which activity begins, peaks, and ends for 43 species, which shows a wide range of thermal possibilities.

Decreasing temperature decreases the metabolic costs of foraging (and presumably also decreases encounter rates with food items), but it makes the ants more "fussy" in that they then reject the less profitable, smaller prey (Traniello, Fujita, and Bowen, 1984).

The ultimate reason why different species are now active at different temperature ranges undoubtedly has to do with the prevailing temperatures over evolutionary time of their environment, especially when food is available. For example, the primitive Australian ant *Nothomyrmecia macrops* is adapted to hunting winged prey in tree tops at night probably because at low nocturnal temperature the prey would be hampered from detecting the ants and slower in escaping from them (Hölldobler and Taylor, 1983). But some food may also be associated with high temperatures.

Most desert ants adjust seasonal and diurnal activity times (Box, 1960; Kay and Whitford, 1978; Briese and MacAuley, 1980; Mehlhop and Scott, 1983; Marsh, 1985a) to avoid the hottest days or times of day, and they stop foraging and retreat to their underground nests at soil temperatures of about 35–45°C (Clark and Connors, 1973; Rogers, 1974; Schumacher and Whitford, 1974; Whitford and Ettershank, 1975; Briese and MacAuley, 1980). However, within each of the very hot, dry deserts of the world there are ants of at least one genus that are strictly diurnal, that seek high temperatures, and that *increase* their foraging activity as temperatures rise above 45°C. These include *Cataglyphis* (Formicinae) of North Africa, *Melophorus* (Formicinae) of Australia, *Dorymyrmex* (Dolichoderinae) of South America, and *Ocymyrmex* (Myrmicinae) of South Africa. As pointed out by Rüdiger Wehner (1987) at the University of Zurich, the reason these diverse ants choose to be active under what appears to be dangerous conditions (numerous of the thermophilic species forage at soil-surface temperatures of up to 70°C) is related to diet: they are all scavengers feeding on arthropods that have died from heat and desiccation. In support of this hypothesis, Wehner documents that *Ocymyrmex velox* from the Namib Desert show greatly increased foraging activity at high (68°C) vs. "low" (51°C) sand-surface temperatures and have greatly increased foraging success at the higher temperature. At 68°C most of the foragers were returning with prey, whereas at 51°C only a few brought prey back to their colony (Wehner, 1987).

One of the most remarkable "thermophilic" ants is probably the silver ant *Cataglyphis bombycina*. This ant forages in the full midday sun in the summer, in the central Sahara, even when surface temperatures reach 63°C. One can assume that this small insect (with a mean body weight of 9.7 mg) would within seconds have a T_b close to T_a. The maximum T_b these ants can tolerate is about 54°C. This temperature is extraordinarily high for an insect, but it is still very close to the upper lethal temperatures the animals should encounter within a few seconds on the sand. The ants prevent themselves from "frying" by pausing frequently to climb dry stalks of vegetation to cool off. If they do not find such refuges they die. That is, they survive very near the upper lethal death point only by exploiting the microclimatological mosaic on their barren sand habitat.

These ants would be remarkable enough for being able to stay alive in such a hot environment, for apparently choosing to be active close to their thermal death point at close to the hottest time of the year and day in an already hot environment. Why not choose to be active when conditions are less severe, like most other insects?

Rüdiger and Sibylle Wehner and Alan Marsh from the University of Namibia (1992) record a remarkable set of observations that shed light on this question. They found, curiously, that *all* foragers leave together in an explosive outburst from their underground nest in the midday sun. The "outbursts" of any one colony last only 3–4 minutes in any one day, and the individual ants then radiate at bursting speed in all directions from the nest. The signal that initiates the mass foraging exodus at a fast run is not known, but it occurs only at very high ambient temperatures following apparent extensive temperature testing by the ants prior to their run to make sure it is hot enough. The Wehners and Marsh speculate that this ant, which feeds on the corpses of heat-killed insects, need not wait for the extraordinarily high temperatures before exiting merely to forage, because the heat-killed prey are already found much earlier, when temperatures are considerably lower. But the ants themselves are prey. They cannot afford to venture out at lower temperatures because the lizard *Acanthodactylus dunerili* that specializes on ants would catch them then. As temperatures rise, however, even the heat-tolerant lizard is forced to retreat underground, and it is precisely at that temperature that the ants burst forth. Were they to exit from the nest a little later, at slightly higher temperatures, they would likely suffer death from heat prostration. In a sense, the lizards and the ants are in an evolutionary arms race, in which the victor is the one that tolerates the highest temperatures or that is best able to thermoregulate. As I have indicated previously, essentially parallel scenarios for predator avoidance may apply to a desert grasshopper and desert tenebrionid beetles. A desert cicada may use the strategy also (Chapter 12).

Extraordinarily high thermal tolerance like that of Saharan *Cataglyphis* ant species are also found in another ant, *Melophorus bagoti* (Fig. 9.2) from the central Australian desert. Keith A. Christian and Stephen R. Morton (personal communication) from the Northern Territory University of Australia report that *M. bagoti* survive for 1 hour at 54°C, and activity in the field does not occur until soil-surface temperatures reach a mean of 56.1°C, at which temperature another Australian arid-zone ant, *Iridomyrmex purpureus*, ceases activity (Greenaway, 1981).

How have the "thermophilic" ants been able to penetrate into the dangerously hot sand environment? First, all have long legs (Wehner, 1983, 1987; Fig. 9.3) that allow them to stilt above the hot substrate. Second, they are extremely speedy. *Cataglyphis fortis* and *C. bombycina* from the Sahara Desert have been timed running at 1 m/s (Wehner, 1983; Wehner, Marsh, and Wehner, 1992), which is near the maximal running speed of the tenebrionid beetle *O. plana*, from the Namib Desert, that cools by forced convection (see Chapter 5). Third, the speedy ants depend on solar and other visual cues for navigation (Wehner, 1984), thus enabling them to return directly to the nest and minimizing the distances they have to travel. In some published pictures (Wehner, 1983; Fig. 9.3) the ants raise their abdomen as though it could serve as a parasol, but the purpose of that posture is likely instead that of facilitating making rapid angular turns; some ants inhabiting the hottest Saharan habitats (namely *Cataglyphis bombycina*) do not show this behavior (R. Wehner, personal communication).

Another behavioral option used by thermophilic desert ants that allows them to hunt for heat-killed arthropods is to make pauses on elevated objects. The Namib Desert ants *Ocymyrmex barbiger* may spend 93 percent of their time directly on the sand at sand temperatures near 51°C (Marsh, 1985b). At higher temperatures,

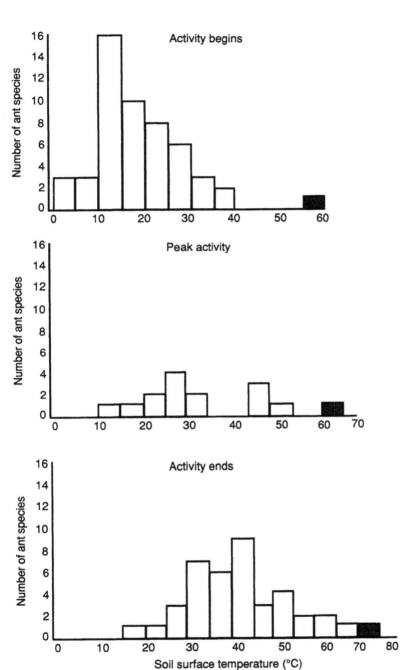

Fig. 9.2 Soil-surface temperatures corresponding to the cessation of activity *(bottom)*, peak activity *(middle)*, and first activity of the day *(top)* for a number of species of ants. Shaded bars = *Melophorus bagoti*. (Figure from Christian and Morton, 1992; data for other ants from Table 10-1, pp. 380–381, in Hölldobler and Wilson, 1990.)

Fig. 9.3 A long-legged ant, *Cataglyphus bicolor*, from the North African desert elevates its abdomen. (From photograph by R. Wehner.)

however, the foragers make frequent pauses (Fig. 9.4) to escape the searing sand surface by climbing up objects, such as grass stalks or goat fecal pellets, where they encounter cooler air (Fig. 9.5). Short pauses occur at all temperatures less than 48°C, but pauses longer than 10 seconds occur only at high temperatures. During particularly hot conditions (>60°C), foraging ants spend more than half of their time away from the nest in thermal refuges (Fig. 9.6) one or more centimeters above the sand, where temperatures are up to 15°C lower than at the surface, and 6–7°C lower than at 4 mm above the surface. Ants experimentally deprived of thermal refuge during foraging at temperatures of 62°C are paralyzed within 60 seconds (Marsh, 1985a,b).

Thermal-respiting behavior has also been observed in certain Chihuahuan Desert ants (Whitford and Ettershank, 1975; Kay and Whitford, 1978). Analogously, desert leafcutter ants, *Acro-*

Fig. 9.4 Relationship between sand-surface temperature and the frequency of thermal-respite behavior of the Namib Desert ant *Ocymyrmex barbiger*. (From Marsh, 1985a.)

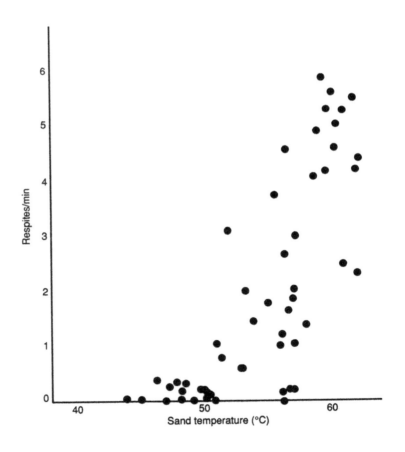

THE HOT-BLOODED INSECTS

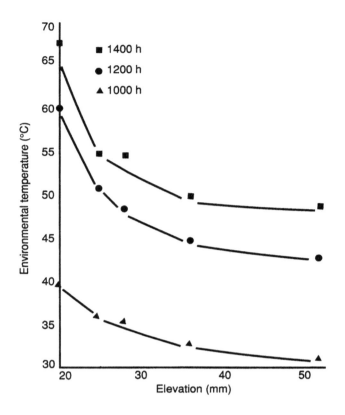

Fig. 9.5 Environmental temperature at different elevations above the sand surface at three different times of day. (From Marsh, 1985a.)

myrmex versicolor, near Tempe, Arizona, at high temperatures (37°C) press themselves onto the cool, shaded ground (at 32°C), presumably to unload heat (Gamboa, 1976).

Although the above behaviors and morphology go a long way to explain why "thermophilia" is possible, another very important consideration is thermal tolerance. Desert ants have extraordinarily high thermal tolerances. The Namib Desert ant *Ocymyrmex barbiger* subsists on arthropod prey found on sand surfaces with temperatures up to 67°C, although the upper temperatures tolerated (until paralysis) by these ants in flasks (where temperatures were raised by 1°C/min) is 51°C (Marsh, 1985b). No ants survived a 30-minute continuous exposure at 51°C, but 80 percent of the ants survived a 30-minute exposure at 51°C if they were given periodic, brief (5 to 30 seconds) respites. The Sahara Desert *Cataglyphis* species, like other desert arthropods (Lubin and Henschel, 1990), have evolved even more extraordinary physiological tolerances to high temperatures. For example, 5 species of *Cata-*

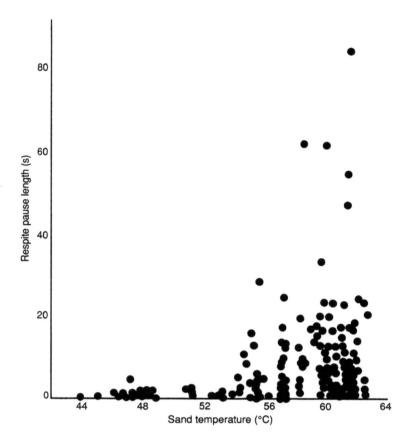

Fig. 9.6 Relationship between length of time *O. barbiger* ants spent in a thermal refuge and sand-surface temperature. (From Marsh, 1985a.)

glyphis exposed to 55°C take 10–25 minutes before they suffer heat coma (Delye, 1967).

Chronic high body temperatures could have costs to an ant colony besides the deaths of individuals due to coma when critical maxima are exceeded. Presumably, longevity is greatly decreased as body temperatures are raised, which could result in long-term costs to the colony as a whole. Although not specifically addressing temperature as a setter of pace, Hölldobler and Wilson (1990, p. 365) speculate that pace or "tempo" in ants could be a product of selection. In general, ants with small colonies and that specialize on relatively evenly distributed or constant food resources are slow-paced, whereas ants having large colonies and that specialize on dispersed and scattered food bonanzas are fast-paced. Even within the same species, the single most consistent determinant

THE HOT-BLOODED INSECTS

of running speeds in two woodland ant species (*Myrmex puncti-ventris* and *Aphaenogaster rudis*) was size. Foragers from large colonies run much faster than those from small colonies (Leonard and Herbers, 1986). Perhaps the ants from small colonies need to be more "careful" or deliberate in their movements so as not to deplete to colony to fatal levels, whereas large colonies can afford to lose more workers, especially if the payoffs may be large. The latter colonies require fast-paced foragers as scouts to find distant food resources. Since temperature is a critical factor affecting pace, the hypothesis predicts that colonies of ants in cool environments should be small while those in hot environments could be larger.

Summary

1. At least during terrestrial locomotion, all ants are poikilothermic.
2. In general, ants have very broad temperature ranges over which they are active, and running pace is a close function of ambient (and hence body) temperature.
3. Different ant species have different thermal optima that depend on the availability of food.
4. Some ants from very hot deserts have evolved extraordinarily high thermal tolerances, and these ants are scavengers specializing on heat-killed prey.
5. Scavengering ants in hot deserts all have very long legs and very fast running speeds. Both features act to reduce body temperatures.

Remaining Problems

1. What are the long-term costs of being active at high temperatures?
2. Ants can choose their pace by choosing the temperatures when they are active. Is that choice related to both immediate foraging success and long-term foraging success as affected by forager longevity?
3. What is the physiological basis of the wide range of temperature adaptation in ants?
4. To what extent are diurnal and seasonal activity shifts driven by temperature effects on the ants themselves, compared with the effects on their prey?

Wasps and the Heat of Battle

W ASPS are another extraordinarily diverse assemblage of insects. They may be parasitic or social, microscopic or surprisingly large. There are "fairy-flies" (mymarids) 0.2 mm in length and the *Trichogramma* (Chalcidae), which are small enough for 70 to grow to maturity within a single butterfly egg. At the other end of the size range is the pompilid spider wasp (*Pepsis* sp.), with a 15 cm wingspread and an 8 cm body that can subdue large tarantulas. There are tens of thousands of graceful, colorful, and agile ichneumon wasps, all of them solitary parasites of other insects. Other kinds of wasps, such as the vespids, have highly evolved social systems.

Thermoregulation

Although most wasps are of small body size and would therefore not likely regulate body temperature independent of ambient temperature, there is every reason to believe from comparative studies that large-bodied species do. However, almost all of the data available that relate to wasp thermoregulation is confined to several social vespids, the hornets and "yellow jackets," from north-temperate regions.

Hornets and yellow jackets (Vespidae) typically build insulated paper nests, and their life cycle is similar to that of bumblebees. After awakening in the spring, the queens make feeding trips as their ovaries begin to develop. Nest initiation starts after 2–3 weeks.

Like bumblebee queens, foundress *Vespula* queens forage for nectar at remarkably low temperatures. Ilkka Teräs (1978) ob-

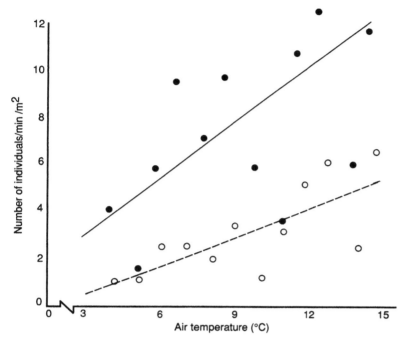

Fig. 10.1 The number of Vespula wasp queens (open circles) and bumblebee queens (filled circles) observed at a study site in Finland at different temperatures. (From Teräs, 1978.)

served them in Finland foraging from whortleberry (*Vaccinium myrtillus*) at temperatures as low as 4°C, which was cool enough to reduce bumblebee activity (Fig. 10.1). Similarly, I have observed *Vespula* queens foraging from blueberry blossoms (*Vaccinium* sp.) on the tundra in northern Alaska in early June at air temperatures near 5°C. Their thoracic temperatures were near 30°C, just above the minimum for flight. As indicated in Chapter 16, *Vespula* queens also generate considerable amounts of heat as they incubate their brood (Ishay, 1973; Makino and Yamane, 1980) in their embryo nests.

Like bumblebees, vespid workers (Fig. 10.2) tend to maintain their T_{thx} relatively independent of air temperature. The workers of *Vespula vulgaris* weigh on the average only 57 mg (near the lower size limit for bumblebee workers), yet despite their small size and their lack of a layer of thick pile, which bumblebees have, they are capable of maintaining T_{thx} up to 25°C above air temperature (Heinrich, 1984). On the average, workers returning to the nest at 7°C have T_{thx} of 28°C, while those returning to the nest at 30°C have T_{thx} of 40°C (Fig. 10.3). Thus, thoracic-tem-

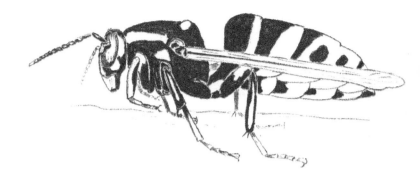

Fig. 10.2 A yellow jacket (*Vespula* sp.) in the alert resting posture.

perature excess at low temperatures is approximately double the excess at high air temperatures. The thermoregulatory mechanism used for halving the temperature excess at high temperatures is not known.

Workers of the white-faced hornet *Dolichovespula maculata* are variable in mass, ranging from 122 to 255 mg. However, on average they are about 3.2 times heavier than *V. vulgaris* workers. As predicted from their size advantage, hornet workers are able to forage at lower temperatures, maintain higher T_{thx}, and show greater independence of T_{thx} from air temperature (Fig. 10.3). These wasps forage already at 2°C, and those returning to the nest have on the average a T_{thx} of 32°C (Heinrich, 1984).

Hornets attacking intruders near the nest have the highest T_{thx} of all. Those leaving the nest to forage (presumably) at 21°C have T_{thx} near 35°C, while those leaving to attack (verified) have T_{thx} of 41°C; internal nest temperature is at least 10°C lower, near 30°C (Heinrich, 1984). These data indicate that although the hornets warm the nest and are warmed in it, they warm up additionally before leaving to attacking or to forage at low air temperatures.

The T_{thx} of the European hornet *Vespa crabro* leaving a feeding dish at 18°C (Stabentheiner and Schmaranzer, 1987) are nearly identical to those of the white-faced hornet leaving the nest to forage at the same temperature.

Curiously, the lower the ambient temperature, the higher the T_{thx} of the hornets exiting the nest (Fig. 10.3). These data suggest that the animals make an extra effort to warm up before confronting low outside air temperatures as well as when confronting enemies, as if to gain additional time for flying before they have to stop to warm up by shivering.

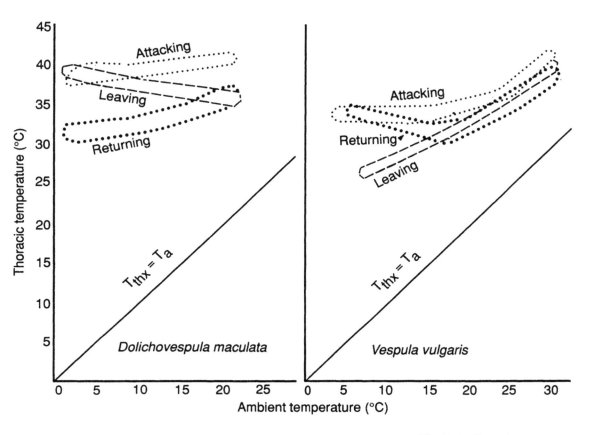

Fig. 10.3 Thoracic temperatures of white-faced hornets (left) and yellow jackets (right) as a function of ambient temperature while attacking a human, leaving the nest to forage, and returning to the nest. In both species the attackers have the highest T_{thx}. Note also that the D. maculata fly at lower ambient temperatures and leave the nest at greatly elevated T_{thx} (nest temperature is maintained near 30°C). (From Heinrich, 1984.)

The mechanisms of thermoregulation in wasps have not been examined. Undoubtedly heat production involves shivering by the flight muscles, as it does in all other endothermic insects. Circumstantial evidence also suggests that some vespid wasps should be able to transport heat into the abdomen. In workers of the European hornet *Vespa crabro* (Fig. 10.4), the circulatory system is superficially similar to that of bumblebees. Like bumblebees, hornets also incubate their brood (Gibo et al., 1977) by direct contact (see Chapter 16), suggesting that *Vespa* also heat with their abdomen. But again, no data are available that demonstrate this. Many solitary wasps have a highly constricted waist (Fig. 10.5), which should work as an ideal counter-current heat exchanger if the animal is endothermic. But the relationship between anatomy and thermoregulatory physiology has not yet been investigated for any wasp species.

Both *Polistes* (Steiner, 1930) and *Vespula* (Weyrauch, 1936) use

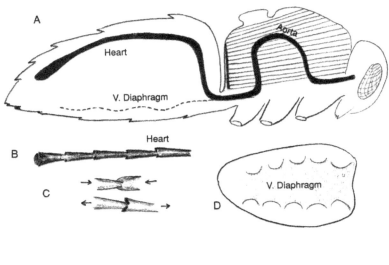

Fig. 10.4 *(A)* The circulatory anatomy of a worker of the European hornet *Vespa crabro;* length = 25 mm. *(B)* The heart in the stretched position, as seen ventro-dorsally to indicate ostia (small openings) along the sides. *(C)* An ostium in the closed position *(top);* it opens *(bottom)* when the heart is stretched. *(D)* The view onto the ventral diaphragm, showing its fenestration. (B. Heinrich, unpublished.)

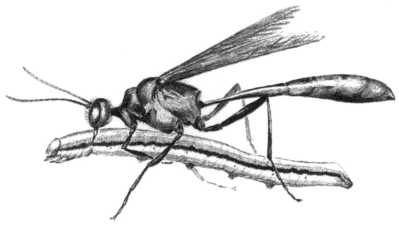

Fig. 10.5 Wasp (*Ammophila* sp.) dragging a caterpillar to its burrow. Its narrow waist is ideally suited for counter-current heat exchange as blood travels between the hot thorax and the cool abdomen, but as of yet the possibility that it acts in this way has not been investigated.

water for evaporative cooling in the nest, and it is therefore at least possible that they do so for individual thermoregulation, as honeybees do.

Behavioral Ecology

The increased T_{thx} of attacking hornets is correlated with impressively increased flight speed as well (Heinrich, unpublished qualitative observations). While thermoregulation obviously enhances the wasps' ability to attack an intruder, it could also function in

THE HOT-BLOODED INSECTS

their pursuit of prey. The white-faced hornets, which feed on other insects, pounce after a short and fast accelerating flight on all suspected prey. Success in hunting would depend in part on an advantage in speed of attack relative to the reaction time of prey. Such a tactical advantage can be gained by maintaining a high T_{thx} and hunting at low air temperatures, when the prey has slow reaction times.

In possible support of the above hypothesis, a large exodus of white-faced hornet workers leave the nest on cold dawns; 2.7 times more workers may be leaving per unit time interval at this time of day than 3 hours later, when it is warmer (Heinrich, 1984). Neither bumblebees nor the closely related but nonpredatory wasps *Vespula vulgaris* leave the nest in such large numbers so early on cool mornings. Instead, in both groups activity increases directly with increasing air temperatures (Fig. 10.6). Perhaps the hornets' thermoregulatory ability allows them more easily to prey on small, winged insects that might be difficult to catch at higher temperatures. The solitary wasp *Crabro monticola* similarly hunts preferentially in the early morning or at low air temperatures (Evans, Kurczewski, and Alcock, 1980). Taken together, the cir-

Fig. 10.6 The numbers of wasps leaving *(filled squares)* and entering *(open squares)* their nest as a function of ambient temperature. (From Heinrich, 1984.)

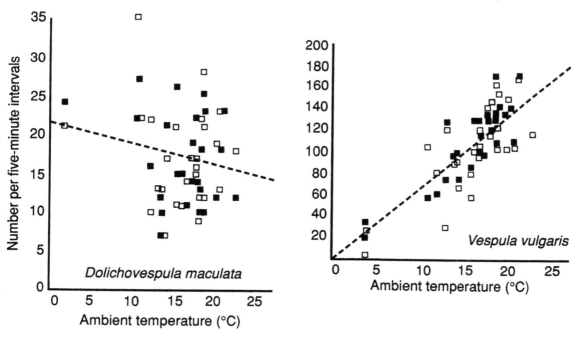

cumstantial evidence for the thermoregulatory hypothesis is so far suggestive, but it has not yet been empirically examined.

Large body mass should facilitate an insect's ability to keep warmed up at low temperatures, but it would work against the insect's endurance in flight at high temperatures, when overheating can occur, unless it is compensated for by heat-loss mechanisms. Some studies suggest that in wasp species having large size variations, individuals of small size have an advantage in hot weather (Willmer, 1985a). Conversely, at least in bees, large size confers a flight advantage earlier in the day at cooler temperatures (Heinrich and Heinrich, 1983; Larsson and Tengo, 1989; Willmer, 1985b). Temperature may explain a significant part of the microhabitat selection that differs between individuals of different body size (O'Neill and O'Neill, 1988) as well as mating advantage relating to size (Larsson, 1989a,b).

Ideally, animals of different mass should be able to be active at different optimal body temperatures, as determined by heating and cooling rates. Biochemistry can then evolve accordingly. However, the biochemical machinery in any one species is necessarily conservative; it cannot vary among individuals as a direct function of specific body mass. Even in bumblebees, whose mass may vary over an order of magnitude, one T_{thx} is required for *all* bees of a species regardless of their mass (Heinrich and Heinrich, 1983) because the mass of each adult bee cannot be predicated by the zygote. Mass variations of adults depend on the vagaries of the food supply and on feeding, which cannot be predicted for each individual; thus the thermal optima are preprogrammed evolutionarily and fixed in the biochemistry of the species.

Some wasps routinely face not low but rather very high temperatures. For example, many bembecid and other solitary wasps dig their nest tunnels (where they provision their eggs with prey) in open sand. They may encounter near-lethal temperatures at the sand surface. Penetration of the hot surface of sand dunes is accomplished as the wasps alternately dig furiously at the surface for a short period of time and then fly about 15–30 cm above the surface of the sand to cool off. As the burrows deepen the wasps encounter lower subsurface temperatures and the flights become less frequent. Chapman and colleagues (1926) clipped off the wings of some of the wasps at a time and in an area where other wasps were alternately digging and flying. The wasps with clipped wings died of heat prostration on the sand surface, whereas the control animals kept on working. The air above the sand thus provides a means for convectively unloading excess heat.

One type of wasp, the mutillids (the females are commonly called "velvet ants") frequent the hot sands where bembecid wasps build their nests—because they parasitize their brood. Mutillid females are wingless, hence they are unable to leave the sand. Their ability to be active on hot sand dunes is not so much to thermoregulation but to physiological tolerance of high temperatures instead. Females of the mutillid *Dasymutilla bioculata*, for example, do not begin activity until 24 °C, and they do not suffer heat paralysis until they encounter temperatures near 52 °C (Mickel, 1928). The winged males can more easily escape the heat, but they have nearly identical (within 1–2 °C) thermal requirements and tolerances. Presumably, thermal tolerances depend on a vast array of the body's biochemistry, not just on a few proteins that can be partitioned to one of the sexes.

Summary

Given their similarities in behavior and morphology, many wasps should have much the same mechanisms of thermoregulation as bees of comparable size. However, almost no physiological studies are available to confirm or refute this hypothesis.

Like other large myogenic flyers, large wasps (at least in the Vespidae) warm up prior to flight. Whether or not T_{thx} is regulated in continuous flight is not known, but at least under field conditions, where both physiological and behavioral mechanisms are simultaneously available, several vespine wasps regulate their T_{thx} during foraging activity.

Some vespine wasps are active in the field near air temperatures as low as those that serve as the lower threshold for activity in bumblebees. Even though they are not insulated with pile, wasps of a size near that of bumblebees have T_{thx} close to those of bumblebees. Like bumblebee queens, foundress vespine queens incubate their brood.

Why have northern vespine wasps not evolved insulating pile? One possibility is that as predators, at least some vespine wasps exploit their thermoregulatory ability at low temperatures to hunt smaller insect prey, which as poikilothermic insects are slow-moving in the cold. The wasps' lack of insulating pile may decrease their aerodynamic drag, which would increase their flight speed and thus their hunting success. Modeling of this problem has not been done.

In a number of species of solitary wasps, in which hunting and mating activity is a function of body temperature, smaller animals have an advantage at high air temperatures because they overheat less readily, whereas larger individuals have an advantage at lower temperatures because they can heat up more easily than smaller individuals can.

Remaining Problems

1. Do large, predatory wasps of hot environments regulate their T_{thx}? If so, how?
2. Do small, fast-paced social wasps have convolutions of the aorta, similar to those in the honeybees, to conserve heat in the thorax?
3. Do some of the wasps that use water to cool the nest evaporatively also use water evaporation to reduce their own body temperatures at high air temperatures?
4. Do large, predatory wasps use their ability to maintain a high T_{thx} at low temperatures to their advantage to hunt smaller, poikilothermic prey?
5. Do large wasps have a counter-current heat exchanger in their narrow "waist" to conserve heat in the thorax, or do they use the narrow waist as a heat radiator?
6. Do large wasps regulate their T_{thx} in continuous flight? If so, how?
7. Why do wasps that fly at the same temperatures that bumblebees fly not have insulating pile?

Flies of All Kinds

THE order Diptera, the "true flies," has been around for a long time. Some of the oldest fly fossils (from Australia) date to the Upper Triassic, about 190 million years ago, and the relatively advanced forms already existing then suggests that Diptera must have appeared much earlier than that, possibly in the Permian period, 220 million years ago (McAlpine, 1979). Dinosaurs emerged at about the same time, but they have been extinct for 65 million years.

The almost infinite variety of fascinating forms and specializations of Diptera are a gold mine for the study of adaptation and evolution. It is estimated that the order includes some 200,000 species worldwide, nearly half of which were recently still unnamed (McAlpine, 1979). The catalogue of the Diptera of America north of Mexico (Stone et al., 1965) recognizes 105 families, 1,971 genera, and 16,130 species. Diptera are not only one of the largest and most diverse elements of our insect fauna, they are also of outstanding importance for their wide-ranging ecological and economic impact. The blood-sucking Diptera (Culicidae, Simuliidae, Ceratopogonidae, and Tabanidae) rate among the world's most significant pests, and they are the vectors of many diseases of terrestrial vertebrates. Larvae of some species, especially the Tachinidae, are parasites of Lepidoptera and are extremely important biological control agents in the management of agricultural crops and in the natural protection of forest trees.

Some flies, particularly Drosophilidae and Calliphoridae, have become key experimental animals for research on genetics, morphogenesis, cold-hardiness, and the mode of action of insecticides. Fruit flies (*Drosophila* sp.), which because of their small size are almost certainly not endothermic, are particularly noteworthy in having served as a model system in numerous studies examining the effects of body temperature on aging (Maynard Smith, 1963), longevity, (Hosgood and Parsons, 1968; Murphy, Giesel, and Manlove, 1983), biochemical strategies of temperature adaptation (Morrison and Milkman, 1978; Alahiotus, 1983; Czajka and Lee, 1990), egg production (Clavel and Clavel, 1969), mating frequencies (Schnebel and Grossfield, 1984), and ecological distribution (Parsons, 1978; Kimura, 1988).

Many species of flies today live in cool or temperate regions and are cold-adapted. Even though the variety of taxa progressively diminishes northward and southward, the relative abundance of flies in the total insect fauna increases in the far north. Diptera, especially the midges (Chironomidae), are more numerous than all other insects in the most severe Arctic conditions and in some other localities of low temperature, such as the Antarctic (Sugg, Edwards, and Baust, 1983).

The Small and the Cold-Adapted

Perhaps the most extreme example of a cold-tolerant midge was found by Shiro Kohshima (1984) from Kyoto University on the Yala glacier of the Nepal Himalayas. Like many other cold-adapted Diptera, the adults of this midge (*Diamesa* sp.) have reduced wings and are unable to fly. Walking activity was observed usually when the snow-surface temperature ranged from 0.0 to −12.8°C, although individuals were found walking on the surface of the glacier at temperatures as low as −16°C, an all-time record for low-temperature activity in an insect. Although the insects preferred to walk about when the sun was shining, they were also observed at night, when temperatures ranged from −10 to −14° C. In contrast to their strong cold tolerance, these glacier flies were very sensitive to higher temperatures; when placed onto a human palm they became hyperactive for a few seconds and were then paralyzed by heat torpor, recovering only if quickly released again onto the snow.

Flies are well known for their ability to fly at low temperatures. Once (January 9, 1990) while I was walking in the woods with snowshoes over deep snow near Weld, Maine, on a cool (−1 to 0°C), overcast day, I encountered an insect slowly flying by and batted it onto the snow. The insect was a winter gnat *Trichocera saltater* (Trichoceridae). Surprisingly, this small (estimated 2–

6 mg), frail-bodied fly flew after a week where temperatures had repeatedly dropped to less than −30°C. One would have guessed that mechanisms of cold-hardening might have reduced flight performance. (Did it have antifreeze?) Another winter gnat, *T. annulata,* has been reported to have a threshold for flight near 5°C (Littlewood, 1966). It can generally be assumed that small flies such as these are poikilothermic. Nevertheless, cryptic behavioral responses may be used to ameliorate at least some of the extremes of the thermal environment (see Jones, Coyne, and Partridge, 1987).

These winter gnats must have been subjected to body temperatures within a few degrees of −30°C. Perhaps they have mechanisms either of supercooling or of surviving being frozen. However, cold shock itself can cause lethal injury in unacclimated insects, even though there is no tissue freezing in those insects that survive in the supercooled state.

The cold-hardening response can occur extremely rapidly, even overnight as temperature is changing. For example, fruit flies, *Drosophila melanogaster,* do not survive acute exposure to −5°C. However, if the flies are first chilled at 5°C for as little as 30 minutes, approximately 50 percent of them subsequently survive exposure to −5°C for 2 hours (Czajka and Lee, 1990). How the winter gnats adapt to be intermittently active as temperatures vary from at least −30°C to 0°C remains to be elucidated.

On another occasion I was cross-country skiing in Vermont (in February 1986) on a late afternoon when snow was falling and the temperature was −3°C. Some 10 inches of fluffy snow lay on top of another foot and a half of solidly packed snow—hardly a place to expect insects to be out and about.

What a shock then, as I was plowing laboriously along, to see a six-legged creature walking nimbly along on the top of the snow. After another mile I spied another, and in 5 miles I had seen about five. On the bright white snow I could have hardly missed a pin, much less a black creature the size of a skinny bee (Fig. 11.1). I put each one in turn into the only thing I had handy, a matchbox. And when I came home, they were alive and well. One pair was *in copula.*

They were "snow flies"—genus *Chionea,* family Tipulidae—a crane fly without wings. George W. Byers at the University of Kansas, the world's authority on this group, has written a valuable monograph (Byers, 1983) of the *Chionea* of North America. It is a treasure trove of information about these fascinating insects.

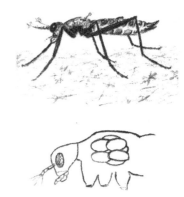

Fig. 11.1 *Chionea vulga* walking on the snow. In the female snow fly, which is wingless, the thoracic volume is used to store eggs. (From Byers, 1969, 1983.)

Apparently these flies live in underground cavities (from rodents' burrows to caves) and their wanderings on the snow may be a dispersal mechanism that prevents inbreeding. Several entomologists have commented on the rapid reaction of males and females of *Chionea* to each other's presence when placed into the same container at subfreezing temperatures. As reviewed by Byers (1983), snow flies are most commonly active on the snow at air temperatures between −4 and −5°C. Apparently their activity is narrowly limited by temperature, because they are rare at even slightly higher temperatures (−3 to 0°C), and their lower temperature limit is near −6°C (Hågvar, 1971).

The insects just described are strict low-temperature conformers. An example of a strict *high*-temperature conformer is the louse fly (*Lynchia* sp.), which lives as an ectoparasite close to the skin of homeothermic birds for its entire life, thus continuously experiencing a temperature within 1–2°C of 40°C in its environmental niche. If one presumes that louse flies choose ambient temperatures to attain specific T_b, then one might consider them to be good thermoregulators. In general, though, it seems more reasonable to suppose that habitat niche determines thermal preferences, and not vice versa. For example, members of some of the most cold-hardy Diptera, such as tipulids, mosquitos, and midges, also occur throughout the tropics.

Winglessness or reduction of wing size is a general phenomenon among insects adapted to very low temperatures (Downes, 1965; Byers, 1969). In addition to *Chionea* crane flies, high-altitude Arctic (Downes, 1965) and Antarctic (Sugg, Edwards, and Baust, 1983) chironomid midges also lack wings, as do the mecopteran *Boreus*, numerous stone flies (Plecoptera), and various Lepidoptera (particularly among the Geometridae active in late fall or early spring) from northern regions. In many species with perfectly formed wings, flight in an area of low prevailing temperatures may be restricted to the few hours when temperatures exceed the flight threshold. For example, few winged stone flies (Plecoptera) are active in midwinter, but in these species flight is restricted to the occasional warm, sunny days when temperatures rise above freezing (Frison, 1935). Many winter insects never fly, because cold durations can be long and their life spans as adults are commonly restricted to only a few days (although the total life cycle tends to be long in winter insects). In some species, presumably those that had rare opportunities to fly, the wings have become useless and hence a burden to be selectively reduced in

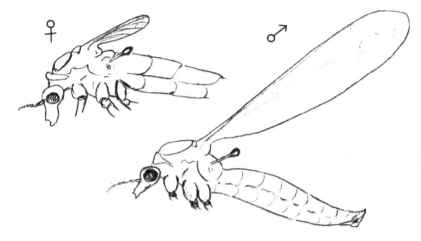

Fig. 11.2 In the Alaskan crane fly *Tipula bergrothiana,* the female is brachypterous and the male, though smaller-bodied, has full-sized wings (from Byers, 1969).

size (Fig. 11.2) or eventually to be eliminated. From there it was only a small step to complete winglessness. In *Chionea* even the thoracic muscles are gone (Fig. 11.1), and the space normally allocated for them is often packed with eggs instead (Byers, 1969).

In most of the tipulid species showing wing reduction or total wing loss, it is the female that is brachypterous while the male is winged (Byers, 1969). The same holds for cold-weather Lepidoptera. Byers (1969) speculates that the males of these species may have been selected to retain the ability to fly in search of mates simply because without the burden of eggs they have a lighter body and can therefore fly more efficiently. Flight by the ovipositing female may also be a waste of energy if she is already near the larval food supply.

The threshold for flight in some temperate and Arctic Diptera can be quite low. In a marshy meadow in the high Rocky Mountains, Byers (1969) saw crane flies *Tipula pendulifera* begin to crawl at 4.4 °C in the early morning to positions of sunlight, though they were still unable to fly at 6.7 °C. As temperatures increased to 7.2 °C they had, presumably in part by basking, achieved sufficient T_{thx} to fly a few inches, and large numbers of them did so nearly at the same time.

Flights of mosquitos in the Arctic can also be quite striking. In northwestern Alaska I observed abundant mosquitos at 6 °C on an overcast day, but there were none to be seen at 4 °C. Byers (1969) reports central Alaskan mosquitos inactivated by 4.4 °C but flying at 7.8 °C. In northern Manitoba, Canada, the two mosquitos *Aedes nigrepes* and *A. impinger* fly at near 4°C (Haufe and Burgess, 1956). Because convection quickly removes all heat in tiny objects, the T_b of mosquitos is within a fraction of a degree of air temperature (Vinogradskaja, 1942; Heinrich, 1974). The lower limit of T_{thx} for flight of northern mosquitos is therefore likely near 4.5°C, which is considerably lower than that of other mosquitos.

Nothing is known about how the necessarily very rapid muscle contractions supporting flight in dipterans at temperatures near 5°C are possible. Myogenic muscles like those of flies have notably reduced sarcoplasmic reticulum (SR), hence the elaboration of the SR proposed for moths at low temperatures (Chapter 1) is not likely a factor. Does it instead involve specific calcium-binding proteins such as parvalbumins (Gerday, 1982)?

In *Aedes aegypti*, the tropical "yellow fever mosquito," the minimum air temperature (and T_b) for flight is 10°C (Rowley and Graham, 1968). In this species, 28°C appears to be optimal for biting activity (Connor, 1924). In a detailed study on the effect of temperature and humidity on *A. aegypti*, Wayne A. Rowley and Charles L. Graham at the U.S. Army Biological Laboratories at Fort Detrick, Maryland, flew female mosquitos (previously maintained with access to 0.3 M sucrose) to exhaustion on flight mills in the laboratory at different temperatures and humidities. Humidity had no effect on flight performance, but temperature effects were marked. The mosquitos flew most willingly and long (averaging 14.4 km) at 21°C, although flight speed peaked (near 28 m/min) at 27–32°C. Flight performances, both in speed and distance covered, decreased markedly at both 35 and 10°C (Rowley and Graham, 1968). Essentially parallel results were found for the sheep blowfly *Phaenicia sericata* (Yurkiewicz, 1968), although these insects may fly nearly twice as long before exhaustion (fuel depletion?) sets in. To what extent these differences seen in Dipterans (see Maynard Smith, 1957; Dingley and Maynard Smith, 1968; Meats, 1973; Thiessen and Mutchmoor, 1967) represent temperature acclimation, as seen in other poikilotherms (Asit and Prosser, 1967; Hochachka, 1965), is not known.

Up to now the studies of dipteran flight as a function of temperature have ignored one important variable: load. Although 4.5°C may indeed appear to be the lower limit of temperature in Arctic mosquitos, that limit could vary dramatically depending upon whether the insect has recently fed. It is not known if the 4.5°C limit represents a physiological or a behavioral threshold;

THE HOT-BLOODED INSECTS

if it is a physiological threshold then only unloaded (unfed) individuals should fly at 4.5°C. On the other hand, perhaps unloaded mosquitos can fly at even lower T_b, but only when they could continue to fly if they happen to get a blood meal. Nevertheless, this seems doubtful, in that I have often observed mosquitos fill up with blood and then become unable to achieve level flight even at relatively high (>20°C) air temperatures.

Load-lifting capacity can have a profound influence on reproductive success. Male dance flies, *Hilara* sp. (Diptera: Empididae), fly in swarms, where they present insect prey as nuptial gifts to females. During prey transfer and initiation of copulation, the males lift not only their own mass but also that of the prey and the female. Males *in copula* have greater relative flight-muscle mass than randomly captured males (Marden, 1989), suggesting at least one strong selective pressure for increased body and flight-muscle mass. However, these small flies (males *in copula* weighed on the average 3.75 mg) are undoubtedly poikilothermic. The results would therefore depend on temperature, because at slightly higher temperatures the flies with a low flight-muscle ratio should be able to overcome their load-lifting disadvantage. If so, the mean flight-muscle mass of males *in copula* should predictably decrease with increasing temperature.

FLOWER BASKING

In the High Arctic the flowers of several species of plants turn their corolla to face the sun and then follow it throughout the day. Primary among them is the Arctic poppy *(Papaver radicatum)*, which follows the sun throughout the entire 24-hour period (Kevan, 1972). Since the poppy flowers are paraboloid in shape, they were presumed to focus heat onto the germ cells. Hocking and Sharplin (1965) measured temperatures of up to 10.5°C above air temperature inside flowers and they speculated that the heliotropism of the parabolic flowers functioned to help ripen the flowers' germ cells. (However, since they used blackened thermocouples they may have been measuring direct solar radiation. The higher temperatures measured would be expected, given that convective heat loss was reduced by the corolla shielding the thermocouple.) They also mentioned that although the poppies secrete no nectar, mosquitos were observed sitting in the corollas, and they suggested that they were basking to accelerate gonadal development.

Peter Kevan (1975) subsequently also examined the solar-tracking of *P. radicatum* and measured the T_{thx} of numerous mosquitos and empidid, syrphid, and calliphorid flies placed into the flowers. The insects showed temperature excesses ranging from 6 to 15° C. Mutilating or removing the corolla from the flowers reduces the temperature excess of the insects, but this does not prove that the insects are heated because of heat focused onto them by the flowers acting as reflectance parabolas, as claimed.

It is feasible that the insects in the flowers were heated by reflected light from the corolla, but they were almost certainly heated primarily by direct solar radiation, as are other Arctic insects. (Would they still have been heated 6–15°C if their bodies were *shaded* while the corolla was left to receive sunlight?)

Cooling is another consideration affecting T_b. Small insects are extremely sensitive to the slightest air movement, and it is not yet clear to what extent the effect of the parabolic petals is due to heating of reflected light as presumed, or to what extent the deep corolla functions as a convection baffle to allow direct solar heating of the insect, which would likely otherwise quickly be eliminated by the smallest air movements. (See also the discussion of reflectance basking of butterflies in Chapter 2.)

In addition, it is also not clear to what extent insects selectively use the flower perches and what biological significance they have. Insects would be very conspicuous on the bright yellow surface of an Arctic poppy. In my own brief stay in the High Arctic of Alaska, I undoubtedly saw millions of mosquitos, but only very rarely was one perched inside one of the very numerous Arctic poppy flowers. Small flies are known to be sensitive to microclimatic variables (Willmer, 1982b). But possibly they use many perches. Do the small Arctic Diptera *preferentially* (and not just incidentally) perch in flowers or open leaves or grass stems, for example? Just because an insect is perched in the sun and thereby enjoys an elevated T_b does not yet mean it is basking. Basking implies a directed behavioral response to seek heat. At temperatures close to the lower thermal threshold of flight (near 5°C), one would have expected the poppy flowers to have become loaded up with mosquitos. But I very rarely saw any there. If perching in heliotropic poppies is a basking response, then the lower the temperature, the higher the predicted incidence of such perching. To my knowledge, no such data are published.

There are other unanswered questions about flower basking as well. If the dipterans perch preferentially in flowers in order to *bask* there, do they prefer those facing the sun as opposed to those that may be experimentally pointed elsewhere? If they perch in both shaded and illuminated flowers, is the duration of their stay longer in flowers in the sun as opposed to those in shade? Do they perch in flowers on days or times of the day when the sun shines? In sum, is the basking by Arctic Diptera in solar-tracking flowers an adaptation or an incidental or accidental behavior? The data do not provide an answer and thus the well-known use of flowers as "solar furnaces" remains a presumption or a speculation.

Thermoregulators

In general, Diptera have low body temperatures and they do not thermoregulate in flight because their small body mass results in rapid convective cooling and prevents any appreciable increase in T_{thx} from either endothermy or solar radiation (Digby, 1955; Willmer and Unwin, 1981). Midges and mosquitos generate a temperature excess of only a fraction of a degree. Even the relatively robust (30 mg) sheep blowfly *Phaenicia sericata* generates a maximum temperature excess of only about 1.2° above air temperature (Yurkiewicz and Smyth, 1966a). Clearly T_{thx} is unregulated in flight; it is, rather, a direct function of ambient temperature, as reflected by the close correlation between wing-beat frequency and air temperature (Sotavalta, 1947; Digby, 1955). As a result, in flies such as the blowfly *Phaenicia sericata*, the rate of oxygen consumption doubles during flight (Yurkiewicz and Smyth, 1966b) as temperature increases from 15°C to 30°C, although the thoracic-temperature excess at the same time only increases from 0.96 to 1.24°C over the same range of air temperature (Yurkiewicz and Smyth, 1966a). (There is no doubling in temperature excess probably because the doubling in heat production is counteracted by increased convective cooling due to faster flight speed.) Endothermy and thermoregulation in continuous flight are likely unavailable to all except possibly the very largest Diptera. However, a fair number of dipterans thermoregulate in the field, where they have numerous behavioral, and sometimes also physiological, options available.

As in other insects, hot-bodiedness in flies is the result of a combination of three factors: high metabolism, large body mass, and availability of direct solar heating. Among those fly species having a very high flight metabolism and availability of sunshine are the syrphids. The males of some *Syrphus* species (Heinrich and

Fig. 11.3 A male of a *Syrphus* species.

Pantle, 1975; Gilbert, 1984) gather in leks (display areas) where they intercept passing females. Their high-pitched whine in flight, and their rapid acceleration as they dart off in pursuit, reminds one of speeding bullets. Although sheep blowflies *Phaenicia sericata* generate a temperature excess of only 1 °C in flight (Yurkiewicz and Smyth, 1966a), syrphids of the same mass (27–30 mg) maintain T_{thx} of 30–38 °C in the field at air temperatures of 10–25 °C (Heinrich and Pantle, 1975).

Syrphid flies make good use of the thermally variable microhabitats in their environment. In the early morning, at temperatures near 10 °C, they perch on the ground, where they bask and are heated to near 28–29 °C. The flies are basking, not just incidentally perching in sunshine, because if they are shaded they immediately fly up and land a few centimeters away in sunshine. They may then also shiver to bring T_{thx} to 30 °C. As the ground is heated to over 35 °C they stop basking and shivering and then perch above ground on vegetation (Fig. 11.3), where it is cooler. Finally, as solar radiation and air temperatures increases so that a fly on the ground would be heated to 45 °C, and one on a leaf to 33 °C, they start hovering in place at the lek, maintaining T_{thx} near 35 °C. Any flight at low air temperatures would, in these small insects, immediately result in cooling of T_{thx} below the minimum for flight (Gilbert, 1984).

The shivering response of these syrphids can (sometimes) be determined by the whine of their partially engaged wings, but it is more reliably determined by simultaneous electrical recordings of T_{thx}, wing vibrations, and the electrical activity of the flight muscles (Fig. 11.4). In the laboratory some flies remain endothermic by shivering up to 20 minutes at a time. Given the very fast passive cooling rate of the thorax (8.6 °C/min at a 10 °C difference between T_{thx} and air temperature), a fly with a thorax weighing 0.0145 g (and having a specific heat of 0.8 cal/g/°C), would have to expend at least (3.35 J/g°C × 0.0145 g × 8.6° C/min) = 0.42 J/min if it stabilized its T_{thx} by endothermy. Flies at their leks on the average hold only about one-quarter of their maximum crop capacity (with sufficient fuel for only 39 minutes of shivering), thereby increasing their flight maneuverability and speed. Lower than maximal loads must compromise endurance at the lek, but at low temperatures basking can greatly prolong the length of stay at the lek since it results in considerable energy savings (see also Gilbert, 1984). One could conclude that at a mere 30 mg or less, syrphid flies are probably one of the most impressive endotherms known, and they lack insulating pile.

The flies are on location in their leks at low temperatures only if they have the opportunity to bask there. Given their rate of warm-up by shivering (Fig. 11.5), however, extrapolations suggest that they might be physiologically capable of warm-up even at temperatures near 2°C. Presumably such impressive shivering capacity would be used only reluctantly and sporadically. It would not likely be used simply to stay warm. Instead, it likely allows the flies to escape shade and seek sunflecks to bask in on cool mornings, rather than having to wait until sunshine strikes them.

Many species of syrphids mimic wasps and bees (Fig. 11.6) in details of color patterns and even behavior. In a study of thermoregulation of 12 species of bee- and wasp-mimicking syrphid flies in Maine ranging in mass from 0.05 to 0.33 g (Morgan and Heinrich, 1987), we found many of the models and mimics foraging from the same flowers at the same time, and they were often difficult to tell apart until in the hand. Even then the defen-

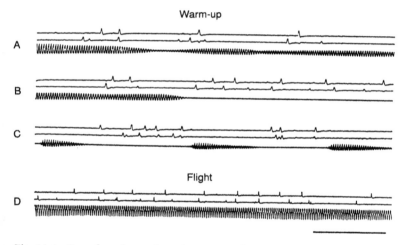

Fig. 11.4 Samples of muscle action potentials and thoracic vibrations during warm-up *(A, B, C)* and during tethered flight *(D)* in *Syrphus* flies. Each of the 4 groups of samples was recorded from the same fly with the electrodes in the same position. The upper record of each group shows the action potentials from the dorso-ventral muscle (indirect wing elevators), the middle record shows the action potentials from the dorsal longitudinal muscle (indirect wing depressors). Note the "cross-talk" from a neighboring muscle unit recorded in the middle trace. The bottom line in each set shows the accompanying vibrations of the thorax. Note that the amplitude of thoracic vibration during warm-up is variable and that shivering can proceed in the absence of external vibrations. Ambient temperature = 16°C during warm-up and 25.5°C during flight. Time mark = 200 ms. (From Heinrich and Pantle, 1975.)

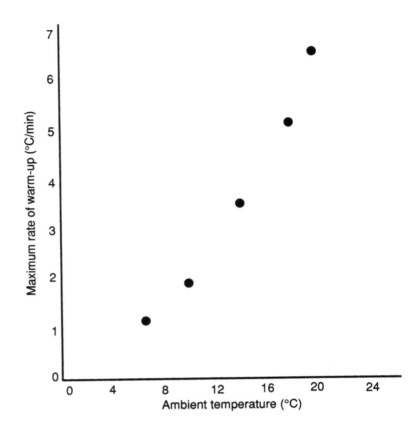

Fig. 11.5 Maximum observed rates of pre-flight warm-up of a *Syrphus* fly at five different ambient temperatures. (From Heinrich and Pantle, 1975.)

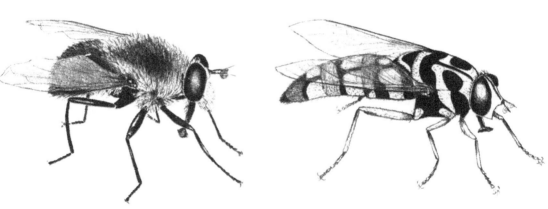

Fig. 11.6 Two flies that mimic stinging *Hymenoptera*. *Left:* the bumblebee-mimicking fly, *Criorhina* sp. (from photograph by E. S. Ross). *Right:* the wasp-mimicking fly, *Milesia virginiensis*.

THE HOT-BLOODED INSECTS

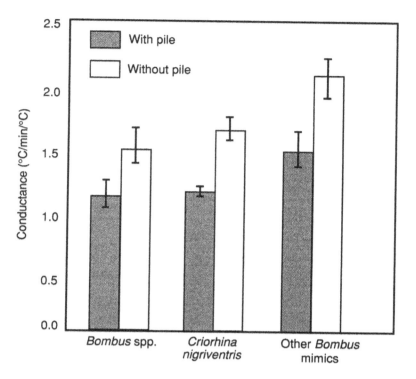

Fig. 11.7 Thermal conductance measured at a wind speed of 5.6 m/s before and after removal of the thoracic pile for *Bombus* species and *Bombus* mimics. The category "Other *Bombus* mimics" includes *Mallota posticata, M. bautias,* and *Eristalis barda,* which have very similar short, dense pile. *Criorhina nigriventris* has longer pile. (From Morgan and Heinrich, 1987.)

sive buzzes of these flies sound like those of their models. The bumblebee mimics are covered with a thick pile, whereas vespid wasp and nonmimetic species are glabrous or have sparse pile. The presence of pile in bumblebee mimics results in a similar decrease in cooling rate, as it does in the models (Fig. 11.7).

The above flies are endothermic when they are active in the field, but over the temperature range of 15–25°C they do not regulate T_{thx} as precisely as their models do. Thoracic temperatures of 4 *Vespula* (yellow jacket wasp)-mimicking syrphids and 1 *Dolichovespula* (white-faced hornet)-mimicking syrphid average 10–13°C above air temperature, depending on the species, whereas both models average T_{thx} 15.5°C above air temperature. The 4 *Bombus*-mimicking syrphids have thoracic-temperature excess similar to the glabrous mimics (averaging 10–15°C), whereas their models have T_{thx} near 18°C above air temperature.

It was not determined by what means (shivering, flight, basking, etc.) the mimics maintain elevated T_{thx} in the field. However, all of them are capable of considerable endothermy through shivering in the laboratory (Fig. 11.8). Some syrphid flies that mimic Hymenoptera increase T_{thx} only a few degrees before takeoff, initi-

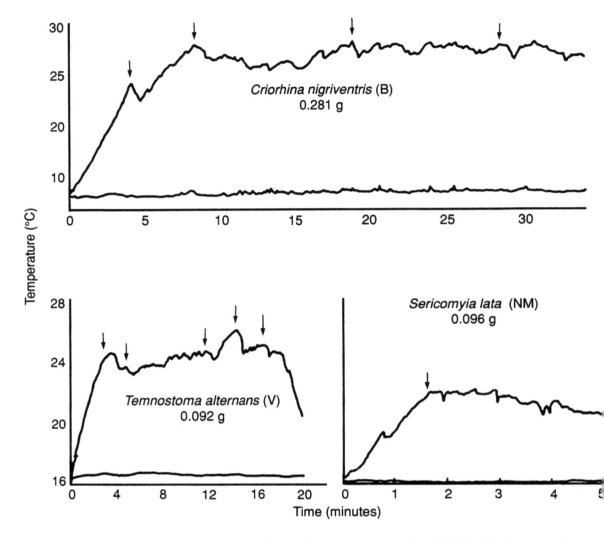

Fig. 11.8 Endothermic warm-up and sustained endothermy in 3 species of syrphid flies. *(B)* denotes *Bombus* mimics, *(V)* denotes *Vespula* mimics, and *(NM)* denotes nonmimetic species. The upper line in each graph represents thoracic temperature; the lower line represents ambient temperature. The arrows indicate flight attempts. (From Morgan and Heinrich, 1987.)

THE HOT-BLOODED INSECTS

ating flight with T_{thx} at near 20°C, some 10°C lower than that required by their models. Other flies increase T_{thx} as much as 15°C above air temperature, taking off at T_{thx} near 25°C and sometimes sustaining endothermy beyond the 2–5 minutes required for T_{thx} to reach flight temperature.

Pre-flight warm-up rates (at 15–18°C) range from 1.2 to 5.5°C/min and are strongly dependent on body mass, with smaller flies having slower rates of warm-up than larger ones. (Warm-up rates of flies weighing 0.1 g average 3.2°C/min, while those weighing 0.3 g average 4.5°C/min.) In general, the trend in endothermic animals is for *larger* animals to have slower warm-up rates than smaller. This general trend seems to hold weakly only for some of the larger insects (May, 1976), or there is no obvious relationship with mass in others (Heinrich and Bartholomew, 1971). I conclude that two factors are operating simultaneously as regards endothermy and size. Smaller insects that are endothermic can warm up faster than larger ones, but the need for endothermy decreases with decreasing size. The smallest insects don't warm up at all, and they therefore cannot be expected to warm up at the fastest rate of all! In many syrphids weighing 0.1 g the effect of small size limiting endothermy may already be in effect.

The relatively low takeoff and flight temperatures of the syrphids relative to their hymenopteran models could be related to their overall lower mass and lower wing loading. Unlike the social hymenoptera they mimic, flies are not transport machines designed to ferry large loads to a nest. Perhaps for them, as for bumblebees, the larger the load carried, the higher the T_{thx} required for takeoff and flight.

In *Bombus, Criorhina nigriventris* (a syrphid with long pile or "fur"), and *Bombus* mimics with short dense pile, removal of the pile increases conductance by a nearly equal amount. Nevertheless, there is no demonstrable difference in T_{thx} between the pubescent *Bombus* mimics and the glabrous flies. (Do the glabrous mimics need to bask more to compensate for less heat retention at high wind speed?) The selective pressure for presence or absence of pile in the syrphids is apparently related more to improving visual mimicry than to thermoregulation. When handled, these syrphids also accurately mimic the buzz of bees, whereas nonmimetic flies lack this buzz.

Other flies in the midsize range (about 135 mg) of syrphids that mimic wasps and bumblebees may also thermoregulate surpris-

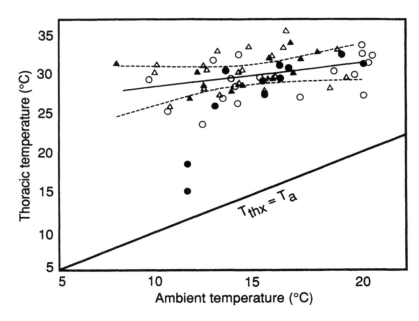

Fig. 11.9 Thoracic temperature as a function of ambient temperature in active *Nowickia* flies. Shaded circles indicate flies sitting in the shade, open circles indicate flies sitting in the sun, shaded triangles indicate flies flying in the shade, and open triangles indicate flies flying in the sun. The regression line does not include the two lowest data points; dashed lines are 95 percent confidence intervals for the regression. (From Chappell and Morgan, 1987.)

ingly well where they are subjected to a wide range of temperatures in the field. For example, the tachinid flies *(Nowickia nitide)* occurring at 3,800 m in the White Mountains of California can shiver like syrphids, and they also maintain (Chappell and Morgan, 1987) a relatively stable T_{thx} (Fig. 11.9) from air temperatures near 5°C in the morning (when there is little solar radiation) to high thermal loads near noon (when operative temperatures—which integrate air temperature, solar radiation, and wind speed—reach 43°C). The relatively narrowly maintained T_{thx} over the large range of environmental conditions reveals an impressive thermoregulation, but within a small temperature range or at any given air temperature, T_{thx} is as variable (±5°C) as it is in seemingly "nonregulating" syrphids at benign thermal conditions.

Heat production by shivering in endothermic flies is continuous during pre-flight warm-up, and rate of oxygen consumption is presumably maximal at all times. When T_{thx} is stabilized and the rate of heat production is not maximal, however, instantaneous measurements of oxygen consumption reveal that oxygen consumption may be rhythmically turned on and off (Fig. 11.10). Since the on-off bouts of oxygen consumption do not correspond with changes of T_{thx}, they are unrelated to on-off heat production for thermoregulation. Instead, they presumably are a form of

THE HOT-BLOODED INSECTS

"discontinuous" respiration that functions in water conservation, a mechanism that was first worked out for moth pupae (Schneiderman and Williams, 1955).

Most other myogenic insects (wasps, bees, beetles) shiver without making any sounds, but many Diptera produce an audible buzz when shivering. Some, like *Syrphus* (Heinrich and Pantle, 1975) or *Nowickia* (Chappell and Morgan, 1987), sometimes but not always accompany warm-up by buzzing. In others, like the horse botfly *Gasterophilus intestinalis* (Humphreys and Reynolds, 1980), the tse-tse fly *Glossina morsitans* (Howe and Lelane, 1986),

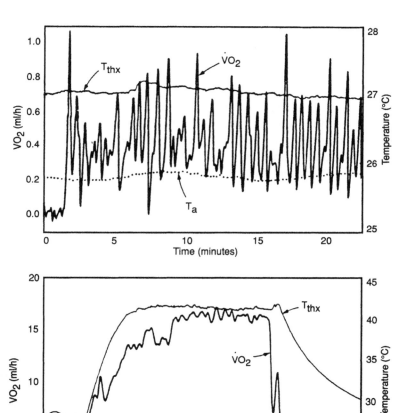

Fig. 11.10 Thoracic temperature and oscillations in the instantaneous rate of oxygen consumption (V̇O2) during a period of sustained but limited endothermy in a *Pantopthalmus tabaninus* fly when motionless *(top)* and during tethered flight *(bottom)*. (From Bartholomew and Lighton, 1986.)

and some syrphids (*Criorhina nigriventris, Sericomyia lata, Sphecomya vittata, Chrysotoxum* sp.), the buzzing seems to be an unavoidable by-product of all shivering warm-up. In still others (*Mallota posticata, M. bautias, Eristalis barda, Temnostoma alternans,* and *T. vespiforme*), there is no audible accompaniment to warm-up (Morgan and Heinrich, 1987). Buzzing and nonbuzzing warm-ups are probably not physiologically different. Rather, it is likely that the difference results from differences in the way the wings are partially engaged (buzzing), fully engaged (flight), or disengaged (silent warm-up) with the very complex wing articulations (Miyan and Ewing, 1985).

The buzzing or humming varies in frequency or pitch directly as a function of T_{thx} (Humphrey and Reynolds, 1980). Since the buzzing frequency is a direct measure of both wing-beat and thoracic-vibration frequency (Heinrich and Pantle, 1975), the strict dependence of buzzing frequency on T_{thx} is expected. In almost all insects so far examined, wing-beat frequency is tightly coupled to T_{thx} during warm-up, and usually also during flight.

At times the "singing" produced by the warm-up of flies can be quite musical and entertaining, especially when there are many animals together. Imagine a "lek" of male syrphids (Heinrich and Pantle, 1975) where each fly, as it lands (having cooled convectively), begins to buzz first at a low frequency and then increasing smoothly to a very high frequency as a high T_{thx} is restored. Other flies join in , "picking up the tune." A female may fly by, and all the males noisily take off in pursuit, to then return and begin the concert all over again.

The tse-tse fly *Glossina morsitans*, after a landing (and after a blood meal before taking off again), will pause for a quick warm-up accompanied by a buzz of rapidly rising pitch. This "post-feed buzz" has been much remarked upon in the literature, and before its thermoregulatory significance was known the conclusion had been made in at least 8 publications that the buzz is used in intraspecific communication (Howe and Lelane, 1986). So far no communication related to the pre-flight warm-up has been demonstrated. However, many bees and wasps (and their dipteran mimics) buzz when handled, and this communicative function in defense behavior has undoubtedly evolved as a modification of the pre-flight warm-up mechanism. Many flies with inaudible pre-flight warm-up nevertheless buzz when handled, showing that the defensive buzz is a derived behavior. The buzzing of a fly can be a very effective predator deterrence even to a great horned owl (see Heinrich, 1988, p. 128), as is the rattlesnake-mimicking buzz made by a burrowing owl when it is disturbed in its burrow (Rowe, 1989).

Regulated Heat Loss

The majority of small insects choose to be active when temperatures are appropriate, and they choose between sunshine or shade and different substrates. Overheating from endothermy is generally not a problem for them, so physiological mechanisms of heat loss need not evolve.

But there are exceptions. Insects normally keep their spiracles closed, except they will open them minimally when needed for respiration. Eric B. Edney and R. Barrass (1962) while at the University College of Rhodesia and Nyasaland demonstrated an exception: the tse-tse fly *Glossina morsitans* continually keeps all 4 thoracic spiracles open at high body temperature. The result is evaporation of water from the tracheal system, which reduces T_b by 1.6 °C in still, dry air. They speculate that since this response occurs at T_b not far below lethal temperatures (45 °C for 5 min) it could have adaptive value possibly when feeding on a hot substrate, such as the black hide on the back of a buffalo in sunshine. In addition, water loss might even be highly beneficial while feasting on blood (most blood-feeding insects have diuretic hormones that aid in ridding their bodies of excess water after a blood meal).

Willmer (1982a) reported a presumed acceleration of cooling in flesh flies, *Sarcophaga* foraging from flat inflorescences, and speculated that these animals maintained T_{thx} at 32 °C by shunting excess heat from thorax to abdomen when they are overheated in sunshine. The data given, however, neither support nor refute this hypothesis. It was observed that in several measurements, both T_{thx} and T_{abd} only approached 32 °C (at air temperatures of 26–28 °C) in flies exposed to sunshine. But the similarity between T_{thx} and T_{abd} is to be expected in *perched* insects exposed to sunshine; no heat transfer need be inferred. When a fine cotton thread was tied between the thorax and abdomen of "some" flies, the rate of thoracic cooling appeared to be slower. However, no conclusions can be made from such experiments. Heat transfer could have been affected by any of a large number of factors, unless the circulatory system *specifically* is ligated. Furthermore, and much more significantly, the presumed mechanism of thoracic cooling *requires* a low T_{abd}, and when flies forage on the exposed surface of flat inflorescences they would not likely be able to orient in such a way as to shade the abdomen specifically, to maintain a low T_{abd} that could then serve as a heat sink. Finally, why should a relatively large-bodied fly like *Sarcophaga* feel "overheated" at 32°C?

The first convincing data showing physiological heat transfer in Diptera were provided by Kenneth R. Morgan and co-workers in two comparative studies of robber flies (Asilidae). In the first (Morgan, Shelly, and Kimsey, 1985), 5 species of sun-loving and 11 shade-loving species were examined on Barro Colorado Island in Panama. Thoracic temperatures of the light-seeking flies ranges from 35.2 to 40.6°C during foraging at 26–30°C. The slope of the regression of T_{thx} on air temperature is not significantly different from 0, suggesting thermoregulation. In contrast, the T_{thx} of the shade-seeking robber flies (Fig. 11.12) averages only 2°C above air temperature at air temperatures of 24.5–30°C (Fig. 11.11), indicating lack of thermoregulation. There is no evidence of shivering in either group of flies. Both groups can initiate flight at ambient (and thoracic) temperatures of 21°C in the laboratory, which is 2°C lower than the lowest temperature recorded in the field during the study period. Only the light-seeking (*Promachus* sp.) flies transfer heat into the abdomen, and in these species the volume of the dorsal aorta for size-matched flies is at least 8 times

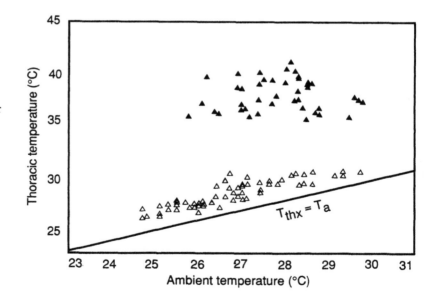

Fig. 11.11 The thoracic temperature of Panamanian robber flies (captured at perches in the field) as a function of ambient temperature. The solid triangles plot the data for 4 light-seeking species, the open triangles for 3 shade-seeking species. All flies weighed more than 0.09 g. (From Morgan, Shelly, and Kimsey, 1985.)

THE HOT-BLOODED INSECTS

Fig. 11.12 The shade-seeking, poikilothermic robber fly *Smeryngolaphria numitor* from Panama. (From photograph by K. Morgan.)

that in the shade-seeking species. Neither species has a thoracic aortic loop, as some other insects have, that can dump heat into the abdomen. However, those which use the abdomen as a heat radiator have a substantial aortal dilation that presumably aids in the pumping of blood (Fig. 11.13).

A follow-up study with two desert robber flies (Morgan and Shelly, 1988) confirmed and extended these results. Of the two species, *Promachus giganteus* (Fig. 11.14) was 4–8 times heavier than the other, *Efferia texana*. Because of their large difference in mass, it might be predicted that one would maintain much higher T_{thx} than the other. However, both species maintained similar T_{thx} (40–45°C) during most of the day, although *P. giganteus* had less difference between T_{thx} and T_{abd} than *E. texana* in the hot portions of the day. The implication is that *P. giganteus* might remain active in sunshine while the smaller *E. texana* is forced to seek shade, because it can regulate T_{thx} by shunting heat via hemolymph into the abdomen.

Laboratory experiments show that *P. giganteus* can indeed transfer impressive amounts of heat into the abdomen, but the abdominal heating is totally abolished when the dorsal aorta is ligated (Morgan and Shelly, 1988). As before, only those species that are

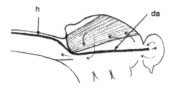

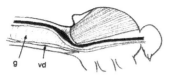

Fig. 11.13 Anatomy pertinent to hemolymph circulation between the thorax and abdomen in the shade-seeking robber fly *Smeryngolaphria numitor (top)* and in a light-seeking robber fly, *Promachus* sp. *(bottom)*. The differences in morphology and relative size of the dorsal vessels are noticeable. The arrows in the top diagram indicate the direction of blood circulation. The bottom diagram includes the gut *(g)* and ventral diaphragm *(vd)* as well as the dorsal aorta *(da)* and heart *(h)*. The dorsal longitudinal flight muscle is shown in both species. (From Morgan, Shelly, and Kimsey, 1985.)

Fig. 11.14 The sun-loving, thermoregulating robber fly *Promachus giganteus* (feeding on a honeybee) from Arizona. (From photograph by K. Morgan.)

heated to or well above 40°C as a result of the combination of solar heating and heat production from flight show physiologically facilitated heat transfer.

Rather than measuring heat flow directly, it is possible to infer it, using thermovision pictures to measure T_b. In one such study (B. Heinrich, unpublished) it appears that the honeybee-mimicking hoverfly *Eristelis tenax*, like honeybees, does not transfer heat into the abdomen (Fig. 11.15). It also does not appear that it transfers heat into the head, as honeybees do. In contrast, the much larger bumblebees transfer heat into both the abdomen and the head.

Like the tropical (Panamanian) robber flies, the desert robber flies (Morgan and Shelly, 1988) show no evidence of shivering thermogenesis. They regulate their T_b in the sun by perching with the long axis of the body at 90° to solar radiation to present a maximal surface area for absorbing heat, or by facing the sun to present a minimum surface area for absorbing heat and to leave the abdomen shaded and available as a heat sink in the larger, more heat-stressed species. A large part of their thermoregulation is also accomplished by selecting perch height. As temperatures increase, they perch increasingly higher above the hot ground (Fig. 11.16). Similar thermoregulatory behavior has also been observed in 3 species of robber flies from the grass prairie of Montana (O'Neill, Kemp, and Johnson, 1990).

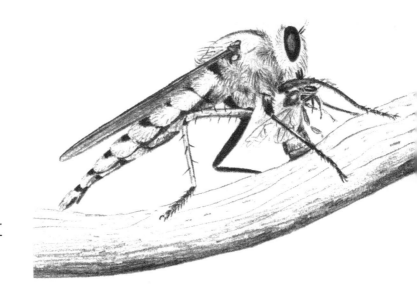

THE HOT-BLOODED INSECTS

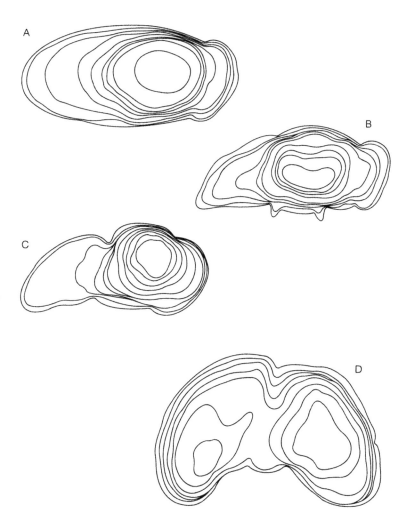

Fig. 11.15 These "thermal contour" maps were drawn from infrared thermovision pictures of insects immediately after they had stopped flying. In the hoverfly *Eristelis tenax* (shown in the dorsal view, *A*, and a lateral view, *B*), only thoracic temperature was very high (the highest temperatures are within the centermost circles); there was little heating of either the head or the abdomen. In the honeybee *Apis mellifera* (shown in a lateral view, *C*), thoracic and head temperature were both high, but little heat had leaked into the abdomen. Significant heat is transferred to the abdomen in the bumblebee *Bombus impatiens* (lateral view *D*), which also shows strong coupling of head and thoracic temperatures. The temperature increases from one isotherm to the next are (*A* and *B*) 0.2 °C, (*C*) 0.8 °C, and (*D*) 1.7 °C. (Thermovision images prepared in collaboration with Cpt. George R. Silver, U.S. Army.)

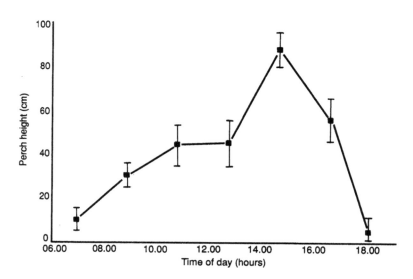

Fig. 11.16 At top is a graph of the body temperatures of dead robber flies *Promachus giganteus,* measured 1 cm *(upper curve)* and 90 cm *(lower curve)* above the ground surface during a single day. The area between the curves thus indicates the expected T_{thx} of live flies (given no thermoregulation) at perches between 1 cm and 90 cm. The squares represent thoracic temperatures of free-ranging flies netted at their perches. The lower graph shows perch height for *P. giganteus* averaged over 2-hour intervals throughout the day. (From Morgan and Shelly, 1988.)

In general, the more northerly the distribution, the smaller the flies (and more numerous). Few flies are large enough to be routinely and obligatorily subjected to high T_{thx} during flight. An exception is some of the neotropical Pantophthalmidae (timber flies), which include probably the largest flies in the world. *Pantophthalmus tabaninus,* a monstrous fly from Panama, weighs up to 2.8 g, and at air temperatures near 25 °C it heats up at 6.0 °C /min and stabilizes its T_{thx} near 40 °C (Bartholomew and Lighton, 1986). If these flies engage in continuous flight (they are reluctant flyers; K. Morgan, personal communication), they would inevitably heat to much higher T_{thx} if it were not for mechanisms for

THE HOT-BLOODED INSECTS

substantial heat transfer out of the thorax, both to the head and to the abdomen (Bartholomew and Lighton, 1986). These large-bodied flies, unlike all others so far examined, are unable to fly with T_{thx} below 25 °C. Obviously they are adapted to operate with high endothermic heat loads and high T_b, and their capacity for endothermic warm-up is well developed (Bartholomew and Lighton, 1986). Even so, their rate of warm-up is similar to that of syrphids of 75 times (Heinrich and Pantle, 1975) lower mass and to that of tachinid flies of 20 times lower mass (Chappell and Morgan, 1987). The latter can cool by convection once in flight, but this option is less available to the large pantophthalmids.

Summary

Flies vary at least 3,000-fold in mass, from less than 1 mg to nearly 3 g. The smallest nocturnal Dipterans are nearly perfect thermoconformers, whereas the largest would likely quickly over-heat to lethal temperatures in continuous flight if it were not for mechanisms of physiological heat transfer. Many medium-sized dipterans (30–300 mg) show little elevation or regulation of T_{thx}, whereas others that are subject to repeated heating because of intense solar radiation or apparent selective pressure for rapid flight regulate their T_{thx} quite well (near 30–40 °C) over a wide range of ambient thermal conditions by a combination of postural maneuvers, shivering, and microhabitat selection.

The larger the body mass and the hotter the ambient thermal conditions, the higher the T_{thx} regulated. Flight, however, is possible with T_{thx} as low as at least 4 °C in some very small species regularly subjected to low temperatures. Adaptation for flight at much lower T_{thx} seems not to have occurred, and those flies subjected to lower temperatures have lost the capacity for flight. Some have even lost their wings entirely, but these species still remain active by walking at very low temperatures (−4 to −16 °C).

Remaining Problems

1. Almost nothing is known about possible thermoregulatory mechanisms in some of the larger dipterans in continuous flight.
2. The extraordinary range of thermal optima of adult dipterans— from near −10 °C to over 40 °C—suggests interesting problems

of biochemical temperature adaptation. What are the underlying cellular mechanisms?

3. To what extent are the shifts to different thermal niches in the same environment by related species related to thermal adaptation *per se* or to niche partitioning?

4. Do these insects acclimate so that different optimum flight T_{thx} occur in the same species?

5. What explains the large variance of T_{thx} often found in any one species displaying great independence of T_{thx} from ambient temperature?

6. Do some Arctic Diptera bask to accelerate gonadal development?

7. Do wasp-mimicking syrphids bask more than *Bombus*-mimicking species?

8. Myogenic muscles are noted for their absence of sarcoplasmic reticulum, and an elaborate SR is thought to be one adaptation permitting either very rapid muscle contraction or contraction at low temperatures in neurogenic flyers. How do myogenic flies manage to have high wing-beat frequencies at low muscle temperatures?

Sweating Cicadas

CICADAS are famous for their "music"—or perhaps one should say they are notorious for it. Their singing has been compared to a chain saw driven at high speed against naked iron, to the frantic thumping of a tin can, and to shrilly, trilling flutes. A single cicada can drive a person to distraction, but when hundreds or thousands chime in, as during an outbreak of the 13- or 17-year cicadas, the noise can become quite irritating. Indeed, the "chorus" of the cicadas is thought to repel their enemies, but in some species, as we will show, a thermoregulatory strategy—not a hammering clamor—is used against predators.

The noise-making instruments of cicadas are the most highly evolved of any insects. Unlike the friction-based sound production of modified joints, legs, or wings that other insects employ, cicadas rely on drums. The males have two drums, each lying in a cavity on the fore part of the abdomen (females are silent and lack drums). The drum head does not vibrate from blows from the outside, but rather from a very rapid succession of brief tugs from within by a powerful muscle that is fastened by a tendon to the lower surface of the membranous drum.

Most of the Homoptera are small insects—whiteflies, leafhoppers, scale insects, and plant lice are examples. Some of these, like the aphids, for example (O'Doherty and Bale, 1985), have remarkably low temperature tolerances. The cicadas, being generally large (over 1 g), are an exception. Like most other Homoptera, they are sedentary insects and suck plant juices. Their noisy and sedentary habits and their predilection for warm weather may not at first suggest an interesting thermal biology. However, in

Fig. 12.1 The desert cicada
Diceroprocta apache. (From
photograph by N. Hadley.)

1970 James E. Heath and Peter J. Wilkin published a by-now classic study on the temperature responses of the desert cicada *Diceroprocta apache* (Fig. 12.1), which spurred much interest in the physiological ecology of insects in general, and on cicadas in particular.

Behavior and Ecology

The desert cicadas *D. apache* inhabit mesquite–palo verde thickets along dry streambeds in the southwestern United States, including near Phoenix, Arizona. Instead of being active at night in the cooler portions of the year, these cicadas have shifted their singing time to nearly the hottest times available: near midday in the summer, when temperatures in the shade may approach limits lethal even to these desert-adapted insects.

The cicadas' day in preparation for singing begins at about sunrise, when the animals move from their nocturnal feeding sites along the large mesquite branches out to exposed twigs and leaves. Here they bask motionless during the early morning hours until their T_{thx} approaches 40°C. Predatory birds and cicada-hunting wasps are active at this time, but although the cicadas are warm enough to sing, they remain silent. As the sun rises and air temperatures increase to 35°C or above, and the birds and wasps escaping the heat retire until dusk, the cicadas begin to sing. Maximum singing intensity is reached near midday, when the desert heat reaches searing intensity. At that time, however, the cicadas fly only 1–10 m when approached, although when disturbed in the morning they may fly as far as 100 m. The insects heat up 5–9°C above ambient temperature in flight, and if they flew for more than 35 seconds even at 39°C at noon they could reach nearly lethal T_{thx}. Thus the short flight range at noon at temperatures above 40°C is probably due to a build-up of metabolic heat.

At air temperatures above 40°C the cicadas seek shade in the mesquite and palo verde, bushes where they remain inactive and where their body temperature is then below that of the air temperature outside the bushes. Air temperatures near Phoenix commonly reach 48°C, and the cicadas lose motor control when T_b rises above 45.6°C. Thus, to stay alive these insects must sometimes maintain their T_{thx} below air temperature (Fig. 12.2). If cicadas in the field allowed their T_b to reach 45–46°C, they would lose motor control and drop to the ground, where surface tem-

THE HOT-BLOODED INSECTS

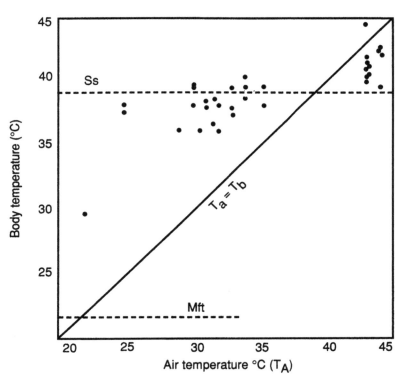

Fig. 12.2 Thoracic temperatures of field-caught *D. apache* plotted against the air temperature measured at 2 m in shade. *Ss* = air temperature where shade-seeking begins; *Mft* = minimum flight temperature. (From Heath and Wilkin, 1970.)

peratures reach 70°C, and a quick death would await them. By choosing summer as their season to be active, the cicadas are thus obliged to balance precariously on a thermal tightrope, where their predators must venture if they are to hunt them.

Birds, the cicadas' major predators, have their heaviest energy demands during the nesting season in the spring, when it is cooler. *Diceroprocta apache* are not to be found in that period, for they restrict their activity to the hottest part of the year, and to the hottest part of the day. They can then persist when the predators are forced to retire because of the heat. The cicadas are thought to be able to exploit the hot, midday period by their ability to locate microclimatically milder perches, such as small flecks of shade under branches (Heath and Wilkin, 1970), which are too small to shelter birds. Additionally, it has recently been revealed that the cicadas may win the thermal battle with their enemies by an impressive capacity for evaporative cooling (see below).

Lloyd and Dybas (1966) argue that cicadas of various species are either solitary and have short life cycles (2–5 years), or they appear in large numbers in striking long-term periodicity. In the first instance the predators or parasites are limited by the scarcity of the prey, and in the second case the cicadas escape predators by "saturating" the appetites of predators because of their large numbers when they do appear (in other words, as more emerge, they face no additional danger of being eaten). The life cycle of the periodical cicadas apparently prevents the parasites from being able to "track" them over time. For example, when developing over 13- and 17-year cycles, they spend most of that time living underground in the nymph form, and then they live only a few weeks in the winged, adult form. The remarkable case of the 3 closely related species of sympatric and synchronous 17-year cicadas (*Magicidada cassini, M. septendecim,* and *M. septendecula*) represents one solution to the predator-parasite problem. But *D. apache* escapes its predators by appropriate timing of its life cycle, with respect to the weather rather than with respect to each other, and it has evolved unique thermal responses to make it possible.

As already mentioned, a "thermophilic" strategy apparently similar to that of *D. apache* has also been observed in desert-dwelling ants (Chapter 9) and in the grasshopper *Trimerotropis pallidipennis* (Chappell, 1983). These grasshoppers hide on open, sunny ground, where lizards do not tread because temperatures there may be dangerously close to their lethal tolerances. The grasshoppers' perching endurance in the heat, after being pursued there by a predator, is a major survival premium. If they stilted to avoid the hot substrate, their cryptic disguise would be destroyed and they could be caught by a lizard, and if they flew (to cool convectively) they could be caught by a bird instead.

The temperature responses of another cicada, the cactus dodger *Cacama valvata*, are similar (Heath, Wilkin, and Heath, 1972). These large insects (1 g) are black dorsally and have white undersides. Cactus dodgers (so named because of their habit of flying erratically among the cactus when startled) spend the night feeding on prickly pear cactus, and then in the morning at air temperatures of near 20°C they bask until achieving T_{thx} near 30°C. By late morning (at air temperatures near 30°C), when T_{thx} approaches 40°C, they finally begin to sing. Singing cactus dodgers take up positions on twigs or shafts of grass with their white belly facing the sun, avoiding exposing an absorbing surface in favor of a more reflecting one.

In grassland habitat of the southwestern United States, where these cicadas are common, shade is restricted to small patches, but by midday, at near 35°C, most of the cicadas have retreated to shade. Like the extremely thermophilic *D. apache*, cactus dodgers enter heat torpor at a relatively low body temperature (near 45°C) in the laboratory, and a T_{thx} rise of 5–8°C in flight probably limits flight activity near noon (Heath, Wilkin, and Heath, 1972). Minimum T_{thx} for flight are surprisingly similar (near 21–24°C) in the cactus dodger *C. valvata*, the 17-year cicada *M. cassini*, and the desert cicada *D. apache*. However, shade seeking occurs at significantly lower T_{thx} (31.8°C) in *M. cassini* than in *C. valvata* (35–37°C) and *D. apache* (39.2°C) (Heath, 1967).

Exercise and Endothermy

Until recently research has stressed the cicadas' behavioral responses to the hot desert environment. It was not suspected that the singing, which is restricted to the hottest part of the day, would create an additional, internal, heat load. Nor was it known that the animals have a physiological cooling mechanism. I shall here first discuss endothermic heat production and then consider these insects' unique cooling mechanism.

Cicadas are unusual among insects both for their extraordinarily vigorous muscle performance while singing and their relatively feeble muscular performance during flight.

Male cicadas produce a calling song by a pair of tymbals, one on each side of the posterior of the thorax. Each tymbal is attached to a large tymbal muscle, which accounts for appreciable endothermy, as shown in the Australian cicadas *Cystosoma saundersii*. These cicadas sing only at dusk, generally at air temperatures near 20°C. They neither shiver before singing nor do they have sunshine available for basking, but the temperature of their tymbal muscles rises rapidly (about 4.4°C/min) during singing and reaches a plateau of about 12°C above air temperature (Josephson and Young, 1979); they can generate a muscle temperature of near 32°C when they sing at 20°C. Motor neurons fire to activate the tymbal muscles at about 40 Hz, and the muscles thus contract, and buckle the tymbals, at this frequency. But much higher activation frequency of the tymbal muscles is possible.

The calling songs of the male cicada *Okanagana vanduzeei* are produced with an extraordinary pulse-repetition frequency of 550 Hz. As in other cicadas, only one sound pulse is produced per tymbal muscle contraction, thus the muscles of *O. vanduzeei* contract at the phenomenal rate of 550 Hz. Furthermore, since as in *C. saundersii* there is a 1:1 correlation between electrical and mechanical activity of the muscle, the muscles are of the synchronous neurogenic type (Josephson and Young, 1985). They are the fastest synchronous muscles known. How is muscle speed this great achieved?

Part of the answer lies in the muscles' ultrastructure. The tymbal muscles are, like other fast muscles, composed of very small myofibrils and the sarcoplasmic reticulum is extraordinarily well developed (Josephson and Young, 1985). The same is true of moths able to generate muscles contractions of sufficient frequency (about 10 Hz) to support flight at T_{thx} below 5°C (see Chapter 1).

In addition, in order to generate the normal high-frequency calling song, the cicadas need to maintain a body temperature of 40–45°C. The performance of the tymbal muscles is highly temperature dependent and the cicadas maintain their tymbal-muscle temperature at an elevated level in part by singing in sunshine and in part by the heat of exercise itself. Although metabolic costs of singing by *O. vanduzeei* are not known, the metabolic cost of singing of the other Australia cicada, *Cystosoma saundersii*, is 98 ml O_2 per gram muscle per hour (McNally and Young, 1981), close to that of flight or stridulation in other insects (Stevens and Josephson, 1977).

Given the high temperature dependence of the tymbal muscle, the insect could theoretically have achieved higher contraction frequencies simply by operating at temperatures higher than 40–45°C. In common with other terrestrial organisms, however, the animals "hold the line" at the common upper temperature setpoint and achieve the higher contraction frequencies by morphological specializations instead.

Although some male cicadas must maintain a high T_b to call at high frequencies and for long durations to attract mates, they have little need for sustained flight and therefore should not normally generate nor need high flight-muscle temperatures. George A. Bartholomew and M. Christopher Barnhart (1984) at the University of California at Los Angeles confirm this idea; they found a very primitive thermoregulation in the cicada *Fidicina mannifera* from the tropical lowlands. On the basis of the large mass alone (3 g), one might have expected well-developed endothermy in this insect. However, these cicadas can fly at the somewhat low (for their mass) T_{thx} of 22°C, near that of other cicadas (Heath and Wilkin, 1970; Heath, Wilkin, and Heath, 1972). Generally, takeoff is preceded by only a relatively feeble warm-up until T_{thx} reaches about 28°C (Bartholomew and Barnhart, 1984).

Metabolic rate in *F. mannifera* increases explosively, some 4.4 times after the initiation of flight following warm-up. In contrast, metabolic rate in other endothermic insects (at a given T_{thx}) is usually very similar between warm-up and flight. Apparently the cicadas are severely limited in their capacity to shiver to warm up. Furthermore, their flight durations are also limited to no more than 100 seconds. There has presumably been little selective pressure for warm-up, because the animals don't fly long enough to heat up, hence the muscles have adapted to operate at the relatively low T_{thx} normally encountered.

The proximal physiological limitation of the cicadas' unimpressive exercise performance during warm-up is apparently their poor gas exchange (Bartholomew and Barnhart, 1984). At rest the O_2 and CO_2 concentrations of the air sacs are 17 and 3 percent, respectively. During warm-up, however, O_2 concentration quickly plummets to as low as 1 percent and CO_2 concentration increases to the extraordinary high level of 21 percent. When flapping flight commences, the O_2 and CO_2 concentrations quickly return to near resting levels, presumably because of the automatic ventilation resulting from the working of the flight muscles (Weis-Fogh, 1967).

But what limits the flight performance to under 100 seconds? Since the respiration is strictly aerobic, Bartholomew and Barnhart (1984) infer that the limited flight endurance is related to the depletion of chemical substrate or fuel in the flight muscles.

Evaporative Cooling

Insects are known to have a variety of adaptations for conserving water, such as having an exoskeleton that is very impervious to water loss (Edney, 1977). In many insects death is accompanied by a several-fold increase in water loss, even when body openings are sealed, suggesting that there is some energy-dependent process that retards water loss through the cuticle. Furthermore, desert insects are much better water-proofed than those from less hot and dry environments.

That there might be something unique about cicadas from hot environments was first suggested by Stacey A. Kaser and Jon Hastings from the University of New Mexico. In New Mexico, the cicada *Tibicen duryi* feeds on the xylem fluid of the pinion pine, or piñon, and although T_b is close to ambient temperature up to 34°C, the cicadas show significant reduction of T_b below ambient temperature at high temperatures and when water is plentifully available to the animals. These results suggested that they cool evaporatively (Kaser and Hastings, 1981).

The intriguing preliminary results with *T. duryi* were later followed up by Eric C. Toolson at the University of New Mexico and Neil F. Hadley at Arizona State University at Tempe. They collected the desert cicada *D. apache* on the Arizona State campus during July and August and immediately examined them in the Tempe laboratory in a series of tests that confirm that these cicadas also cool themselves evaporatively.

Live *D. apache* were subjected to 45.5°C in dry air, and after 1 hour of exposure to this temperature (Fig. 12.3) they maintained T_b more than 2.9°C below air temperature (Toolson, 1987). Subsequent studies showed T_b was regulated at 37–38°C, even at ambient temperatures up to 42°C (Hadley, Quinlan, and Ken-

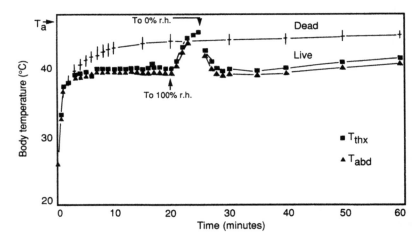

Fig. 12.3 The effect of evaporative cooling on thoracic and abdominal temperature in a *D. apache*. The animal was exposed to dry air (relative humidity less than 5 percent) at an ambient temperature of 45.5°C and then transferred to a saturated atmosphere (relative humidity 100 percent) for 20 minutes before being returned to the dry air. Summary data for three dead (control) animals under the dry conditions are plotted by the top line. Note that T_b of the live animal (both thorax and abdomen) are about 4°C cooler than the temperatures of the dead animals, and that T_b rises immediately when an animal is placed into a humid environment, when evaporative cooling is abolished, and then cooling resumes in dry air. (From Toolson, 1987.)

THE HOT-BLOODED INSECTS

nedy, 1991). Only evaporative cooling can bring T_b below ambient temperature. Furthermore, dead animals did not show reductions of T_b, indicating that water loss for cooling is an actively controlled process. Since the cicadas may in the field be subjected to even higher heat loads (higher air temperatures, solar radiation, and metabolic heat production from singing), the above data probably underestimate the amount of evaporative cooling possible. When Toolson (1987) increased the relative humidity around a live cicada to 100 percent, its T_{thx} quickly reached the temperature of control (dead) animals. The T_b recorded in the laboratory were similar to those that Heath and Wilkin (1970) had previously measured in the field at the same air temperatures, suggesting a far greater role for physiological temperature regulation as opposed to or in addition to selection of temperature microhabitats (as previously supposed) that allow *D. apache* to maintain T_b below ambient thermal conditions. In contrast to the usual situation in other insects (where desert-adapted forms are best at water conservation), the cicadas best adapted to deserts, such as *D. apache*, have *greater* rates of water loss than a species, such as *Tibicen dealbatus* (Toolson, 1984), from a more mesic environment.

Hourly water losses of 30–35 percent of the desert cicadas' body water were observed. Undoubtedly such high water losses cannot be sustained unless the insects have a means of replenishing the water. Even though living in the desert, the cicadas indeed do have a readily available source of water: they feed from the xylem of plants, such as that from mesquite trees in the case of *D. apache*. The evaporative-cooling mechanism has only a very modest metabolic cost (Hadley, Quinlan, and Kennedy, 1991).

Toolson and Hadley (1987) glued small capsules onto the cicadas' cuticle directly. A humidity sensor was placed inside the capsule so that water loss could be measured as a function of the cicadas' temperature (Fig. 12.4). Typically, at T_b of 39°C and below there was very little water loss, although sometimes a transient peak of up to 9 mg H_2O/cm^2 body surface area/hr occurred at that temperature. However, when the cicadas were heated to 41°C they showed a very rapid rise of water loss, and at 43°C rates of water loss as high as 70 mg H_2O/cm^2 body surface area/hr were observed. Only live cicadas showed the high and often cyclical changes of water-loss rate. Cicadas injected with an aqueous solution of sodium cyanide showed an immediate disappearance of the cyclical fluctuations of water loss from the cuticle, as well as a rapid decline of that water loss to levels typical of dead controls.

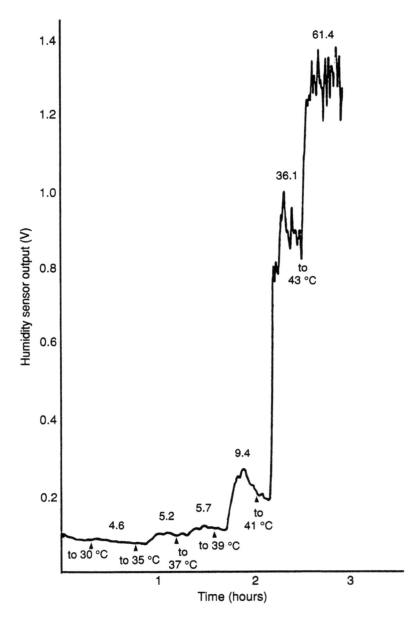

Fig. 12.4 The effect of temperature on cuticular water loss in a live *D. apache*. The vertical axis represents humidity sensor output (in volts). Temperatures were ramped upward in discrete jumps (indicated under the line). Numbers above the line indicate average water loss (in mg H_2O /cm^2/hr) during each interval. (From Toolson and Hadley, 1987.)

Although many details of the mechanism underlying the evaporative water loss in cicadas remain unclear, Toolson and Hadley (1987) have identified dermal pores as the likely site through which the water moves through the cuticle. Microscopic views of the dorsum of the thorax of *D. apache* reveal three tracts of pores, one along the midline, and one on each side of the midline (Fig. 12.5). Additional large pores (7 μm average diameter) are also

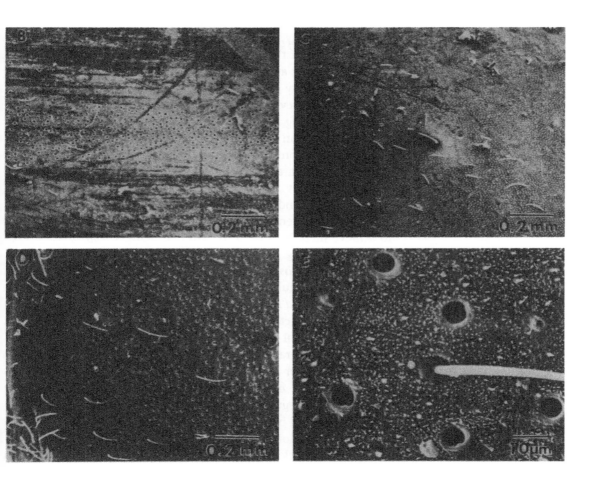

Fig. 12.5 The dorsal surface of the desert cicada *Diceroprocta apache* as revealed by low-magnification scanning electron microscope: through these pores water is lost for evaporative cooling. *(B)* In this view (51×) of the mid-line thoracic area the large pores of the center of the tract are visible. *(C)* On the lateral side of the thorax (also 51×), the pores are absent. *(D)* On the dorsal side of the abdomen (51×), the large pores are distributed uniformly. *(E)* In this higher-magnification view, both large and small pores are visible, as is the basal portion of a bristle. Scale bars in *(B)–(D)* represent 0.2 mm; in *(E)* the bar equals 10 μm. (From Hadley, Toolson, and Quinlan, 1989.)

distributed over the abdominal surface (Hadley, Toolson, and Quinlan, 1989). Water-loss rates in these areas of the body have been examined, and the regional water-loss rates reflect measured pore densities. Both males and females exhibit the water-loss response at high ambient temperature.

Summary

Cicadas are large, relatively sedentary insects. They fly only briefly and generate relatively modest (for their size) temperature excess in flight, and they have therefore evolved to be able to fly with relatively low (for their size) T_{thx}. They exhibit either no pre-flight warm-up or only a feeble warm-up. Warm-up is apparently limited by ventilatory gas exchange, suggesting that in those insects that have had to evolve pre-flight warm-up, ventilation would be a primary limiting factor since during warm-up ventilation (due to the deformation of the thorax from the wing beats) would no longer occur automatically, as in flight.

Although flight, and flight-muscle endothermy, is brief (usually restricted to a few seconds), male cicadas sing for many minutes without pause and their tymbal muscles heat up. One species generates neurogenic contraction frequencies of over 500 Hz, and to achieve these high rates the animals achieve T_{thx} of 40–45 °C, in part by basking.

Possibly the most comprehensive and detailed information on cicada thermoregulation is available for one species, the desert cicada *Diceroprocta apache*. This insect from the arid southwestern United States is active at the hottest part of the year, and sings at the hottest part of most days at body temperatures of 42–44 °C and at air temperatures up to 48 °C. It becomes uncoordinated due to heat torpor at a body temperature of 45.6 °C, yet manages to remain active on days when air temperatures in the shade reach 48 °C. It regulates its body temperature by basking at low temperatures and by microhabitat selection and evaporative cooling at high heat loads.

Cicadas, by sucking plant juices from the xylem tissues, have a means of replenishing their water supplies, and they have evolved a unique system of accelerated water loss for evaporative cooling. Their "sweating" response is activated by T_b of 41 °C or above, and it allows these insects to survive and be active (sing) at ambient thermal conditions too extreme for their major predators (birds and wasps).

Remaining Problems

1. What is the mechanism that accounts for the accelerated evaporative water loss in the desert cicadas? How is it turned on and off?
2. How general is the high-temperature predator-avoidance strategy among the cicadas?
3. What is the relationship between the optimum temperature of the tymbal muscles and the temperature excess generated by these muscles during singing in different species?

THIRTEEN

Warm Caterpillars and Hot Maggots

I T I S difficult to imagine two kinds of animals more different from each other than the larva and the adult of a holometabolous insect—an insect that undergoes complete metamorphosis in its life cycle. Consider a caterpillar and a moth. The larva of a *Cocytius* moth, a sphinx moth, is a huge, slug-like creature that weighs up to 36 g (D. Janzen, personal communication). It eats green leaves, and like most other insect larvae, it is presumably poikilothermic. But the adult form of the same individual weighs about 6 g and has wings and a scaly covering. It is highly endothermic, and it regulates a T_{thx} near 46°C while in hovering flight, harvesting nectar from flowers.

Not all the insects discussed in this book undergo complete metamorphosis. Cicadas and dragonflies emerge from eggs as nymphs, which are similar in form to the adults but are smaller and wingless. For them, metamorphosis is incomplete. Still other insects, such as grasshoppers and hemipteran bugs, do not metamorphose at all but leave the egg as a smaller version of the adult. In this chapter we will ignore these other forms of insects and focus instead on those that pass through a larval stage—in particular, caterpillars (the larval form of moths and butterflies) and maggots (the larval form of flies).

Throughout this book I have repeatedly stressed how thermoregulation is in the ultimate evolutionary sense related to overheating, which in most species occurs only during flight. No larval insect flies, and one might assume therefore that none thermoregulates. There are "exceptions," however, and these exceptions illuminate the more fundamental theory that the *evolution* of thermoregulation is ultimately related not so much to flight but to inevitable heating. But another phenomenon is also operating in larvae. Larvae, unlike all adults, have the capacity to grow. Growth is temperature dependent, and to a limited extent some larvae do choose their T_b, or they have that temperature controlled by the adults' choice of oviposition site (Kingsolver, 1979) or place of residence.

In general, eggs and pupae, the other stages in complete metamorphosis, have no thermoregulatory options. Their temperatures are a function of the environment where they find themselves, and that environment is generally chosen by the adults and larvae, respectively, with little or no concern for temperature, even though ambient thermal conditions have profound ecological consequences (Huey, 1992). Larvae can be active over a wide range of T_b, and their broad temperature tolerances for activity (in comparison with their endothermic adult counterparts) is associated with few thermoregulatory options. Aside from the potential of avoiding extremes of heat or cold, only a tiny percentage of insect larvae regulate their T_b (I speculate far less than 0.1 percent). Here I review some of the rare examples of thermoregulation in larvae.

Caterpillar Body Temperatures and Growth

As pointed out elsewhere (see Chapter 14), silk growers and other rearers of Lepidoptera have long known that caterpillars grow best at certain temperatures. Paul W. Sherman and Ward B. Watt (1973) at Stanford University were the first to make an analysis of the thermal ecology of caterpillars by comparing two very closely related species. One, *Colias eurytheme*, originated from a warm environment (Merced County, California) and the other, *C. philodice eriphyle*, came from a cool habitat (in Delta County, Colorado). Both species feed on vetch, *Vicia*, and feeding rate is strongly temperature dependent.

The optimal rearing temperature differs between the two species. Feeding rate already declines at 25°C in *C. p. eriphyle*, while it does not even peak until 27°C in *C. eurytheme* (Fig. 13.1). Growth rates as a function of temperature approximately parallel feeding rates. The egg-to-adult duration in *C. eurytheme* is approximately 1,150 hours at 20°C and 425 hours at 30°C (Fig. 13.2). Thus, merely by being 10°C warmer the insects can nearly triple their growth rate (and potentially their reproductive output). Behaviors to control body temperature, which presumably translate to increased growth and reproductive rates, are observed. Results

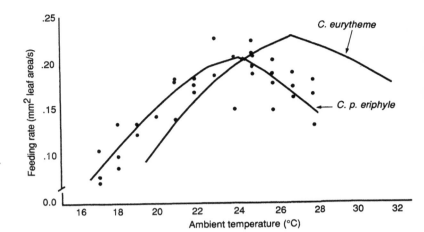

Fig. 13.1 Feeding rate as a
function of temperature in *Colias philodice eriphyle* and *C.
eurytheme* caterpillars. (From
Sherman and Watt, 1973.)

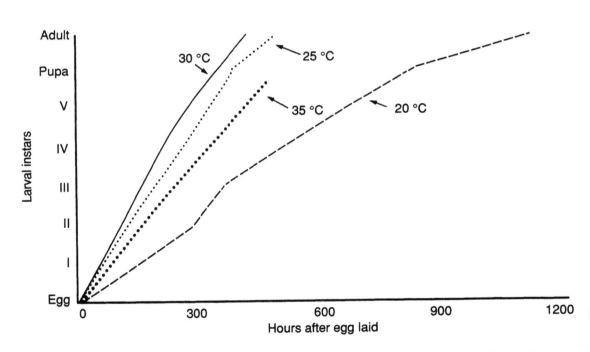

Fig. 13.2 Developmental rates of *C. eurytheme* caterpillars at different rearing temperatures. (From Sherman and Watt, 1973.)

THE HOT-BLOODED INSECTS

similar to these have now been reported for some dozen species of caterpillars (Stamp and Bowers, 1990).

Clearly, the temperature optima for growth in larvae differ greatly from the temperature optima for activity in flying adults, because the two experience very different T_b. But why do the optimum temperatures differ between the two caterpillars, whereas they are the same for the adults of both species? Sherman and Watt (1973) speculate that the ultimate answer may lie with predator pressure, which prevents the larvae from freely choosing their own T_b. The cryptic green *Colias* larvae have dorsal stripe patterns mimicking highlights on plant stems. The effectiveness of this disguise was determined by Gerould (1921), who showed that sparrows failed to find normal larvae, although they readily found a blue mutant form. Predator pressure for cryptic coloration has probably prevented larvae from evolving both darker forms (as the adults have done to enhance solar heating rates) and forms that expose themselves to sunlight (and predators). The predators may thus have driven the larvae to optimize their biochemical machinery to operate at temperatures specific to their habitats and hence lower than that of the sun-loving adults, which generate considerable temperature excess in flight. Both larvae probably choose temperatures that maximize growth efficiency within the range of temperatures now genetically fixed. But selection for growth efficiency in two very closely related species should not have been maximized at two *different* temperature optima unless there is an ecological reason for it.

The larvae, curiously, have relatively *narrow* temperature optima, and to maintain these optima they are forced in the morning to climb up and feed on those parts of the plant where they can achieve optimum T_b, and they are forced to retreat to shade and cease to feed as temperatures rise above the optimum. Why not simply evolve to be active at a broader temperature range? The answer probably resides in biochemistry (Hochachka and Somero, 1973). As argued elsewhere (Heinrich, 1977; Huey and Hertz, 1984), perhaps the metabolism cannot be organized to operate at maximal efficiency over a broad range of temperature. If there is selective pressure to *maximize* growth rates (to reduce the length of the larval stage and thereby reduce exposure to predators), then there is a need to "choose" a specific T_b. Functions *can* evolve to operate at temperature-independent rates over a broad range, but likely by the *inhibition* of the peaks (Hochachka and Somero, 1973). And in caterpillar development, it seems likely that the major selective pressure has been to accelerate growth rate as much as possible within the constraints posed by parasites and predators, rather than to develop temperature insensitivity (but see Casey and Knapp, 1987).

There is, however, an important point to be kept in mind when comparing temperature optima of larvae vs. adults. Larvae can live at any of a broad range of temperatures and grow well, even though the "optimum" temperature range—the temperature where growth is most rapid—is narrow. In the adults, on the other hand, the cutoff between all activity (flight) and no activity can be all-or-none at a 1–2°C difference. And that cutoff will not be reflected in a curve of optimum enzyme levels or muscle performance vs. temperature.

The foregoing example of two *Colias* species from different habitats and with different thermal optima contrasts with another study of two species of sphinx moth (Sphingidae) caterpillars. In this case the two species co-occur in the same environment, the Mojave Desert (although they also occur widely outside deserts over most of North America). In the Mojave Desert both sphinx moth caterpillars are subjected to air temperatures ranging from near 10°C to at least 36°C, with the possibility of intense solar radiation added.

The caterpillar of the white-lined sphinx moth *Hyles lineata* regulates its T_b relatively well (Fig. 13.3) by orienting parallel or perpendicular to the rays of the sun, depending on the air temperature, and by shuttling between the ground and the plant. The other, the *Manduca sexta* caterpillar, feeds on the underside of large leaves (Fig. 13.4). It is a relatively passive thermal conformer that does not bask nor leave its food plant (Casey, 1976). Both species have similar body-temperature-dependent biting rates and upper lethal temperatures of 45°C (Casey, 1976), although in *M. sexta* growth rate is already severely depressed at 35°C (Casey, 1977).

Why is it that one of the caterpillars basks and regulates its T_b and one does not? Casey (1976) argues that *H. lineata* regulates a high T_b to allow it to maximize feeding rate, which may be critical because some of the desert annuals upon which these caterpillars feed are available for only a short time. From the perspective of proximate mechanisms that explanation makes indisputable sense. However, there is no reason to suppose that *M. sexta* would *also* not attempt to maximize feeding rate. Indeed, its ability to efficiently manipulate leaves of complex shapes (Heinrich, 1971) suggests that feeding rate is also at a premium in this species.

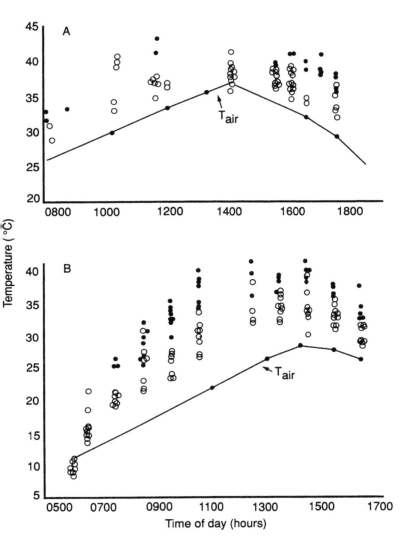

Fig. 13.3 Body temperature of *Hyles lineata* caterpillars during the day on a warm day *(A)* and on a cool day *(B)*. Caterpillars on the ground *(filled circles)* and on vegetation *(open circles)* were sampled. (From Casey, 1976.)

Although at the present time *H. lineata* may indeed have to bask at low temperatures, a quite different question is the evolutionary one: why hasn't it evolved to do the same job at a lower body temperature? The answer, once again, probably is related to predator pressure and micro-environment. The two relatively closely related animals (both Sphingidae) occur almost side by side at the same time, but they necessarily experience widely different thermal environments. The caterpillars of *M. sexta* occur on dense, bushy, broad-leafed jimsonweed, *Datura metelloides*, where they are able to feed almost continuously, day and night, while remaining suspended underneath large leaves (Heinrich, 1971) and thus shielded both from avian predators (Stewart, 1969; Thurston and Prachuabmoh, 1971) and from direct solar radiation (Fig. 13.4). They generally complete their entire larval development on one plant. The larvae of *H. lineata*, in contrast, feed primarily on small, creeping, ground vegetation, and they are forced to wander in search of food over the hot, open desert soil, often in direct sunlight, especially during the recurrent caterpillar outbreaks. In other words, *Hyles lineata* are forced to confront very great heat loads, which *M. sexta* like all or almost all other sphinx larvae can avoid. *Hyles lineata* is a unique sphingid not only because its larvae wander over the desert, but also because of its high thermal "preferences." These preferences probably are a biochemical consequence of its being exposed throughout evolutionary time to higher average temperatures. *Hyles lineata* has had to adapt to operate at a higher T_b, thus necessitating basking at low temperatures.

The pattern for cryptic, palatable caterpillars, like *M. sexta*, to avoid harsh sunlight (Capinera, Wiener, and Anamosa, 1980) and

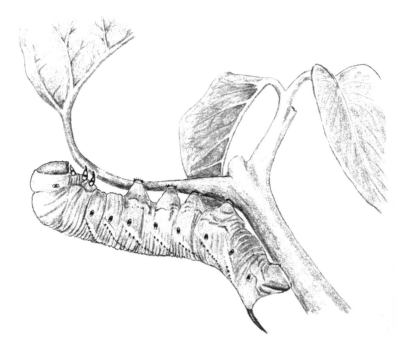

Fig. 13.4   Tomato hornworm caterpillar *Manduca sexta* on jimson weed (*Datura* sp., Solanacea), its food plant in the Mojave Desert.

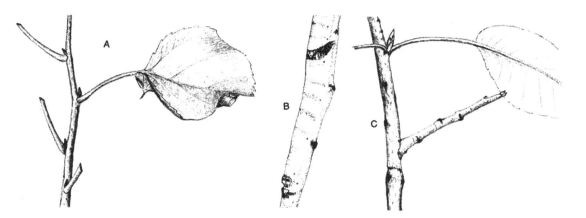

Fig. 13.5 Three cryptic caterpillars that are prevented by predator-avoidance postures from engaging in thermoregulatory behavior: the poplar sphinx *Pachysphinx modesta* *(A)*, which remains hidden under leaves while feeding; the underwing moth caterpillar, *Catocola* sp. *(B)*; and a large geometrid moth caterpillar on balsam poplar *(C)*. Neither of the two latter species move from the spot all day long.

thus to to be "thermoconformers" because they need to remain hidden (Fig. 13.5) is probably the much more general pattern. For example, of 32 New England forest-dwelling sphingids, saturniids, notodontids, and noctuids observed in a series of feeding studies (Heinrich and Collins, 1983), none were observed basking.

Huey (1982) has previously described a possibly analogous case in tropical forest–dwelling lizards that, like forest caterpillars, tend to be poikilothermic. He has proposed that in lizards the cost of finding suitable basking sites may be prohibitive, or that predators could learn to return to specific good basking sites, making basking an excessive risk. The same logic may apply to forest caterpillars.

Although most caterpillars palatable to birds (Heinrich, 1992) hide in shade and under the foliage, there is an interesting exception that may prove the rule, in that it also supports the hypothesis that unavoidable heating is a major evolutionary cause of thermoregulation.

The larvae of *Papilio glaucus* mimic bird droppings during early instars, and then small green snakes after they reach the final two instars. Bird droppings are found on the *surface* of leaves, hence potentially in sunshine, as are the caterpillars that mimic them. One would expect that these mimics would more likely be subjected to higher temperatures, on the average, than caterpillars hidden in shade under the leaf. Grossmueller and Lederhouse (1985) show that *P. glaucus* females oviposit preferentially on sunlit leaves, where the larvae later experience T_b excess of up to 8°C in sunshine near noon. The oviposition behavior alone indicates thermal "preference" imposed on the larvae by their

Fig. 13.6 The monarch butterfly caterpillar *Danaus plexippus*, a typically poisonous caterpillar, often feeds while exposed to direct sunshine.

mother. In the *proximate* sense the higher thermal "preference" results in a shorter life cycle, making possible two broods per year rather than just one (Grossmueller and Lederhouse, 1985). But I speculate that, as in the other example, the high thermal preference can ultimately also be explained by the thermal load imposed by the particular anti-predator strategy employed: in other words, if the caterpillars were always subjected to low T_b while hiding, they would evolve to have faster growth rates at lower T_b.

For the most part, only those caterpillar species that do *not* need to hide from predators and that are therefore consistently seen in sunshine (Fig. 13.6) are spiny, hairy, or poisonous (Rawlins and Lederhouse, 1981). One example, the eastern tent caterpillar *Malacosoma americanum*, shows large elevation of T_b above air temperature (Knapp and Casey, 1986). Nevertheless, the pattern of T_b as a function of air temperature (see Fig. 13.7) shows very little thermoregulation. Indeed, at temperatures from 10 to 25°C, T_b ranges from 15 to 40°C. Even with sunshine available, T_b can vary by as much as 15°C.

Although one may quibble whether or not *M. americanum* "thermoregulate," they clearly heat up and one may ask why. The most immediate reason is that it is advantageous for them to heat up because growth rates are very depressed at T_b from 15 to 25°C (Fig. 13.8), but then growth rates more than double from 25 to 30°C. Since these insects are active early in the spring, when it is still cool, it is therefore advantageous for them to heat up above the prevailing temperatures (generally less than 25°C). In addition to this proximate physiological answer, however, there is likely also an ultimate or evolutionary explanation that needs

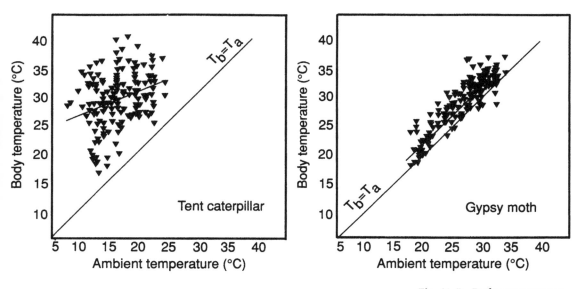

Fig. 13.7 Body temperature as a function of ambient temperature for tent caterpillars *Malacosoma americanum* and gypsy moth caterpillars *Lymantria dispar*. (From Knapp and Casey, 1986.)

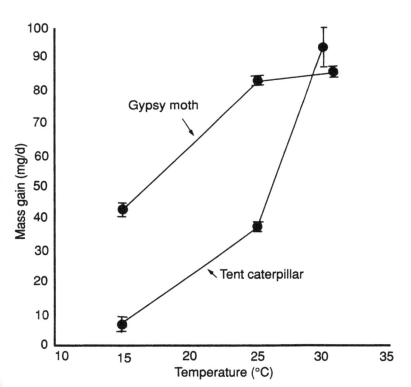

Fig. 13.8 Growth rates of late-instar tent caterpillars and gypsy moth caterpillars with respect to ambient temperature in the laboratory. (From Knapp and Casey, 1986.)

to be distinguished. For example, the gypsy moth *Lymantria dispar* is active at higher temperatures later in the season and is able to grow quite well at low temperatures; while *M. americanum* gains only 9 mg/day at 18°C, *L. dispar* can grow 45 mg/day at the same temperature (Knapp and Casey, 1986). Why does *M. americanum* grow rapidly only at high T_b, for which it heats itself up? Why does it bother with all the tent building and thermoregulation to achieve faster growth rates when a faster growth rate is apparently a viable evolutionary option at low T_b?

It is first necessary to answer *how* they heat up. As their name implies, tent caterpillars communally spin silken tents into which they retire when not feeding (Fitzgerald, 1980; Casey et al. 1988). The tents act as solar greenhouses (Wellington, 1950; Joos et al., 1988) in that they allow solar heating and retard convection from the metabolically produced heat of the larvae within. Additionally, the crowding of the caterpillars inside retards cooling (Knapp and Casey, 1986). As a result, temperature excesses in these tents are commonly 20–25°C, and up to 43°C above air temperature (Joos et al., 1988). The tent warms (and often overheats) the caterpillars, and a common supposition is that the animals make the tents *in order to* heat up and thermoregulate, and the heating up (or thermoregulation) allows them to exploit the cool environment of spring (Joos et al., 1988).

The above scenario may *now* be the case, because without the use of tents these caterpillars might not survive in the spring. However, given that *L. dispar* and other caterpillars without tents can grow rapidly at considerably lower T_b than *M. americanum*, the thermoregulatory idea is not an adequate *evolutionary* hypothesis. Even if they were unable to evolve to be able to grow at the same T_b as *L. dispar* caterpillars, they could simply emerge slightly later at a warmer season—there is no evidence that they cannot feed on June foliage. Also, like some other larvae, they could switch activity cycles to the warm part of the day. For example, *Halisidota argentata* switches from being nocturnal in summer to diurnal in winter (Edwards, 1964).

An alternate argument that can be made is that the tent evolved, originally, as a means of defense from predators and parasites, a function it may still serve (Morris, 1972a,b; Damman, 1987; Casey et al., 1988). (Parasites have evolved to specialize even on tent users, but are *unspecialized* parasites, such as tachinid flies and small hymenopterans, less likely to oviposit on caterpillars that are covered by the multi-layered webbing of the tents than on

Fig. 13.9 A "tent" of the eastern tent caterpillar *Malacosoma americanum* exposed to full sunshine. The larvae here are forced to leave the tent because of overheating, and they dangle in the shade beneath the tent.

those which are uncovered?) Once they have escaped from originally unspecialized parasites by taking shelter in the tents, the caterpillars will have had to endure T_b at times up to at least 40°C (Casey et al., 1988), and such temperatures are highly inimical to growth in most caterpillars (see Chapter 14). When T_b threatens to exceed 40°C the caterpillars are forced to evacuate their tent to cool convectively outside it (Fig. 13.9). Obviously, the higher the T_b they can endure, the more their tent can serve as a protective refuge from predators and parasites. Therefore, given the temperature constraint imposed by predators and parasites, one obvious evolutionary option is to shift the biochemical specialization of the tissues to operate at higher temperatures.

Support for the above hypothesis exists in a parallel situation for the caterpillars of the butterfly *Euphydras aurinia* (Nymphalidae). The first three larval instars of this species do not bask; they spin a communal web and remain inside it to feed. At the fifth or sixth larval instar the larvae change color from brown to jet black, and now on sunny days they bask in sunshine, achieving temperature excesses of up to 30°C (Porter, 1982, 1989). Like *M. americanum*, they commute between food and basking clusters, thus splitting feeding and digestion times. Porter (1982) observes that the thermoregulatory behavior allows them to develop faster. This is to be expected, because (as in all caterpillars) once adjustment to a *high* thermal load has occurred, the animals cannot then develop equally well at low temperatures. Again, however, tent building is not likely to have originally evolved to aid thermoregulation. There are dozens and perhaps hundreds of species of caterpillars, including the Cuculiinae, the winter moths (Schweitzer, 1974), that hatch, feed, and grow at very low temperatures (<10°C) on early spring buds, long before *M. americanum* or any other caterpillars emerge. This shows that growth is not obligatorily restricted to a high T_b, as is available in a tent.

Although I have here emphasized the effect of temperature on growth rate, molting may be even more affected by temperature—at least at low temperatures. For example, at 12°C the larvae of *Galerucella sagittariae* (Chrysomelidae) require 3.5 days to grow through the penultimate instar, and 3.6 days to then molt into the final instar. At 6°C the larvae are still able to grow, but they are incapable of completing a molt (Ayres and MacLean, 1987). The caterpillars of *Espirrita autumnata* (Geometridae) are similarly highly sensitive to temperature in their molt. Low temperature retards molting more than it does growth.

Hairy Caterpillars

A caterpillar hair is called a "seta." Many thousands of caterpillars have setae, which are often associated with poisons and/or are sharp and bristly. The main function of setae is therefore probably anti-predator defense. For example, chickadees *(Parus atricapillus)* accept without hesitation smooth-skinned heterocampid, geo-

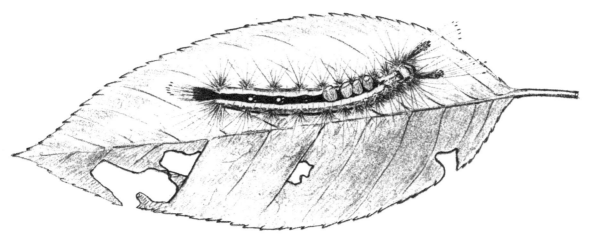

Fig. 13.10 Tussock moth caterpillar *Hemerocampus leucostigma*, showing its highly
ornamental but noxious tufts
of "hair." Note the posterior-
directed tuft and the 2 ante-
rior-directed tufts of long thin
hairs of different lengths, the 4
decorative hummocks of short
white hairs on the dorsum,
and the tufts of lateral hairs.
This caterpillar is highly distasteful to birds, and it exposes
itself on the surface of leaves
in sunshine.

metrid, and sphingid caterpillars, but bristly caterpillars—such as those of the tussock moth *Hemerocampus leucostigma* (Fig. 13.10), most arctiids, including *Halysidota maculata*, the mourning cloak butterfly *Nymphalis antiopa*, or the saturniid *Anisota rubicunda*— are not even pecked at (Heinrich and Collins, 1983). When hairy caterpillars are (rarely) taken by the red-eyed vireo *(Vireo oliva-ceus)*, the birds required a long time to knock off the spines (Hein-rich, 1979) and these caterpillars are therefore only marginally palatable.

Some of the hairs of caterpillars also serve as sound receptors (Markl and Tautz, 1975), suitable for detecting wasp predators or parasites and allowing evasive actions (Minnich, 1925).

Aside from serving as "ears" and mechanical anti-predator de-vices, caterpillar hairs in some species have also taken on a ther-moregulatory function. In the gypsy moth caterpillars *Lymantria dispar*, the setae along the side of the body are long and soft, while those on the dorsal surface are short and stiff. Pubescence reduces the rate of convective cooling without retarding the rate of radia-tive heating (Casey and Hegel, 1981). Are these characteristics useful to the animal?

In the field, gypsy moth caterpillars are thermoconformers (Fig. 13.7). Early instars bask in the sun, but later instars spend most of their time in the shade (except possibly during caterpillar out-breaks, when they feed both day and night). Body temperatures of gypsy moth caterpillars are generally within about 2°C of air temperature from at least 17°C to 35°C (Knapp and Casey, 1986), indicating that although the pubescence offers a great potential for thermoregulation in these animals, it is not used. Growth rates peak at 25°C, and the animals hatch late in the summer (mid-May to July, in New Jersey). Temperatures are normally high enough so that *L. dispar* do not need to rely on behavioral thermoregulation to complete their larval development (Knapp and Casey, 1986). Thus it appears doubtful that the hairs actually aid in thermoregulation. Either the setae are useful for heating up those animals occurring at the northern edge of their range, or else they function mainly for predator defense.

There is, however, one example where caterpillar fur makes a crucial difference in the thermal biology of two closely related caterpillars. Two relatives of the gypsy moth, *Gynaephora rossii* and *G. groenlandica* (Lymantriidae), both of which occur in the High Arctic tundra, are covered with long, reddish-brown to black pubescence both dorsally and laterally (Fig. 13.11). At Lake Hazen (81° 49′N) on Ellesmere Island, late-instar *G. rossi* maintained T_b of up to about 10°C above air temperature in the field while shaved animals only heat to 6.9°C above air temperature (Kevan, Jensen, and Shorthouse, 1982). At Alexandra Fiord (78° 53′N), *G. groenlandica* body temperatures in the midnight sun (at 1–4°

Fig. 13.11 The Arctic caterpillar *Gynaephora groenlandica.* The fur is an essential component of its thermoregulatory strategy; it reduces the convective loss of heat gained from the sun by basking.

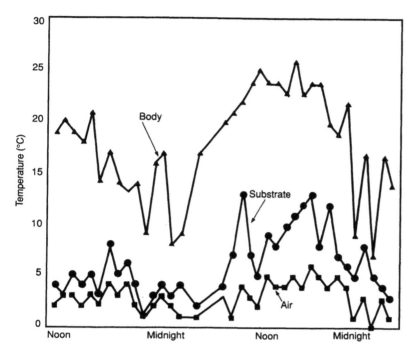

Fig. 13.12 Diel changes in
mean body temperature of 8
Gynaephora groenlandica caterpillars relative to air and
substrate temperatures as recorded bihourly over 2 days
(an overcast day followed by a
sunny day) on the open tundra of Ellesmere Island. (From
Kukal, Heinrich, and Duman,
1988.)

C) range near 10–17°C (Fig. 13.12), while at noon (near 5°C)
T_b increases to near 25°C (Kukal, Heinrich, and Duman, 1988).
The caterpillars have a daily pattern of activity; at midday larvae
either bask, move, or feed, but around midnight most of them
bask. Basking caterpillars are motionless and align themselves
perpendicular to the sun's rays. Their body temperature declines
4–6°C when they start moving or feeding. Setae significantly
retard cooling rates, as seen by comparing cooling rates of shaved
vs. unshaved caterpillars in a wind tunnel.

The *Gynaephora* caterpillars apparently use all of the solar radiation they can get. In their environment the angle of the sun in
mid-June (at 81°N) varies from 33° at noon to 10° at midnight.
Low ambient temperatures (from below 0°C to about 6°C) and
wind prevail. Additionally, the caterpillars already curtail their
activity in midsummer, apparently a result of severe selective
pressure to escape tachinid and ichneumonid parasitism (Kukal
and Kevan, 1987), and possibly also because the *Salix arctica* upon
which they feed has decreased caloric and nutrient content and

increased concentrations of tannins and phenols later on in the season (Kukal and Dawson, 1989).

The problem of temperature adaptation in Arctic caterpillars to maintain energy balance is intricate because these insects live "on the edge." They need to store energy reserves for overwintering, yet the higher their T_b, the greater their rate of energy expenditure in maintenance metabolism. High T_b can also be an energy drain unless it is counter-balanced by rapid food intake and rapid assimilation. Larvae assimilate food most rapidly at 15°C. At 30°C assimilation efficiency decreases because of the increased maintenance metabolism (Kukal and Dawson, 1989). Presumably a high T_b enhances food handling, or digestion, but in the absence of feeding it is economically advantageous for organisms to maintain low T_b.

In the winter, with no food intake, the caterpillars greatly reduce the energy drain by being frozen and degrading their mitochondria (Kukal, Duman, and Serianni, 1989). Given the constraints by parasites, temperature, and food quality, the caterpillars are forced to remain up to about 14 years (Kukal and Kevan, 1987) in the larval stage undergoing repeated freezing and thawing (Kukal, 1988). There is no indication that the caterpillars regulate their T_b at or around a particular setpoint; instead, they appear to attempt to attain the maximum T_b they can at *all* times before midsummer (Kevan et al., 1982; Kukal, Heinrich, and Duman, 1988). Therefore, the extra elevation of T_b due to the setae (about 5°C) is critical to their growth and survival above the Arctic Circle, likely reducing the duration of their larval stage by at least 2 years. Undoubtedly, these animals have evolved from ancestors adapted to much higher temperatures, and thus in the High Arctic they are living close on the thermal edge. The Arctic lymantriids are noticeably furrier than temperate lymantriids.

Some temperate arctiids are also very furry, even though the effect of their fur on possible thermoregulation has scarcely been examined. However, Fields and McNeil (1988) report that in the fuzzy arctiid *Ctenucha virginica* (occurring from Pennsylvania to Labrador) the autumn and spring larvae have black and yellow setae, while the summer larvae have light yellow setae, and they propose that the seasonal color change may help promote thermal balance. Critical heat-balance models to test this hypothesis are lacking.

The need for direct solar heating (and hence setae to help

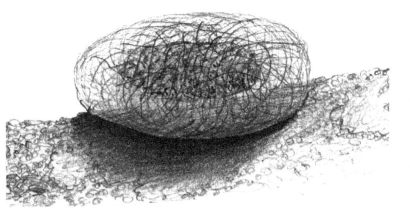

Fig. 13.13 "Basking" cocoon of *G. groenlandica* on exposed rock in sunshine. The transparent outer tent is thin enough to admit sunlight yet thick enough to retard convection.

conserve the solar heat gained) in both *G. rossi* and *G. groenlandica* can be seen by the thermoregulation posture that the caterpillars impose on the cocoons they spin. In other lymantriids the cocoon is generally an opaque structure hidden in some shady crevice or under bark. In the two Arctic lymantriids, however, the cocoons are surrounded by a secondary diaphanous structure, or tent, that s thin enough to admit sunlight (Fig. 13.13). Cocoons are prominently placed atop hummocks or rocks or on flat ground, where they experience direct solar radiation for 68 percent of their pupal life (Kevan, Jensen, and Shorthouse, 1982), which is completed in one summer. At Alexandria Fiord most (70 percent) of the cocoons are oriented in a N–S direction, and when the larvae are restrained in a container and placed into a room with only one window they spin their cocoons at a right angle to the incident light. This orientation is consistent with maximizing interception of light, or heat. In the field, the pupae experience an average temperature excess of about 14°C (Kevan, Jensen, and Shorthouse, 1982). Curiously, at Atkasook, Alaska, where there are "strong prevailing northeast winds" (when cocoons were spun, or observed?), the cocoons are aligned primarily in the NW–SE direction, presumably to "minimize convective cooling" (Kevan, Jensen, and Shorthouse, 1982). Contrary to this hypothesis, however, such orientation is precisely opposite (90° off) what might be expected if they were trying to minimize convective cooling. Alignment *head-on* rather than broad-side to the wind should minimize convective cooling. (I tested an oblong pupal-shaped piece of round rubber heated under a heat lamp in a wind tunnel when facing and when parallel to the wind. When facing the air stream of 5 m/sec the rubber reached a temperature excess of 8.6 °C; perpendicular orientation to the air reduced the excess to 6.0 °C.)

So far, then, the orientation of the *Gynaephora* pupae remains unexplained, because to compromise between an orientation for maximum heating (N–S) and one for minimum convection (NE–SW), the pupae should have been oriented east of north and west of south. Nevertheless, regardless of their precise alignment, they are conspicuously exposed to sunshine, unlike, to my knowledge, the pupae of any other lepidopterans. An example of sunlight-exposed pupae exists in the Diptera also: maggots of the Arctic carrion fly *Borealus atricepts* are reported to pupate on the upper surfaces of carcasses, so that the pupae "bask" throughout their development (Downes, 1965).

Measurements (near Cambridge Bay, Northwest Territories, Canada) by Bruce Lyon and Ralph Carter (personal communcations) show that the tents of the *Gynaephora* indeed speed up development time, as predicted. However, the caterpillars pay a price for spinning the large tents—increased predation by birds.

Endothermic Caterpillars

For the most part caterpillars are small and thermally isolated from each other, and they subsist on food that is of low caloric content. Furthermore, they do not have sufficient muscle mass to produce heat faster than they lose it. They are therefore not endothermic. It was long ago reported (Girard, 1865), however, that wax moth larvae *Galleria cerella* may elevate the temperature of honeybee hives that they infest after a swarm dies. Within hives where the bees have died, *G. cerella* as well as *G. mellonella* frequently occur in crowds of hundreds or thousands, and they feed on concentrated energy resources (honey and wax) while being insulated from convective cooling. (They may also occur in small numbers in weak hives.)

Measurements by Stephen L. Buchmann and Hayward G. Spangler (1992) at the Carl Hayden Bee Research Center in Tucson, Arizona, confirm that when *Galleria mellonella* larvae infest hives that are devoid of bees they generate hive and hence body temperatures of up to 43 °C at air temperatures of near 25–28 °C.

The elevated temperatures persist for many days as the larvae mature. Are the high temperatures generated specifically for heating, or are they the result of physical factors such as crowding in an enclosed space, with heat produced only by normal activity and resting metabolism?

The immediate result of the elevated temperatures should be an acceleration of growth rate, and I speculate that, unlike most other caterpillars, *Galleria* will not grow at T_b below 10°C, but they will likely not be inhibited from growth at 40°C. It is tempting to speculate that the heating has evolved into a strategy of an "arms race" among different egg clutches to utilize the rich resources of a hive. Presumably, chance determines which female moth finds an empty hive filled with vast resources. But the more eggs she lays, the more likely she will convert these resources to offspring before a competing female finds the same resources. Groups of larvae that do not heat up will be out-competed for by others in the race to use the finite resources of the enclosed hive.

Thermoregulation in Other Larvae

There is a relatively large literature on thermoregulation in caterpillars, our subject so far. Thermoregulation by larvae of Diptera, Hymenoptera, Coleoptera, and the other orders is less frequently studied and can be summarized quickly.

MAGGOTS

There is very little information available on the thermal responses of fly larvae. Presumably most fly larvae have few thermoregulatory options. Nevertheless, when in large numbers, as when infesting a carcass, they may generate considerable amounts of heat that is not readily dissipated. Already in 1869 M. Girard reported that a box full of the *Lucilia caesar* maggots (Calliphora) generated a temperature excess of 32°C. Most likely at high T_b the maggots would tend to disperse, lowering the temperature excess of the mass as the individuals at least temporarily escape to more agreeable quarters.

As possibly the first and only published confirmation of Girard's observations of some 120 years ago, I offer the following data showing that maggots can indeed be responsible for generating heat. I put a deer and a calf carcass side by side in the open air in the second week of November in western Maine, after the first snowstorm and after the customary nocturnal frosts. The road-killed deer had been fly-blown two weeks earlier during a few warm days. The calf was without maggots.

The deer, by the time it was converted to a writhing mass of maggots, was steaming in the chilly autumn air. Inserting a thermometer to a depth of 10 cm into it, I measured temperatures (on two days) of 24°C and 29°C, while the calf's temperature was only 4°C and 9°C, respectively. These measurements were taken late at night and at dawn following days when there was no sunshine. Therefore, the activity of the fly larvae and bacteria accounted for a 20°C temperature excess in the deer. (The larval thermogenesis undoubtedly facilitated the growth of the bacteria—the prey for the larvae—and the bacteria, once warmed, were likely also metabolizing at a higher rate and generating more heat.)

HYMENOPTERANS

Most hymenopteran larvae are thermally passive because they live in an environment that is thermally controlled (the eusocial Hymenoptera, see Chapter 16), or they are at the mercy of the thermal responses of their hosts (the parasitic Hymenoptera). As far as I know, only one study exists of one of the individually free-living hymenopteran larvae. This is a study of sawflies (Tenthredinidae).

Sawfly larvae feed on foliage and resemble caterpillars closely enough to be confused with them by nonentomologists. Many of them are gregarious. Roger S. Seymour (1974) from Monash University in Australia reported a unique set of observations on thermoregulation in the gregarious larvae of the sawfly *Perga dorsalis* that feed on *Eucalyptus* trees. At low temperatures the larvae are heated by the sun, and by aggregating they diminish convective heat loss. The larvae form a compact unit at low temperatures, but under intense sunshine they avoid contact with each other and they raise their abdomens into the air to facilitate convective cooling.

Seymour heated larval aggregations in an incubator and subjected them to increasing temperatures and radiation with a heat lamp. Abdomen lifting (Fig. 13.14) occurs at a body temperature of 30°C. When heated further the larvae wave their abdomens about, and at T_b of 37–40°C they void watery fecal pellets from the anus and smear it over their bodies. In a short time the larvae become quite wet and considerable evaporative cooling occurs. For example, at 45–48°C, larvae maintain T_b between 39° and 42°C (Fig. 13.15).

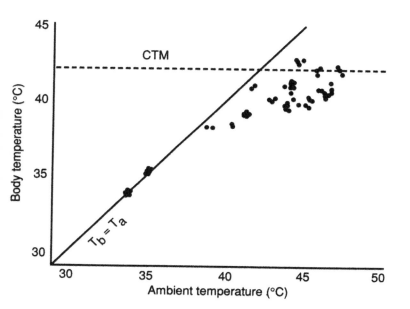

Fig. 13.14 Larva of the sawfly *Perga dorsalis*, resting at room temperature *(left)*. It raises its abdomen in response to heat stress *(middle)*; at even higher temperatures it spreads rectal fluid onto the ventral surface of the abdomen *(right)* for evaporative cooling. (From photographs by Seymour, 1974.)

Fig. 13.15 Body temperatures of *P. dorsalis* as a function of air temperature. The critical thermal maximum (CTM) that the larvae can tolerate is 42 °C. (From Seymour, 1974.)

TURRETS, PITS, AND SHIELDS

Another interesting example of thermoregulation concerns the larva of a tiger beetle, *Cicindela willistoni* (Coleoptera: Cicindelidae). *Cicindela* larvae are predators that construct vertical burrows in the ground. They then wait at the mouth of the burrow to snatch any small moving object coming within reach of their extended body and large mandibles. Throughout its range in the central and southwestern United States, *C. willistoni* adults and larvae occupy hot saline-alkali flats with little or no vegetation. Unlike other *Cicindela* larvae, *C. willistoni* construct turrets on top of the burrow entrance, 1–4 cm high, and then wait for prey in the top of the turret (Fig. 13.16). C. Barry Knisley and David L. Pearson (1981) from the Pennsylvania State University tested several different hypotheses to explain the functional significance of this curious behavior of turret construction and use. They conclude that the turrets are thermoregulation aids.

In the tiger beetle larvae's environment in southeastern Arizona, temperatures near the top of the soil commonly reached 50°C at noon, but air temperatures 4 cm above the soil surface average 38°C in sunshine. These temperatures also correspond to temperatures within C. *willistoni* burrows; without the use of its turret, a hunting larva remaining at the surface to hunt would be soon "cooked" at 50°C. Within the turret, however, the larva is subjected to lower temperatures and greater convective air flow, and it can remain there on station to hunt throughout the day. Additionally, prey also ascend or land on the turrets to escape the heat, which means that the structure also serves as thermal "bait."

Turret construction is a widespread phenomenon that has evolved independently in several groups of ground-dwelling arthropods, including ants, spiders, and solitary bees, but for other purposes. For example, in the gregariously nesting bee *Diadasia bituberculata*, turrets are made facing downhill on sandy slopes, where they may function primarily to prevent infilling of burrows by loose dirt and debris (North and Lillywhite, 1980), or to foil predators and parasites.

Larvae make other structures that have one primary function but could also function in another way, such as the defensive "fecal shields" of cassidine (Chrysomelidae) larvae. Tortoise-shell beetle larvae carry tight packets of cast skins and feces on a fork held over their back, and these shields are effective in blocking the bites of ants (Eisner, von Tassel, and Carrel, 1967). When pinched, these larvae will tilt their shield toward the site stimulated. Additionally, irritation from overheating in sunshine causes the same response (Heinrich, personal observation). Perhaps, therefore, the shield may also act as a thermoregulatory aid, but no studies have examined this hypothesis.

The sand pits (Fig. 13.17) of ant lion larvae (Neuroptera: Myrmeleontidae) are also thermal devices. The larvae dig pits in loose sand for prey capture, and in these they orient on the cooler sides of their pits on hot days and seek thermal refuges in the sand below their pits (Green, 1955; Youthed, 1973). They may also

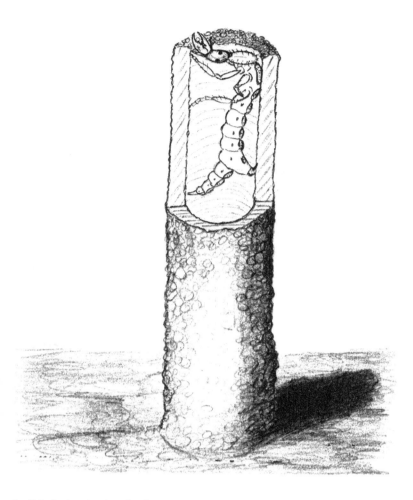

Fig. 13.16 A turret built by a third-instar tiger beetle larva *Cicindela willistoni*. The larva is in foraging position at the mouth of the burrow. (Adapted from Knisley and Pearson, 1981.)

build their pits in shade away from lethal temperatures (Heinrich and Heinrich, 1984).

For the most part, however, the larvae are chronically subjected to high temperatures, and hunting duration will be a function of their thermal tolerances. Various species of ant lions have tolerances for very high T_b (some species tolerate T_b between 47 and 53 °C for 5 hours), which allows them to extend their activity time throughout the day (Youthed, 1973). For example, *Cueta trivirgata*, a species whose larvae construct pits in the dry Kuiseb River bed in the Namib Desert, have death points near 53.4 °C. Their very high thermal tolerance permits these predators to capture the thermophilic ants *Ocymyrmex robustior* (Marsh, 1987).

Indeed, these ants comprise 65 percent of the prey biomass consumed by these ant lions. Most of the ants' activity is at surface temperatures of 45–65 °C (Fig. 13.18), and to capture these ants it is necessary for the ant lion larvae to expose themselves to near-lethal temperatures; they remain vigilant in their pits until pit temperatures approach 53 °C, when the ants are still abundant on the sand surface.

BEHAVIORAL STRATEGIES AND ECOLOGY

Many poikilothermic vertebrates display diel variations in their thermal preferences that presumably reflect switching between active (energy acquisition) and passive (energy saving) phases (Regal, 1967; Brett, 1971). As shown in the largely nocturnal aquatic larval crane flies (Diptera: Tipulidae), a rhythmic thermoregulation may persist on a circadian basis; when the animals were tested to a temperature gradient, they preferred temperatures

Fig. 13.17 An ant lion larva excavating a pit. This pit is approximately half done. The larva flings sand while moving from one pit site to another, leaving a conspicuous trail, as seen here.

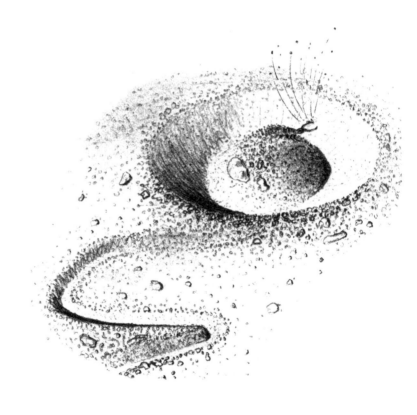

THE HOT-BLOODED INSECTS

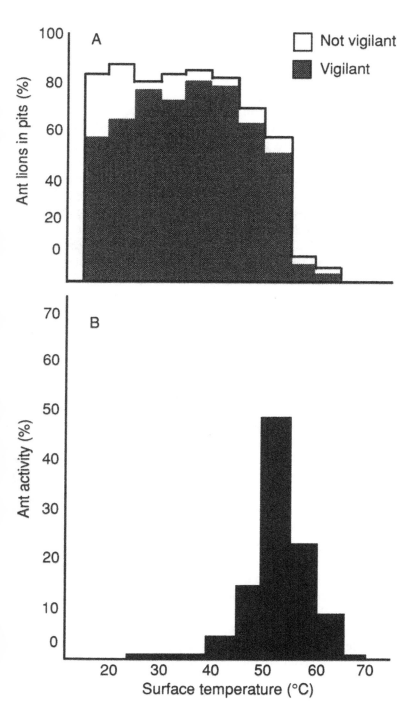

Fig. 13.18 The relationship between sand-surface temperature and *(A)* the proportion of ant lion larvae *C. trivirgata* that are vigilant within their pits and *(B)* the proportion of the total numbers of their prey, the ants *Ocymyrmex robustior,* active on the sand surface. The ant lion larvae are vigilant (near top of sand surface in their pits) and in their pits even at about 50 °C, when ant activity is at its peak. (From Marsh, 1987.)

of 16–18°C in the dark and 12–14°C in the light (Kavaliers, 1981). In the Canadian subarctic, mosquito larvae aggregate in the warmest sectors of partly shaded pools, and they travel around the pool clockwise, as thermal conditions change during the day (Haufe, 1957).

In my estimation, one of the more clever experiments showing how behavioral plasticity can counteract climatic effects was done by Jones, Coyne, and Partridge (1987), who worked with a temperature-sensitive eye mutant of *Drosophila melanogaster*. Flies with the sex-linked recessive allele white-blood (W^{bl}) have plum-colored eyes when raised at 15°C in the early pupal stage, but eye color becomes lighter at higher temperatures, fading to pale yellow at 30°C. By scanning eye color, 2°C differences in the temperatures that pupae had experienced can be detected. The researchers released flies at two sites (at 400 and 1,000 m elevation) having air-temperature differences of at least 4.5°C. From recoveries of W^{bl} males it was determined that the mean temperature *experienced* by the flies released at high elevation was only 1.3°C rather than 4.5°C lower. Thus, either the larvae exercised thermal choice in their pupation niche, or (more likely) their mothers exercised the choice for them.

The profound effect on temperature selection of the adult on the ecological adaptation of an insect was illustrated by Joel G. Kingsolver (1979) in a classic paper on the pitcher plant mosquito *Wyeomyia smithii*. In northern Michigan bogs *W. smithii* can oviposit either in pitchers in the shade or those exposed to sunshine. Controlling food and larval density, Kingsolver determined that fourth-instar larvae developed 26 percent faster in sun-exposed pitchers. In the bog, sun-exposed pitchers held *W. smithii* in more advanced developmental stages than the larvae found in shaded pitchers. Kingsolver concludes that larvae in shaded pitchers in northern Michigan develop so slowly that they are forced to diapause after one generation, whereas those in sun-exposed pitcher plants can complete two generations in one season.

The fact that different thermal niches, and not different life stages as such, determine T_b can be seen in the Colorado potato beetle *Leptinotarsa decemlineata* (Chrysomelidae). Both adults and larvae feed on the same foliage simultaneously, and both are aposematically colored and unpalatable to most predators (Hsiao and Fraenkel, 1969). Like aposematically colored caterpillars (Heinrich, 1979), the beetles and their larvae are commonly found on the leaf surfaces, and they maintain T_b 1–10°C above air temperature, depending on solar intensities (May, 1982). Both larvae and adults seek shade at 32 °C to maintain T_b below 36–38 °C. At 15 °C both adults and larvae have T_b of 23–24°, and at 28 °C T_b are 34–36 °C. These temperatures are presumably close to those of dead animals in the same positions. One may quibble that the temperature excess at 28 °C is *slightly* less than that at 15 °C, and that other passive factors may act to increase T_b at the higher temperatures, but I think most observers will agree that the dependence of T_b on air temperature (in other words, the *lack* of thermoregulation) is more conspicuous than *in*dependence of T_b from air temperature.

Summary

Insect larvae are such a heterogenous group that few except the simplest generalizations about them are possible. Most larval forms likely have very different (generally lower) thermal preferenda from those of the flying adults. Thermal preferenda within individual species are nevertheless specialized, and they are suited to the particular thermal environment that the larvae experience. It is likely that all larvae show increased locomotor (escape) responses at high temperatures and become sedentary in zones of comfort. An extension of this behavior is shuttling between sun and shade, provided this activity is not constrained by predator pressure. A further refinement is postural positioning to increase surface area for the interception of solar radiation. Structures made by larvae—such as turrets made by some tiger beetle larvae, pits made by ant lion larvae, and "tents" made by some caterpillars—act as thermoregulatory aids though they may have evolved originally to serve other functions.

Considerable research has been done on thermoregulation in caterpillars. Except for caterpillars that live confined in large numbers in a small space (such as *Galleria*), no caterpillars produce enough heat to affect body temperature. Although a few caterpillars bask, most of them avoid sunshine and do not thermoregulate. Caterpillar thermoregulation in sunshine can best be understood in terms of advantages (accelerated feeding and growth) and costs (increased visibility to predators and parasites). Whether a species evolves basking behavior involves a very unstable equilibrium, because feeding in sunshine can immediately expose larvae to avian predators while at the same time reducing the *duration* of exposure (by accelerating growth and therefore shortening the larval stage). Nevertheless, for the most part only those caterpillars bask that are protected by poisons or spines or that are forced to endure high T_b. The spines or setae act primarily as anti-predator devices, but in some caterpillars they are also essential thermoregulatory aids.

Remaining Problems

Mechanisms of thermoregulation, where they exist, tend to be rudimentary in larvae. Most revolve around behavior, and temperature adaptation revolves around biochemistry. To understand the behavior it is necessary to understand the sensory physiology. Nothing is known about temperature sensing of larvae as compared with that of adults. Possible similarities and differences of given enzymes that function at different temperature optima in larvae and adults also remain unexplored.

Fever

T H E idea that troublesome organisms can be killed by heat treatments originated with Louis Pasteur. Most bacteria, viruses, and protozoa have inactivation temperatures that are much higher than 45 ° C, the upper limit for many insects. Indeed, heat treatment (pasteurization) of dairy products for tuberculosis and other disease organisms typically involves heating at 62 ° C for 30 minutes, or 72 ° C for 16 seconds. Since the lethal temperatures of most disease organisms are considerably higher than those of their hosts, it seems therefore not at all obvious that animals would have invented heat treatment long before Louis Pasteur thought of it. Nevertheless, it is now clear that fever, one form of heat treatment, is a common survival mechanism (Kluger, 1979).

Already in 1936 L. O. Kunkel from the Rockefeller Institute for Medical Research at Princeton, New Jersey, used heat therapy to cure viral infections of peach trees. Heat therapy was also used in the control of viral diseases of other plants (Maramorosch, 1950; Bawden, 1964). This kind of "treatment" sometimes occurs naturally, because some plants are already naturally protected from certain viruses by the very high summer temperatures where they grow (Bawden, 1955, as cited by Tanada, 1967).

With the possible exceptions of endothermy during metabolic flare-ups in flowering spathes—flowers that dispense volatile substances that attract pollinators (Meeuse, 1966; Skubatz et al., 1990)—plants have no control of their body temperature so they cannot combat microorganisms by generating heat. In endothermic vertebrate animals, however, the development of fever during certain viral infections has long been associated with combating disease (Lwoff and Lwoff, 1958). Fever is now well recognized as an adaptive response in both endothermic vertebrate animals and in some of those that regulate T_b by behavioral means (Kluger, 1979). However, the details whereby fever exerts its effects are still largely unknown.

Self-Defense against Microscopic Agents

It has only recently been learned that some insects regulate their body temperature. Not surprisingly, therefore, fever in insects has so far barely been investigated. But studies of the effect of high temperature on insect pathogens, and the use of heat treatment to combat pathogens while rearing insects, have a long history. L'Héritier and Sigot (1946) observed that *Drosophila* reared at 30 ° C lost the symptoms of infection caused by the sigma virus. De Lestrange (1954, 1963) investigated the phenomenon in detail and found a continuous decrease of the virus content in infected flies after they were placed at 30 ° C.

That many insects are protected from lethal viral infections when reared at high temperatures is now well known. Temperature suppression of virus has been reported for a granulosis virus in *Pieris rapae* reared at 36 ° C (Tanada, 1953); a nuclear-polyhedrosis virus in a number of insects, including *Diprion hercyniae* at 29.4 ° C (Bird, 1955), *Trichoplusia ni* and *Heliothus zea* at 39 ° C (Thompson, 1959), *Bombyx mori* at 36 ° C (Tanada, 1967), and *Pseudaletia unipuncta* at 37 ° C (Watanabe and Tanada, 1972); a cytoplasmic-polyhedrosis virus in *Colias eurytheme* at 35 ° C (Tanada and Chang, 1968) and *Bombyx mori* at 36–37 ° C (Tanada, 1967); the flacherie virus in *B. mori* at 32–35 ° C (Aruga and Tanaka, 1968) and 37 ° C (Inoue and Tanada, 1977); non-inclusion viruses in *Sericesthis pruinosa* at 28 ° C (Day and Mercer, 1964) and *Galleria mellonella* at 30 ° C (Tanada and Tanabe, 1965).

In most of these examples larval-stage hosts were infected by preparing a viral inoculum derived from infected animals and spraying it onto the food. The animals were then reared under constant environmental conditions, and no attempts were made to determine if they would choose warmer temperatures when infected than when not infected.

Almost nothing is known of the natural thermal preferences, which are needed for comparison in order to determine if these insects develop a "fever." Nevertheless, in some species the temperatures sufficient to combat the lethal viral infections are readily available through basking (see Chapter 13) and other means. For example, the greater waxmoth caterpillars *Galleria mellonella* feed on wax in the darkness of beehives (generally abandoned ones) and thus they do not have the option of basking. Large numbers of these caterpillars living in abandoned hives often clump together and achieve temperatures of 40 ° C (Chapter 13). Similar heating in a related species, *Galleria cerella*, has long been known (Girard, 1865). Thus, waxmoth caterpillars are heated to well above the 30 ° C required to combat the lethal viral infection of the *Tipula* iridescent virus that could infect them (Tanada and Tanabe, 1965). And this heating apparently results as a by-product of the crowding—it is not necessarily an endothermic response to pathogens.

Deliberate thermal therapy has practical application for mass-reared species, such as in the culture of larvae of the commercial silkworm *Bombyx mori*. The most important disease of the silkworm in Japan is a virus described as the "flacherie" virus (Inoue and Tanada, 1977). Heat therapy at 35 ° C for this viral infection was first reported by Suzuki, Kimura, and Suzuki (1963), although 35 ° C does not kill all of the virus. Nevertheless, the longer the larvae are reared at high temperature, the greater the probability of their survival (Miyajima, 1970).

Inoue, and Ayuzawa, and Kawamura (1972) followed up these initial observations to examine the effect of temperature on the multiplication of this virus in *B. mori*. Virus-infected cells were visually labeled with an anti-flacherine virus globulin created in the rabbit that was then conjugated with a fluorescent stain. Newly hatched larvae were infected by feeding them on mulberry leaves smeared with the virus. Subsequently, thin sections of the infected caterpillars were treated with the fluorescent stain, which, because of the attached anti-flacherine globulin, found and attached to the viral-infected cells on the sections. The results clearly show the time-course of the infection. At the same time, infected larvae subjected to different temperature regimes were ground up and supernatant pieces were diluted and sprayed onto new mulberry leaves, to examine their infectivity and compare it to the incidence of virus as determined by the fluorescent method above.

In controls reared at 27 ° C there are about 480 fluorescent (virus-infected) cells per 6 μ section per larva at 96 hours after infection. At 120 hours these increase to over 600 in those larvae that continue to be kept at 27 ° C and decrease to 200 in those heated to 37 ° C. At 168 hours, infection in the controls (27 ° C)

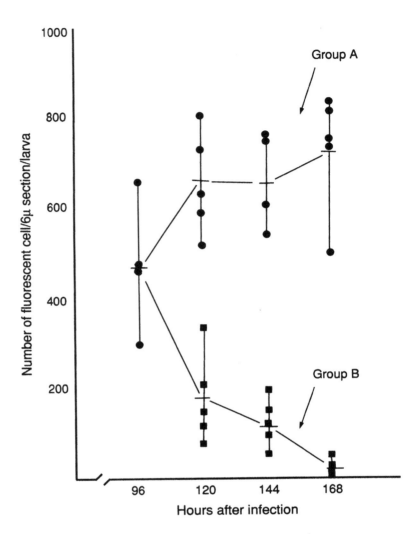

Fig. 14.1 Number of fluorescent (virus-infected) cells in 6 μ sections from *Bombyx mori* larvae. Group A was maintained at 27° C. Group B was held at 27° C for 96 hours after virus inoculation, then moved to 37° C. All larvae of group A died, and none of group B died. (From Inoue, Ayuzawa, and Kawamura, Jr., 1972.)

has increased to 700 fluorescent cells, while in the experimental (37 ° C) larvae it had declined to less than 10 (Fig. 14.1). Infectivity of the spray of infected larvae that was fed to uninfected larvae closely parallels the incidence of fluorescence; it is strictly dependent on dose. Since dissociation of the viral polyhedra by high temperature occurs in live but not in dead larvae (Yamaguchi, Iwashita, and Inoue, 1969), this suggests that the high temperatures do not directly deactivate the virus, but that the viral deactivation is in some way mediated indirectly through some mechanisms by the host.

High rearing temperatures adversely affect other pathogens besides viruses. High temperatures are detrimental to immature stages of some insect parasites. Certain hymenopterous parasites of aphids are not able to survive within their hosts at temperatures greater than 32–35 ° C, depending on the species (Force and Messenger, 1964). At ambient temperatures above 35 ° C the braconid parasitic wasp *Apanteles medicaginis* does not survive in its host, the alfalfa caterpillar *Colias philodice eurytheme* (Allen and Smith, 1958). (The hosts survive without apparent harm containing the heat-killed parasites.) The eggs of another braconid wasp, *A. militaris,* are encapsulated (and killed) when their hosts, the cornborer larvae *Pseudaletia unipuncta,* are exposed to 35 ° C for 24 hours (Kaya and Tanada, 1969). Rearing the larvae after parasitization at 23 ° C for 2 days and then at 35 ° C for 1–2 days does not prevent the parasite eggs from developing.

Although high T_b can be associated with certain benefits, there are also costs associated with it that could preclude it from being adaptive. First, caterpillars that bask expose themselves to visually oriented avian predators. In addition, although higher temperatures result in faster growth rates, temperatures slightly above the optimum result in decreased feeding, growth, and assimilation rates (Mathavan and Pandian, 1975). For example, although both the armyworm *Pseudaletia unipuncta* (Watanabe and Tanada, 1972) and the commercial silkworm *Bombyx mori* (Inoue and Tanada, 1977) are protected from certain *viruses* at 37 ° C, in both species the growth rate is adversely affected at that temperature (Guppy, 1969), and both suffer from a higher incidence of *bacterial* infections at the higher temperatures (McLaughlin, 1962). *Colias eurytheme* butterfly larvae greatly increase their growth rate from 20 ° C to 30 ° C, but when kept at 35 ° C they die (Sherman and Watt, 1973). If their pathogens are not killed until 37 ° C, then the larvae have little to gain unless they strictly regulate the timing and durations of fever. (They can survive short durations at 37 ° C.) Additionally, certain *essential* symbiotes are also adversely affected when their insect hosts are reared at high temperatures (Huger, 1954, 1956; Schneider, 1956; Brooks, 1963).

The evolutionary perspective of the parasite should also be considered. Pathogens differ in their sensitivity to temperature (Ewald, 1980). Could a pathogen evolve to raise its thermal optimum so that it is not harmed by the host? Could it exploit the host's defensive response to speed up its own developmental rate?

Given the different costs and benefits of heat treatment and the fact that effective heat treatment against one pathogen may be totally ineffective against another, it is not at all clear whether or not a general, indiscriminate febrile response by insects to any one pathogen would be adaptive. Another major difficulty is that, at least in the laboratory, insects are not generally able to survive infections when the temperatures alternate between high and low (Tanada, 1967), as occurs in the field and as is a necessity in a heliothermic organism which cannot bask continuously (except in the High Arctic in the summer).

Apparently, the high temperatures of the heat treatment do not act analogously to pasteurization in killing the pathogens directly. Instead, the pathogenic organisms are somehow suppressed, but only for as long as the temperature remains elevated. If the temperature remains elevated long enough, the pathogen may stop multiplying and eventually be diluted out. A febrile response could presumably nevertheless be effective at relatively low, physiological temperatures, provided it is activated very quickly after the inoculation. However, it is not clear if, when, and how "fever" in insects occurs. Additionally, since even a primitive arthropod, a millipede, produces an anti-bacterial protein in response to a bacterial infection (Van der Walt et al., 1990), it would appear that there is an alternative to fever to combat infection. The above background information points to a plethora of potentially interesting problems that relate to fever as a functional response for fighting disease organisms.

Behavioral fever as a response to bacterial endotoxins has been observed in a variety of ectothermic vertebrates, including amphibians, lizards, teleost fish, and neonatal mammals (Kluger, 1979). It has also been observed in response to prostaglandins and bacteria in arthropods, including crayfish *(Cambarus bartoni)* (Casterlin and Reynolds, 1977), lobsters *(Homarus americanis),* shrimp *(Penseus durorarum),* horseshoe crabs *(Limulus polyphemus)* (Casterlin and Reynolds, 1979) and the scorpions *Buthus occitatus* and *Androctonus australis* (Cabanac and Le Guelte, 1980). The examples of the marine organisms seem particularly surprising because under natural conditions these animals have no opportunity to significantly alter their body temperature.

The terrestrial environment is complex and varied, and most insects have both the means and capability to vary their body temperature with ease. However, to my knowledge only four studies of fever in insects have been published to date.

The first study concerns the Madagascar cockroach *Grampha-*

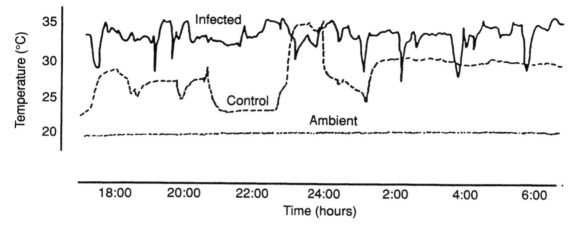

Fig. 14.2 Sample of body temperatures of a cricket, *Gryllus bimaculatus*, infected with *Rickettsiella grylli* vs. a control injected with sterile Ringer solution. (From Louis, Jourdan, and Cabanac, 1986.)

dorhina portentosa, which shows a febrile response when injected with suspensions of heat-killed *E. coli* (Bronstein and Conner, 1984). Cockroaches injected with the bacteria prefer temperatures 3.6 °C higher than controls. Lipopolysaccharide purified from *E. coli* has a similar effect to the *E. coli* themselves. The study by Louis, Jourdan, and Cabanac (1986) from France demonstrates that the cricket *Gryllus bimaculatus* injected with *Rickettsiella grylli*, a natural pathogen of crickets, also shows a behavioral fever in a temperature gradient (Fig. 14.2). Furthermore, animals kept in a temperature gradient where they are free to choose their own T_b have only 10 percent mortality after 21 days, whereas those kept at a constant 28 °C show a 50 percent mortality at the end of that time. Stephen M. Boorstein and Paul W. Ewald (1987) show that the American migratory grasshopper *Melanoplus sanguinipes* has a similar response. Its preferred temperature increases 6 °C when it is infected (by ingestion through food) by the spores of the protozoan *Nosema acridophagus*. This study also shows increased survival of the febrile animals, although febrile temperatures of *uninfected* grasshoppers negatively affects growth.

These three studies focused on insects that do not normally regulate a high body temperature. Yet, the infected animals all choose body temperatures near 35–36 °C, about 4 °C or more above their normally preferred T_b. Only one study has been published to date on an insect that even when not febrile maintains a high (>35 °C) body temperature. Field and laboratory studies by Elizabeth McClain and colleagues (1988) from the University

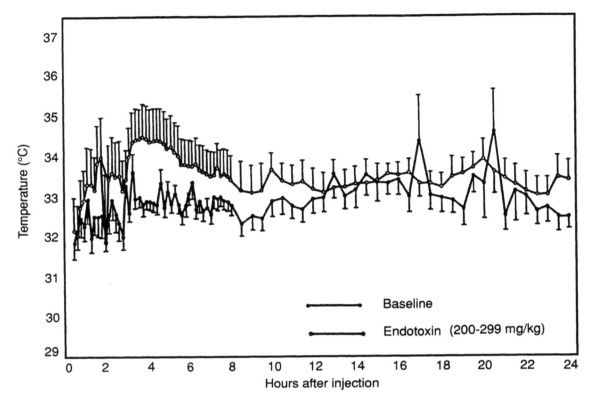

Temperature (°C)

Hours after injection

Baseline
Endotoxin (200-299 mg/kg)

Fig. 14.3 Thermal preferences (means and standard errors) of the desert tenebrionid beetles *Onymacris plana* injected with lipopolysaccharide isolated from *E. coli* vs. uninjected beetles that had previously been shown to have thermal preferences similar to those injected with sterile saline solution. (From McClain, Magnuson, and Warner, 1988.)

of Namibia indicate that the desert tenebrionid beetle *Onymacris plana* normally maintains a T_b between 37 and 38 ° C when active. But beetles injected with highly purified lipopolysaccharide isolated from *E. coli* sought an even higher temperature in a temperature gradient than control animals injected with saline solution. As in the other organisms so far investigated, the response of *O. plana* is dose-dependent. The elevation of thermal preference (up to about 1.5 ° C) is expressed within 1 hour of pyrogen injection and body temperature remains elevated above controls for at least 12 hours (Fig. 14.3). It is of interest that whereas in insects the pyrogen consists of a lipopolysaccharide, in at least some *Arum* lilies, such as the vodoo lily *Sauromatum guttatum*, the "calorigen" that functions to trigger the endogenous heat production is salicylic acid (Raskin et al., 1987); salicylic acid is the active ingredient of aspirin, which depresses body temperature in febrile vertebrate homeotherms.

Raymond I. Carruthers and colleagues (1992) from the USDA

Plant Protection Research Unit at Ithaca, New York, have published highly relevant data on the thermal ecology of the rangeland grasshopper *Camnula pellucida* in relation to its fungal parasite, *Entomophaga grylli*. I believe this study is one of the best yet of the role of temperature in the etiology of an insect host, even though their grasshoppers do *not* show "fever" (elevation of T_b when infected), probably because they already bask to achieve T_{thx} near 38–40 ° C nearly at *all* times possible. At the researchers' main study site near Alpine, Arizona (altitude about 2,450 m), a temperature-dependent development model shows that basking and the maintenance of a high T_b near 38–40 ° C is required for the animals to complete their life cycle from the first to the fifth instar within the year. However, protection from the fungal parasite is a second clear advantage of basking.

Optimal conditions for fungal pathogen development is typically between 25 and 30 ° C, with upper lethal limits under constant incubation conditions being near 35 ° C (MacLeod, 1963; Roberts and Campbell, 1977). In the fungal parasite *E. grylli*, fastest growth is near 25 ° C (Fig. 14.4), and both *in vitro* and *in vivo* in the grasshoppers, there is no growth at 33–35 ° C. Grasshoppers having as little as 4 hours' exposure to 40 ° C can adversely affect survival of their *E. grylli* fungal infections (Fig. 14.5), and the infections are almost totally eliminated in grasshoppers given the opportunity to bask. The simulation results of Carruthers and

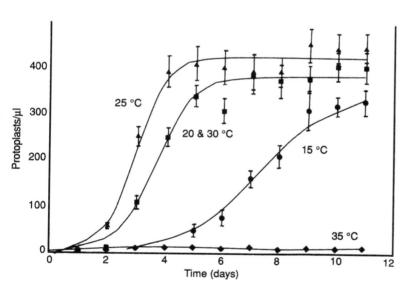

Fig. 14.4 Growth curves of *Entomophaga grylli* fungal protoplasts at 15, 20, 25, 30, and 35 ° C in insect tissue culture medium. Note that the optimal temperature for development is near 25 ° C, and the upper thermal limit is near 35 ° C. (From Carutthers et al., 1992.)

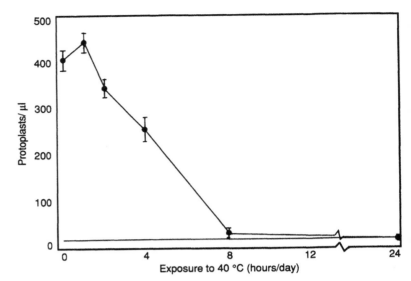

Fig. 14.5 Concentration of *in vitro E. grylli* protoplasts grown at 20 ° C and then following 6 days of exposure to 40 ° C for either 0, 1, 2, 4, 8, or 24 hours per day. (From Carruthers et al., 1992.)

Protoplasts/ µl

Exposure to 40 °C (hours/day)

colleagues predict the prevalence of the disease in the field depending on thermal conditions; at the hotter sites the *E. grylli* infections are eliminated.

Many Davids Slay a Goliath

The examples cited in this chapter all refer to hosts killing microscopic pathogens by heat, possibly in part indirectly through activation of the immune response. There is at least one example of an insect killing predators larger than itself by heating them to death.

In Japan a vespid wasp, the hornet *Vespa simillima xanthoptera*, is a primary predator on the native honeybees there, *Apis cerana japonica*. It hunts bees at the hive entrance, but it is penalized severely if it gets too close. Individually, the bees are powerless against the wasps, but they have evolved a social counter-strategy that, from the perspective of a colony response, is analogous to "fever." The bees attack a hornet by attaching themselves around it to form a ball. Such balls usually involve 180–300 bees (Ono, Okada, and Sasaki, 1987). As soon as a balling reaction starts, temperatures inside the ball increase rapidly and reach 46 ° C (average maximum = 45.2–47.0 ° C) within 4 minutes. Determinations of upper lethal temperatures showed that the Japanese honeybees tolerated 48–50 ° C, but the hornets only 45–47 ° C.

Temperatures within balls are maintained at near 46 ° C for 20 minutes. After 30 minutes the number of workers balling the wasp gradually decreases and the temperature inside the ball declines. The dead wasp is released after about an hour. In the 36 ballings that M. Ono and co-workers at the Laboratory of Entomology at Tanagawa University in Tokyo observed, the balled hornet was always killed. No honeybee stings were found in the corpses. These results strongly suggest that Japanese honeybees can kill their primary predators, the hornets, by heating rather than sting-ing them to death. (The researchers also observed a milder balling reaction of the same wasps by the introduced European honebees, *A. mellifera*. However, the average temperature inside their balls was only 42.8 ° C and the killed hornets usually had 2 or 3 stings embedded in their intersegmental membranes.)

Summary

There is a large literature showing that heat treatment helps insects to survive infections of viruses, fungi, bacteria, and other insects that have invaded the body as parasites. Many insects are now also known to regulate their body temperature, both through behavioral and physiological means. A great potential exists for many interesting studies, but the surface has barely been scratched. So far, however, the evidence is in that some insects indeed change their thermal preferenda when subjected to dan-gerous pathogens or their products, and that this behavior can aid their survival.

Remaining Problems

This area has been so little explored that it is probably still pre-mature to point out specific questions. For the time being, a larger data base is needed on a great variety of insects in order to determine if fever is a widespread phenomenon. It is necessary to discover in which type of insects—those with behavioral or phys-iological mechanisms for thermoregulation or those with none—fever occurs, what its effects on both the hosts and the parasites are, and how the effects are realized.

FIFTEEN

Cold Jumpers

FOR very small animals endothermy is an extravagantly costly way of remaining instantly ready for rapid locomotion. Always being prepared for making a quick getaway from a predator uses a lot of energy. Heliothermy also has obvious disadvantages for predator avoidance, because the presence of sunshine is locally unreliable, and when the animal does bask it is exposed to potentially hungry visual scanners; furthermore, once a very small animal moves, convection almost instantly cools it again.

Except for a very few isolated cases, elevations of body temperature in insects are functionally related to activity. Activity is the prize. In order to attain it despite the thermally varying environment, insects have, as we have seen throughout this book, evolved physiological and behavioral mechanisms of heating and cooling. They have evolved anatomy and covering to promote either heat retention or heat loss, and they have evolved specialized muscle morphology and biochemistry. All of the adaptations I have discussed are related to promoting activity independently of ambient temperature by regulating body temperature.

Given that temperature has had such an extraordinary impact on several levels of organization, it seems possible that evolution might also find a way to promote activity *independently* of body temperature—to short-circuit the process of temperature adaptation. A triumph over the constraints of ambient temperature, if possible, is clearly not "thermoregulation." But it is a triumph over temperature nevertheless, and not a defeat, for it is as a war won without fighting a battle. It is winning without appearing to try, because it skirts the battlefield entirely. In the insects' war against temperature, this victory is achieved by the use of intricate and diverse energy-storage mechanisms. Using these mechanisms, obligate poikilotherms have won some of the prizes gained by endotherms.

Flea Beetles

The flea beetles, Alticinae, are the largest of the leaf beetle (Chrysomelidae) subfamilies. Approximately 560 genera and some 8,000 species of flea beetles have been described. Most species are small, weighing only several milligrams, and many (but not all) are good jumpers. The scientific name of the type genus, *Altica*, is derived from the Greek *haltikos*, which means "good at jumping," and some species may jump at least 100 times their body length. Alticinae are easily distinguished from other Chrysomelidae and most other beetles by their greatly enlarged hind femora, which contain their jumping mechanisms (Fig. 15.1). The mechanism works independently of temperature, allowing beetles to jump as far at 4°C as at room temperature (Ker, 1977).

The jumping mechanism of alticinine beetles is not clearly described in the literature, probably because major details are still unknown. In principle, however, the mechanism is elegantly simple. The model I here summarize and synthesize is undoubtedly an oversimplification, but I hope it is consistent with the available evidence without merely being comprehensible at the expense of accuracy.

Literature on the jumping mechanism of flea beetles dates from over sixty years ago (Maulik, 1929) to papers in press (Furth and Suzuki, 1991). However, most of the later literature by David Furth at Harvard University relates to the use of a part of the jumping anatomy as a tool in taxonomy (Furth, 1982; Furth and Suzuki, 1990b) without reference to the precise steps that result in the generation, storage, and release of energy for the jump.

Rudolf Barth (1954) first correctly inferred that the flea beetle *Homophoeta sexnotata* gets its energy for the jump by contracting the massive muscles in its thick hind femora. These muscles pull apart a stiff, S-shaped structure, the metafemoral spring (Fig. 15.1), that was earlier described by Maulik (1929). The function of this spring is best understood from considerations of flea beetle taxonomy.

The shape of the metafemoral spring varies greatly among the beetles (Furth, 1982, 1988)—enough so to be a valuable taxo-

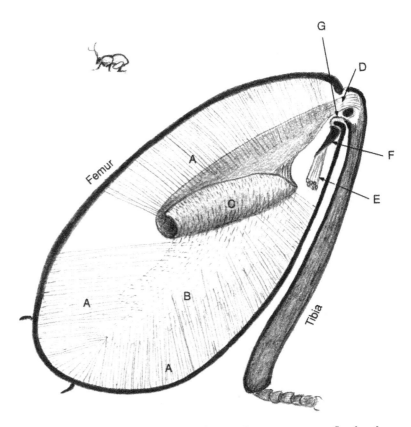

Fig. 15.1 The jumping mechanism of *Homophoeta sexnotata,* a flea beetle. In the hind leg: *A* = tibial extensor muscle; *B* = cuticular sheet for partial attachment of *A; C* = metafemoral spring; *D* = tibial extensor ligament; *E* = tibial retractor (flexor) muscle; *F* = "trigger" sclerite; *G* = tibial retractor ligament. In this and all of the following diagrams, I have tried to highlight and interpret the structures, not necessarily to draw all of the anatomy "as is." I have also tried to minimize lines by simplifying. (Adapted from Barth, 1954.)

nomic tool in differentiating about 200 of among the approximately 560 genera of Alticinae.

Comparative studies (Furth, 1982, 1988; Furth and Suzuki, 1990b) indicate that the spring arose as a modification of the tibial extensor tendon. It consists of protein together with fibers of the polysaccharide chitin (Furth, Traub, and Harpaz, 1983). Possibly in the first primitive jumpers the tendon could stretch and store a small amount of energy, but stretching would have quickly reached a limit given that compactness of the leg is at a premium.

Miniaturization in leg length and packing of much muscle into a small space became possible by stiffening the tendon, curving it, and ultimately recurving it, as a compound bow is made. Energy could then be stored in the tendon by bending it a short distance much as a bow is bent, rather than stretching it for a long distance, as a rubber band in a slingshot must be pulled. The anatomical studies suggest, but do not prove, that contraction of the dorsal or secondary leg-extensor muscle creates tension in the upper portion of the S-shaped coil, while the contraction of the primary extensor muscle (which is attached to a cuticular sheet on the upper lip of the S-shaped coil) stretches the coil apart at the bottom. Presumably, the coil is maximally distorted when both muscles are contracted. In a sense, then, the beetles have utilized probably for millions of years a technology developed only centuries ago by humans in our most advanced bows.

In order for a beetle to initiate a jump, the energy of the contracted extensor muscles and the uncurled spring must be released all at once. The release mechanism, however, is not understood. Most likely, continued contraction of the tibial retractor muscle, through a mechanical advantage gained by the leverage of the "trigger" sclerite, keeps the tibia flush against the femur. In that position the force of the spring (via the tibial extensor ligament) and the force of the tibial retractor muscle (via the tibial retractor sclerite) are more or less parallel with each other. The lever of the powerful spring is short and very close to the fulcrum, while the lever of the much weaker flexor is long. Consequently, as suggested by Ker (1977), the forces in this "dead-center" catch mechanism are adjusted so that flexure of the tibia is maintained. A slight tension of the tibial retractor muscle can oppose a *major* tension of the spring because of this large mechanical advantage. Again, this arrangement parallels the principle employed in the compound bow. In order to jump, the beetle needs to relax the retractor muscle, allowing the trigger sclerite to rotate, to upset the balance of forces. The tension of the spring is then free to act on the tibia, violently extending it. An analogous mechanism is used by bees as the dorsal longitudinal flight muscles oppose the dorso-ventral muscles in a tetanus during shivering, but here one is contracted more strongly than the other to attain the mechanical advantage that prevents the initiation of contraction oscillations.

As in the bee's flight system, the beetles have no catch mechanism, as far as is known, for storing the energy of the spring as the extensor contracts. Therefore, the flexor muscles must continually remain in tension prior to the jump in order to bend the metafemoral spring, in the same way that an archer must supply continual muscle power to keep the bow bent. As already mentioned, however, mechanical arrangements minimize the tension required; a stable position can be held by a minimum of force exerted by the tibial retractor on the trigger sclerite.

In the beetle *Longitarsus rubiginosus*, the jump performance (but not walking) is unaffected by temperatures from 4°C to room temperature (Ker, 1977), which strongly suggests that a spring is released during the jump itself. The beetles need not always make long jumps of some 100 times their body length, of course. They can also make short, "controlled" jumps. This argues for an absence of a catch mechanism such as that proposed for the flea (see below). In other words, as in a bow, the degree of muscle contraction of the tibial extensor determines the tension of the spring, which determines the length of the jump. The spring acts merely to store and release the tension of that muscle all at once.

Although the jumping organ is best known for the Alticinae, it is also found in numerous other unrelated coleopteran families (Furth and Suzuki, 1990b, 1991). Its distribution argues for separate independent evolution of the same or very similar mechanisms.

Click Beetles

Click beetles (Elateridae) are small- to medium-sized but slow-moving beetles, and they are therefore probably poikilothermic most or all of the time. Nevertheless, they are capable of prodigious vertical leaps that probably act to startle potential predators. The click beetles' jump does not involve legs. Instead, a click beetle initiates its jump while lying on its back (Fig. 15.2).

Just before jumping, the otherwise very stiff-bodied beetle arches its back slightly at the only major articulation of its body (aside from appendages), between the pro- and mesothorax. While in this position, the beetle tenses its dorsal intersegmental muscles located within the prothorax (Fig. 15.2). These large muscles, accounting for 6–9 percent of the beetles' weight, may be maintained in a state of tension for seconds, or minutes, until that tension is suddenly released to propel the insect upward in leaps of up to 0.3 m (Evans, 1972).

A 40 mg *Athous haemorrhoidalis* at 25°C has a takeoff velocity of 2.4 m/s, and the jumping action is completed in about 0.6 ms

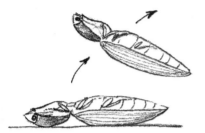

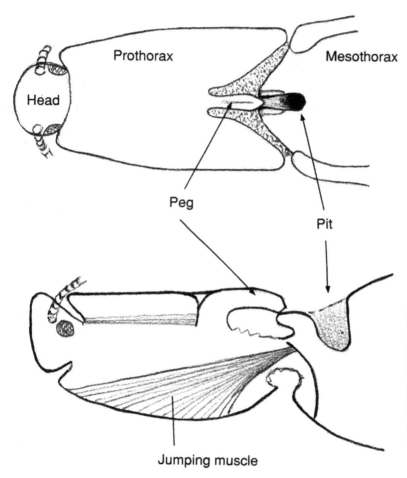

Fig. 15.2 The jumping mechanism of a click beetle. The jumping muscle in the prothorax contracts, forcing a peg against the tight lip of a pit in the mesothorax. The momentum of the accelerating head and prothorax propels the beetle upward when the peg suddenly pops into the bottom of the pit. (From Evans, 1972.)

(Evans, 1973). Total energy of the jump is $2-4 \times 10^{-4}$ J, with a power output during the jump of between 80 and 180 W/kg muscle. As in other poikilothermic insects, such unusually high but extremely brief power output from unheated muscles is only possible when energy is available in storage. The energy is released only when the peg slips through its narrow neck into the mesothoracic pit.

When the beetle is not jumping or preparing to jump, the peg fits partway into a groove anterior to the narrow neck leading into the pit. The pro- and mesothorax are thereby normally fitted tightly together. When preparing to jump, however, the beetle arches its prothorax and withdraws the peg from the groove until the tip of the peg is jammed against the anterior lip of the groove. As a consequence of some not yet clearly understood process involving other muscles (Evans, 1972), a slight buckling of the cuticle creates a "catch" that prevents the peg from sliding back down the groove and popping into the mesothoracic pit. The peg may be prevented from slipping into its pit in part by friction. But friction alone is probably not sufficient.

Following tensing of the powerful jumping muscle, the jump is initiated when the catch is released, presumably by the reverse mechanism that sets the catch. The trigger mechanism that causes the peg to lose its hold at the lip of the track, and thus to slam into the pit, is not clearly known. But Evans (1972) suggests several possibilities. It could potentially involve any of several muscles that slightly depress the anterior mesonotum or that elevate the prothoracic peg. Either way, the peg should then slip over its hold.

How does the rapidly *upward* accelerating head and prothorax lift the beetle off the substrate? From working with a model of a click beetle constructed out of plywood and rubber bands (to simulate muscles), Evans (1972) deduced that the mass of the very rapidly upward-accelerating head and prothorax, when abruptly stopped short by the peg in the pit (which causes the well-known "click" sound of the click beetle), transfers its kinetic energy in an upward vector and then propels the body upward in a tumbling, rotating motion.

Springtails

Springtails are members of a primitive and widely distributed order of wingless insects Collembola, certain species of which

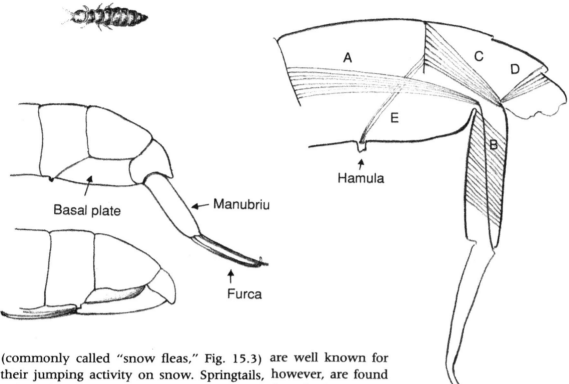

Basal plate

← Manubriu

↑
Furca

Hamula

A

E

C
D

B

(commonly called "snow fleas," Fig. 15.3) are well known for their jumping activity on snow. Springtails, however, are found in both tropical and temperate climates, and they also live on the soil, on the surface of ponds, and on the top of the water in the tidal zone (they do not break the surface tension of the water).

Unlike any other insects, these minute insects have a special springing organ, a "tail" (or manubrium) having two extensions or a fork, the furca. When not in use, the tail folds under the abdomen, where it is caught by a special hook, the hamula. When set free, the tail lashes back against the substrate and propels the animal upward and forward. The details of the jumping mechanism, however, are not entirely clear.

Manton (1972) believes that the muscles of the springing organ are concerned merely with "recovery movements and with stabilization." She maintains that the propulsive thrust of the tail against the substratum is almost entirely hydrostatic in origin. According to this view, "a small longitudinal contraction of the body . . . gives a sudden hydrostatic force blowing out the springing organ and swinging it strongly away from the body."

The muscular system of the thorax and five abdominal segments and the morphology of the jumping apparatus have been thor-

Fig. 15.3 The tail of a snow flea, with its manubrium and furca extended and folded against the basal plate *(left)*, propels the insect upward. The sketch at right shows the major jumping muscles. See text for possible functions of muscles *A* to *E*. (From Manton, 1972.)

oughly described for three species of the genus *Tomocerus* by Gerhard Eisenbeis and Signe Ulmer at the University of Mainz. Despite their amazingly detailed and thorough studies, the authors still conclude that we do not have a full explanation of the physiology of the collembolan jumping mechanism. Although Manton (1972) had previously proposed that the jump was entirely of hydrostatic origin, they conclude instead that the jump is primarily of direct muscular origin, because animals can jump even when the body cavity has been punctured. However, the length of the jump is probably regulated by the hydraulic mechanism, because punctured animals do not jump as far as unpunctured ones.

The collembolan jumping mechanism apparently evolved from specializations of the intersegmental body musculature (Eisenbeis, 1978; Eisenbeis and Ulmer, 1978), and Eisenbeis and Ulmer (1978) suggest that the strongly developed polysegmental longitudinal muscles in the rump (Fig. 15.3) indicate a pressure-generating hydraulic system. However, other direct muscles work antagonistically on the furca. For example, it is likely that prior to the jump a large flexor muscle contracts antagonistically to extensor muscles.

Subsequently, Christian (1979) found that the movements of the organ involve a click mechanism; there are only two stable positions for the furca: fully extended or fully flexed under the abdomen. High-speed cinematographic analysis reveals that take-off velocities are about 1.4 m/s, and energies of the jump amount to 10^{-9} to 10^{-7} J.

Snow fleas likely use an energy-storage mechanism. They conspicuously crouch a second or so just before jumping, when they are presumably contracting their jumping muscles and storing energy. When the springing organ is drawn up, it is held by the hamula fitting in a small groove on the base of the furca. As one possibility, perhaps when in the flexed position, contractions of muscles B (see Fig. 15.3) within the manubrium could be contracting against muscles A and C, which extend the organ and are attached at its base. Contraction of both muscle groups should hold the tail in place with relatively little counter-force, provided it remains tightly folded against the abdomen. Thus, the contracted muscles A and B could act like a stretched rubber band around a solid object (the basal plate). Force could be released by a trigger (such as unhinging of the hamula by contracting muscle E and/ or by contracting muscle D, to initiate the rotation to allow the release of the tension).

Collembolans at high elevation (Mani, 1968) and high latitude

(Agrell, 1941; Healey, 1967) can be active in the field at temperatures down to $-10°C$. Temperatures in the winter in New England, where I have for years observed these insects, often dip to -30 and sometimes even $-40°C$. But there are also "balmy" days of near $0°C$, and on such days one is often immediately rewarded with the appearance of the snow fleas, *Hypogastrura nivicolus*, peppered all over the surface of the snow.

The habits of *H. nivicolus* are not well known, but these insects are conspicuously capable of prodigious leaps at low temperatures. Because of their small size (length = 1–2 mm), it is safe to assume that, at least in the absence of sunshine, their T_b is nearly identical with ambient temperature at the snow surface. Thus, by measuring ambient temperatures near them (in shade), one knows their T_b and can then examine their behavior as a function of T_b.

In one set of observations (on January 9, 1990, at $-2.0°C$ in the shade on the snow in Weld, Maine), I noted the insects peppered by the hundreds on the snow near tree trunks in sunshine as well as in shade. On the next day, during a fall of sleet and snow, they were now (near $0°C$) crawling on the exposed surfaces of tree trunks. The following day temperatures dropped to $-3.8°C$ on the snow surface in shade following an even colder night. The snow fleas were again jumping on the snow and they appeared first in large numbers on the snow around the trunks of trees, then they gradually spread out uniformly (but patchily) over the snow surface in the forest. On January 12 the snow was melting, and now thousands could be seen uniformly spread over the snow, with no pattern with respect to tree trunks. Finally, on January 13, temperatures dipped to -6 and $-7°C$ on the snow (both in shade and sunshine) and not a single snow flea was then visible on the snow. However, many were now jammed into and under cracks of bark at the base of tree trunks. Thus, locomotion on the snow appears to involve travel to and from tree trunks, at body temperatures down to near $-4°C$ but not $-6°C$.

Locomotion consists of two modes, crawling and jumping. I watched over 200 individuals on tree trunks at 0 to $1.0°C$, and almost all of them were continuously crawling. Mean crawling velocity of 20 individuals was 0.88 mm/s. I saw not a single jump. On the other hand, *all* of the individuals I observed on the snow surface moved forward primarily if not exclusively by jumping. One individual that I focused on moved over 10 cm on each of over five successive jumps (at $0°C$), with only 5–10 seconds between each jump.

No animals jumped, and none could be induced to jump at

temperatures of $-6.0\,^{\circ}$C. Indeed, at this temperature all the snow fleas ceased movement. At only slightly higher temperatures $(-3.8\,^{\circ}$C), however, they were already jumping on the average 5.40 cm per jump. At slightly higher temperatures $(-2.0\,^{\circ}$C), jump distance on the same surface (snow) had nearly doubled, to 9.67 cm. And at $0\,^{\circ}$C (melting snow in shade), they jumped 10.4 cm. In other words, even if crawling on the extremely bumpy (for a snow flea) surface of snow were as feasible as on a smooth tree trunk, the insect can, by jumping, accomplish in a fraction of a second what it might take 110 times as long if it crawled instead.

Jumps, however, are not continuous. There is at least a several seconds' delay between individual jumps. As in other insects, collembolans presumably jump by gradually contracting muscles, storing the energy, and then releasing it all at once. Thus, the limiting factor in the jump is not so much the speed of muscle contraction (which is presumably dependent on temperature) but the ability to contract at all. The data cited above appear to suggest that these 1.5-mm-long animals can jump some 50 body lengths even when their T_b is within $2\,^{\circ}$C of cold torpor. Jumps allow them to travel rapidly at sub-zero temperatures, and even at very low temperatures, barely above the point when jumps are finally extinguished, distance per jump is still impressive.

Bristletails and Other Body Flexers

The members of another primitive group of very small-bodied insects, the bristletails (Thysanura), are also poikilothermic because of their very small size and sedentary habits. They include *Atelura* species, which thrive by the thousands in the warmth of some ant and termite mounds, and the firebrat *Thermobia domestica*, which is active even at $50\,^{\circ}$C and prefers the environment of boiler rooms and the warmest or hottest parts of our dwellings (Edwards and Nutting, 1950).

Other species, such as the seashore bristletail *Petrobius maritimus*, carry out their normal activities at or near $0\,^{\circ}$C (Makings, 1973). This species, at least, has a jump mechanism similar to that found in malacostracan crustaceans and some fly larvae.

Jumping fly larvae, such as cheese skippers (Piophilidae) and various fruit fly larvae, such as *Dacus* (Olroyd, 1964), arch the body until the mandibles grasp the caudal end on a papilla, and when the tension of the muscles builds up, the mandibles let go and the animal straightens and leaps up like a spring unbending.

In contrast, the crustacea and bristletails lack an external mechanism for holding while muscular tension builds up.

The thrust of the jump in *Petrobius* is achieved by pushing off from the abdomen, which is against the ground (Manton, 1972). Prior to the jump there is a strong downward bend, to produce an abdominal concavity. The jump is caused by the rapid straightening of the flexed abdomen with a simultaneous curvature of the thorax. This action slams the posterior end of the abdomen and the tail against the substrate and propels the animal upward and forward.

The jumping mechanism of bristletails appears to involve two antagonistic abdominal muscle systems—the longitudinal dorsal muscles and the deep oblique, plus the longitudinal ventral muscles (Manton, 1972). The dorsal longitudinal muscles have a "twisted rope" configuration and are thought to be used primarily for storing energy when the animal is cocked or ready for the jump. The abdominal straightening that causes the push of the jump is due to the rapid relaxation of the dorsal abdominal muscles.

Bristletails need only a very short time (11–86 ms) to store sufficient energy for a jump. Jump directions are apparently random, but the erratic behavior presumably aids in predator avoidance.

Jumping by making quick contractions of long, lateral muscles to give momentum to an elongate body may be a fairly general phenomenon in insects, including the famous Mexican "jumping bean." I once observed near Playa Hermosa (in Costa Rica) a green object rapidly moving across a hot dirt road in the sunshine in a series of quick consecutive hops. I stopped to "catch" this object and was surprised to see that it was a rolled-up leaf, held together with silk. Inside the leaf was a 1.5-cm-long pyralid moth larva. In this case the jumping behavior was not erratic, because the caterpillar and its "house" moved in a relatively straight line. And the behavior was more likely related to heat avoidance than to predator avoidance.

Jumping Homoptera

Homopteran insects suck plant juices and are generally known for their sedentary life-styles. They include cicadas, aphids (several families), and particularly the scale insects (also a number of families), which plug into the phloem tissue of plants on a relatively long-term basis. However, there are also at least four families of Homoptera of at least 20,000 species that lead more free existences, and some of these are superb jumpers. They include (Fig. 15.4) the jumping plant lice (Psyllidae), planthoppers (Fulgaroidae), treehoppers (Membracidae), leafhoppers (Cicadellidae), and froghoppers or spittlebugs (Cercopidae). Animals from these groups are some of the most beautiful and bizarre forms of the insect world. Some tropical planthoppers have monstrous heads decorated with markings that resemble teeth and that mimic crocodile heads. Treehoppers (Fig. 15.4) have bizarre structural extravaganzas on their prothorax that defy description. Leafhoppers, on the other hand, are elegantly streamlined, and they are ornamented with brilliant colors. All of the jumpers have slender legs, and in all of them the muscles powering the jump are within the thorax.

The homopteran jumping mechanism(s) is only very sketchily known from a single, preliminary publication on *Pyrilla perpusilla* (Fulgoridae) from India (Sander, 1957). The jump is powered by two bundles of jumping muscles that are apparently modified

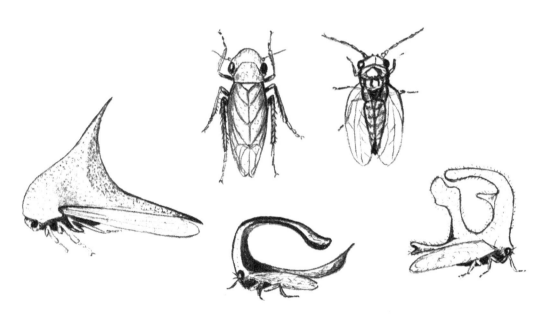

Fig. 15.4 Some jumping homopterans. *Clockwise from top center:* a leaf-hopper, *Tetigella* sp. (Cicadellidae); a plant louse, *Psylla* sp. (Psyllidae); and three treehoppers (Membracidae).

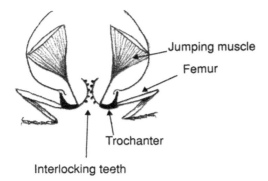

Jumping muscle

Femur

Trochanter

Interlocking teeth

Fig. 15.5 The jumping mechanism of the planthopper *Pyrilla perpusilla.* Note the interlocking teeth on trochanter. (From Sander, 1957.)

mesothoracic pleural muscles. Anchored dorsally to a highly sclerotized plate that is extensively braced, they terminate in strong tendons that insert on the trochanter (Fig. 15.5). Rigid struts within the thorax (from the dorsal plate and down to the trochanter) prevent the compression of the thorax so that the force of the contracting jump muscles is exerted on the trochanters (Weber, 1929). Rotation of the trochanters is mechanically translated to a downward and backward kick of the (hind) legs that propels the animal on its jump.

The jumping muscles are presumably relaxed and inactive when *Pyrilla* is walking, since the trochanters remain immobile. Thus, walking and jumping are separate mechanisms, and the jump can be initiated from any of a variety of different leg extensions found in normal walking motions (Sander, 1957).

Walking motions are presumably coordinated solely by the nervous system. However, *Pyrilla perpusilla* has an interesting mechanical device that aids in achieving a synchrony in the kick of the two hind legs. As the trochanters are rotated following contraction of the thoracic jumping muscles, they twist inward and their inner surfaces contact each other. Further rotation between the two (and hence right and left leg extension) are precisely synchronized because of a wheel-like structure with teeth on the trochanter's inner margins (Fig. 15.5). The teeth of one trochanter fit into the depressions of the other, so that the two are mechanically interlinked like two interlocking gears in a watch.

To my eye the leaps of some leaf- and treehoppers are every bit as impressive as the leaps of fleas and flea beetles. Without having measured these jumps, I would guess they are well over 100 body lengths, and they persist at low temperatures. It seems obvious, though, that the picture of the homopteran jumping mechanism I've just described must be very incomplete, because as it now stands it implies that the jump is powered by a *twitch* of the powerful jumping muscles. Almost certainly, however, there is an opposing muscle, an energy-storage mechanism, a catch mechanism, and probably also a release. The overall mechanism could be superficially similar to that of fleas (see later section), but if so then it is not likely homologous with it because, unlike homopterans, fleas are flightless and have some of their flight muscles highly modified into jumping muscles.

Orthoptera

All grasshoppers, katydids, and crickets jump (and they can do so during all instars) using their metathoracic legs, which are specialized to store energy. The jump has been studied in detail in the locust *Schistocerca* (Acrididae), from the standpoint of neuromuscular physiology (Hoyle, 1955; Godden, 1969; Usherwood and Runion, 1970; Heitler and Bräunig, 1988), energetics and forces (Gynther and Pearson, 1986; Hoyle, 1955; Brown, 1967; Bennet-Clark, 1975; Gabriel, 1985) and morphology (Heitler, 1974; Bennet-Clark, 1975).

In the locust, the jump results from the contraction of the large extensor tibiae muscle in the femur (Fig. 15.6). This muscle has numerous insertions throughout the length of the femur, adding to the great force of its contraction over a short distance. The muscle gradually contracts for up to 1 second prior to the jump. The tibia nevertheless remains motionless and flexed against the femur, in part because of the simultaneous contraction of the much smaller flexor tibiae muscle. Only a small force is exerted on the tibia when the extensor tibiae muscle acts nearly in a parallel direction with the tibia, when it is flexed; minor forces can hold the tibia in place. Extension of the tibia is prevented by a catch formed by the flexor "tendon" or apodeme and a cuticular knob inside the femur (Heitler, 1974). When the flexor tibiae muscle is relaxed, the energy from the contracted extensor system is released, first gradually and then at great force, as the tibia starts to extend. The extensor system delivers power not only from the muscle itself, but also from a short piece of elastic cuticle, the

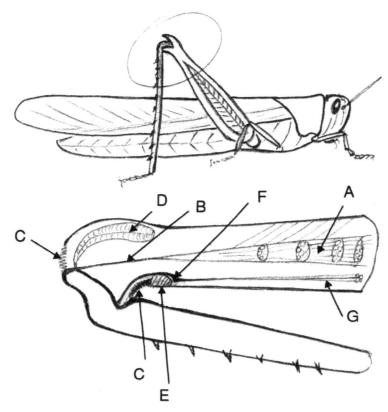

Fig. 15.6 The jumping mechanism of acridids, as seen in the desert locust *Schistocerca*. The circle indicates area of blow-up below. *A* = extensor tibiae muscle; *B* = suspensory ligament or extensor apodeme; *C* = flexible cuticle; *D* = semi-lunar process; *E* = Heitler's lump; *F* = pocket in flexor apodeme; *G* = flexor tibiae muscle. (From Bennet-Clark, 1975.)

semi-lunar process, at the end of the extensor apodeme near the femur-tibial joint.

How can the tibia remain flexed by the contraction of the relatively tiny flexor tibiae muscle working against the simultaneous contraction of the large and powerful tibial extensor muscle? As already suggested, the balance is apparently due to the mechanical advantage of levers; the flexor apodeme is attached a relatively long distance from the suspensory ligament connecting the tibia to the femur, while the powerful extensor muscle (which is adapted for short contraction at high force) has less leverage because it is more closely attached to the pivot of the two forces. The proximal muscle insertion results in more speed in tibial extension per given length of muscle contraction.

The contraction of the flexor muscle prevents the extensor muscle from releasing its force by pulling against it with a mechanical advantage, but also by pulling the cuticular knob (Heitler's lump) away from the wide terminal portion of the femur into a constriction of the femur. It here pushes the semi-lunar process to the dorsal side, thereby acting as a wedge that mechanically helps to maintain the mechanical advantage of the flexor muscle acting against the extensor; that is, it aids in keeping the forces of the extensor tibiae muscle pulling in parallel with the orientation of the tibia.

All acridids have the same principal anatomical features found in the locust *Schistocerca*, although they differ in detail (Bennet-Clark, 1975). For example, crickets *(Gryllus)* run and dig actively but jump poorly, and their semi-lunar processes are less developed. In the katydids, or long-horned grasshoppers (Tettigoniidae), the metathoracic legs are relatively longer than those of *Schistocerca*, and the femur contains most of the muscle in an expanded proximal region while extensor apodemes or tendons pass through a long, narrow distal part of the femur. There are apparently no bow-like semi-lunar processes, and it appears that most of the energy for the jump is stored instead in the extensor apodeme. In any case, because of the jumping mechanism in their long hind legs, orthopterans enjoy perhaps a unique mode of locomotion that should yield considerable energy economy in that it bypasses endothermy, which would otherwise be required. The locomotion is highly discontinuous, but that may not be a problem for grazers.

Fleas

The jump of fleas *(Siphonaptera)* has long been regarded as an interesting puzzle because it could not be explained in terms of the known properties of muscle. The rabbit flea *Spilopsyllus cuniculi*, leaping 3.5 cm or 23 of its body lengths (1.5 mm), requires 2.25 ergs, and this energy must be delivered in less than a millisecond over a distance of 0.37–0.5 mm (Bennet-Clark and Lucey, 1967). Normally in insects, very high muscle temperatures are required to speed up the rate of muscle shortening but endothermy in *Spilopsylla*, weighing 0.2 mg, is clearly out of the question, as it is in all other jumpers or their parts (hind legs of grasshoppers). Fleas are generally prodigious jumpers not only when they are already lodged on their warm-blooded hosts but also, and perhaps more irritatingly, when they reach them from a cool substrate.

The mystery of how fleas make their impressive leaps was first reported by H. C. Bennet-Clark and E. C. A. Lucey (1967) at the University of Edinburgh. They analyzed high-speed films taken of jumps (at 1,000 frames/second) and made deductions from general anatomical features worked out previously (Snodgrass, 1946).

Before a flea jumps it arranges its legs so that the hind femora are nearly vertical and the hind tarsi and tibial spurs are in contact with the substrate. It then remains stationary for at least 0.1 second while the fore and middle legs are folded back and the front end of the animal is pushed up. It is at this time that the forces for the jump are generated internally (Fig. 15.7). At takeoff the femora snap downward through 90° to 120° (full depression occurring in less than a thousandth of a second) and the rotating femora drive against the hind tibia, which are anchored onto the substrate. Thus, the flea is propelled upward and forward, usually at a trajectory angle of 50° to the horizontal although some jumps are nearly vertical. The propulsion comes from the hind pair of legs.

According to the model proposed by Bennet-Clark and Lucey (1967), jump preparation is initiated when the trochanter is raised by contraction of the massive trochanter elevator muscle (Fig. 15.7), that nearly fills the bulging hind coxae. The epipleural muscle (likely a modified wing depressor muscle) in the thorax contracts along the same lines of force along with the continuing contraction of the trochanter levator, and the trochanter-femur is raised nearly vertically. The lines of force acting on it are only slightly in back of the pivot with the coxal joint, thus minimizing the torque. The lack of strong torque then allows the trochanter depressor muscles (thick dorso-ventral thoracic muscles that are likely modified wing elevator muscles) to contract and exert their force via a long tendon attached to the other (anterior) end of the pivot about the coxal-trochanter joint. Again, as in the other jumpers discussed previously, when the flea is in the cocked, prejump position, the lines of force are relatively parallel and close to the pivot joint; both the leg extensor and levator muscles are contracting simultaneously and torque is prevented by their alignment over the pivot.

Dorso-ventral compression of the insect is resisted by strong

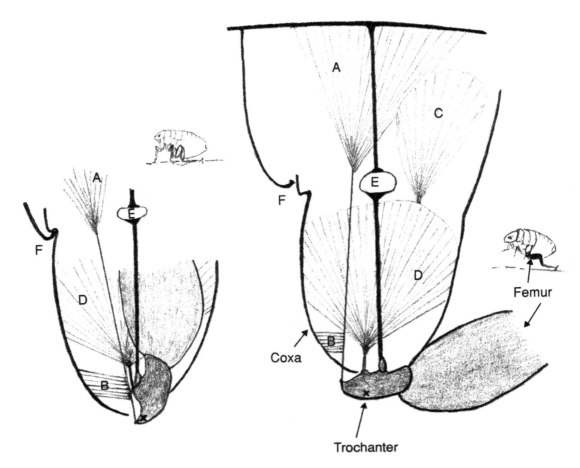

Fig. 15.7 The main muscles and primary anatomy responsible for the flea's jump. At left, a flea is shown in the "cocked" position with the resilin pad *(E)* depressed and the thoracic catch *(F)* engaged after the trochanter depressor *(A)* and the trochanter levator *(D)* and the epipleural muscle *(C)* have simultaneously contracted. At right, the thoracic catch *(F)* has been released, and the trigger muscle *(B)* has forced the trochanter depressor *(A)* off-center from the pivot point *(X)* on the trochanter. Now the resilin can expand and depress the trochanter-femur for the jump. (From Bennet-Clark and Lucey, 1967, and from various papers by Rothschild and colleagues.)

internal struts (Snodgrass, 1946), but the combined dorso-ventral contraction of the two sets of antagonistic muscles serves to compress a rubber-like pad of resilin (see Weis-Fogh, 1960) within the stiff struts. The compressed resilin then stores the energy of the contraction.

Resilin is a rubber-like protein originally discovered in the wing-hinge elements of the locust, where it minimizes the inertial losses of energy in flight (see Weis-Fogh, 1961; Neville, 1965) during the wing upstroke. In the flea the resilin pad is a modified wing hinge that has apparently been retained and modified from a winged mecopteran-like ancestor that became parasitic and lost the powers of flight. It now stores a major portion of the energy for a jump instead of a wing beat. Different flea species have different jumping abilities, and in general the size of this resilin store is a good predictor of a flea's jumping performance (Bennet-Clark and Lucey, 1967). That, however, says nothing about whether the resilin alone provides all of the power. Other features presumably vary in parallel.

Because of the heavily braced thorax in the flea, the continuing contraction of the two sets of muscles on either side of the pivot, the compressed resilin stores energy from *both* the leg depressors and leg elevators. According to this model, the jump is initiated by a twitch of the "trochanter depressor" muscle B (Fig. 15.7), which acts as a trigger. This muscle is used to release the stored energy by altering the pivot of the coxal-trochanter joint, so that leverage and torque can now be brought to bear on that joint. The muscle can alter the pivot because it is attached laterally to the tensed trochanter depressor tendon, and upon contracting it thus pulls this tendon anteriorly so that the trochanter-femur pivots slightly until the forces of the stored energy are released in a snap. It is necessary, however, that there is simultaneous relaxation of the trochanter levator and epipleural muscles so that the trochanter can violently flip over the pivot.

Despite its alluring simplicity, this mechanism has some weak points. First, it requires that the trochanter levator and depressor muscles are *continually* contracting so long as the flea remains "cocked" to jump. This would be an energetically very costly mechanism. Second, it requires that the trochanter levator muscles be relaxed "instantly," even though muscle relaxation is generally not an instant process.

A second model of the flea's jump has been proposed by Miriam Rothschild and colleagues (1972, 1975; Rothschild and Schlein, 1975) from England. This model has many features in common with the previous one, but it differs in essential details based on additional studies of the flea's anatomy. It differs from the first mainly in that the energy of the compressed resilin is not stored by continual isometric muscle contraction, but instead the flea is "cocked" when the coxae of the dorso-ventrally contracted flea are secured by a system of hooks and catches to the thorax. This mechanism presumably allows for the relaxation of the muscles after *one* quick contraction only, and after that the flea remains jump-ready because the resilin should remain compressed without additional muscular work.

According to a much simplified version of this model, as in the other, the trochanter levator, the epipleural, and other muscles contract, compressing the resilin pad and engaging a medial thoracic hook that keeps the resilin compressed as trochanter levator and epipleural muscles relax. The flea is then cocked for the jump. The trochanter depressor (called the "dorso-ventral muscle" by Bennet-Clark and Lucey, 1967) is responsible for the takeoff, and it is thought to have a dual function at this time. It supplies energy for the jump during its contraction, and it triggers the whole process of the jump by releasing the central curved hook or catch. Rothschild and Schlein (1975) suggest that the jump is initiated by a rapid twitch of this muscle. The tension of the muscle twitch presumably expands the thorax laterally sufficiently to disengage the medial hook and it simultaneously rotates the trochanter (with attached femur) about its pivot.

Which mechanism is correct? I speculate that the jump of the flea probably involves features of both models. Both incorporate the click mechanism with energy storage in resilin. As suggested by Bennet-Clark and Lucey (1967), it seems logical to propose that both the trochanter levators and depressors are contracted prior to the jump to store energy. Subsequently, as suggested in Rothschild's studies, the muscles could relax because of the engaged catch mechanism. The "trigger" that releases the click has not been elucidated, but any of a number of muscles or combination of muscles could be involved. It could be a twitch of the trochanter depressor, as proposed by Rothschild and Schlein (1975), to disengage the medial catch via a thoracic deformation. However, a twitch of the muscle attached to the lower portion of the tendon of the trochanter depressor could also be involved in order to achieve a more favorable leverage of the pulled tendon on the trochanter (Bennet-Clark and Lucey, 1967). Indeed, both muscle contractions could occur simultaneously, accomplishing both functions at once.

Whatever the precise mechanism, however, it will likely remain unproven because the animal's small size precludes direct experimentation. The function, however, is obvious. The jump permits these insects to accomplish a very important part of their life cycle relatively independent of temperature: namely, to hop onto their hosts. For example, Ioff (1950) describes how the flea *Vermipsylla alakurt* jumps from freezing High Arctic tundra soil (where it pupates) onto passing ungulates. Similarly, many other fleas must accomplish prodigious leaps to secure a host, after emerging from their pupae in very cool environments.

Not all fleas jump, and there is great variation in jumping ability among different species. For example, various fleas parasitizing bats and arboreal rodents or birds have secondarily lost their jumping ability, in the same way that many insects inhabiting small windy islands have lost their ability to fly. In general, the larger the host, the better the jumper.

To return, after all that work, to the point: "The uniqueness of the flea's jump lies not in its speed, acceleration, trajectory or sustained repetition, but rather in the subtle artistry by which natural selection has parsimoniously turned back the flight mechanism and incorporated it into the new jumping mechanism" (Rothschild et al., 1973). My brief overview of the flea's jump was gleaned from reading 77 turgid pages of 5 manuscripts by two researchers who are violently at odds with each other. After reading the material I confessed to some confusion, which I presumed was caused by my inability to master the plethora of details. Through lengthy correspondence I got even more detail, but I remained disappointed because I could not arrive at a full understanding of the mechanism of how fleas jump. But even one of the principal researchers on the flea jump wrote me saying, "I am *by no means* confident we know how fleas jump," and was moved to ask, "Why are you interested in this obscure subject?" The answer seems obvious: To understand why some insects are endothermic one needs to know why others are not. Most insects are too small to be able to afford the enormous energy costs of staying warm and flight-ready on a continuous basis. It seems fascinating, therefore, that by a "simple" mechanical device some have bypassed the whole ecological problem of endothermy. The mechanism has made them independent of temperature limitations (and thermoregulation).

Wing-Click Mechanisms and Energy Storage

The examples discussed so far refer to single jumps, but they also apply to locomotion in consecutive jumps. Furthermore, similar click mechanisms involving energy storage have been identified in the normal wing-stroke cycles of many insects (Pringle, 1957).

Their net effect is to economize the flight energetics by reducing the work load per wing-beat cycle. As a result, performance is possible at lower T_{thx}, given specific work loads.

The click mechanism of the wing movements is now thought to apply to a variety of insects (Pringle, 1957), but it was first demonstrated in blowflies *(Sarcophaga)* by Edward G. Boettiger and Edwin Furshpan at the University of Connecticut. Boettiger and Furshpan (1952) observed that when the flies are anesthetized with CO_2 the wings can easily be pushed into many positions. Under carbon tetrachloride (CCl_4) anesthesia, however, flies held their wings either up or down. The researchers demonstrated that the pleurosternal muscle contracts on treatment with CCl_4, and the resulting inward pull of the pleuron flips the wings either up or down because of stiffness in the sides. A complex wing-hinge mechanism was described (Pringle, 1957, pp. 12–18) whose elements ensure that very large amplitude wing beats result from very small contractions of the flight muscles. Since the rate of muscle shortening is very much dependent on temperature, insects with the click mechanism can achieve high-amplitude wing beats that would otherwise be possible only at much higher T_{thx}.

In these click mechanisms, some of the energy of the wing stroke acts against the resistance of the air. However, the decelerating wing also transfers energy into stretching of the antagonistic muscles. In other words, the antagonistic upstroke and downstroke muscles serve alternately as energy-release and -storage mechanisms.

Energy can also be stored in elastic tendons containing the rubber-like protein resilin. This substance was discovered by Torkel Weis-Fogh (1960) of the University of Cambridge. In the locust *Schistocerca*, it is found in the wing hinge, and it stretches in the upstroke movement of the wing stroke. It functions to capture the inertial component of the energy near the top of the wing stroke, releasing it again during the downstroke. Resilin is a nearly perfect rubber, releasing nearly 97 percent of the energy stored after stretching.

Energy-storage mechanisms such as that in the resilin pad of the wing hinge in locusts (Weis-Fogh, 1960) reduce the inertial component of the energy expenditure of the wing movements during flapping flight. Thus, by reducing the power requirements they could indirectly affect body-temperature regulation, since the reduced power requirement can be achieved at reductions in thoracic temperature. However, catch mechanisms are obviously not helpful during *continuous* locomotion, where the rates of muscle contractions must be quickly repetitive.

Energy-storage mechanisms like those of the locust wing hinge are common in mammals, but all use such storage mechanisms during locomotion (consisting of repetitive leaps) and none use catch mechanisms. Classic examples of mammalian energy-storage mechanisms include the ligaments in the fore and hind hooves of horses (Camp and Smith, 1942) and the Achilles tendon of dogs (Alexander, 1974) and kangaroos (Dawson and Taylor, 1973). In all of these cases, the kinetic energy of the jumps is stored in the elasticity of the stretching tendons as the limbs are decelerated upon striking the ground. The same principle is increasingly being applied in the design of running shoes (Wolkomir, 1989).

Summary

A number of insects that are too small to be endothermic (fleas, collembolans, flea beetles, numerous homopterans) or that are large yet relatively poikilothermic with regard to body (or leg) temperature during terrestrial locomotion are nevertheless capable of releasing large amounts of energy in quick bursts. They do so by gradually contracting muscles prior to the jump and conserving the energy by a catch mechanism until ready for release. In several cases preparations for the jump involve simultaneous isometric contraction of two sets of powerful opposing muscles mechanically acting analogously to the mechanism of pre-flight warm-up. But the result of the mechanism is the ability to dispense with warm-up. The variety in the morphology of the jump mechanisms indicate multiple evolutionary origins. For example, the jump mechanism of fleas involves a modification of muscles otherwise used for flight in other insects. In acridids it involves modifications of the third pair of legs otherwise used for walking. In collembolans the terminal portion of the abdomen has a special appendage used to hurl the animal forward. In click beetles the jump mechanism involves the articulation between thoracic segments, whereas in flea beetles it involves the use of the hind legs (as in acridids and fleas) but by an entirely different mechanism, one that is highly analogous to the compound bow. The thoracic anatomy for wing movements in insects involves principles of energy storage similar to those that apply to the jump, but because wing beats must be continuous there are obviously no catch mechanisms.

Remaining Problems

1. Although the broad outlines of several jumping mechanisms are known, in no case (except that of grasshoppers) have all the details been worked out. In particular, there are not even models for the release mechanisms of the energy stores in some groups, and the existing models in others are untested speculations.
2. Predictably the "mouse-trap" mechanisms should result in temperature-dependent rates of energy storage, and temperature-independent energy release. However, few relevant data are available.
3. To what selective pressures are the jumping mechanisms biologically relevant adaptations, and how do they currently function? For example, to what extent and to what distances do snow fleas travel by means of leaping, and how does locomotion by leaping vs. crawling relate to temperature?

Social Thermoregulation

T H E nests of ants, termites, and social bees and wasps serve as incubators for raising the immatures and as refuges from enemies and temperature extremes. The importance of these thermal refuges, where the microclimate is often rigidly controlled, cannot be overemphasized in any study of the life strategies of social insects. Indeed, most treatises on the social life of insects discuss numerous facets of this fascinating thermoregulatory behavior at length. Heat generation of honeybee colonies has been well known for at least 250 years (Réaumur, 1742), and an overall review of thermoregulation by insect societies is available (Seeley and Heinrich, 1981). Work on ants has recently been updated (Hölldobler and Wilson, 1990). Given that much information on social thermoregulation is available elsewhere, I here refrain from covering the topic in detail and attempt only a summary of the main features and a limited critique of some controversial viewpoints.

The Ants

Ants are strongly thermophilic organisms. But unlike social bees and wasps, ants lack the thoracic musculature for flight, which can help provide the colony with a rapid influx of energy supplies, and for shivering, which can also result in rapid energy expenditure. Both processes are prerequisites for significant nest heating through shivering. Nevertheless, given the great range of thermal environments on earth where ants occur, the workers of many species maintain their brood at a relatively high and narrow range of temperatures despite their lack of flight musculature.

One of the major generalizations of the thermal biology of ants is that nest temperature is seldom if ever physiologically regulated. Instead, ants choose where to build their nests or where within the nest to deposit their brood according to temperature. In the tropical rainforest, with its relatively stable air temperature, soil-dwelling species are rare, and most ants nest either above the ground rather than in it or in small pieces of rotting wood (Wilson, 1959). In contrast, in deserts having tremendous temperature fluctuations, most species nest deep underground where temperature extremes are moderate. For example, in the Sahara Desert of Algeria, most ants nest some 50 cm underground, where temperatures range annually between 15 and 38°C whereas surface temperatures range between 0 and 70°C (Délye, 1967). In northern Florida the ants *Prenolepis imparis* are only active on the ground surface in the winter (Tschinkel, 1987). In the summer they seal themselves into their vertical nest shaft as much as 3.6 m deep. Laterally to the shaft the ants excavate horizontally domed chambers where temperatures never exceed 24°C. Here they spend the summer because the relatively low temperatures conserve their energy supplies during the 7–8 months when no foraging occurs.

The majority of ant species in warm temperate regions nest at or in the soil surface, whereas in colder temperate regions, such as the boreal forests, they nest at sites warmed by the sun. Nest shading (by growing vegetation) is often lethal to northern colonies (Brian, 1956).

Colony distribution is determined by choice of nest site by the founding queen as well as by differential colony mortality. For example, colony-founding queens of *Formica fusca* and *Myrmica ruginodis* in Britain settle on a warm (30°C) substrate under experimental conditions (Brian, 1952), and in the field queens of *Lasius niger* and *L. flavus* alight predominantly on patches of bare, sunlit soil before shedding their wings and burrowing into the ground to start their colonies.

Sunlight is important in creating favorable microhabitats for colony growth. In northern Europe founding queens of *L. niger* and *L. flavus* were seeded in glass tubes at open and in shaded sites in grassland and woodland locations (Pontin, 1960). After one summer the broods of the shaded queens had not advanced beyond the larval stage, whereas the broods in sunlight had pupae. Similarly, within a shrinking glade of a spruce-pine plantation, the ant community dwindled from 45 nests of 4 species to 9 nests of 3 species in 4 years (Brian, 1956), whereas in colonies of *M. ruginodis* that were in open, sunlit habitat brood development time advanced by 1 month relative to the shaded colonies (Brian and Brian, 1951). Most important, the shaded colonies never reared reproductives, thus effectively committing genetic suicide. In general, shading mounds in habitats undergoing succession is accompanied by mound extinction (Waloff and Blacklith, 1962; Pontin, 1963).

Aside from merely selecting between sunlit vs. shaded sites, ants may select very specifically warmed sites. Perhaps the most common example of nest-site adaptation for increased nest warmth is the habit of living under stones lying on the soil surface. In the Swiss Alps at over 1,500 m, many species of ants live under flat stones. At these elevations nightly frosts occur regularly and the damp ground remains cool because of its high heat capacity and the evaporation of water from it. Because of their low specific heat, however, dry stones heat quickly in sunshine, and when stones serving as nest covers are heated by the sun the ants quickly move their brood up out of the ground and into cavities directly beneath the stones (Steiner, 1926, 1929).

A small minority of ants, the mound builders, have evolved to make structures that capture solar heat, rather than relying on preexisting structures. Mounds often consist of excavated soil. But in some other species, such as the red wood ants *Formica polyctena* and *F. rufa* (Fig. 16.1), the mounds are constructed almost entirely out of organic materials (such as twigs and conifer needles) gathered from surrounding areas. Mounds range in height from a few centimeters to the impressive hills of a meter or more of the widespread Eurasian ants *F. polyctena* and *F. rufa*. Mound-building species occur in the myrmicine genera *Atta, Acromyrmex, Myrmica, Pogonomyrmex,* and *Solonopsis* of the tropics and warm temperate regions, in the dolichoderine genus *Iridomyrmex* of Australia, and in the formicine genera *Formica* and *Lasius* of Europe, Asia, and North America (Wilson, 1971). Mounds are most often found in sunny clearings and forest margins with southern and eastern exposures (Elton, 1932; Dreyer, 1942; Wellenstein, 1967; Sudd et al., 1977).

Already in 1810 Pierre Huber suggested that the mounds act as solar collectors, and his work was elaborated on by Forel (1874), who proposed the "Théorie des Dômes." The idea has now been tested by a long line of European, American, and Japanese investigators. All of the detailed studies (Andrews, 1927; Wellenstein,

Fig. 16.1 A large mound of the wood ant *Formica polyctena* in Finland. The mound is heaped against the southern side of a tree. Pine needles, bits of leaves, small twigs, and other organic debris provide thatching for the mound. (From photograph by B. Hölldobler.)

1928; Zahn, 1958; Steiner, 1924, 1926, 1929; Kato, 1939; Raignier, 1948; Kneitz, 1964, 1970; Scherba, 1962; Heimann, 1963; Coenen-Stass, Schaarschmidt, and Lamprecht, 1980; Horstmann and Schmid, 1986; Rosengren et al., 1987; Etterschank, 1971; MacKay and MacKay, 1985) confirm that mound structures provide higher and often more stable temperatures than the adjacent soil or surrounding air.

What is probably a typical pattern of temperature dynamics for a large mound of *F. rufa* or *F. polyctena* was recorded by Raignier (1948). As air temperatures rose from 17 to 22°C from 7:30 to 11:30 A.M., mound temperature at a depth of 30 cm increased from 29 to 32°C. Although the sky was overcast for the rest of the day, the mound interior remained steady near 32°C, finally dropping to 29°C at night. It took 20 hours for the mound to lose the heat that it had acquired in 4 hours' exposure to the morning sun.

A number of features of mounds probably contribute to their ability to serve as solar heat collectors. Most mounds have faces sloping to the south and southeast, thus increasing the amount of sunlight intercepted (Forel, 1874; Andrews, 1927; Linder, 1908; Scherba, 1958; Hubbard and Cunningham, 1977). The ants apparently foster sun exposure by removing vegetation from the south-facing surfaces (Dreyer, 1942; Linder, 1908) and also by making mounds taller (Raignier, 1948); decreasing the intensity of lamp illumination on nests of *F. rufa* and *F. polyctena* causes the ants to rebuild their mounds in pointed forms that intercept more light (Lange, 1959).

The mounds of some species may be more than passive heat absorbers. To test whether mound architecture is an instrument of nest heating, Horstmann and Schmid (1986) heated *Formica polyctena* mounds internally. The ants responded by reducing the height of the mound, enlarging the exits on the mound surface, and by dispersing both workers and brood away from the heated center. These measures resulted in a lowering of mound temperature. In addition to increasing mound height to achieve heating, some ants (especially *Pogonomyrmex, Iridomyrmex,* and *Cataglyphis*) add pebbles and pieces of charcoal to the surface of their mounds. These dry materials heat rapidly in the sun and may serve as solar traps (Harkness and Wehner, 1977).

Large mounds promote population growth because of the heat they store, and the large populations of 100,000 or more in turn may add significant metabolic heat (Kneitz, 1964; Steiner, 1929; Rosengren et al., 1987). The first evidence for metabolic heating in large *F. polyctena* colonies comes from colony-kill experiments (Raignier, 1948). The temperatures of two similarly sized colonies were alike for one month. One of the colonies was then killed (by pouring 3.5 l of carbon disulfide over it) and subsequently the dead mound's temperature at a depth of 30 cm declined by 7–9°C even though its surface temperature averaged 3°C *above* that of the live nest.

Despite these seemingly convincing data for metabolic heating within mounds, the idea has not always been accepted. Other possibilities of nest heating include the Wärmeträgertheorie, or "warmth-carrier theory," by Zahn (1958). Red wood ants *F. rufa* often crowd upon the mound surface in the sun, where they undoubtedly absorb heat, and some of this heat could be carried within their bodies to the mound interior. Undoubtedly, this phenomenon occurs, but it cannot account for heating under overcast conditions. Furthermore, as pointed out by Rosengren and colleagues (1987), ant basking is characteristically observed on weak ant mounds and on cool mounds, not on vigorous colonies already having warmed mounds.

An earlier idea (Wasmann, 1915) held that the mound interiors are heated by microbial decay, as in a compost heap. Using sensitive electronic calorimeters, Coenen-Stass, Schaarschmidt, and Lamprecht (1980) confirmed that considerable heat is produced by material isolated from the nests of wood ants *(F. polyctena)*, and they show that aeration of this experimentally isolated material causes an immediate rise in heat production. Oxygen-consumption measurements also confirmed aerobic metabolism from microbial activity, particularly in the nest material taken from the center of the nest. The researchers conclude that seasonal fluctuations in ant activity and heat production in mounds coincide because the ants may aerate the nest material or otherwise cause optimal conditions for microbial heat production.

More recent studies by Rosengren and colleagues (1987) in Finland do much to clarify the relative contributions of various sources of heat within red ant mounds. Their observations indicate that in colonies of *F. rufa* with 1 million or more workers, intra-nest temperatures already reach 30°C in early spring, when the ground temperature is near 0°C and snow is still on the ground (Fig. 16.2). Furthermore, there is a highly significant trend for nest temperatures to *increase* at lower air temperatures. For example, at an air temperature of −5°C nest temperatures were near 30°C, whereas when air temperature increased to 15°C average nest temperature declined 27.5°C (Fig. 16.3). Nests pro-

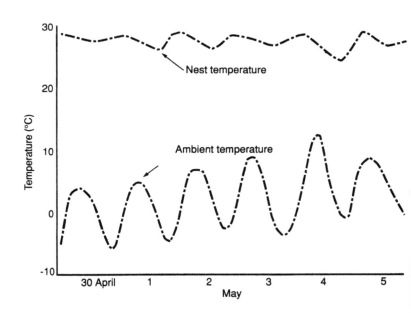

Fig. 16.2 The relation between diel variation in inner nest temperature and outside air temperature at the beginning of May 1978 in a *Formica rufa* colony in Finland. (From Rosengren et al., 1987.)

THE HOT-BLOODED INSECTS

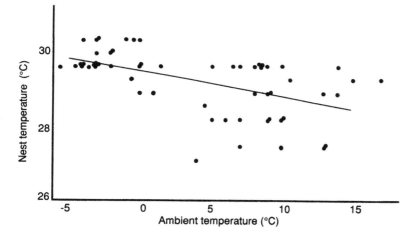

Fig. 16.3 The regression of inner *F. rufa* nest temperatures at noon and at midnight vs. outside air temperature for May 1978. The negative correlation is statistically significant ($p < 0.001$). A decrease in the external temperature regularly coincides with an increase in temperature within the mound. (From Rosengren et al., 1987.)

ducing sexuals maintain higher temperatures than those producing only workers. These observations can be explained neither by the Wärmeträgertheorie nor by the effect of microbial fermentation alone. Instead, these authors convincingly demonstrate that considerable nest warming occurs by metabolic heat production of the ants themselves. Metabolic heat production in nest warming has also been suggested for carpenter ants (*Camponotus* spp.) nesting in hollow trees where decay is not a contender for heat generation. In northeastern Ontario, Canada, when nests contained growing brood (June) they generated temperatures up to 16 °C higher than control parts of the same trees without ants (Sanders, 1972).

The data do not necessarily indicate "alternative" hypotheses of mound heating. Instead, various mechanisms are involved simultaneously, and the magnitude of the different mechanisms vary from one species and from one growth stage of the mound to another. In general, solar heating predominates in smaller colonies, in part because of their larger relative surface area. Intermediate-sized colonies may in addition be affected by microbial action, and in very large colonies the metabolically produced heat of the nest occupants assumes an increasingly larger role in raising internal nest temperature.

Metabolically produced heat is the primary source of heat in the temporary nests (bivouacs) of the neotropical army ants *Eciton burchelli*. These surface-dwelling ants continuously alternate between a nomadic phase roughly 15 days long and an approximately 20-day-long statary phase. During the nomadic phase the ants make daily raids to feed the growing larvae, and following the completion of the day's raid they change their nest site. When the larvae finally pupate the colony enters the statary phase and it then remains in the same bivouac while the brood metamorphoses to adults and the queen then lays new eggs.

Each nest or bivouac (in *E. burchelli*) consists of from 200,000 to 600,000 workers that are hooked together by their tarsal claws into a single mass, like a honeybee swarm. The largest of these ant bivouacs can be 80 cm across and 40 cm in breadth and depth. Temperature inside the massed ant clusters with brood is consistently 2–5°C higher than that of the surrounding air, and it is much more stable as well (Schneirla, Brown, and Brown, 1954; Jackson, 1957). Temperature increase is an inevitable consequence of the trapped heat from the metabolism of the half-million or so ants, but temperature increase, as such, says nothing about thermoregulation.

Nigel R. Franks of the University of Bath shows, however, that the temperature of the clusters of *E. burchelli* is to some extent regulated (Fig. 16.4) by the control of heat loss. Frank's study was conducted on a statary bivouac hanging 0.5 m above the forest floor from a horizontal tree trunk in the lowland tropical rainforest of Barro Colorado Island, in Panama. As in honeybee swarms, the core of the cluster was warmer than the mantle. Core temperature was maintained within 2°C (at 27–29°C), independent of environmental temperature changes that amounted to about 6–7°C over a day. Calculations indicate (Franks, 1989) that the basal metabolic rate of the army ants in the bivouac generates all the heat they need to keep warm when air temperature drops. Indeed, as suggested by similar calculations for honeybees (Heinrich, 1981a), the problem for the ants is not how to keep warm but rather how to lose sufficient heat. Presumably the thermal challenges faced by neotropical army ants are potentially even greater for African driver ants (*Dorylus* spp.), where a single colony can contain more than 20 million workers rather than only half a million.

Army ants appear to regulate cluster temperatures in the same way honeybees regulate swarm temperatures at modest air temperature. As air temperature drops, the bivouac contracts (Jackson, 1957), presumably as cooler outside ants crawl into the interior, plugging ventilation channels. Such channels have long been known to exist. Thomas Belt (1874) in describing *Eciton*

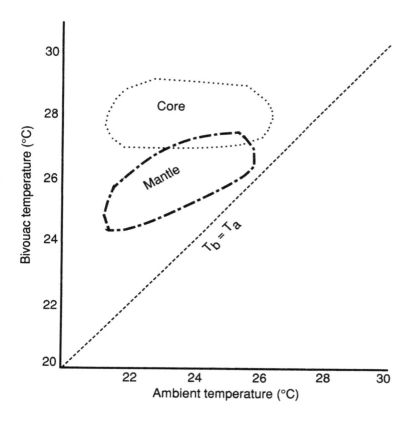

Fig. 16.4 Temperatures of army ant bivouacs in Panama as a function of ambient temperature. (From Franks, 1989.)

bivouacs wrote: "I was surprised to see in this living nest tubular passages leading down to the centre of the mass . . . I thrust a long stick down to the centre of a cluster, and brought out clinging to it many ants holding larvae and pupae, which were probably kept warm by the crowding together of the ants."

As previously mentioned, ants have evolved a mechanism, mound building, for trapping solar heat. But they have very limited or no capacity to regulate mound temperature. Ants do, however, regulate brood temperature fairly precisely. Rather than regulating the temperature of the nest where the brood are fixed in place (as their relatives the wasps and bees do), ants pick up their brood and carry it around to those parts of the nest where it experiences the proper thermal environment. Usually a colony's brood is placed in the warmest chambers available, although when selecting a proper microenvironment in a nest's network of galleries and chambers, ants commonly distinguish among those appropriate for themselves as adults, eggs, and larvae and pupae (Ceusters, 1977). Pupal chambers are generally the warmest (Wilson, 1971).

The upper temperature limit to which ants subject their brood varies somewhere within 20–40°C, depending on the species (Fielde, 1904; Brian, 1973; Ceusters, 1977). Workers of *Lasius flavus*, for example, begin to evacuate their brood from beneath stones when the temperature approaches 28°C, whereas *Formica fusca* delay until the brood chamber warms to about 36°C (Steiner, 1929). Shifts of brood for temperature control can occur on an hourly basis; in the evening *L. flavus* ants brought the colony's brood to a depth of about 10 cm in its mound, and as the mound cooled in the night, they carried the brood down further to 20–30 cm, to keep them warm. In the morning the brood was brought back up again to within 2 cm of the sun-warmed surface of the mound, and at high solar heating in the afternoon the ants brought their brood back down to 5 cm (Steiner, 1929). By heating the stones used by *L. flavus* with a flatiron or shading the mound from the sun, Steiner (1926, 1929) showed that the ants' fine-tuning of brood-temperature regulation through brood transport was related directly to temperature changes and not to time of day, as such. Differential use of the nest is therefore seasonal as well as diurnal, and Kondoh (1968) has examined in some detail the vertical within-nest migrations of *Formica japonica* that correspond closely with seasonal temperature changes.

Another well-known example of brood transport for thermoregulation is that of weaver ants *Oecophylla longinodis*. Weaver ant nests are strictly arboreal, and any one colony generally has numerous domiciles on a tree. Nests are created by lashing several leaves together with silk produced by the larvae. Weaver ant colonies near the equator construct domiciles on northern and southern sides of the trees, alternately with the seasons. They shift their nest positions to keep the nests exposed to the sun (Vanderplank, 1960).

Another ant with silken nests, *Polyrachis simplex*, moves seasonally between cavities in streamside cliffs and vegetation during the summer. Migration to the streamside enables the ants to avoid the high temperatures of the cliffs in summer (Ofer, 1970). Numerous other "fugitive" ants (Wilson, 1971) that have small colonies nesting in temporary shelters, such as holes in twigs, in acorns, under small stones, or under a few fallen leaves, depend on frequent colony emigration to attain a favorable nest environment. Nest emigrations begin with scouting for a new nest site by a small fraction of a colony's workers, followed by recruitment to the new nest site (Wilson, 1971; Abraham and Pasteels, 1977; Meudec, 1977; Möglich, 1978).

So far ant-nest temperatures have been presumed to be related almost exclusively to the purpose of promoting brood production. Brood thermoregulation has undoubtedly helped make it possible for ants to invade into and diversify in a wide variety of habitats and geographical areas. It is possible, however, that there are other physiological correlates to nest-temperature regulation. For example, does nest-temperature regulation in leaf-cutter ants (*Atta* sp.) help in growing food on the fungus gardens? Does the choice of low and high temperatures in a nest allow reserve workers to choose low temperatures in times of food shortage and thereby conserve energy supplies and prolong longevity?

Social Wasps

Our knowledge of thermoregulation in social wasps is based on studies of the paper wasps (*Polistes* sp.), which have small colonies on naked hanging combs, and on the yellow jacket and hornets (subfamily Vespinae), which have much larger colonies and whose combs are enclosed in multilayered paper envelopes or are enclosed in cavities in tress or in the ground.

Polistes wasps are primarily of tropical distribution, and lowland tropical *Polistes* build nests on vegetation in the forest with probably little concern for temperature in nest-site selection. But many species also occur in temperate regions, where a large part of the colony's thermal strategy is to place the comb in a warm and sheltered microenvironment. The wasps do not incubate their brood, and nest heating is often accomplished by direct solar radiation.

In Germany *Polistes gallica* and *P. biglumis* nest along southern and eastern sides of barns and other buildings, avoiding north- and west-facing surfaces (Steiner, 1932). Their nests consist of dark-grey carton and although these combs warm to suitable temperatures in the sun, they are sometimes subject to overheating. However, the *Polistes* wasps have a well-developed fanning and water-sprinkling response that prevents the killing of the larvae should overheating threaten (Fig. 16.5). Fanning begins as nest temperature reaches 31–36°C, and if nest temperature continues to rise to 34–38°C then water is transported and regurgi-

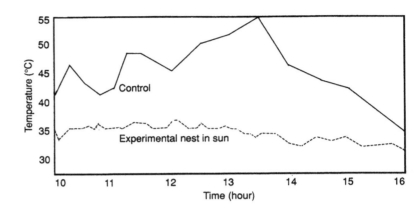

Fig. 16.5 Temperatures of two open-combed wasp *(Polistes gallica)* nests, in sunshine for 6 hours on July 31 in Germany. One nest *(solid line)* was uninhabited, and the other *(dashed line)* had an attending queen that brought water for evaporative cooling and fanned. (From Steiner, 1930.)

tated in droplets onto the brood and onto the back of the comb containing the brood.

When the queen is alone during the early stages of colony founding she may during times of potential nest overheating devote herself exclusively to nest thermoregulation, making 8 or more water-collecting trips in 15 minutes (Steiner, 1932). After her workers eclose, she still assumes most of the water-carrying chores although these daughters participate in fanning. The adults can receive fluid secretions from their larvae (Morimoto, 1960) but they apparently do not use these secretions as a quick source of fluid for evaporative cooling, as do some vespine wasps (see below).

The effectiveness of the queen's apparent thermoregulatory behavior was shown by Steiner (1930), who caged a queen *Polistes* wasp while her nest was left *in situ* to receive sunlight. The temperature of the unattended nest soared from 35°C to 55°C, while the temperature of a nearby control nest with a queen did not exceed 38°C. At the control nest the queen was making 6–11 water-collecting trips every 15 minutes.

The queen's crop volume is 30–35 μl. At a maximum frequency of water-collecting trips of 52 per hour, a queen therefore transports about 1.8 ml water to the nest in an hour. The evaporation of this volume of water would dissipate 1,050 cal (or 4,396 J) at 35°C. A typical nest weighs 15.4 g (Steiner's estimate), and thus its temperature should plummet by 68°C if all of the water were evaporated instantly. Undoubtedly not all of the 1,050 cal cools just the brood, but Steiner's calculations indicate that nest cooling by evaporation is a feasible phenomenon. The placement of water droplets on nests on hot days confirmed that evaporative cooling occurs in nests of *Polistes gallicus, P. biglumis,* and *P. foederata* (Weyrauch, 1936).

The nests of hornets and yellow jacket wasps (Fig. 16.6) differ from those of *Polistes* in that they are located in cavities or have a multilayered paper jacket covering of the combs (Fig. 16.7). The paper carton covering the combs provides considerable insulation and allows nests to be placed in shaded localities. The wasps are distributed throughout the Northern Hemisphere, and unlike *Polistes*, these insects can generate (and retain) metabolic heat of their own to warm the nest. Nest temperature in large nests may be regulated with precision. Himmer (1927) measured nest temperatures of 26–36°C in a large *Vespa vulgaris* nest for over a month during which air temperatures ranged from 9 to 34°C, averaging 18.4°C. The wasps' nest temperature averaged 30.7°C.

The annual vespine wasp colony starts with a single overwintered queen in the spring. During nest initiation, the queen builds a small pedicel onto which a paper comb with several cells is attached. Eggs are then laid into the cells, and the queen then also initiates construction of 2 or 3 paper envelopes around the comb. Soon after the first eggs are laid the foundress queen curls herself about the pedicel (Janet, 1895; Brian and Brian, 1948) and produces heat that conducts through the petiole into the brood cells with eggs and small larvae (Makino and Yamane, 1980). The queen usually spends more than half of each day curling, when she increases brood temperature by 2.5–4.0°C (Makino and Yamane, 1980). In the small foundress nests, heat loss is very rapid, and heat is more efficiently conducted to the immatures via the cell wall than by the air surrounding the cell. Some studies report little evidence of nest heating by foundress queens (Gibo et al., 1977). In these studies, however, the thermometer probe was *adjacent* to the comb, not *in* it.

The nest population grows through successive broods of workers, and then it dies out at the end of the season after the new reproductives are produced. Throughout the annual colony cycle, as the colony population grows, nest thermoregulation continually improves (Roland, 1969; Gibo et al., 1974a,b, 1977). Small, newly initiated nests show poor thermoregulatory capacity, as do old, declining nests.

The larvae themselves are reported to become major heat producers (Ishay and Ruttner, 1971; Ishay, 1973; Gibo et al., 1977; Makino and Yamane, 1980). For example, in one nest of *Vespa*

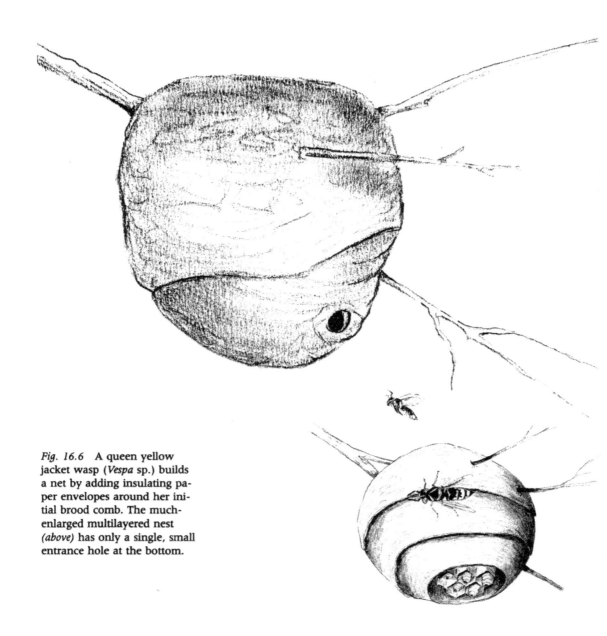

Fig. 16.6 A queen yellow jacket wasp (*Vespa* sp.) builds a net by adding insulating paper envelopes around her initial brood comb. The much-enlarged multilayered nest *(above)* has only a single, small entrance hole at the bottom.

simillima temperatures of a larva were maintained at 31.0 to 30.5°C for 45 minutes while ambient temperature outside the nest remained at 24°C (Makino and Yamane, 1980). Since these data were given to show heat production of the larvae, it must be assumed that adult wasps had been removed from the nest, although this was not explicitly stated. Other data (Fig. 4 in Ishay,

1973) also suggest modest heat production by larvae of *V. crabro*, but the "T_a" (ambient temperature) to which brood temperatures were subjected are not critically described (how and where measured?) in the publication, making independent analysis of the results not possible. The brood temperatures of *Vespa arenaria* that were reported to show heat production are nearly identical to ambient temperature (Fig. 3 in Gibo et al., 1977), making another claim of heat production by larvae problematical. The mechanisms for the presumed heat production also remains unknown, inasmuch as larvae lack muscles for shivering thermogenesis. Nest-temperature regulation by vespine wasp larvae remains a possibility, but it seems to me that it has not been demonstrated. Nest temperatures are much better regulated as larval numbers increase, but this increase in larval numbers is also associated with a combination of other factors, such as larger thermal inertia (due to mass increases of the nest), better nest insulation, and more workers who heat the nest. All of these factors would act to

Fig. 16.7 Cross-sections of vespine nests. At left a queen white-faced hornet *Dolichovespula maculata* is shown in her initial nest; note the long entrance tube. At a later stage *(right)*, the nest holds two large brood combs.

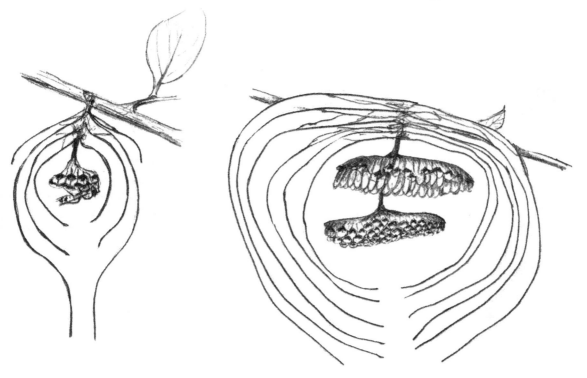

stabilize larval temperatures independent of possible larval thermoregulation.

Another thermoregulatory response emerges when the colony contains pupae. Workers heat pupae by pressing upon them with their bodies, both when the pupae are in the comb and removed from it (Ishay, 1973). Incubated pupae can increase in temperature by about 10°C in 5–7 minutes (Ishay, 1972).

Incubation of brood among the yellow jackets and hornets or Vespinae (Ishay and Ruttner, 1971; Ishay, 1973) is directed at pupae in response to a pheromone as well as to the physical characteristics of the pupae inside the cell. For the European hornet *Vespa crabro*, the incubation pheromone is cis-9-pentacosene (Veith and Koeniger, 1978) and the physical characteristics of the pupa inside a cell that stimulate incubation can be mimicked by a 350–400 mg pebble slipped into a cell (Koeniger, 1975; Fabritius, 1976). It is not known if the wasps sense the temperature of the brood before incubating it, and whether they attach themselves to it because they are attracted to it and then produce enough heat until their own (and the brood's) temperature reaches appropriate levels.

Heat production by shivering to maintain a high nest temperature (Stussi, 1972a,b) at low air temperature requires high rates of energy expenditure (Gibo et al., 1977). Unlike bees, however, wasps do not have combs filled with honey that can serve as energy reserves to fuel the heat production. During days of good weather, incubating wasps can refuel by foraging, but at night or during inclement weather they tap another source; they rely on sugary secretions from their larvae. In a sense, the larvae serve as honeypots that are otherwise lacking in most (but not all) wasps. Thus, the larvae are *indirectly* essential for sustained elevation of nest temperature by endothermic heat production.

Food exchange between adult wasps and their larvae (Réaumur, 1742; Janet, 1895) is triggered by a tactile response. A light touch on a larva's mouthpart by either an adult wasp or an experimenter's pencil triggers the larva to lean backward, spread its mandibles, and disgorge a droplet of clear liquid. This liquid contains less amino acid than the larva's blood, but it has about 4 times its carbohydrate concentration (Ishay and Ikan, 1968a; Maschwitz, 1966). By one "milking" of a larva a worker wasp is calculated to gain sufficient food energy to support its metabolic rate for a half day (Maschwitz, 1966). Although wasps feed on nectar, they are primarily predators and feed their larvae meat. The larvae then synthesize the sugars fed to their adults from their meat diet (Ishay and Ikan, 1968b). An apparent mutual dietary dependency between larvae and adults results that presumably ensures brood care, but of course many social insects have adequate brood care without such mutual dependency.

The larvae provide a rather limited energy reserve. Commonly there is a drop in wasp-nest temperature late at night and early in the morning, when food reserves decline (Gaul, 1952a,b; Roland, 1969). In one colony of *Vespa vulgaris* monitored by Roland (1969), nest temperature each night declined from 28–30°C to 18–20°C. But when he reprovisioned the nest with honey, nest temperature again rose to 28–30°C. Apparently colonies "turn down their thermostat" at night when they run low on fuel.

The second major mechanism of nest thermoregulation by vespine wasps involves nest cooling by evaporation of water, if ventilatory fanning is inadequate to prevent overheating. Water evaporation for nest cooling has been reported for *Vespa vulgaris, V. germanica,* and *Dolichovespula silvestris* (Himmer, 1932), *V. arenaria* and *D. maculata* (Gaul, 1952), and *V. orientalis* (Ishay, Bytinski-Salz, and Shulov, 1967). Workers in overheated colonies of *V. vulgaris* and *V. germanica* may even turn to larvae for liquid secretions that are then applied as droplets and smears onto the combs and nest (Weyrauch, 1936). The tapping of larvae, however, is probably only an emergency measure when water is not immediately available.

An experimentally overheating colony of *V. crabro* did not start water collection (Himmer, 1931). Instead, as the nest temperature rose to above 39.°C the hornets deserted the nest. Aside from this experimental situation, *V. crabro* may rarely be overheated in the field because they occupy hollow trees. They are there well insulated from potential overheating, and may therefore not have evolved responses that are necessary in those colonies commonly subjected to direct solar radiation.

The maintenance of a high nest temperature speeds up the colony cycle, and it also promotes its proper development. In *V. crabro,* pupae reared at 20°C produced many deformed adults, whereas those reared at 32°C developed normally (Ishay, 1973).

Social Bees

Most of the world's thousands of bee species are solitary. Solitary bee females provision a nest with a pollen lump sweetened with nectar, lay their eggs on it, and then leave their offspring to their fate. Presumably nest sites are chosen at least in part with consideration to suitable temperature, although this has not been systematically investigated. In social bees, the female stays with her eggs until they hatch, and since she reuses the nest for subsequent broods of offspring, overlapping generations are reared together, and now she qualifies for the title of "queen."

Of all the major social insects, bees present the greatest diversity of levels of sociality and of range of control mechanisms for social thermoregulation. But nest-temperature regulation has been studied only spottily, and most of our insights of thermoregulatory physiology and behavior come from data from bumblebees and honeybees.

SWEAT BEES

The sweat bees (subfamily Halictinae) are small-bodied bees with a worldwide distribution. Social organization ranges from solitary species to those with colonies of several hundred individuals. In all species the females dig a shaft into the soil and excavate a number of lateral chambers or cells. Each cell is provisioned with pollen and nectar and a single egg before it is sealed (Michener, 1974).

As in all social insects, temperature is apparently of prime importance in brood development, and the placement of brood cells is varied in accordance with suitable temperature conditions. Like many ants, most halictines avoid thick, shaded plant cover and choose for their nest sites bare soil surfaces warmed by the sun (Sakagami and Michener, 1962; Michener et al., 1958; Sakagami and Hayashida, 1961).

To more finely tune the temperature of their brood, halictines apparently dig their burrows to depths where appropriate temperatures are encountered. For example, in *Lasioglossum inconspicuum* in Kansas, the cells are on the average at a depth of 7.7 cm in April and May, but they average 41 cm deep (range 7–63 cm) in the much hotter weather of August and September (Michener and Wille, 1961).

BUMBLEBEES

Most bumblebee species live within the temperate zones of North America and Eurasia, although several are found above the Arctic Circle. Bumblebees have annual life cycles like those of temperate *Polistes* and vespine wasps. The colony is initiated in the spring by a single overwintered queen, and following colony growth throughout the season, the colony dies in the fall (Free and Butler, 1959; Alford, 1957; Heinrich, 1979a). As in vespine wasps, nest temperature is maintained near 29–32°C in vigorous colonies, but the level of temperature control changes along with the colony cycle (Himmer, 1933; Vogt, 1986a).

The majority of bumblebees nest in the vacated nests of small mammals, such as those of mice and squirrels, which supply existing insulation. In some species the queen may also pull together any locally available fibrous material to fashion a tightly knit hollow mass the size of a tennis ball. Later in the colony cycle, as the nest cavity enlarges, bees under cold stress (Plowright, 1977) add a waxen canopy into which they incorporate a variety of handy fibrous materials that they may drag or carry to the nest. Bird nests are sometimes also used. In the Arctic, *Bombus polaris* queens sometimes evict snow buntings *(Plectrophenax nivalis)* from their feather-lined nests to build their own brood clumps on the deserted birds eggs, after fashioning a roof over the otherwise open nest (Kukal and Pattie, 1988).

Inside the fuzzy ball of the initial nest material, the queen builds a brood cell that holds her eggs and developing brood, as well as a waxen honeypot that is daily filled with nectar. From the time the eggs are laid until the workers eclose, the queen steadily incubates her brood clump (Fig. 16.8), only occasionally pausing to feed from her honeypot at night and to forage in the daytime. She is able to orient to the brood clump in total darkness with the aid of a pheromone that she herself deposits to "mark" the brood location (Heinrich, 1974a). In captivity, queens occasionally deposit pheromone marks on inappropriate locations (such as a spot on the cage wall), which I have seen them then incubate for weeks on end.

As shown in the field at Lake Hazen on Ellesmere Island, Canada (Richards, 1973), and in the laboratory (Heinrich, 1974b), the temperature of the initial brood clump of a single nest-founding queen is regulated at 30°C or above while the bee is in the nest. Brood-clump temperature immediately declines when the queen no longer contacts the brood, as when she leaves to forage.

Direct temperature measurements by means of thermocouples implanted in a brood clump of a *B. vosnesenskii* queen in the laboratory under *ad libitum* food supply show that even an uninsulated brood clump can be maintained at near 20°C above air

Fig. 16.8 A resting bumble-bee *(bottom)* keeps its body elevated above the substrate. In contrast, an incubating bumblebee *(top)* presses her extended abdomen against the brood. The queen is perched on a brood clump containing pupae and a packet of eggs (near her right hind tarsus).

temperature even at air temperature near 5°C. At near 25°C, the brood clump is maintained at 32°C (Heinrich, 1974b).

During incubation, the bees (old and new queens, workers as well as drones) press themselves onto the brood and remain there for many minutes to hours. Their only obvious motion is the rapidly pumping abdomen, as their shivering flight muscles are ventilated. Incubating bees regulate their T_{abd} at levels slightly

THE HOT-BLOODED INSECTS

higher than that of the brood (Heinrich, 1974b). It is therefore not clear whether the mechanism of brood-temperature regulation involves simply a regulation of abdominal temperature, or whether brood temperature as such is monitored directly by means other than its thermal effect on T_{abd}. In support of either hypothesis, the metabolic rate of incubating bees rises steeply with decreasing air temperature (Heinrich, 1974b; Vogt, 1986b), to approach that observed in flight at the lower air temperatures.

What does the queen's endothermy for brood incubation mean in terms of foraging effort? The queens of *B. vosnesenskii* mentioned above were foraging from *Arctostaphylos* flowers in early spring. One flower contains on the average 0.4 mg sugar, or the equivalent of 6.3 J (or 1.5 calories). While incubating uninsulated nests in the laboratory at 5°C, each bee consumed approximately 30 ml O_2/g body weight/hr. This is the equivalent of an energy expenditure of 628 J, which would be available in 100 *Arctostaphylos* flowers (Heinrich, 1974b). Since the bees in the field visit approximately 15 flowers of this species per minute, a minimum foraging time of 7 minutes must be needed to supply the fuel for one hour's incubation at 5°C. Having filled her honey-stomach (about 0.2 ml) with a 30 percent sugar concentration (*Arctostaphylos* nectar), a bee should be able to incubate uninterruptedly for 1.4 hours before refilling. However, a filled honeypot (0.5 ml) would extend her incubation duration for 4 hours at 5°C. Nevertheless, these times are minimum estimates. In the field the bees' incubation duration would be considerably greater because nests are highly insulated.

When a bee exhausts her food supply she enters torpor and ceases to incubate. Sladen (1912) observed: "Occasional periods of semi-starvation, lasting for a day or two do not harm a colony of bumblebees: the bees simply become drowsy, remaining in a state of suspended animation."

Suspended animation may not cause immediate harm, but it can have long-term consequences. The demographic aspects of colony thermoregulation have been examined by F. Daniel Vogt of the University of Plattsburgh, New York. Vogt (1986a,b) founded colonies in the laboratory with wild-caught queens in the spring and followed their behavior and their fate throughout their entire colony cycles while subjecting them to experimental manipulations, including exposures of air temperatures varying from 3°C to 38°C. The cost of brood-temperature regulation is high. At low air temperatures not only do the bees (*B. impatiens*

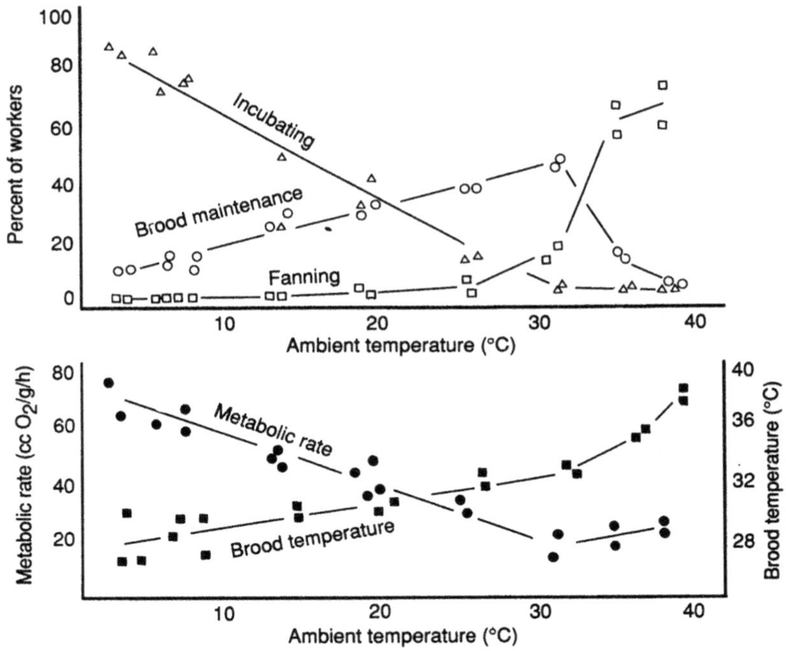

Fig. 16.9 Thermoregulation in a *Bombus impatiens* colony with 42 workers. *Upper graph:* Percentage of worker population engaged in brood incubation, wing fanning, and brood maintenance as a function of outside ambient temperature. *Lower graph:* Metabolic rate and brood temperature as a function of ambient temperature. (From Vogt, 1986b.)

and *B. affinis*) pay a high energetic cost, they also pay a high cost in time, since nearly 90 percent of a colony's worker force may be occupied with incubation as opposed to brood maintenance. Similarly, at air temperatures above 35°C the major colony activity is fanning, again depleting brood-maintenance activity (Fig. 16.9). The bees spend a minimum of energy and time in nest thermoregulation near 30–32°C, and correspondingly they then have the most time available for foraging and brood maintenance.

The effects of freeing bees for brood maintenance (and presumably foraging) are indirectly evident through insulation and colony size. Colonies reared at 15°C with nest insulation grow as rapidly and produce as many workers as those reared at 25°C. However, colonies reared without nest insulation at 15°C require that the colony members spend an increasingly greater proportion of their time incubating, and such colonies produce approximately one-third as many workers as are produced in an insulated nest (Fig. 16.10) and no new queens (Vogt, 1986b). Undoubtedly the laboratory results greatly underestimate the tremendous cost of the lack of nest insulation, because the lab bees had a constant food supply whereas bees in the field must leave the colony to forage for pollen and nectar.

Large numbers of workers are an immediate thermoregulatory advantage to the colony. As the queen is joined by her daughter workers, the nest temperature becomes much more stable; nest temperature now does not decline when the sole member of the colony leaves the colony to forage (Richards, 1973). Large numbers of individuals ensure that one or several bees will always be available to incubate the brood. Additionally, as bee numbers

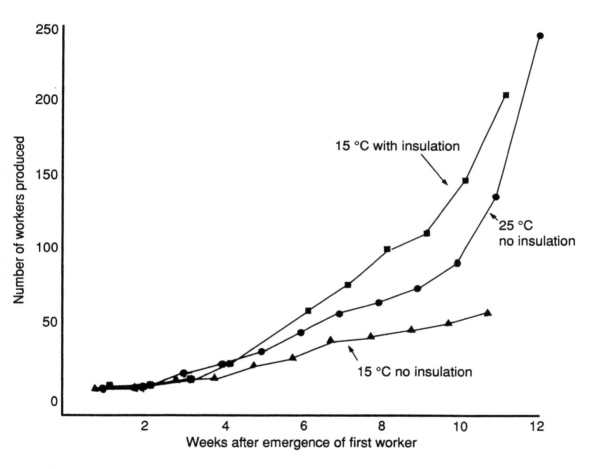

Fig. 16.10 Mean number of workers produced in colonies reared at 25 °C (6 colonies), 15 °C with nest insulation (3 colonies), and 15 °C without nest insulation (3 colonies) as a function of the number of weeks after the emergence of the first worker in the colony. (From Vogt, 1986b.)

increase the whole nest gets heated as a by-product of their activity, and except at very low air temperatures, *no* or very few bees now need to stop their other activities in the colony in order to incubate. Increased food stores may also improve the otherwise precarious energy base for thermoregulation. The pattern of increasing control of nest temperature throughout colony development, and its subsequent decline with the decline of the colony at the end of the cycle, has now been documented in a number of *Bombus* species (Nielsen, 1938; Cumber, 1949; Fye and Medler, 1954; Hasselrot, 1960; Wójtowski, 1963; Vogt, 1986b).

Nest overheating is probably not a common problem of bumblebee colonies in the field. However, experimental overheating of the nests of a variety of bumblebee species causes fanning, abandonment of the nest, or the formation of ventilation holes in the ceiling above the brood (Lindhard, 1912; Jordan, 1936; Postner, 1951; Hasselrot, 1960; Vogt, 1986a).

STINGLESS BEES

The stingless honeybees (tribe Meliponini) are a strictly tropical and subtropical group. Some species, such as *Trigona spinipes*, have colonies with up to 100,000 individuals (Lindauer and Kerr, 1958), and the complexity of their social behavior is probably comparable to that of the more familiar honeybee, *Apis mellifera* (Sakagami, 1971).

Very little is known of mechanisms of thermoregulation in this group (Wille, 1976), although several features of their biology suggest that temperature control is likely. First, large colonies and flight muscles should make heat production both inevitable and feasible. Second, most species of Meliponini nest either in tree cavities or in other enclosed nest sites, such as among tree roots, in abandoned ant or termite nests, or even in walled-off chambers in occupied ant and termite nests. A few species build aerial nests like those of vespine wasps, and these nests are insulated and protected with layers of mixtures of their own wax, plant resins, and sometimes mud or vegetable matter (Michener, 1974).

T. spinipes, one of the few meliponine bees whose nest temperatures have been studied, has nests containing some 100,000 individuals built in tree branches, where they are exposed to sun. The brood nest is protected by numerous layers of hard material, and temperature in the nest is maintained at 33–36°C as air temperature varies from 8 to 30°C. Nevertheless, the brood nests of 6 other species of stingless bees show very little or only moderate nest-temperature regulation (Zucchi and Sakagami, 1972).

Similarly, in colonies of *Melipona rufiventris* and *M. seminigra* in Brazil, temperatures within the nest's cavity space tend to fluctuate with air temperature (varying from 23 to 30°C over a day, which is equivalent to the total temperature fluctuations throughout the year). Brood-temperature fluctuations are low, however, and brood temperature is maintained at 31–32°C (Roubik and Peralta, 1983). Roubik and Peralta conclude that the larvae supply most of the heat for the nest, but no experimental evidence is provided to either support or refute this claim; the brood was not isolated from the hive occupants to test the hypothesis. The authors also maintained that there is "no evidence" that *Melipona* or other stingless bees carry water to the nest for evaporative cooling, but whether these bees use evaporative cooling is a fully open question; no evidence for or against evaporative cooling as a mechanism in stingless bees is likely to be forthcoming until experimental overheating of nests to over 40°C have been done to test the idea.

Some species adjust nest temperature by alterations in nest architecture. An African species, *Dactyluvina staudingeri*, has sticky, resin-lined pores, approximately 1 mm in diameter, through the nest covering, but the pore size is apparently adjustable (Darchen, 1973). At 9:00 A.M. on 3 days when air temperature was about 23°C, the average density of open pores was 0.5/cm². At 3:00 in the afternoon on these 3 days, when air temperature had increased to 27–28°C, the pore density had more than doubled to an average density of 1.1/cm². The pores are all closed at night. In addition to the pores and the main entrance, the nests of *D. staudingeri* have a second large opening near the base of the nest. This opening is left open or closed off, depending on temperature and/or humidity.

HONEYBEES

Any discussion of thermoregulation in honeybees must be approached with both bravado and trepidation. In no area of insect thermoregulation has there been so much research. As a consequence, our "literature" knowledge is good, but the same facts and ideas are endlessly repeated. Unfortunately, the continuing froth of activity sometimes threatens to obscure what is already known from what is not known, particularly as in some cases the issues have become more obfuscated than clarified.

Like other *Apis* species, the "European" honeybee *A. mellifera*, with its now nearly world-wide distribution, probably originated in the tropics. The perennial nature of its colonies and its reproduction by swarming are common features among tropical social bees, features that distinguish it from most social bees endemic to cold temperate regions. Through the evolution of advanced techniques of colony thermoregulation, *Apis mellifera mellifera* and the other European races of the honeybee have subsequently penetrated to colder climates. Winter is still the time of greatest mortality of even those races now adapted to northern climates, and colony thermoregulation is a critical feature of its biology.

No area of insect thermoregulation has received as much attention as honeybee nest-temperature regulation, ever since René Réaumur (1742) and John Hunter (1792) discovered the phenomenon over 2 centuries ago. Hunter originally suggested that the warmth generated by the bees kept the wax soft so as to allow them to shape it into cells. The ductility of beeswax is indeed uniquely optimized at the temperature that is regulated within the nest (Hepburn, Armstrong, and Kurstjens, 1983). But the ductility of wax at precisely 30°C is undoubtedly not a given in nature, any more than is flight with a T_{thx} of 40°C. Both are effects.

The growth and development of the eggs, larvae, and pupae also require a high temperature. If the colony is raising brood, then at least part of the brood nest must be maintained between 32 and 36°C, coincident with the temperature sensitivity of the brood. Larvae and pupae reared in an incubator at 28–30°C have a high incidence of shriveled wings and other malformations. Those reared at 32–36°C are normal, and those reared at 38–39 °C all die (Himmer, 1927). In the absence of brood (such as in swarms or in the hive in fall and early winter), the temperature of the bees at the nest periphery must not fall below the chill-coma temperature of near 10°C (Free and Spencer-Booth, 1960). Thus, in the proximate sense, the bees indeed regulate nest temperature "in order to" maintain suitable ductility of the beeswax and suitable temperatures for the immatures. However, the chemistry underlying the temperature characteristics of the wax and the temperature sensitivity of the larvae are undoubtedly both products of natural selection due to the temperatures that they

Fig. 16.11 A honeybee swarm (10,000 to 30,000 bees) that has left the hive and is looking for a new home.

are subjected to as a consequence of many bees crowding together.

The broodless, waxless cluster. The simplest situation for a brood-less cluster of bees is a swarm that has just left the colony. A swarm consists of an old queen and about 20,000–60,000 bees that have left the parent hive and a new queen behind. The bees of a swarm typically coalesce into a beardlike mass on a tree branch (Fig. 16.11). They hang there with interlocking tarsae and remain virtually immobile except for a few scouts that leave it and search for a suitable living space, such as a hollow tree. If the swarm is allowed to hang for an hour or more and is then shaken off its branch or other attachment, most of the bees drop to the ground, because their T_{thx} is too low for flight. Yet, when the time comes to move to a new home, all of the bees of a swarm warm up and the entire swarm is airborne in less than a minute.

If the bees do not find a suitable nest site within a few days, they eventually begin to forage to replenish their food supplies and they then begin to build combs inside the swarm cluster; the swarm site becomes the nest site. But in north-temperate areas such colonies do not survive the winter. In tropical honeybees, such as in the giant honeybee *A. dorsata* or the dwarf honeybee *A. florea*, swarm cluster and hive or nest cluster are more closely equivalent, since the swarm settles where the nest will grow. The combs of these two tropical species are always in the open, and the colony as a whole may eventually move on to areas where food is more abundant.

Colony thermoregulation can be viewed from both the response of the swarm as a whole and the behavior of individual bees within it. In general, the temperature characteristics of the whole swarm provide the "environment" in which the individuals find themselves, and their actions in turn affect not only themselves but also that environment of which they are an integral part. Understanding colony thermoregulation requires a consideration of both what the individuals perceive and respond to and what the effects of their responses are on their immediate environment.

The temperatures throughout swarms as a whole are not uniform. Core temperatures remain near 35–36°C over a wide range of air temperature and over a wide range of swarm sizes (Büdel, 1968; Nagy and Stallone, 1976; Heinrich, 1981a,b), but temperatures of the swarm mantle fluctuate directly with air temperature at air temperatures from about 30°C to 15°C. From air temperatures of 15 to 0°C, on the other hand, mantle temperature stabilizes just above 15°C (Fig. 16.12), which is near the lower T_b threshold where bees can spontaneously warm up from shivering. In between the hot core near 36°C or above and the mantle at 15°C, the temperature of the bees grades gradually from the one temperature to the other.

Fig. 16.12 Core *(circles)* and mantle *(triangles)* temperatures of captive swarms in the laboratory as a function of ambient temperature. *Open circles:* Swarms of 2,000–10,000 bees. *Filled circles:* Swarms of 15,000–17,000 bees. Note the one swarm core temperature of over 46°C at an ambient temperature of 0°C. (From Heinrich, 1981b.)

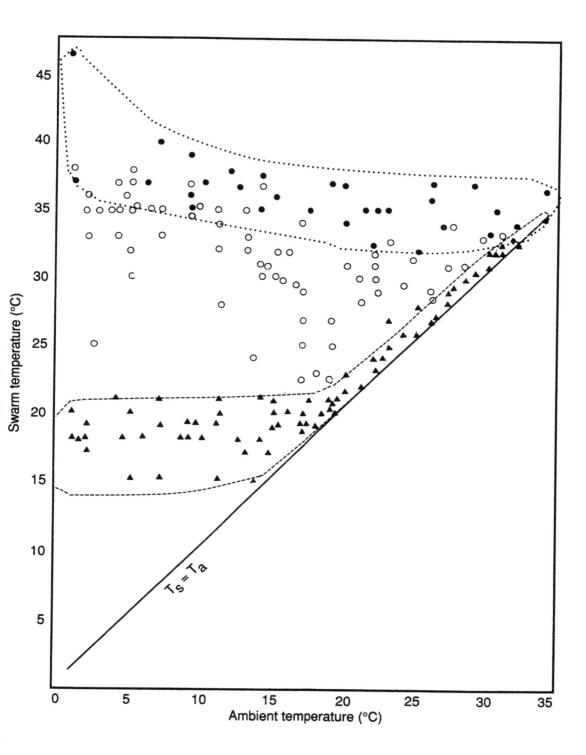

Individual bees, if given a choice, seek out environmental temperatures near 35°C (Heran, 1952), close to the thoracic temperature they regulate outside the hive while in flight (Heinrich, 1979b), while caged in small groups (Cahill and Lustick, 1976), as well as while living in the summer hive with brood (Himmer, 1932; Simpson, 1961). In one very large swarm at an air temperature of 1°C, however, bees at the core attained temperatures of 46°C (Heinrich, 1981a,b). Did the bees in the core of this swarm selflessly increase their metabolic rate to heat themselves up to near lethal temperatures in order to keep the bees on the swarm mantle from getting cold? Numerous workers (Gates, 1914; Hess, 1926; Phillips and Demuth, 1914; Himmer, 1926, 1933; Southwick and Mugaas, 1971; Ritter, 1978) had also observed previously that at air temperatures less than about 0°C, there is in large groups of bees often an increase of temperature at the core as air temperature is *lowered*.

Data like the above have almost universally been interpreted to show that the individuals do not regulate their T_b as such, but that they instead regulate colony temperature instead. Such "superorganism" models (for the *mechanism* of thermoregulation) presume that the bees in the center of the swarm shiver even though they are *already* warm, in order that the mantle bees will be heated to at least 15°C so that they do not become incapacitated! The superorganism model predicts that the lower the air temperature, the more the well-situated core bees shiver to protect their colony mates at the periphery. Nagy and Stallone (1976) measured CO_2 concentrations of 0.5 percent inside a swarm cluster at an air temperature of 15°C, and when air temperature was lowered to 5°C, the CO_2 concentration increased three-fold, to 1.5 percent. Since the rate of CO_2 production is indeed a reliable way to measure metabolic rate, these data were taken as proof that the core bees were raising their metabolic rate at low air temperatures, just as the superorganism hypothesis predicts. However, the primary presumption of the hypothesis, which was *not* tested, was that the bees inside the swarm cluster "know" (and care) when the bees on the mantle are getting cold so that they can respond appropriate to aid them. A second assumption was that the rate of CO_2 *loss* from the swarm core remains constant.

Although the "superorganism" model (Himmer, 1933; Simpson, 1961; Southwick and Mugaas, 1971; Southwick, 1982, 1983, 1987, 1988, 1990) is superficially seductive, additional studies of swarm thermoregulation support an alternative model for the

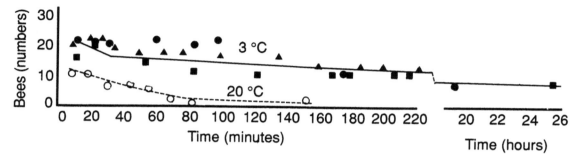

Fig. 16.13 Numbers of marked bees seen on the mantle of a captive swarm as a function of time after being paint-marked while on the swarm mantle. Different filled symbols indicate experiments at 3 °C, and open circles indicate one experiment at 20 °C. Swarm size was 16,600 bees. (From Heinrich, 1981b.)

mechanism of thermoregulation that is based primarily on the actions of the individual bees. In this model (Heinrich, 1981c, 1985) all bees act independently and no bee needs to know what any other bee is doing, and it generates the identical results of swarm temperature and swarm metabolism that the superorganism model predicts.

A variety of experiments (Heinrich, 1981a,b) fail to support the assumption that there is communication of T_b or the thermal environment between mantle and core bees and/or between the queen near the center of the swarm and the rest of the bees. First, swarms with and without queens have similar thermal gradients through them, and they have similar core and mantle temperatures. Second, at low air temperatures individually marked bees on the mantle remained in place for hours or days; they do not appear to exchange places with those in the center (Fig. 16.13). Most of the exchange of bees between mantle and core takes place only at high air temperatures, when bees are milling about for presumably various reasons. When the exchange of bees between the core and the mantle of a swarm is prevented by a physical barrier (a screen sock), thermoregulation of the swarm remains unaltered (Fig. 16.14). Clearly, there are no "messenger" bees shuttling about between the core and the periphery, telling the bees inside the swarm when to increase their heat production. Third, air (with presumptive pheromones) pumped from the center of one swarm located in the warm into a swarm located in the cold, and vice versa, fails to alter the thermoregulatory response, indicating that the bees are not communicating possible thermal needs by pheronomal messages. Fourth, perhaps the buzzing of chilled bees conveys information that is heeded by core bees. However, sound recordings from the periphery made with

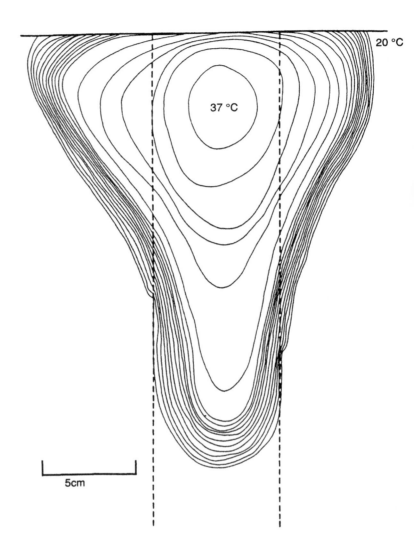

20 °C

37 °C

5cm

Fig. 16.14 Temperature isotherms (1 °C) in a swarm of 8,000 bees that is partially built in a gauze cylinder (and attached to a plexiglass foothold). (From Heinrich, 1981b.)

a microphone about which the bees were clustered also have no effect on altering core temperature. In summary, the core bees do not respond to messages relating to thermoregulatory needs that they receive from mantle bees, nor do they themselves travel to the mantle to check on external temperatures in order to gauge their response when they return to their positions and their attendant thermoregulatory duties at the core.

As in any homeothermic organism, the metabolic rate of bee groups increases at decreasing air temperatures (Woodworth, 1936; Roth, 1965; Heinrich, 1981a,b,c; Southwick, 1982, 1983,

THE HOT-BLOODED INSECTS

1984, 1985, 1987; Southwick and Heldmaier, 1987). However, this superficial resemblance between homeothermic organisms and bee groups says nothing about a superorganism mechanism. The response could more economically result from individual bees increasing their metabolic rate only to increase their own body temperature. Additionally or alternately, individuals are forced to have a higher metabolic rate (resting rate) if they are forcibly subjected to increasing body temperature, as when they are over-heated in the core of a large swarm.

In large swarms most bees are inevitably deep in the interior, shielded from low air temperatures. These bees are unavoidably warmed by the dense crowding, and they cool slowly. Indeed, based on cooling rates of bees inside the cores of heated *dead* swarms, calculations of how much heat core bees need to produce if they were shivering to keep warm indicate that even their *resting* metabolism (which they cannot shut off and which approximately doubles with each $10°C$ increase) is about ten times *more* than needed to keep warm at air temperatures near $0°C$ (Heinrich, 1981a,b). In other words, live swarms unavoidably heated and respiring must have active mechanisms of dissipating heat from the core. The observed metabolic rates of large swarms at air temperatures above $0°C$ can be best understood in terms of pas-sive production of heat by heated bees, rather than in terms of the bees' production of heat by shivering to keep themselves or peripheral bees warm.

If the core bees don't regulate their metabolic rates to keep the mantle bees warm, how then do the mantle bees manage to have T_b above air temperature? First, composition of the swarm is not uniform. The young bees that do not yet have the full shivering and heat-generating response (Allen, 1959; Fahrenholz, 1986) and that prefer higher air temperatures tend to crawl into the center of the swarm where it is warmer. The older forager bees, which have the full shivering response and lower temperature preferences even when alone, tend to aggregate near the periphery where it is cooler (Allen, 1959; Meyer, 1956). But as air temper-ature near them is lowered they push themselves into the swarm (Fig. 16.15), to seek warmer microclimates and to be heated by conduction by other bees. Second, the peripheral bees shiver, but after prolonged exposure to low temperatures they acclimate and tolerate lower T_b before shivering (Heran, 1952; Free and Spencer-Booth, 1960).

As the bees on the mantle crowd ever closer to keep warm, the

Fig. 16.15 Bees on the mantle of a swarm at an air temperature of 25 ° C *(top)* don't crowd each other, as they do at 3 ° C *(bottom)*. (From Heinrich, 1981b.)

THE HOT-BLOODED INSECTS

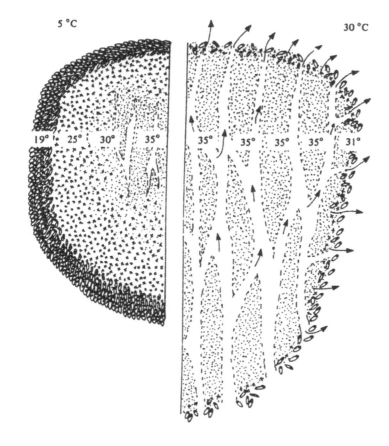

5 °C   30 °C

19° 25° 30° 35°   35° 35° 35° 35° 31°

Fig. 16.16 Diagrammatic model summarizing thermoregulation of the swarm cluster at low *(left)* and high ambient temperature *(right)*, indicating positions of bees, channels for ventilation, heat loss *(arrows)*, areas of active *(crosses)* and resting metabolism *(dots)*, and approximate temperatures. (From Heinrich, 1981b.)

swarm contracts and the air channels in it through which heat could escape get plugged (Fig. 16.16). In effect the mantle bees, by regulating their individual T_{thx}, inadvertently start forming an insulating shell that aids the core bees by trapping heat around them. As predicted by these as well as the superorganism hypothesis, as air temperature is experimentally lowered, the core temperature of large swarms (Heinrich, 1981b) and broodless winter clusters (Corkins, 1930, 1932; Fahrenholz, 1986) may even *rise* as the mantle bees crowd ever inward and closer, but the latter hypothesis offers no mechanism to account for the data.

If heat is trapped then CO_2 will be trapped as well, accounting for the high CO_2 concentrations that were thought to be the result of increased metabolism of the core bees to heat the mantle bees (Nagy and Stallone, 1976). Paradoxically, there may indeed be a

SOCIAL THERMOREGULATION   481

small increase in metabolic rate of the core bees, but it is because their *resting* metabolism rises with the increased T_b that crowding has forced upon them; their increased metabolic rate is not a superorganism response to help the mantle bees. Instead, the core bees are victimized by the mantle bees.

Although the core bees escape the cold, they may at times be subjected to overheating. How then do they regulate their T_b? Thermoregulation by the core bees occurs primarily through the converse response of the mantle bees subjected to cooling. As temperatures near them rise they work their way to the cooler periphery of the swarm. The behavioral tendency to seek cooler ambient temperatures has the effect of creating spaces or channels within the cluster as the cluster expands. As a result there is increased heat loss from the swarm because (1) its surface area is increased, (2) the steepness of the temperature gradient from the swarm to the air is increased, and (3) air channels are created whereby heat is released from the swarm interior. These passage-ways serve as ventilatory ducts and allow bees to pass freely from the hot interior to the cooler outside. Bringing a large swarm from room temperature to near 0°C results in the mantle bees imme-diately crowding close, after which there is an immediate over-compensation of core temperature to near 40°C. Conversely, suddenly taking a swarm from a low to a high air temperature results in an immediate reduction in core temperature as the mantle bees respond by crowding less loosely, thus opening the insulating shell and allowing heat to escape (Heinrich, 1981b).

The behavioral processes of the mantle bees attempting to main-tain a minimum T_{thx} for themselves at very low air temperatures is complementary to the action of the core bees attempting to maintain a preferred T_b near 35°C. There is a conflict, since behavioral thermoregulation by the mantle bees can result in overheating of the core bees. An equilibrium results, however, from the opposing actions, since as the core bees are trying to cool themselves, they inevitably release some of the core heat that then helps heat the cool mantle bees.

From individuals to the winter cluster. Measurements of over 50 years are consistent with the idea that the above model of ther-moregulation based on individual responses of bees in swarms also applies to bees in other broodless clusters. For example, as air temperature is lowered bees in the winter hive also begin to cluster, and undisturbed colonies that are killed in winter reveal an outer shell of tightly packed bees and an inner core where the bees have room to move about. Also as in a swarm cluster at low

THE HOT-BLOODED INSECTS

air temperatures (Fig. 16.15), the mantle bees of broodless winter clusters are arranged with their heads pointing inward (Farrar, 1943) and temperatures are relatively low and fluctuate widely (Himmer, 1926; Fahrenholz, 1986). In general, at air temperatures above 14 °C (which is outside the low-temperature danger zone) cluster temperature fluctuates with ambient temperature. The bees defend a lower temperature near 13–15 °C, as do swarms, but winter bees are often subjected to considerably lower environmental temperatures than swarms. As air temperatures fall the cluster shrinks, and it reaches a lower size limit near 0 to −5 °C. At still lower ambient temperatures the cluster increases heat production by shivering thermogenesis.

That bees in a winter hive sometimes increase their metabolic rate as air temperature is lowered has long been known (Corkins and Gilbert, 1932; Simpson, 1961; Free and Simpson, 1963). However, several problems have plagued much of the data that make interpretation of possible mechanisms of thermoregulation often difficult if not impossible. First, unlike as has been commonly assumed in studies where metabolism as a function of temperature has been measured, insects do more than just regulate their T_b at any air temperature. In many studies of the metabolic rate of hives (sealed), the bees themselves were not visible and individuals (necessarily confined) were possibly very active (especially at high temperatures), struggling to escape or to go forage. The metabolism of such few but very active individuals could mask the response of the cluster to temperature. Second, and perhaps more important, many different units have been used to describe metabolic rate, and no attempts have previously been made at standardization into common units so that studies can be compared and integrated to solve common problems. I here convert a variety of metabolic-rate measurements (such as cal/g /hr, ml O_2/g/hr, J/kg/min) to W/kg so that one may draw together some of the diverse data and then compare and evaluate the figures (Fig. 16.17).

As expected, the highest metabolic rates and the greatest effect of temperature on metabolic rate are found in individuals (or small groups of them). With groups of just 10 workers, the regression of metabolic rate on temperature shows that the average body temperature of these workers is near 35 °C (A, Fig. 16.17). These data indicate that all of the bees of the group are acting as individuals, regulating the same T_{thx} that they do in colonies (Fig. 16.18).

In a group of 300 bees (B, Fig. 16.17), metabolic rates at low

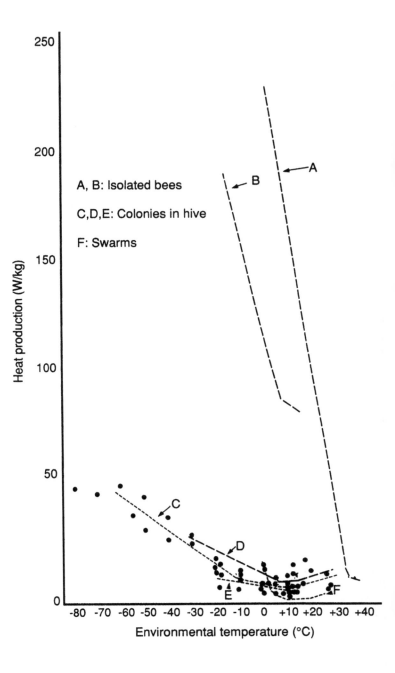

Fig. 16.17 Metabolic rate (heat production) of different honeybee groups as a function of ambient temperature as converted to watts/kg from a variety of sources in the literature (*A* = Cahill and Lustick, 1976; *B* = Southwick, 1990; *C* = Southwick, 1988; *D* = Southwick and Heldmaier, 1987; *E* = Southwick, 1983; *F* = Heinrich, 1981a.) Data are from intact colonies within hives with at least 10,000 bees (*C, D, E*), from groups of 10 (*A*) or 300 bees (*B*), or from resting swarms of 1,800 to 16,600 bees (*F*). Data points for the lines (fitted by eye) are included only for *C*. (The swarms had been long at rest, and caution was taken to exclude measurements when individual bees were active outside the swarm cluster, because such bees could generate as much or more heat than the cluster itself.) Note that in the large groups the regression of metabolic rate extrapolates to considerably lower temperatures than it does in individuals and in small groups.

THE HOT-BLOODED INSECTS

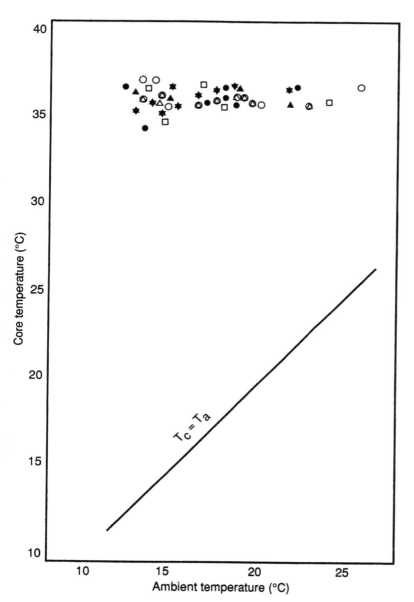

Fig. 16.18 Temperatures recorded from the brood nest of a hive *(filled circles)* and the stabilized core temperatures of free swarms of various sizes from 250 g to 3,400 g (different symbols represent different sizes) over a range of ambient temperature (<15° C, at night; >15° C, in daytime). (From Heinrich, 1981b.)

air temperatures are still high, indicating that most of the bees still feel the direct effects of the environment. However, the individuals now shiver only half as vigorously at any one low air temperature, and they can now exist at considerably lower environmental temperatures. As group size increases to massive (many thousands) numbers in swarms, metabolism drops rapidly to very low values (F, Fig. 16.17).

The combined data suggest that there are probably three entirely separate reasons for the massive reduction in metabolic rate as group size increases. First, like a bagful of beans assembled into one uniform mass, cooling rate is necessarily reduced and the individual "beans" now cool at only a small fraction of the rate they did when they were isolated from each other. Consequently, as the animals try to keep warm at very low air temperatures, the weight-specific rate of heat production is reduced accordingly. But there is another major effect. Clumping by itself with *no change in* T_{thx} would result in a lower slope of the heat production vs. temperature line, but the line should still extrapolate to the same temperature as in individuals, near 35–36°C. To the contrary, however, although in very small groups of bees the regression does extrapolate to near 35°C, in groups of 300 a very interesting thing happens: it now extrapolates to near 30°C. In other words, these bees have a much lower average T_{thx}. Furthermore, in still larger groups of several thousands, it extrapolates to 10°C or less (Fig. 16.17). These data are entirely different from what would be expected in differently sized vertebrate endotherms, in whom, regardless of size, the regression would always extrapolate to near 37–40°C as long as the animal remains homeothermic.

What happens in the case of the bee groups? I speculate that the most reasonable explanation is that the *average* T_{thx} of the bees in the clusters is considerably lower than it is in individuals or small groups of them held alone. In effect, the bee cluster does *not* behave like a classical homeotherm, because if it did, it would merely change its conductance (that is, the *slope* of the regression line) with the intercept remaining constant regardless of mass. The intercept itself varies probably because as group size increases, the bees in the outside layers of the cluster vastly reduce their temperature. As a consequence, they loose even less heat than predicted by their being part of a larger mass alone.

There is, at the present time, no adequate theory to explain why bees in very large groups tolerate much lower T_{thx} than they do in small groups or as individuals. I suggest here merely that it

may be a "comfort" reaction. In groups of bees, any small distur-
bance causes an immediate jump in oxygen consumption. Indeed,
any bees that leave the cluster always warm up first before doing
so, and as long as they remain out of contact with the cluster they
continue to shiver as necessary to keep warmed up. In a sense,
honeybees away from the cluster are in an "alarmed" or alert
state where they shiver nearly constantly. Most nonsocial insects
remain poikilothermic most of the time. I suggest that in individual honeybees, the poikilothermic "rest" or comfort stage, as ob-
jectively determined by the oxygen consumption measurements,
is reserved for times when the animals are in cluster with other
bees.

In hives or nests of bees additional phenomena complicate
metabolic-rate measurements as a function of air temperature. As
already mentioned, in a hive that is sealed up for metabolic-rate
measurements at higher air temperatures, bees may be attempting
to exit for foraging trips. These diurnal individuals, with very high
metabolic rate, would make it appear that there is a diurnal
fluctuation of hive metabolism. To some extent diurnal changes
of metabolic rate (Southwick, 1982) could be related to such
activity. Additionally, unlike in swarms, some bees in a hive are
also active with hive duties, and these individuals are therefore
not likely locked into a cluster for heat conservation, except at
very low air temperatures.

Finally, in winter clusters, continuous measurements of hive
temperatures and hive metabolism do not necessarily reflect the
steady-state physiology. The bees of a winter cluster must occasionally break up to move to neighboring honeycombs when local
energy sources become exhausted. During the periods when bees
move at low ambient temperature, there may be great transient
increases in overall hive metabolism (Southwick, 1982). Temporary increases of metabolism and food consumption are to be
expected, as many individuals, without the benefit of the cluster,
are each attempting to regulate their own body temperature by
increasing their rate of heat production (Allen, 1959; Esch, 1960;
Cahill and Lustick, 1976). As a consequence of extraneous activities, the metabolism of hives is higher and more variable at any
one air temperature than it is for swarms (C, D, Fig. 16.17).

Metabolic rates of bees within hives probably extrapolate to
higher temperatures than do bees in swarms, because in hives the
walls and empty comb provide insulation so that at any one
ambient temperature there is a higher average T_b of the bees

(because those on the outside of the cluster experience higher T_b than do bees in exposed swarms). However, if air temperatures continue to drop from $-10\,°C$ to $-70\,°C$ the winter hive bees, as those in a swarm, become more and more committed to maintaining a cluster. The mantle bees at first try (and eventually fail) to maintain their T_{thx} above the T_{thx} threshold where shivering is possible.

It is astounding that a colony can still keep increasing its metabolic rate as air temperatures decline to near $-70\,°C$. However, from the graph (Fig. 16.17) it might be tempting to conclude that, since the weight-specific metabolism of individual bees can increase nearly 5 times *above* that observed of a *colony* at $-80\,°C$, the bees of the colony at $-80\,°C$ have a large metabolic "reserve" to allow them to withstand even lower air temperatures than $-80\,°C$. So far the lower temperature limits have not been tested, but the data suggest that the metabolism of bee groups at near $-70\,°C$ is already plateauing out near an apparent maximum (Fig. 16.19).

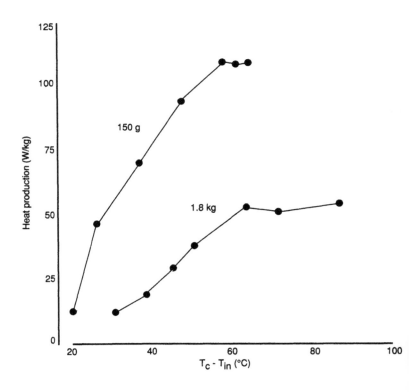

Fig. 16.19 Cold-induced heat production in small and large clusters of honeybees. $(T_c - T_{in})$ is the temperature difference maintained by the bees between the cluster core and the air outside the cluster. (From Southwick, 1987.)

THE HOT - BLOODED INSECTS

The reason for the apparent dichotomy in there being reserve capacity yet no further metabolic increase is easy to explain: At the *very* low air temperatures, cooling of the mantle bees to less than 15°C is inevitable, and the lower their T_b, the lower is their maximum rate of heat production. Indeed, at very low air temperatures only the core bees will remain warm enough to retain any capacity to produce appreciable amounts of heat by shivering. After the mantle bees have cooled to become passive insulators, the core bees also begin to cool, and overcooled bees can produce much less heat by shivering than fully warmed bees.

A major difference between swarms and winter clusters concerns the physiology of the individual bees. Following prolonged exposure to cold, bees acclimate (Heran, 1952). Bees from the winter nest, in temperature-preference tests in a temperature gradient, choose temperatures averaging 32.8°C but as low as 28°C, while summer bees choose temperatures near 35–38°C (Heran, 1952). Furthermore, winter bees have lower chill-coma temperatures than summer bees (Free and Spencer-Booth, 1960). Presumably these differences could alone account for both the lower core and mantle temperatures observed between winter hive clusters and summer swarms.

The largely individually based responses noted above are likely modified by natural selection in such a way that in addition to regulating individual T_b they also aid colony thermoregulation (Brückner, 1975, 1976). Superorganism concepts apply sociobiologically to considerations of a community of bees. But with respect to the mechanism of a *specific* function, such as thermoregulation, it would apply only if that function is accomplished through communication, to which individuals respond by fulfilling the thermal need of the colony as a whole, *as opposed to* individuals acting to regulate their own T_b.

An organism has exterior and interior temperature sensors, a centrally located thermostat, information on temperature and responses to it that are integrated by communication and central processing. Nothing like it occurs in a honeybee cluster, so the analogy breaks down. So far most of the observed data and mathematical modeling (Omholt, 1987) of colony thermoregulation at low air temperatures are consistent with the idea that bees regulate their own T_b independently of knowledge of, or response to, optimal temperatures of the group. The idea that temperatures of groups of bees are the result of superorganism responses is either an assumption or an admission of ignorance about the mechanisms that cause the observed behavior of the group.

Brood incubation. One of the major responses of bees that is unrelated to their seeking of direct individual comfort relates to brood. A dramatic change occurs in the colony response after brood rearing begins in the spring. Brood rearing can occur in the coldest weather, as well as the hottest, at air temperatures from $-40°$ to $40°$C or more. Over this wide range of temperature the bees maintain the temperature of the brood nest between 30 and 35°C (Gates, 1914; Hess, 1926; Himmer, 1927; Corkins, 1930; Dunham, 1929, 1931, 1933; Farrar, 1943; Ribbands, 1953; Simpson, 1961; Wohlgemuth, 1957; Büdel, 1960, 1968; Fahrenholz, 1986).

During brood incubation (Ritter, 1978; Kronenberg and Heller, 1982), the bees shiver where they might otherwise allow their T_b to decline. The details of thermoregulation of brood are not clear, but bees are attracted to clusters of capped brood (Koeniger, 1978; Ritter and Koeniger, 1977), where they have a higher metabolic rate than at combs with honey (Kronenberg, 1979; Kronenberg and Heller, 1982). The cues that cause both attraction and shivering may be both chemical and tactile. For example, workers ignore empty queen cells, but replacing a queen pupa shortly after its removal with a stone restores the attractiveness of the cell (Koeniger, 1975). Extracts of queen and worker pupae also help restore the attractiveness of long-empty queen cells. It is not known if the bees respond directly to brood temperature (but see Kronenberg and Heller, 1982). If there is a tight coupling between their own T_b and that of the brood, then they could potentially regulate their own T_b in the presence of brood so that brood-temperature regulation results secondarily. However, the workers have "cold" receptors on the last five antennal segments whose impulse frequency increases with decreasing temperature (Lacher, 1964). The bees therefore have the capacity to monitor brood temperature directly.

Brood cluster thermoregulation and geographical distribution. A problem of practical interest is the northward spread of the Africanized honeybee from Central America to North America. Colony thermoregulation is a primary feature thought to limit this spread. Edward E. Southwick and colleagues (1990) address this issue by comparing thermoregulation in groups of Africanized and European honeybees in Panama. As had previously been concluded for individuals (Heinrich, 1979b), groups of Africanized honeybees consistently have higher rates of energy expenditure than the European race (on a per-weight or per-individual basis)

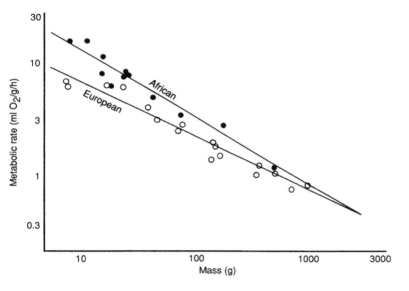

Fig. 16.20 The relation of bee-cluster mass and minimum maintained oxygen consumption achieved by groups of Africanized and European honeybees held at a constant air temperature of 2° C overnight. (From Southwick, Roubik, and Williams, 1990.)

when they are exposed to low air temperature (Fig. 16.20). The difference in metabolic rate is pronounced in small groups, but it disappears in large (>10,000 bees) groups. In 26 groups of bees (300 Africanized bees vs. 230 European), the Africanized bees showed 46 percent higher mass-specific metabolic rates. Greater mass-specific rates might be expected simply because of smaller body size, but at the individual level, the smaller Africanized bees still consumed, on the average, only 21 percent more oxygen than individuals of their larger European counterparts. The 46 percent higher metabolic rate of individuals in small groups is likely due to greater "nervousness," and individuals in small groups have a lesser tendency to form clumps. Even at 2°C groups of 50 Africanized bees apparently do not form a tight group because then all fall into cold coma in 2–6 hours, whereas European bees in similarly sized groups enter cold torpor only after 15 hours of exposure at that temperature (Southwick, Roubik, and Williams, 1990).

The increased activity levels of the Africanized bee might aid it in exploiting new food resources when they become available, but it could also result in a large winter loss of bees who leave the hive to attempt to forage (does it lead to the correlation of increased pace with decreased longevity?) or in exhaustion of food reserves. Smaller size, in turn, greatly reduces the ability of a winter colony to withstand further cold stress and hampers the colony's build-up in late winter in time for the spring honey flow. As a consequence of their increased activity levels, the Africanized bee will likely remain limited to a southerly distribution when they arrive in the United States (Southwick, Roubik, and Williams, 1990).

Curiously, the mass-specific rates of oxygen consumption of Africanized and European honeybees at 2°C become indistinguishable at large group sizes (Southwick, Roubik, and Williams, 1990). Thus, although the data suggest that Africanized bees differ by having less dense clusters than European bees, this can only be true at lower bee densities. I conclude from the data available so far that the "comfort factor," which is very important in honeybee colony thermoregulation at low air temperatures, has a higher population threshold in African than European bees. European bees, by having a lower threshold, clump earlier at lower bee densities and thus thermoregulate better as a group at lower energy cost.

Cooling the nest. When Martin Lindauer from the University of Würzburg (1954) placed a honeybee hive in full sunlight on a lava field near Salerno, southern Italy, the hive held its core temperature at 36°C, even though outside temperatures rose to 60°C. Such an impressive feat of honeybee colony thermoregulation depends on a highly coordinated social response that involves both water evaporation and fanning (Bruman, 1928; Chadwick, 1931; Lensky, 1964). Both processes are also related to honey-making, and they probably evolved from it.

Nectar is deposited into combs above the brood nest, and rising warm air from the brood nest would tend to evaporate water from the nectar and help make it into honey. However, air must be circulated through the hive to maintain its relative humidity low enough for evaporation to proceed, and this air circulation is accomplished by fanning (Hazelhoff, 1954). When nest temperature reaches 36°C fanning is greatly accelerated (Hess, 1926; Wohlgemuth, 1957; Southwick and Moritz, 1987) as the bees now rid the hive not only of CO_2 and water vapor but also of excess heat.

Fanning bees tend to align themselves along existing airflows (Fig. 16.21). As a result they produce strong unidirectional currents. Possibly after convection currents are created inside the nest, some of the air escapes through an entrance following lines of least resistance, and these lines of flow then attract additional

Fig. 16.21 Honeybee fanners aligning themselves at the hive entrance.

fanners who augment the existing air flow. Gangs of fanners stir the air in the brood nest. But entrance fanners expel the warm air currents (giving the pleasant honey and flower scent to the air near hives on warm summer days!), and without the entrance fanners there would likely be little air exchange between the nest atmosphere and that of the environment.

It is not known how the fanning response evolved. Possibly the simplest scenario is to suppose that it is derived from an escape behavior from noxious stimuli. But in the social setting the "flying" bee remains in place in the hive. At this time in evolution, however, fanning bees are not merely flying in place, because they create different patterns of air movements with their wings than do freely flying bees (Neuhaus and Wohlgemuth, 1960). Are fanning bees hotter than other bees? Perhaps they originally aligned along existing air streams to enhance convective cooling of their own bodies. However, the act of fanning should generate increases of T_b (how much?). Do the bees minimize potential thoracic overheating by seeking existing air streams in which to fan?

Fanners do not respond to hive temperature pe se. For example, foragers, particularly those carrying pollen, often fan before entering the hive (Neuhaus and Wohlgemuth, 1960). One might suppose they respond to warm air being blown out of the hive, but the temperature of the air in the vicinity of the antennae bearing the thermoreceptors (Heran, 1952) is apparently not effective in stimulating fanning (Neuhaus and Wohlgemuth, 1960). But body temperature itself may be relevant. At least in the Arizona desert, most of the pollen carriers returning to the hive have higher T_{thx} (average = 43°C) than nectar foragers, and Neuhaus and Wohlgemuth (1960) and Cooper, Schaffer, and Buchmann (1985) report that these pollen carriers "almost always stop and fan upon alighting before entering the hive." Since the fanners presumably have no knowledge of temperature within the hive, it is difficult to rationalize their behavior except in terms of a response to their own body temperature. The returning pollen foragers fan only for short durations. It would be of great interest to know what their body temperatures are when they stop fanning. If the bees indeed fan in part to affect their own body temperature, then it is nevertheless significant that they have evolved to do it primarily in the vicinity of the hive, where it immediately benefits the colony.

Alternatives to fanning to augment existing air currents are to work against existing air streams or to fan in random directions. Both responses are obviously counterproductive since both would increase heat production in the individual bee and in the hive without promoting convective cooling either of themselves or of distant portions of the hive.

Fanning by driving air unidirectionally through the hive could be relatively simple when the hive has two or more entrances. Many hives have only one small entrance hole, however. In order to examine how such hives ventilate, Southwick and Moritz (1987) used small hives containing about 2,000 bees and they measured ventilatory air movements at the one small (2 cm diameter) hole in the hive bottoms. As air inside the nest was artificially raised to near 40°C, bees moved outside the entrance and began fanning at or near the entrance. The air at the entrance hole was continually monitored for temperature amd for CO_2 and O_2 content. Since air leaving the entrance hole had a high temperature, a high CO_2 content, and a low O_2 content, and air entering was vice versa, in-out airflow could be monitored. Curiously, the hives "breathed" in-out at a frequency of about 3 times per minute during the daytime. Apparently periodic, active fanning moves an air current out, which is followed by a passive influx of air. The mechanism of the nest breathing so far remains unknown, inasmuch as the responses of the individual bees that cause the on-off fanning have not been examined. Also, normally colonies have lateral entrance holes. Do such colonies show the same "breathing" behavior?

Fanning to cool can work by itself only if air temperature is less than hive temperature. At high air temperature evaporation of water, and fanning to drive off the water vapor to make more evaporation possible, is effectively employed. A limited amount of cooling results from the evaporation of water from the nectar or honey stores. But if heating of the hive continues despite fanning, then the bees bring water into the hive and regurgitate it in small droplets into the cells, sometimes smearing it in a thin layer onto cells, as well as evaporating it from the tongue via "tongue-lashing" (Lindauer, 1954; Kiechle, 1961).

Tongue-lashing is also a common mechanism of honeybees and other bees for the evaporation of dilute nectar to produce honey. Each time the proboscis is extended back and forth, the bees exude a regurgitated droplet of fluid from the mouth and spread it with the proboscis into a film, which has a large evaporatory surface. Tonge-lashing for evaporative cooling is a mechanism whereby individual bees regulate their own T_{thx} under hot conditions outside the hive (see Chapter 8). When employed inside the hive, the same mechanism would also remove heat from the environment where it is used, such as near or in the brood nest.

As first demonstrated by Lindauer, a remarkable communication system is used to regulate the colony's water collection (Kiechle, 1961) that feeds the colony's thermoregulatory capacity at high temperatures. The system is based on a division of labor between the "water sprinklers," generally the young nurse bees in the central brood nest, and the water carriers, who are older forager bees that respond not to temperature as such but to the need for water.

Different foragers bring in nectar of a great range of concentrations, and some bees bring in water. Foragers bringing in concentrated nectar are normally unloaded much faster than those bringing back dilute nectar or water, but when a water shortage exists, those bees bringing back dilute nectar or water get a stormy reception near the hive entrance. Hive bees rush up to them, relieve them of their forage, and these bees then dance and quickly recruit followers. When there is again less need for water or dilute nectar, then the foragers cannot unload their water in 60 seconds and their eagerness to continue collecting decreases and they also do not dance to recruit more foragers. As a result of this feedback loop involving division of labor, the colony's water input is precisely calibrated to needs. The same feedback exists for the regulation of intake of other products (sugar and pollen), from which the thermoregulatory response is undoubtedly derived.

The tropical Asian giant honeybee *Apis dorsata* also uses evaporative cooling to maintain nest temperature at about 33.5°C (Mardan and Kevan, 1989). In addition, under humid conditions when heat is not easily lost by evaporation, the giant honeybees leave their open comb *en masse,* and mass defecation during these flights also helps dissipate excess heat.

Nest-site selection. All of the above thermoregulatory responses probably apply in varying degrees to all the different *Apis* species. However, only *A. mellifera* and other European races of the honeybee have penetrated cold and highly seasonal climates. Probably no other aspect of the colony response has allowed this penetration more than the choice of a nest site. Nest-site choice is the most critical step that ultimately determines nest temperature and hence survival during the winter. Winter is by far the most vulnerable time—most of the nest mortality occurs then.

Strictly tropical or subtropical species of *Apis* (*A. florea* and *A. dorsata*) nest in the open. Only the two cavity-nesting species, *A. mellifera* and *A. cerana,* the Himalayan honeybee, have ranges that have expanded beyond the tropics (Ruttner, 1968). And of the first species, *A. mellifera adansonii,* the African subspecies of the honeybee, is fairly indiscriminate in nest-site selection, sometimes even nesting in flimsy shelters such as holes in the ground (Darchen, 1973).

As a first step in nest-site selection, European races of the honeybee choose a cavity of proper volume, rejecting those below approximately 20 liters. Measurement of a cavity's volume involves the bee scout's walking about the inner surfaces and then integrating information on the distances and directions of the walking movements (Seeley, 1977). The nest-site entrance is also evaluated during site selection. Bees prefer small south-facing entrances that lie at the base of the cavity (Seeley and Morse, 1978).

All of the factors critical in nest-site preference have a common denominator of facilitating nest thermoregulation. A large cavity is required to store enough honey to fuel the heat production in the winter and to house the large bee population necessary to survive; small groups cannot maintain sufficient warmth long enough (Southwick, 1987). Southward entrances are probably warmer than northern entrances in the northern hemisphere, and a small, low entrance hole helps keep the nest warm by inhibiting loss of the rising warm air.

Bees improve the protection provided by the nest cavity by sealing all unnecessary openings with propolis, a resinous material. Within the nest the brood combs are enclosed by combs devoted to honey and pollen storage (Fig. 16.22). These combs serve to insulate, especially after they are emptied. Taken together, the many individual thermoregulatory responses of *Apis mellifera* conspire to make this the only insect species that survives northern winters without migrating or becoming torpid.

The honeybees of the genus *Apis* are all of tropical origin, and they have likely become too specialized for living continually in a complex society at a high body temperature to allow specialization in the opposite direction—namely, the dissolution of the colonial way of life and tolerance of freezing temperatures, as in northern bees. Instead, they have ever improved their social organization as a means of combating the cold. The refinements in choosing a nesting site, the critical timing of colony growth (Nolan, 1925) and reproduction (Seeley and Visscher, 1985), which involves the striking habit of starting brood rearing in midwinter so that swarms can leave in spring that will have enough time to accumulate 20 or more kg of honey for the winter, and the improved social response of regulating its temperature—all are key elements in the adaptations that enabled *Apis mellifera* to invade northern Europe and North America.

As shown for the Himalayan honeybee *Apis laboriosa*, migration is also an option for dealing with temperature extremes. The Himalayan honeybee suspends a single large comb in the open under a cliff overhang. A living curtain of bees several layers thick encloses this comb and protects it from predators and helps to maintain the proper nest environment for brood rearing. Although

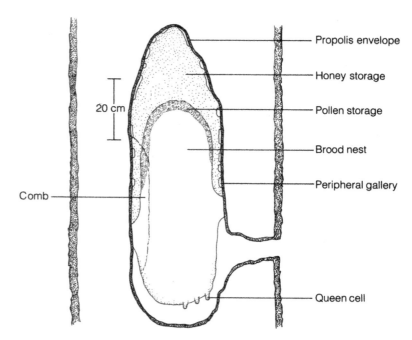

Fig. 16.22 Vertical cross-section of a honeybee nest inside a tree cavity. A large portion of the nest is devoted to storage of honey, the colony's food reserve that provides the fuel to heat the nest throughout the winter. (From Seeley and Morse, 1976.)

Propolis envelope

Honey storage

Pollen storage

Brood nest

Peripheral gallery

Comb

Queen cell

these bees have been exploited by humans of the remote mountainous regions of Bhutan, China, India, and Nepal for thousands of years, little was known of their biology. Using rock-climbing techniques and equipment, Benjamin A. Underwood (1990) made a detailed study of these bees on their high vertical cliffs in Nepal. One of his surprising conclusions was that entire colonies migrate intact up and down the mountains with the seasons.

Nests in the subalpine (above 2,800 m) are occupied for a maximum of only 4 months in the summer, while those within the warm temperate zone (1,200–2,000 m) are used for 10 months of the year. Due to the steepness of the terrain, colonies need not migrate more than 10–20 km to reach the warm temperate zone.

After abandoning their nests on cliffs and moving down the mountains, the bees survive the cold winter by huddling near the ground, and in spring they migrate back up the mountain to the cliffs to resume their yearly cycle. Undoubtedly part of the advantage for seasonal migration is related to seasonal availability of flowers. For example, the African honeybee likely emigrates for food. At high elevations in Colombia, South America, colonies of Africanized bees nesting at over 2,000 m in the wild leave after

THE HOT-BLOODED INSECTS

about 4 months of poor foraging conditions or rain (Villa, 1987). However, *A. laboriosa* ceases coming to feeders provided for them in fall, prior to migration (Underwood, 1990). Unavailability of food is thus likely not their only reason for leaving when nightly temperatures dip below 0°C and maximum daily temperatures are near 12–15°C.

Termites

Termites are a primarily tropical and subtropical group of some 2,000 species that, except for caste differences, are similar morphologically. But their nests vary enormously. The nests of damp-wood termites (Termopsidae) are mere galleries in wood, for example, but others are marvels of the insect world. At 6 or more meters in height, they are imposing for the sheer enormity of their size and for the sophistication of their construction, which among other advantages results in largely automatic, precise climate control. Although termites show a great array of nest forms (Frisch, 1974), they all live in enclosed spaces more or less isolated from the external environment. The elaboration of colony thermoregulation is largely paced by and made possible by nesting habits.

Termites evolved from cockroach-like ancestors that began to live directly within wood largely because of a partnership with gut symbionts able to digest cellulose. Harnessing the cellulose-digesting ability of microbes made feeding and tunneling directly within the wood possible. In turn, it not only gave these animals a nearly limitless supply of food, it also gave them a safe haven. Consequently, a key point that led to their evolution of sociality was likely the unattractive alternative the offspring would face in leaving their save cornucopian haven; young stayed, paying the price of being dominated by their parents, and their fitness benefits became indirect, to help their parents rear winged siblings that would disperse and die (only one in millions lives to reproduce).

Many termites still live directly within their food supply, although most species, now with massive populations, have evolved to harvest cellulose food energy beyond the confines of the nest structure. They use their own excrement as a binder of soil, sand, and wood particles to make cement- or carton-like building material; first they enclose their food and build tunnels to it, and ultimately they build complete nests. These adaptations have allowed them to rely even more on foraging for distant food sources, resulting in possibly even greater colony size.

More than any other social insects, some termites have reached great heights in adapting nest architecture for microclimate control. (One general complication in studying the relationship between nest form and temperature homeostasis is that temperature, humidity, and carbon dioxide are all simultaneously affected by nest design, and it is often impossible to tell, given current knowledge, which factors are primary and which vary concurrently.) However, most termite species exhibit no colony thermoregulation. Instead, as in ants, temperature fluctuations are buffered by nest placement and by the ability to move from one nest chamber to another as changing nest-substrate temperatures demand (Snyder, 1926; Skaife, 1955; Coaton, 1948; Hill, 1942). Soil temperatures near the surface may fluctuate hugely, but in colonies that have escaped to deep underground chambers or built large nests, relatively stable nursery temperatures can result from thermal inertia alone (Ebner, 1926; Cowles, 1930; Holdaway and Gay, 1948; Greaves, 1964; Josens, 1971). The effects of exposure to extreme temperature oscillations and low humidity are perhaps most obvious in many species of *Anacanthotermes*, *Psammotermes* and *Amitermes*, which sink deep subterranean nests in the Sahara Desert (Grassé, 1949). Alternately, like some ants in cooler climates, some termites select sunny sites or they live primarily under stones exposed to the sun (Lee and Wood, 1971).

Like ant mounds, the above-ground mounds of termites provide a warm, solar-heated nest space. These may also serve as a refuge from flooding. For example, *Cubiotermes glebae* in East Africa builds subterranean nests on the dry Masai plateau but switches to small (30 cm) mounds in wetter parts of its range. *Microcerotermes parvulus* likewise shows a transition from strictly underground nests in dry areas to mound nests in wetter areas (Bouillon, 1970).

Perhaps the most well-known example of termite mounds built to escape flooding are the wedge-shaped mounds, 3–4 m tall, of *Amitermes meridionalis* in generally swampy floodplains of northern Australia (Gay and Calaby, 1970). These termites are, in effect, denied underground thermal refuge while being subjected to intense solar heating, but their peculiar nest architecture serves specific thermal regulatory functions. A conspicuous feature of the *Amitermes* "compass" or "magnetic" termite mounds is their strongly wedge-shaped architecture with the axis of elongation in the approximately north-south direction (Fig. 16.23). Grigg and Underwood (1977) measured the orientation of 248 mounts of *A. laurensis* and found that all were aligned between 11° W of north and 30° E of north.

Fig. 16.23 Two views of the "compass mounds" of the Australian desert *Amitermes* termite. The design facilitates the maintenance of a high and stable internal temperature.

As had previously been supposed (Hill, 1942; Gay and Calaby, 1970), it was confirmed by Gordon Grigg (1973) of the University of Sydney, Australia, that the nest's shape and orientation function to minimize temperature fluctuations by reducing the interception of solar heating at midday. During the hot midday, only a narrow end of the mound intercepts the sun's rays. In northern Australia, the sun is somewhat north throughout the day, so that a flat termite mound in the east-west orientation would have one side facing the sun the entire day. Grigg (1973) experimentally created such a compass mound by sawing it off at its base and then pivoting it 90° to an east-west orientation. Before the mound was rotated its core temperature plateaued at near 30–35 °C each day between 10:00 A.M. and 5:30 P.M., but in the experimental mound temperature kept on rising steadily throughout the day to 40–42 °C, then again declined in the afternoon. In contrast, *Cubitermes* species living in areas of possible flooding with heavy rainfall, such as in rainforest, construct a mushroom-like cap on the top of the nest (Fig. 16.24) that protects the nest from rainfall (Emerson, 1938, 1956). The mounds of some species have a cap in wet habitats but are found without a cap when nesting in dry habitats. Therefore, the caps function as umbrellas and not sunshades.

How can the termites construct a mound such as the Australian compass termites do, in such a shape that it regulates its own temperature? One nest-building response that could in part result in the observed architecture is depositing nest material at the hottest spots, as has been observed in the laboratory in a Panamanian termite, *Nasutitermes corniger* (Stuart, 1977). Building in the hot spots would put in place a shield that protects the termites from the immediate heating effect of the sun's rays. (This would be a response similar to that of building in response to other

Fig. 16.24 The *Cubitermes* mound from lowland forests of West Africa is designed to shed rain.

noxious stimuli, such as light, which results in plugging up holes.) Thus, at noon, the colony would grow taller, and throughout the day it would keep being gradually elongated in the northerly direction as the animals "plug" the hot walls with moist building material. Alternately, perhaps termites orient their building activity to lines of magnetic force directly. Grigg, Jacklyn, and Taplin (1988) buried bar magnets in the sides of the bases of four new *Amitermes* colonies, and they placed nonmagnetized iron bars near four control colonies. Seven years later, in 1986, all of the four "treated" colonies had died, but all of the four colonies whose magnetic environment had not been altered still survived. (This result may be typical of biological research in general, although it is probably not illuminating of the specific question asked.)

Another architectural feature that influences the temperature in some mounds is a system of air ducts. Ventilatory systems may consist of well-like depressions or chimneys that dead-end deep in the mound and so cannot be used to create regular air currents (Grassé, 1944, 1949). Since the walls separating the chimneys from the nest galleries are thin and porous, however, they may nevertheless function primarily for ventilation in the deeper parts of the nest (Grassé and Noirot, 1958).

Mounds of *Macrotermes subhyalinus* on the Serengeti Plains of Tanzania may have a more sophisticated air-duct system (Weir, 1973). These mounds have pits near their base as well as funnels atop the mound (Fig. 16.25). The pits lead into tunnels that connect underground to form a network of ducts, which converge in a space under the nursery region of the nest. Air flows into the basal pits, passes up through the nursery, and then exits via the top funnels. Weir (1973) suggests that air is drawn through the nest by venturi forces generated by wind blowing over the rimmed mound-top funnels. Experimentally stoppering all the basal pits in one small mound resulted in an immediate 10°C temperature increase in the mound interior relative to nearby control mounds. But perhaps more important, within 48 hours of stoppering the mound, its walls became damp and soggy and started crumbling because of the increase of nest humidity.

As first elucidated by Martin Lüscher (1951, 1955, 1961), perhaps the ultimate in temperature homeostasis within a termite nest occurs in the immense nests of *Macrotermes bellicosus* (formerly *M. natalensis*). Continuous records from nests for several weeks showed nursery temperature maintained within 3°C of 30°C while air temperature outside the nest varied by 23°C (between 15 and 38°C) (Lüscher, 1961; Ruelle, 1964; Noirot, 1970).

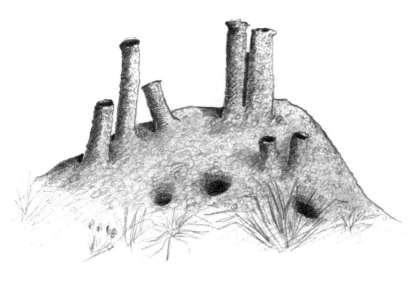

Fig. 16.25 A mound of the *Macrotermes subhyalinus* termite, from East Africa, showing chimneys and air-intake ducts for cooling.

The large (some 2 m tall) mounds of *M. bellicosus* are heavily fluted and have projecting ribs running down their sides (Fig. 16.26). These ribs enclose a system of 2–3-cm-diameter ventilatory tubes that interconnect air spaces above and below the fungus gardens (Sands, 1969; Martin and Martin, 1978) cultured on grass that the termites gather. A colony of 2 million individuals may consume some 240 liters of oxygen per day (Lüscher, 1955); this gas supply and the dissipation of the carbon dioxide produced could require more than passive diffusion. Air circulation begins as air is heated in the nest core by the metabolism of the millions of occupants and their fungus gardens (Holdaway and Gay, 1948; Ruelle, 1964; Lüscher, 1951; Geyer, 1951). The air then raises into the central chimney above the nest, thereby drawing up cool air from the nest basement. In the nest attic the air is distributed into a dozen or so radial canals of about 10 cm diameter, and then further channeled into the numerous small channels in the outer ribs mentioned above. Here, in the lateral ribs, gas and heat exchange take place with the external environment. Air then descends back into the cooled cellar, thus completing the circle through and around the nest.

The full circularity of the "thermosiphon" system of *M. bellicosus* nests was initially only inferred by Lüscher, but it has subsequently been confirmed experimentally and elaborated on by Ruelle (1964) and Loos (1964). Ruelle showed that complete circulation was possible by tracing a nest's air-duct system after pouring

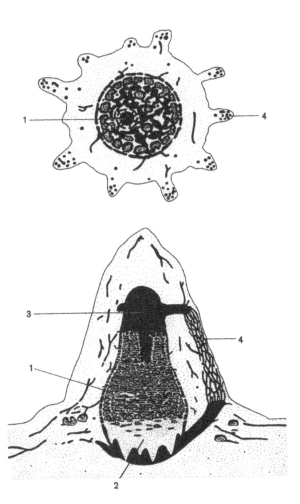

Fig. 16.26 A mound of *Macrotermes bellicosus* as seen from the top and the side. Both are partly transected to show the internal organization. The fungus gardens are dotted and the air spaces and air passages are indicated in black. (Adapted from Lüscher, 1961.)

400 kg of cement into a nest and then washing away its walls. Loos tested the thermosiphon idea directly by inserting a periscope-like microanemometer directly into a nest to measure air currents. As predicted, on cool days warmed air flowed upward from the cellar into the nest attic and downward in the peripheral galleries leading back into the nest cellar. However, in the afternoon at higher temperatures and with the outside walls of the termitarium being heated by the sun, the air currents were reversed and started flowing upward in the peripheral galleries. Loos (1964) also registered reversals of air currents in the nest's outer galleries in response to outside air gusts. It appears that the thermosiphon system operates sometimes, but not always. Air-flow

directions are not altered for nest-temperature control, but the system allows for automatic dissipation of heat from the interior, and it also permits nest warming when the equatorial sun strikes the sides of the steep mound during the early morning and evening. It is not known to what extent the stabilization of nest temperature at the core is due to the mere thermal inertia of the large nest, the effects of the air currents, or the possible aggregation of the termites within the nest (Holdaway and Gay, 1948; Greaves, 1964, 1967). At least in *Nasutitermes triodiae*, on very hot days the occupants apparently create temporary openings in the roofs of mounds (Lee and Wood, 1971) that allow hot air to escape. At low temperatures, an air space between the colony's armor-like covering and the inner nest in species such as *Cubitermes acinaciformis* (Lee and Wood, 1971) and *Macrotermes bellicosus* (Nye, 1955) serves as insulation. Several of the grass-storing species fill the outer galleries with loosely packed grass fragments (Lee and Wood, 1971), which could also serve as insulation as well as a buffer for the inconstancy of the food supply.

Summary

In each of the major groups of social insects there is a gradation from little or no control of nest temperature in some species to elaborate temperature control in others. In general, the degree of nest-temperature regulation is related to the total mass of colony occupants.

Most ants have little or no control over the temperature of their nests. No ants produce heat by shivering to increase nest temperature, although in some nests temperature is increased in part by the metabolism of the many occupants. In northern climates some species either build mounds or nest under flat rocks, both of which function as solar heat collectors. An ant brood is liberally and frequently moved around to exploit shifting temperatures within different portions of the nest as they occur because of solar heating and subsequent cooling. As a result, the brood is kept at a relatively specific range of temperatures.

Social wasps that have invaded northern climates exhibit both nest heating and cooling. *Polistes* wasps in cool environments build their open combs when they receive exposure to the sun, and they cool the nest evaporatively when overheating threatens by depositing water droplets onto the comb. Vespine wasps enclose their combs in layers of insulating paper and produce metabolic heat by shivering to heat the brood. Pupae are incubated directly by body contact. In addition, some *Vespa* species deposit water in the nest for evaporative cooling, as *Polistes* wasps do.

Bees have a tremendous range of sociality, and their thermoregulatory responses reflect that range. The primitively social bees, such as sweat bees (Halictinae), rely primarily on nest placement, capitalizing on the environment to achieve suitable nest temperatures. Fine-tuning of nest temperature is achieved by digging their nests in burrows to depths at which they find appropriate temperatures.

Bumblebees nest in and modify the nests of mice and other animals. Brood temperature is controlled by modifying the nest insulation and by incubating the brood by shivering. Eggs, larvae, and pupae are incubated by queens workers, and drones by direct body contact. In the initial colony-founding stage, the brood is heated primarily by direct contact with the queen's abdomen. But after the colony becomes large, the whole nest space becomes heated as a result of the activity and resting metabolism of all of its members. Direct shivering and incubation are no longer necessary at this stage except at unusually low temperatures.

In the "European" honeybees, *Apis mellifera,* as in other honeybees, colony reproduction is by fission, so that population size of a colony is always high. In these bees, brood "incubation" consists of shivering in the vicinity of the brood rather than heating it by direct body contact (conductive heating), as vespine wasps and bumblebees do.

Of the several species of honeybees, only *Apis mellifera* (several races) are found in northern climates. All honeybees (genus *Apis*) are incapable as individuals of withstanding freezing temperatures (unlike many other bees), and several responses have evolved in *Apis* that maintain elevated body temperature in the winter, even to temperatures near $-70\,^\circ$C. Colony-level responses that aid in thermal balance for winter include elaborate nest-site selection and large population size combined with huddling, shivering, and storage of large energy reserves accumulated during the summer months.

The tight clustering of tens of thousands of individuals in a swarm retains the heat produced by the resting metabolism of the warmed bees in the core of the cluster. The bees on the mantle (on the outside of the cluster) are more exposed to temperature fluctuations, but because of the inevitable higher temperatures inside of the cluster, the mantle bees automatically help regulate the cluster's core temperature by crowding inward as the external temperature is lowered. The heat retention that results communally then helps them to maintain T_{thx} within the range where shivering response is still possible, should ambient temperature decline still further. Conversely, if the mantle bees cluster too tightly (to keep warm) by crawling inward and blocking channels for heat loss, the core bees then become potentially overheated. Overheated core bees attempt to crawl to the periphery to reduce their T_b, thus automatically creating channels for heat loss from the core. The operation of these two simultaneous processes explains the data on metabolic rates and temperature profiles reasonably well.

At very low ($<0°C$) ambient temperatures or small cluster sizes, heat from resting metabolism alone is not enough. During initial cooling the bees crowd together and conserve body heat. Additionally, however, the effect of the crowd has a calming effect that reduces energy expenditure, making the animals more thermally passive; they tolerate T_{thx} below that normally required for flight and activity. In a swarm cluster with resting bees, most of the metabolism can be accounted for by passive or resting metabolism. As air temperatures decline below $0°C$, however, the bees on the outside of the cluster are cooled sufficiently to activate their shivering response, while the bees in the core continue to maintain high T_b by resting metabolism alone. If ambient temperatures continue to drop to $-20°C$ to $-80°C$, as may happen in winter, the outermost layer of bees eventually lose their ability to produce heat and are killed by the cold. (Cool bees that shiver produce little heat. Only warm bees produce much heat by shivering.) Heat production by shivering is then required by ever more centrally located bees. Ultimately, at the very lowest temperatures tolerated by the group, most of the heat is produced by shivering by the core bees as they fight to keep their T_b elevated. There are no absolute temperatures at which these shifts of behavior occur. They are a function of physical heat-flow dynamics, and hence they are determined by the size or mass of the bee clusters and the insulation around them, which varies in different situations— for example, there is no insulation in swarms, but empty combs in the winter hive insulate the core.

Although honeybees, unlike most other insects, regulate their T_b to resist individual cold torpor (near $10°C$), they regulate much higher temperatures (near $36°C$) in the presence of brood. Brood production usually starts in late winter, when stored food reserves are used; thus, some of the highest nest temperatures may be maintained at the coldest part of the year, down to near $-40\,°C$.

Honeybees prevent overheating in their nests by a combination of fanning and evaporation of water. Water carrying and distribution in the nest involves a feedback system using both the communication system normally used for foraging for nectar and an elaboration of the systems used both to regulate individual T_{thx} at high ambient temperature and to make honey out of the nectar. Nest fanning also involves a social response that is possibly an evolved elaboration originally used for individual cooling, inasmuch as the fanners align themselves in preexisting air flows where they can be cooled by convection. As a result, unidirectional flow through the hive (as opposed to mere air stirring) results.

Strictly tropical honeybees nest in open combs and they experience little variation in ambient temperature. At least one species that lives at high elevations in the Himalayan Mountains, however, regularly encounters cool temperatures in the winter. These bees have met the challenge of seasonality in that their colonies annually migrate en masse between lower and higher elevations.

Termites also show a range of control over nest temperature. Only those species with the largest mounds show nest-temperature regulation beyond that of microhabitat choice and temperature choice within the nest. Climate regulation is largely a function of nest morphology. Mounds are constructed as both solar heaters and as sun minimizers; temperature control and gas exchange are achieved in some mounds by circulatory ducts. Elaborate ducts in mounds of the African termite *Macrotermes bellicosus* serve as a thermosiphon system in which heat generated by the central fungus gardens and nest inhabitants rises to the top and is dissipated through external cooling fins.

Remaining Problems

1. What are the cost-benefit balances of increased nest temperatures, in terms of the decreased longevity of the nest occupants vs. the increased growth and production of offspring?
2. Do ants utilize climate differences in the nest for functions (such as regulating longevity) besides brood incubation?
3. To what extent does clustering of ants in large mound-building colonies (such as *Formica rufa*) contribute to nest-temperature regulation?

4. Do the African driver ants have temperature control equivalent to that of New World army ants?

5. Do any social insects determine the temperature of their brood as such in order to regulate its temperature, or are they merely attracted to the brood and then regulate the brood's temperature secondarily by regulating their own T_b near the brood?

6. Is the insulation of an expanding (growing) wasp colony varied? If so, what are the proximate cues and responses?

7. Do some wasp larvae generate their own heat?

8. Do any stingless bees (Meliponini) regulate nest temperature at low air temperatures? If so, how?

9. Do fanning bees preferentially align themselves in cool or in hot air? Which bees fan, and why? How is fanning related to T_b?

10. What are the individual responses of termites that together result in the construction of simple tunnels in some species and to the erection of elaborate "castles" with thermosiphon systems in others?

Summary

D EPENDING on different points of view, the class Insecta is divided into some 25 to 35 orders. Thermoregulation has so far been examined in only a few families of some of these orders. From these studies, however, a set of principles has emerged, noting both similarities and distinctions among the various insects, concerning the evolution of mechanisms and the ecology of thermoregulation. The insects show us that endothermy is usually not an advantage that some species may acquire and others do without. Instead, it is something that cannot be avoided, given a certain size and life-style. The sometimes intricate mechanisms of thermoregulation that have evolved along with endothermy are less easily thought of as proof of evolutionary advancement than as necessary adjustments that are apparently easy to acquire. Much progress has been made over the last 20 years in elucidating the mechanisms of insect thermoregulation, but the literature that has accumulated in this time is plagued by numerous misconceptions. This book is a synthesis of the physiology, behavior, ecology, and evolution of insect thermoregulation written with the aim of exposing controversies, confusions, and questions. It is not an unbiased review of all that has appeared in print on the subject.

Energetics

The biochemical rates of all processes underlying activity in insects, as in all living things, are strongly dependent on temperature. A 10°C increase in temperature (in the physiologically relevant range for the species) generally leads to a doubling of biochemical rates. This temperature dependence is a two-edged sword. When an animal is inactive, a low body temperature is highly desirable because it conserves energy resources. When the animal needs to accomplish certain tasks that place a premium on high rates of performance, however, then a high body temperature can be a great advantage. If flight performance is limited by temperature (and minimum flight temperature is narrowly defined in most species), then it becomes a necessity.

Because of their small size, insects enjoy a tremendous advantage over many other animals in their ability to cool quickly and to occupy cool microenvironments where they can conserve energy resources. Also because of their small size they can exploit the other edge of the "sword"—they can utilize microclimatological differences to heat up quickly. For example, by crawling from the shade under a leaf into the sunshine, an insect can heat up virtually "instantly" and take flight. Physiological warm-up, too, is rapid because of small size (although this trend is obscured or superseded at very small body mass by the fact that the smallest insects don't warm up at all endothermically).

Flight and Endothermy

Insect flight is metabolically the most demanding activity known. It unavoidably results in rates of heat production that are often over a thousand times greater than during rest at low body temperature. Endothermy (as such) in flight is not an evolutionary adaptation. Instead, it is an unavoidable biophysical consequence of the relationship between body size and sustained high rates of energy expenditure. Nevertheless, in the proximate sense an elevated thoracic temperature (T_{thx}) is now necessary for some insects to generate sufficient power for takeoff, and this necessity is largely an evolutionary consequence of the muscles having become adapted to operate at the high temperatures they normally generate during flight.

In general, endothermy in flight is necessarily closely correlated with body mass. As has been systematically demonstrated in flies, beetles, bees, and moths, the smallest insects fly with a T_{thx} close to air temperature, whereas the largest generate thoracic-temperature excesses of 20–35 °C above air temperature.

The energy expenditure of flight can be supported over a wide range of muscle temperatures in different species, inasmuch as many insects fly with a T_{thx} well below 15 °C. Regardless of phylogenetic origin, each species has muscles adapted to operate within ranges, often narrow, of temperature, and the high muscle temperatures normally generated by some of the larger insects are far above the optima for some of the smaller ones. In general, the larger the insect, the higher and also more narrow is its evolved range of T_{thx} suitable for flight.

Within any one species, the maximal work output that the muscles can now perform is directly related to their temperature. The specific temperatures required vary not only among different species, however, but also sometimes among the different muscles of the same species. For example, the muscles used for both singing and flight in some male katydids have much higher temperature optima than those used only for flight. Furthermore, the females, which do not sing, have lower temperature optima for the same muscles. In contrast, the temperature optima of the flight muscles in bumblebee females, whose body mass (and cooling rates) can vary over an order of magnitude, are independent of size; as a consequence, the small bees (because they cool rapidly) are restricted to flying in warm weather. Apparently evolution, and not developmental or other proximate factors, has tailor-made the different species-specific, and even sex-specific, temperature optima for the specific muscle temperatures encountered by the insects.

Endothermy is not only an immediate consequence of flight but it is now also an evolved necessity for it. Not surprisingly, various features of insects have evolved that promote endothermy. First, thoracic endothermy is enhanced by air sacs surrounding the muscles in the thorax (found principally in dragonflies), and air sacs also insulate the thorax from the abdomen (in most insects). Second, counter-current heat exchangers of the circulatory system act to retard heat flow to head and abdomen (in some moths and bees). Finally, a variety of insects (many bees, moths, and some beetles and flies) also have insulating pile on the thorax. The thoracic "scales" of moths have become long and hair-like; the insulation more than halves the rate of convective heat loss in some species. In contrast, the fuzziness of bumblebees is due to dense setae, which have a different evolutionary origin, but this insulation is as effective as that of moths. Except in those few cases where flying insects have extended flight activity to the wintertime (some winter moths fly near $0°C$ with T_{thx} near $30°$ C), however, those insects that have special features that retard heat loss also have mechanisms to accelerate it. This allows activity over a broad range of ambient temperatures, since those insects which operate at a high T_{thx} also necessarily operate in a narrow T_{thx} range. Either T_b is physiologically controlled to allow continuous activity, or activity itself controls T_b.

In butterflies, T_b is apparently not physiologically controlled in flight. Instead, it is behaviorally regulated in the interflight intervals, but mass determines subsequent cooling rate in flight.

Body mass is, then, of critical importance in determining activity patterns. Large-bodied butterflies when they fly tend to be "continuous" flyers; that is, they may fly several minutes without stopping. Small-bodied butterflies from temperate climates that have adapted to a high T_b often have only a limited flight range (seconds) because they cool rapidly by convection when in flight. In them T_b fluctuate from a temperature generally well above the minimum for flight at the initiation of a flight to a much lower temperature at the end of the flight. Since the distance or range per flight is temperature dependent, individuals thermoregulate by basking and making intermittent short flights at low ambient temperatures, and by basking less or not at all but making long flights at high temperatures. The greater the body mass, the more continuously long flights are possible without the intervening stops for basking. Some of the large, nocturnal, well-insulated moths are the epitome of this strategy; they can fly continuously by relying solely on their internally generated heat to maintain an elevated (often higher than 45°C) thoracic temperature.

Physiological Mechanisms of Temperature Stabilization in Flight

Many large insects have the potential problem of overheating from their flight metabolism after a minute or two of flight. Since they have little or no option of reducing energy expenditure if they are not to stop flying, they must deal with a mounting temperature excess if they fly at temperatures slightly higher than the minimum for flight. Several solutions to this problem have evolved. One solution is the tolerance of an even higher muscle temperature (up to 45–46°C in some bees and some moths). Second, in some insects with very large wing areas relative to body mass (some dragonflies and some butterflies), it is possible (evidence is skimpy) that heat production is reduced by increasing the incidence of soaring. In sharp contrast, in those insects with small wings and hence very rapid wing beats in flight (some bees) where heat production is *independent* of flight speed, increasing flight speed (at high air temperatures or at increasing T_{thx}) can automatically result in T_{thx} stabilization because it accelerates convective cooling. Fourth, physiological mechanisms of heat loss are used to reduce temperature excess.

The largest and most active flyers of several groups of insects (moths, bees, true flies, dragonflies) help stabilize T_{thx} during continuous free flight by shunting excess heat to the abdomen by way of the blood. The abdomen, in contrast to the thorax, is lightly insulated, so it serves as an effective window for convective heat loss. Since the cooling mechanism ultimately relies on convection from a thermal window, it is only effective when a relatively large temperature gradient is generated between T_{thx} and air temperature. Head temperature varies passively with T_{thx} (in most moths) or it may be regulated (in some bees), in which case the head secondarily acts to stabilize T_{thx}.

Several insects use modest amounts of evaporative cooling. For example, blood-sucking tse-tse flies open their spiracles and cool evaporatively when they are subjected to intense heating while perched in direct sunshine to feed. Desert cicadas have body pores that exude evaporating fluid ultimately derived from the plentifully available water from the xylem of plants that they feed from. But only one insect, the honeybee, has been demonstrated to be able to fly at air temperatures higher than 40°C because of an effective evaporative cooling mechanism that dissipates all of the large amounts of metabolic heat generated by its flight metabolism. Honeybees regurgitate fluid from the honey-crop, and this fluid cools the head (and helps convert the nectar to honey) and then, by accelerating heat transfer toward the head, the thorax.

Very modest evolutionary changes, such as a slight morphological change in the shape of the loop of the dorsal vessel, a change in the length of the setae or scales, or a behavioral alteration in orientation to sunlight, may have drastic consequences for the T_b experienced. In contrast, nearly every biochemical constituent of the body would have to be altered in order for an insect to evolve a different optimal T_b for activity. It is therefore likely that T_b would vary relatively little across ecological or geographical ranges within species, whereas the factors that might help to maintain the same T_b despite environmental changes should be the prime targets of evolution. The data from insects strongly support this supposition. Thermoregulatory mechanisms of social insect colonies are also conservative. For example, the pre-flight shivering response has been used as a pre-adaptation in all flying social insects as their mechanism for maintaining a high nest temperature (and/or individual high T_b) at low ambient temperatures.

Similar conservatism is seen in heat-loss mechanisms. Bumblebees use the abdomen as a heat radiator during flight to prevent overheating of the thorax. And they use the same mechanism for heat "loss" to incubate their brood comb (and possibly to incubate their ovaries to accelerate egg production) via the abdomen. Honeybees make honey by regurgitating nectar and evaporating water. The same mechanism used in flight to prevent the body from overheating at high temperatures is also used within the colony, where it acts to prevent the colony from overheating under heat stress. Many of the responses of social clumping and dispersion (that help in colony thermoregulation) are in part the responses of individual animals seeking higher or lower ambient temperatures to help regulate their own T_b.

Heat production from flight metabolism can be a limiting factor on continuous flight, provided the insect exceeds certain size limits. In general, insects weighing more than 0.2 g have physiological mechanisms of heat dissipation that are activated even at relatively modest ambient temperatures (<25°C) during continuous flight. Because of the physiological constraints, it can be predicted that very large (>2–3 g), lowland tropical insects would have to make intermittent stops (rather than fly continuously), fly at the cooler parts of the year or the night, and have well-developed cooling mechanisms.

Pre-Flight Warm-Up

As anyone knows who has ever tried to swat a housefly or a mosquito, many (or most) insects do not need to spend time or energy in pre-flight warm-up. The endothermic ones (the large-bodied flyers), however, need to increase the temperature of their thoracic muscles above the average prevailing ambient temperatures before they can generate sufficient power and lift for takeoff. Such warm-up is achieved by basking, shivering, or by both basking and shivering. Basking is almost universal in diurnal insects, and it has been extensively described for butterflies, grasshoppers, and some beetles, flies, cicadas, and some species of dragonflies.

Basking differs from mere perching in sunshine in that the insect assumes different body or wing positions when orienting itself to the solar radiation. First, in all species the body is oriented at right angles to the incident solar radiation for basking, thus maximizing the amount of body surface area exposed. The first priority in basking is to maximize body exposure. The second consideration is to conserve as much as possible the heat that is gained by this exposure. Convective heat loss in small bodies can be extremely rapid even in gently moving air, and both basking butterflies and basking dragonflies use their wings as convection baffles that reduce convective heat loss during basking. In some butterflies and in dragonfly baskers, the laterally extended wings are depressed around the body; the wings trap warm air against the body, which reduces convection from the ventral and lateral body surfaces. In other butterflies (principally those basking above rather than directly on a warm substrate), the wings are instead partially raised and provide convection baffles over the *dorsal* rather than over the ventral body surface.

Most butterflies are highly conservative in their warm-up mechanisms. They usually have only one type of basking posture. Many that bask never shiver (although some that shiver also bask). Still others neither shiver nor bask. The variety of responses in the different species shows that warm-up has independently evolved numerous times, even within the same taxonomic family. Multiple independent evolution of warm-up within the sphinx moths is evident from an electrophysiological analysis of the details of the muscle activation during shivering.

Shivering warm-up is probably found in most nocturnal lepidopterans larger than about 100 mg. It is a crude form of flight-like behavior, or a behavior evolutionarily derived from flight, in which wing movement is minimized by activating the upstroke and downstroke muscles synchronously rather than alternately. Once slow flight is initiated, however, the flight itself can act as means of further warm-up. Indeed, relatively cool insects that are experimentally suspended will initiate wing flapping almost indistinguishable from normal flight, and they then gradually increase their wing-beat frequency as they warm up from their flight metabolism.

In all insects so far examined, wing-beat frequency is a direct function of muscle temperature, and during shivering warm-up the wing-beat frequency also increases directly with T_{thx} until that appropriate for liftoff is achieved.

In the "higher" insects, the Diptera, Coleoptera, and Hymenoptera, shivering warm-up has become most advanced. In most of these forms (but not all) the wings can be fully uncoupled from the wing muscles so that there are no detectable external wing vibrations except for minute movements of the scutellum. In some Diptera the wings may be seldom or never totally uncoupled and a buzz may then be heard during warm-up (in tse-tse flies). In some syrphids and in bees, however, coupling (and buzzing) is facultative and shivering is only sometimes accompanied by audible signals. No beetles have been reported to buzz during warm-up, but mechanical probes placed into the thorax vibrate, thus showing that the muscles are mechanically active during warm-up. During warm-up (but not in flight) in myogenic flyers the flight muscles act in a neurogenic mode, showing twitches associated with individual action potentials. For the most part, however, the flight muscles of bees (and presumably other myogenic flyers) are in near tetanus during shivering. The pre-flight warm-up mechanism shows a close physiological resemblance to the pre-jump preparation of small poikilothermic insects.

Pre-flight warm-up is primarily a function of body mass and ambient temperature. In general, it is most pronounced in large insects at low temperatures. However, it is sometimes also observed in relatively small (<10 mg) insects provided temperatures are low enough, and it may also be required at relatively high temperatures provided the insect is large enough. Thus, in the lowland tropics almost all except the largest insects can be expected to be able to initiate at least clumsy flight without warming up.

The Function of Thermoregulation

Perhaps more obviously for insects than for other organisms, temperature is time. On a millisecond scale each degree difference in T_b has a direct effect on the rate of muscle contraction. The rate at which a flight muscle contracts immediately affects wing-beat frequency, which translates into power in flight. In turn, small differences in T_{thx} affect the rate of foraging and the outcome of contests for mates and for food resources. It affects flight speed and the ability to outmaneuver rivals and predators.

On a larger time scale, an insect's attainable muscle temperature affects whether or not flight is possible at all. And since in most adult insects foraging, mating, oviposition, and dispersal all depend on flight, one hardly needs to elaborate on the obvious consequences of an insufficiently high T_{thx}. Heat "buys" time for all of these activities, none of which are possible without it. Finally, temperature is time especially for social insects whose nests serve as incubators for the young, whose growth rate and development are strictly temperature dependent. Short development times are necessary in seasonal environments if the social life cycle is to fit within the seasonal time constraint. But what are the specific temperatures "chosen," and why?

The many flying species of large body mass, and nonflying species from *hot* environments, all regulate thoracic temperature near the same common ceiling of 40–45°C. This temperature range is near the upper lethal temperature for most endothermic organisms, and it is presumably a consequence of biochemic constraints that apply to all complex multicellular organisms that are not under severe hydrostatic pressure.

Except for those insects experiencing the apparent upper ceiling of near 45°C, there is an unmistakable pattern in thermal optima among species: temperature optima of different species are closely related to those temperatures the animals most commonly experience in nature *when they are active*, when the premium is on performance. Thus, in flying adult insects, the temperature experienced is largely dictated by the rates of heat production unavoidably generated in flight, and the rates of heat loss determined by body size and insulation. It has been concluded in innumerable publications of insect thermoregulation that a higher body temperature enables the insect to accomplish more than a lower body temperature will. This conclusion is justified, proximally. But it does not imply that the *evolution* of regulation of a high T_b was necessary to accomplish the advantages now derived.

In order to try to answer the much more complex question of why insects have *evolved* to regulate a high T_b, it is useful to distinguish between those insects that are small- and large-bodied, and between those that fly and those that do not fly. Very small-bodied insects fly with T_{thx} close to air temperature, and in them T_{thx} is almost a direct function of ambient thermal conditions. In several diverse taxa (Diptera, Coleoptera, Lepidoptera, Hymenoptera, Homoptera, and others), it is possible for individuals to fly with very low (<5°C) muscle temperature. This raises a question: If flight muscles can evolve to operate at muscle temperatures near 0°C, then why should some insects (of the same taxa) evolve the very costly (in terms of eneregy and time) requirement of warming up to over 40°C and the mechanisms of thermoregulation in flight?

Like other animals, insects are presumably under selective pressure to be able to fly over a broad range of ambient temperatures. Assuming that biochemical limitations evolutionarily constrain high rates of activity to the range of T_b from near 0 °C to 40 °C, then a small insect automatically experiences a potentially favorable T_{thx} at most temperatures encountered. However, a large insect may be heated to temperatures of 25 °C or more above air temperature within a minute of flight; it may approach lethal T_b within a minute of starting flight in the same way that a flightless insect adapted to a body temperature of 20 °C would be stressed, or killed, by venturing onto a sand surface at 40 °C. Thus, vigorously flying insects are under strong selective pressure to evolve mechanisms of shunting excess heat away from their working flight muscles.

What have been the selective pressures operating on the temperature constraints of the biochemical machinery? I presume that in insects that fly there have been three: to operate at high or maximal rates, to continue flight for sustained durations, and to have T_b independent of temperature fluctuations in the environment. The three have mutual incompatibilities. For example, if rates are made independent of temperature by having three copies of the biochemical machinery (the isoenzymes, for example) each adapted to a different temperature, then temperature independence could theoretically be achieved as these isoenzymes turn on and off throughout the temperature range. But if *maximization* were required, then it would be better to have the three presumptive sets of machinery combining forces to operate at the same temperature (in other words, to have no isoenzymes), and to use physiological and/or behavioral mechanisms to maintain the cellular environment at that specific temperature selected where all of the available machinery can operate at the *same* time. Undoubtedly, the greater the ease of controlling body temperature, and the greater the need for very high rates of activity, the less the likelihood of finding broad temperature optima. And as a matter of observation, the temperature requirements during the insect's maximum metabolic demands—flight—are very narrow, suggesting that flight activity has selected for *maximization* of rates of muscle performance.

The temperature specificity is exacerbated in flying insects because of a mechanical constraint; a specific minimum (rather high) metabolic demand is required to become and remain airborne. A walking insect, in contrast, can utilize the low end of its biochemical activity curve—it can simply walk slower as it cools, limited by its biochemical capabilities but little or not at all by a mechanical constraint to be supported above the substrate.

Aside from the likely selection for maximization of metabolic output over a relatively narrow temperature range, there is likely also selection to keep going once in flight. There is little point to evolving the capability for powered flight if it can be sustained only for a few seconds at a time. This seemingly trivial point, however, is extremely important in differentiating between the proximate reason for regulating a high T_{thx} and the reason for evolving thermoregulation at a high T_{thx} with all the costs this entails. The problem can be illustrated with a hypothetical example.

Suppose a bumblebee existed which miraculously at one stroke had each of the thousands of units of its biochemical machinery changed to operate optimally at 20°C (as millions of other insects do) rather than at near 40°C (as in other bumblebees). When this hypothetical mutant regulates its T_{thx} by shivering at 0°C, it now pays only half the metabolic cost of warming up and of keeping warmed up when not in flight. Furthermore, at 20°C the bee can now take off "instantly" without any warm-up at all, and it would now be operating at maximal efficiency and speed without any investment of time or energy, *immediately* upon takeoff.

So far so good. But after 30 seconds of flight this "cold" fast bee will no longer be cold. It will have heated up to near 30°C, some 10°C above its thermal optimum. It will lose power or have to stop flight to cool off. If it somehow still keeps on flying, however, it will quickly approach the thermal death point. (If the bee is simultaneously small, say no bigger than a housefly, it could keep right on flying because it would never heat up.) Eventually a T_{thx} optimum would evolve that is a compromise between the biochemical constraints, the need for specific body size, rates of activity, endurance, and the temperatures normally encountered when active.

A similar logic of optimum T_{thx} applies to insects that never fly but that are regularly subjected to high T_b because of their thermal environment (rather than by endothermy). But in them the T_b regulated should be relatively independent of body mass and/or metabolic rate and instead be more closely related to operative temperatures in the environment (Operative temperatures have unfortunately generally referred only to *inactive* insects. But to gain a broader, more unifying perspective, they should refer to the insects' experience while active.)

There is an almost infinite variety of solutions for thermal balance, including temporal and geographical patterning of activity. Presumably only minor factors and minor mutations—sometimes no more than the behavior to orient the wings in a certain way—can change body temperature over tens of degrees. However, a very large number of mutations are required to change all of the proteins and all of the rest of the biochemical machinery if temperature adaptation is from within, via biochemistry. Hence, as a result of the various behavioral and physiological options, there will likely be great conservatism in body temperature, which probably explains the minor difference in T_{thx} optima of the endothermic insects as a function of geographical distribution.

The number of different "solutions" to thermal balance in any one species is extremely large. In reality, however, we usually treat a large number of features as relatively fixed (for example, type of habitat, body mass, geographical range) or invariant (for example, body posture, the shape of a blood vessel, the amount of fur, the diurnal activity time, flight duration). Yet all are probably modifiable, to a certain degree. And a given thermal shift may change a few, or all, features to varying degrees through evolutionary time, depending on the lines of least resistance. Thus, it is difficult to think in terms of *an* adaptation to temperature in insects. The animal's ecology, life cycle, morphology, physiology, and behavior are all intertwined in an integrated response. These responses include fever and rare cases of mate attraction (singing) that are possible only at high T_b (some katydids). They include predator-avoidance strategies that appear to be contests in tolerating the highest ambient (and body) temperatures (some ants, beetles, grasshoppers, and cicadas); one case, for example, has resulted in the evolution of a unique evaporative cooling mechanism in an insect that can tap water where no water exists for others.

One is led to conclude that insects, which until recently were considered primitive "poikilotherms," may be the paragon thermoregulators of the animal kingdom. As a group their sophistication in dealing with temperature as a variable in the environment is unmatched. And we still have much to learn from them.

Future Research

Research in thermoregulation in insects has already started to reach the stage of redundancy. Studies are being published that corroborate what is already well known for other species (and sometimes even the same species) but that do not report any new discovery. A few years ago a random examination of the thermal biology of almost any insect may have been the ideal strategy to root out the unanticipated and thus the most interesting, but the probability of random searching turning up new discoveries is now greatly diminished.

In order to provide some direction and to reduce duplication, we must explicitly consider what studies probably need not be done and what research may yield new insights. Obviously such judgments will be perceived as bold assertions in the face of the unpredictability of science. There are still likely to be surprises, and these surprises, because they defy predictions, will still be most interesting. Therefore, the prognosis I offer here is meant as a personal one. It describes what would guide my own research, and what I would advise graduate students if they asked.

Some of the studies that are probably now redundant include those whose conclusions are that (1) another insect species shivers, basks, does neither, or any combination thereof; (2) small insects fly at lower or higher air temperatures than large insects; (3) large insects show better thermoregulation than small ones; (4) a high T_{thx} results in faster flight; (5) color affects body temperature; (6) body temperature affects flight activity and vice versa; (7) body temperature affects mating success, foraging rate, or predator avoidance; (8) there is a tradeoff between maintenance of high vs. low body temperatures; (9) sunshine and/or convection affects T_b; (10) energy- and heat-balance budgets affect behavior.

This list covers only very general topics because each item reflects a broad range of studies that have already been done. My "wish list" of questions that are not yet answered tends to be more specific, as they concern immediately doable research projects that are not aimed at answering ultimate questions (in keeping with my view that the aim of research is to make discoveries). Ultimately, unifying concepts will emerge not because of a plan but because discoveries supported by empirical results will create a pattern.

The following questions are, in a sense, all potential thesis problems that I would consider if I were searching for a topic now. (1) Can some of the world's largest endotherms, the beetles that weigh over 20 g, regulate their T_{thx} in continuous flight, and if so, by what mechanism(s)? (2) What is the relationship between T_b, viability of pathogenic organisms, and behavior and physiology? This is still a very large and open topic that needs to be explored in a variety of different kinds of adult and immature insects before patterns will become apparent. (3) What are the precise jumping mechanisms of any of the small-bodied jumpers that can be active down to 0°C or below? The basic jump mechanisms have been elucidated. Now it's time to focus on some specifics. For example, what are the catch mechanisms for energy storage? How are they released? (4) What is the explanation for the enzyme profiles of some insects (principally bumblebees) that superficially suggest the presence of nonshivering thermogenesis? Are they, instead, related to the promotion of *low* body temperature (as discussed in the text)? (5) What allows different muscles to maintain the same activity rates at different temperatures? In other words, what features allow muscles to have different temperature optima? (6) Are there temperature "setpoints" in the body in terms of neurological networks with controls and feedback mechanisms? (7) How much blood is circulated during preflight warm-up and during flight at low and high T_a? Is blood flow shut off during warm-up? (8) Do very large tropical insects have special heat-dissipation mechanisms? (9) Do African dung-ball-rolling beetles adjust their T_b to affect the speed of ball making and ball rolling in response to competition from other beetles? (10) Do the large *Pepsis* wasps with their narrow petiole (waist) regulate their T_b, and if so does their narrow waist play a role in their thermal adaptations?

This list is by no means meant to be exhaustive. Perhaps the numerous studies that have been explored in this text will have raised many other questions. I hope they will be sufficient to stimulate at least some readers to start a publishable research project that will contribute to a developing field.

References

Prologue

Alahiotus, S. N. 1983. Heat shock proteins. A new view on the temperature compensation. *Comp. Biochem. Physiol.* 75B:379–387.

Bartholomew, G. A. 1981. A matter of size: An examination of endothermy in insects and terrestrial vertebrates. In *Insect Thermoregulation,* ed. B. Heinrich. New York: Wiley.

Bennett, A. F., K. M. Dao, and R. E. Lenski. 1990. Rapid evolution in response to high-temperature selection. *Nature* 346:79–81.

Casey, T. M. 1981. Insect flight energetics. In *Locomotion and Energetics of Arthropods,* ed. C. F. Herreid II and C. R. Fourtner. New York: Plenum.

——— 1988. Thermoregulation and heat exchange. *Adv. Ins. Physiol.* 20:119–146.

Cloudsley-Thomson, J. L. 1970. Terrestrial invertebrates. In *Comparative Physiology of Thermoregulation,* vol. 1, *Invertebrates and Nonmammalian Vertebrates,* ed. G. C. Whittow, pp. 15–77. New York: Academic Press.

Cossins, A. R., and K. Bowler. 1987. *Temperature Biology of Animals.* New York and London: Chapman and Hall.

Gates, D. M. 1980. *Biophysical Ecology.* New York, Heidelberg, Berlin: Springer-Verlag.

Gates, D. M., and R. B. Schmerl, eds. 1975. *Perspectives of Biophysical Ecology.* New York, Heidelberg, Berlin: Springer-Verlag.

Heath, J. E. 1964. Reptilian thermoregulation: Evaluation of field studies. Science 146:784–785.

Heath, J. E., and M. S. Heath. 1982. Energetics of locomotion in endothermic insects. *Ann. Rev. Physiol.* 44:133–143.

Heinrich, B. 1974. Thermoregulation in endothermic insects. *Science* 185:747–756.

——— 1975. Thermoregulation in bumblebees: II. Energetics of warm-up and free flight. *J. Comp. Physiol.* 96:155–166.

——— 1979. *Bumblebee Economics.* Cambridge, Mass.: Harvard University Press. 245 pp.

——— ed. 1981. *Insect Thermoregulation.* New York: Wiley.

Heinrich, B., and C. Pantle. 1975. Thermoregulation in small flies (*Syrphus* sp.): Basking and shivering. *J. Exp. Biol.* 62:599–610.

Hochachka, P. W., and G. N. Somero. 1984. *Biochemical Adaptation.* Princeton, N.J.: Princeton University Press.

Huey, R. B., E. R. Pianka, and J. A. Hoffman. 1977. Seasonal variation in thermoregulatory behavior and body temperature of diurnal Kalahari lizards. *Ecology* 58:1066–1075.

Huey, R. B., W. D. Crill, J. G. Kingsolver, and K. E. Weber. 1991. A rapid method for measuring heat or cold resistance of *Drosophila* or other small arthropods. Unpublished manuscript.

Kammer, A. E., and B. Heinrich. 1978. Insect flight metabolism. *Adv. Insect Physiol.* 13:133–228.

Kimura, M. T. 1988. Adaptations to temperate climates and evolution of overwintering strategies in the *Drosophila melanogaster* species group. *Evolution* 42:1288–1297.

Lee, R. E., Jr., and D. L. Denlinger. 1991. *Insects at Low Temperature.* New York and London: Chapman and Hall.

May, M. L. 1979. Insect Thermoregulation. *Ann. Rev. Entomol.* 24:313–349.

———— 1985. Thermoregulation. In *Comprehensive Insect Physiology, Chemistry and Pharmacology,* vol. 4, ed. G. A. Kerkut and G. I. Gilbert, pp. 491–552. Oxford: Pergamon Press.

Parry, D. A. 1951. Factors determining the temperature of terrestrial arthropods in sunlight. *J. Exp. Biol.* 28:445–462.

Parsons, P. A. 1978. Boundary conditions for *Drosophila* resource utilization in temperate regions, especially at low temperatures. *Am. Nat.* 112:1063–1074.

Porter, W. P., and D. M. Gates. 1969. Thermodynamic equilibrium of animals with environment. *Ecol. Monogr.* 39:96–103.

Stone, B., and P. G. Willmer. 1989. Endothermy and temperature regulation in bees: A critique of "grab and stab" measurement of body temperature. *J. Exp. Biol.* 143:211–223.

White, E. B., P. deBach, and M. K. Garber. 1970. Artificial selection for genetic adaptation to temperature extremes in *Aphytis lingnanensis* Compere (Hymenoptera in Aphelinidae). *Hildegardia* 40:161–192.

Wieser, W. 1973. *Effects of Temperature on Ectothermic Organisms.* New York, Heidelberg, Berlin: Springer-Verlag.

Willmer, P. G. 1982. Microclimate and the environmental physiology of insects. *Adv. Insect Physiol.* 16:1–57.

1. Night-Flying Moths

Adams, P. A. 1969. How moths keep warm. *Discovery* 4:83–88.

Adams, P. A., and J. E. Heath. 1964a. An evaporative cooling mechanism in *Pholus achemon* (Sphingidae). *J. Res. Lepid.* 3:69–72.

———— 1964b. Temperature regulation in the sphinx moth, *Celerio lineata.* *Nature* 201:20–22.

Bachmetjev, P. 1899. Über die Temperatur der Insekten nach Beobachtungen in Bulgarien. *Z. Wiss. Zool.* 66:521–604.

Bartholomew, G. A., and T. M. Casey. 1973. Effects of ambient temperature on warm-up in the moth *Hyalophora cecropia*. *J. Exp. Biol.* 58:503–507.

——— 1978. Oxygen consumption of moths during rest, pre-flight warm-up, and flight in relation to body size and wing morphology. *J. Exp. Biol.* 76:11–25.

Bartholomew, G. A., and R. J. Epting. 1975a. Allometry of post-flight cooling rates in moths: A comparison with veretebrate homeotherms. *J. Exp. Biol.* 63:603–613.

——— 1975b. Rates of post-flight cooling in sphinx moths. In *Perspectives of Biophysical Ecology*, ed. D. M. Gates and R. B. Schmerl, pp. 405–415. New York: Springer-Verlag.

Bartholomew, G. A., and B. H. Heinrich. 1973. A field study of flight temperatures in moths in relation to body weight and wing loading. *J. Exp. Biol.* 58:123–135.

Bartholomew, G. A., D. Vleck, and C. M. Vleck. 1981. Instantaneous measurements of oxygen consumption during pre-flight warm-up and post-flight cooling in sphingid and saturniid moths. *J. Exp. Biol.* 90:17–32.

Bennett, A. F. 1985. Temperature and muscle. *J. Exp. Biol.* 115:333–344.

Brocher, F. 1920. Etude expérimentale sur le fonctionnement du vaisseau dorsal et sur la circulation du sang chez les insectes. *Arch. Zool. Exper. et Gen.* 60:1–45.

Casey, T. M. 1976a. Flight energetics of sphinx moths: Power output during hovering flight. *J. Exp. Biol.* 64:529–543.

——— 1976b. Flight energetics of sphinx moths: Heat production and heat loss in *Hyles lineata* during free flight. *J. Exp. Biol.* 64:545–560.

——— 1980. Flight energetics and heat exchange of gypsy moths in relation to air temperature. *J. Exp. Biol.* 88:133–145.

——— 1981a. Energetics and thermoregulation of *Malacosoma americanum* (Lepidoptera: Lasiocampidae) during hovering flight. *Physiol. Zool.* 54:362–371.

——— 1981b. A comparison of mechanical and energetic estimates of flight costs for hovering sphinx moths. *J. Exp. Biol.* 91:117–129.

——— 1981c. Insect flight energetics. In *Locomotion and Energetics in Arthropods*, ed. C. F. Herreid and C. R. Fourtner, pp. 419–452. New York: Plenum.

Casey, T. M., J. R. Hegel, and C. S. Buser. 1981. Physiology and energetics of pre-flight warm-up in the Eastern tent caterpillar moth *Malacosoma americanum*. *J. Exp. Biol.* 94:119–135.

Casey, T. M., and J. R. Hegel-Little. 1987. Instantaneous oxygen consumption and muscle stroke work in *Malacosoma americanum* during pre-flight warm-up. *J. Exp. Biol.* 127:389–400.

Casey, T. M., and B. A. Joos. 1983. Morphometrics, conductance, thoracic temperature, and flight energetics of noctuid and geometrid moths. *Physiol. Zool.* 56:160–173.

Church, N. S. 1959a. Heat loss and body temperature of flying insects. I. Heat loss by evaporation of water from the body. *J. Exp. Biol.* 37:171–185.

——— 1959b. Heat loss and body temperature of flying insects. II. Heat conduction within the body and its loss by radiation and convection. *J. Exp. Biol.* 37:186–212.

Dorsett, D. A. 1962. Preparation for flight by hawk-moths. *J. Exp. Biol.* 39:579–588.

Dotterweich, H. 1928. Beiträge zur Nervenphysiologie der Insekten. *Zool. Jb. Abt. Allg. Zool. Physiol. Tiere* 44:399–450.

Ellington, C. P. 1984. The aerodynamics of hovering insect flight. VI. Lift and power requirements. *Phil. Trans. R. Soc. Lond.* B305:145–181.

—— 1985. Power and efficiency of insect flight muscle. *J. Exp. Biol.* 115:293–304.

Esch, H. 1988. The effects of temperature on flight muscle potentials in honeybees and cuculiinid winter moths. *J. Exp. Biol.* 135:109–117.

Franklemont, J. G. 1954. Cuculiinae. In *Lepidoptera of New York and Neighboring States*, part 3, ed W. T. M. Forbes, pp. 116–164. Ithaca, N.Y.: Cornell Experimental Station Memo 329.

Hanegan, J. L. 1973. Control of heart rate in cecropia moths; response to thermal stimulation. *J. Exp. Biol.* 59:67–76.

Hanegan, J. L., and J. E. Heath 1970a. Mechanisms for the control of body temperature in the moth, *Hyalophora cecropia. J. Exp. Biol.* 53:349–362.

—— 1970b. Activity patterns and energetics of the moth, *Hyalophora cecropia. J. Exp. Biol.* 53:611–627.

—— 1970c. Temperature dependence of the neural control of the moth flight system. *J. Exp. Biol.* 53:629–639.

Heath, J. E., and P. A. Adams. 1965. Temperature regulation in the sphinx moth during flight. *Nature* 205:309–310.

—— 1967. Regulation of heat production by large moths. *J. Exp. Biol.* 47:21–33.

Hegel, J. I., and T. M. Casey. 1982. Thermoregulation and control of head temperature in the sphinx moth, *Manduca sexta. J. Exp. Biol.* 101:1–15.

Heinrich, B. 1970a. Thoracic temperature stabilization by blood circulation in a free-flying moth. *Science* 168:580–582.

—— 1970b. Nervous control of the heart during thoracic temperature regulation in a sphinx moth. *Science* 169:606–607.

—— 1971a. Temperature regulation of the sphinx moth, *Manduca sexta.* I. Flight energetics and body temperature during free and tethered flight. *J. Exp. Biol.* 54:141–152.

—— 1971b. Temperature regulation of the sphinx moth, *Manduca sexta.* II. Regulation of heat loss by control of blood circulation. *J. Exp. Biol.* 54:153–166.

—— 1974. Thermoregulation in endothermic insects. *Science* 185:747–756.

—— 1977. Why have some animals evolved to regulate a high body temperature? *Am. Nat.* 111:623–640.

—— 1987a. Thermoregulation by winter-flying endothermic moths. *J. Exp. Biol.* 127:313–332.

—— 1987b. Thermoregulation in winter moths. *Sci. Am.* 255:104–111.

—— 1992. Maple sugaring by red quirrels. *J. Mammal.* 73:51–54.

Heinrich, B., and G. A. Bartholomew. 1971. An analysis of pre-flight warm-up in the sphinx moth, *Manduca sexta*. *J. Exp. Biol.* 55:223–239.

Heinrich, B., and T. M. Casey. 1973. Metabolic rate and endothermy in sphinx moths. *J. Comp. Physiol.* 82:195–206.

Heinrich, B., and T. P. Mommsen. 1985. Flight of winter moths near 0°C. *Science* 228:177–179.

Jacobs, M. E., and J. E. Heath. 1967. Temperature regulation, heat production, and wing beat frequency of the sphinx moth, *Protoparce sexta*. *Am. Zool.* 7:399.

Joos, B. 1983. Carbohydrate metabolism during pre-flight warm-up in the tobacco hornworm moth, *Manduca sexta*. *Am. Zool.* 23(4), no. 501.

Josephson, R. K., and R. D. Stevenson. 1991. The efficiency of a flight muscle from the locust *Schistocerca americana*. *J. Physiol.* (in press).

Kalmus, H. 1929. Die CO$_2$-produktion beim Fluge von *Deilephila elpenor* (Weinschwärmer). Baustein zu einer Energetik des Tierfluges. *Z. Vergl. Physiol.* 10:445–455.

Kammer, A. E. 1967. Muscle activity during flight in some large Lepidoptera. *J. Exp. Biol.* 47:277–295.

—— 1968. Motor patterns during flight and warm-up in Lepidoptera. *J. Exp. Biol.* 48:89–109.

—— 1970. A comparative study of motor patterns during pre-flight warm-up in hawkmoths. *Z. Vergl. Physiol.* 70:45–56.

—— 1971. The motor output during flight in a hawkmoth, *Manduca sexta*. *J. Insect Physiol.* 17:1073–1086.

Kammer, A. E., and S. C. Kinnamon. 1979. Maturation of flight motor pattern without movement in *Manduca sexta*. *J. Comp. Physiol.* 130:29–37.

Kammer, A. E., and M. B. Rheuben. 1976. Adult motor patterns produced by moth pupae during development. *J. Exp. Biol.* 65:65–84.

Marden, J. H. 1987. Maximum lift production during takeoff in flying animals. *J. Exp. Biol.* 130:235–258.

May, M. L., P. J. Wilkin, J. E. Heath, and B. A. Williams. 1980. Flight performance of the moth, *Manduca sexta*, at variable gravity. *J. Insect Physiol.* 26:257–265.

McCrea, M. J., and J. E. Heath. 1971. Dependence of flight on temperature regulation in the moth, *Manduca sexta*. *J. Exp. Biol.* 54:415–435.

Moran, V. C., and D. C. Ewer. 1966. Observations on certain characteristics of the flight motor of sphingid and saturniid moths. *J. Insect Physiol.* 12:457–463.

Newport, G. 1837. On the temperature of Insects, and its connexion with the Functions of Respiration and Circulation in this class of Invertebrated Animals. *Phil. Trans. Roy. Soc. Lond.* 127:259–338.

Oosthuizen, M. J. 1939. The body temperature of *Samia cecropia* Linn. (Lepidoptera, Saturniidae) as influenced by muscular activity. *J. Entomol. Soc. S. Afr.* 2:63–73.

Ploye, H. 1979. Endothermy and partial thermoregulation in the silkworm moth, *Bombyx mori*. *J. Comp. Physiol.* 129:315–318.

Sargent, T. D. 1966. Background selection of geometrid and noctuid moths. *Science* 154:1674–1675.

Schweitzer, D. 1974. Notes on the biology and distribution of Cuculiinae (Noctuidae). *J. Lepid. Soc.* 28:5–21.

Stevenson, R. D., and R. K. Josephson. 1990. Effects of operating frequency and temperature on mechanical power output from moth flight muscle. *J. Exp. Biol.* 149:61–78.

Wasserthal, L. T. 1981. Oscillating haemolymph "circulation" and discontinuous tracheal ventilation in the giant silk moth *Attacus atlas* L. *J. Comp. Physiol.* 145:1–15.

Wilson, D. M. 1968. The nervous control of insect flight and related behaviour. *Adv. Insect Physiol.* 5:289–338.

Zebe, E. 1954. Über den Stoffwechsel der Lepidoptera. *Z. Vergl. Physiol.* 36:290–317.

2. Butterflies and Wings

Arnold, J. W. 1964. Blood circulation in insect wings. *Mem. Entomol. Soc. Can.*, no. 38.

Ayres, M. P., and S. F. MacLean, Jr. 1987. Molt as a component of insect development: *Galerucella sagittariae* (Chrysomelidae) and *Epirrita autumnata* (Geometridae). *Oikos* 48:273–279.

Bartlett, P. N., and D. M. Gates. 1967. The energy budget of a lizard on a tree trunk. *Ecology* 48:315–322.

Birket-Smith, S. J. R. 1984. *Prolegs, Legs and Wings of Insects.* Entomonograph, vol. 5. Copenhagen: Scandinavian Science Press, Ltd.

Blau, W. S. 1981. Latitudinal variation in the life histories of insects occupying disturbed habitats: A case study. In *Insect Life History Patterns*, ed. R. F. Denno and H. Dengle, pp. 75–95. New York, Heidelberg, Berlin: Springer-Verlag.

Bogert, C. M. 1949. Thermoregulation in reptiles, a factor in evolution. *Evolution* 3:195–211.

——— 1959. How reptiles regulate their body temperature. *Sci. Am.* 200:105–120.

Brower, L. P., J. van Zandt Brower, and F. P. Cranston. 1965. Courtship behavior of the queen butterfly, *Danaus plexippus berenice* (Cramer). *Zoologica* (New York) 50:1–37.

Chai, P., and R. B. Srygley. 1986. Associations of flight patterns and thermal biology of butterflies to their palatability. *Am. Zool.* 24:98A.

——— 1989. Predation and the flight, morphology, and temperature of neotropical rainforest butterflies. *Am. Nat.* 135:748–765.

Clark, J. A., K. Cena, and N. J. Mills. 1973. Radiative temperatures of butterfly wings. *Z. Angew. Entomol.* 73:327–332.

Clench, H. K. 1966. Behavioral thermoregulation in butterflies. *Ecology* 47:1021–1034.

Cooper, W. 1874. A dissertation on northern butterflies. *Can. Entomol.* 6:9–96.

Cowles, R. B., and C. M. Bogert. 1944. A preliminary study of the thermal requirements of desert reptiles. *Bull. Am. Mus. Nat. Hist.* 85:261–296.

Digby, P. S. B. 1955. Factors affecting the temperature excess of insects in sunshine. *J. Exp. Biol.* 32:279–298.

Douglas, M. M. 1981. Thermoregulatory significance of thoracic lobes in the evolution of insect wings. *Science* 211:84–86.

Douglas, M. M., and J. W. Grula. 1978. Thermoregulatory adaptations allowing ecological range expansion by the pierid butterfly, *Nathalis iole* Boisduval. *Evolution* 32:776–783.

Downes, J. A. 1964. Arctic insects and their environment. *Can. Entomol.* 96:279–307.

——— 1965. Adaptation of insects in the Arctic. *Ann. Rev. Entomol.* 10:257–274.

Dumont, J. P. C., and R. M. Robertson. 1986. Neural circuits: An evolutionary perspective. *Science* 233:849–853.

Ehrlich, P. R., and D. Wheye. 1984. Some observations on spatial distribution in a montane population of *Euphydras editha*. *J. Res. Lepid.* 23:143–152.

Findlay, R., M. R. Young, and J. A. Findlay. 1983. Orientation behaviour in the Grayling butterfly: Thermoregulation or crypsis? *Ecol. Entomol.* 8:145–153.

Flower, J. W. 1964. On the origin of flight in insects. *J. Insect Physiol.* 10:81–88.

Gibo, D. K., and M. L. Pallett. 1979. Soaring flight of monarch butterflies, *Danaus plexippus* (Lepidoptera:Danaidae), during the late summer migration in Southern Ontario. *Can. J. Zool.* 57:1393–1401.

Gossard, T. W., and R. E. Jones. 1977. The effects of age and weather on egg-laying in *Pieris rapae* L. *J. Appl. Ecol.* 14:65–71.

Gould, S. J. 1985. Not necessarily a wing. *Nat. Hist.* 10:13–25.

Grodnitzky, D. L., and M. V. Kozlov. 1991. Evolution and function of wings and their scale covering in butterflies and moths (Insecta: Papilionida = Lepidoptera). *Biol. Zent. Bl.* 110:199–206.

Guppy, C. S. 1986. The adaptive significance of alpine melanism in the butterfly *Parnassius phoebus* F. (Lepidoptera:Papilionidae). *Oecologia* (Berlin) 70:205–213.

Heinrich, B. 1972. Thoracic temperatures of butterflies in the field near the equator. *Comp. Biochem. Physiol.* 43A:459–467.

——— 1977. Why have some animals evolved to regulate a high body temperature? *Am. Nat.* 111:623–640.

——— 1986a. Thermoregulation and flight activity of the satyr, *Coenonympha inornata* (Lepidoptera:Satyridae). *Ecology* 67:593–597.

——— 1986b. Comparative thermoregulation of four montane butterflies of different mass. *Physiol. Zool.* 59:616–626.

——— 1990. Is "reflectance" basking real? *J. Exp. Biol.* 154:31–43.

Herter, K. 1953. *Der Temperatursinn der Insekten.* Berlin: Dunker & Humbolt.

Hidaka, R. 1973. Logic of mating behavior of Lepidoptera. *Ann. N.Y. Acad. Sci.* 223:70–76.

Hidaka, T., and K. Yamashita. 1975. Wing color pattern as a releaser of mating behavior in the swallowtail butterfly *Papilio xuthus* L. (Lepidoptera Papilionidae). *Appl. Entomol. Zool.* 10:263–267.

Hoffmann, R. J. 1974. Environmental control of seasonal variation in the butterfly *Colias eurytheme:* Effects of photoperiod and temperature on pteridine pigmentation. *J. Insect Physiol.* 20:1913–1924.

———— 1978. Environmental uncertainty and evolution of physiological adaptation in *Colias* butterflies. *Am. Nat.* 112:999–1015.

Kammer, A. E. 1968. Motor patterns during flight and warm-up in Lepidoptera. *J. Exp. Biol.* 48:89–109.

———— 1970. Thoracic temperature, shivering, and flight in the monarch butterfly, *Danaus plexippus* L. *Z. Vergl. Physiol.* 68:334–344.

———— 1971. Influence of acclimation temperature on the shivering behavior of the butterfly, *Danaus plexippus* (L.). *Z. Vergl. Physiol.* 72:364–369.

Kammer, A. E., and J. Bracchi. 1973. Role of the wings in the absorption of radiant energy by a butterfly. *Comp. Biochem. Physiol.* 45A:1057–1064.

Kevan, P. G., and J. D. Shorthouse. 1970. Behavioral thermoregulation by High Arctic butterflies. *Arctic* 23:268–279.

Kingsolver, J. G. 1983. Thermoregulation and flight in *Colias* butterflies: Elevational patterns and mechanistic limitations. *Ecology* 64:534–545.

———— 1985a. Thermal ecology of *Pieris* butterflies (Lepidoptera: Pieridae): A new mechanism of behavioral thermoregulation. *Oecologia* 66:540–545.

———— 1985b. Thermoregulatory significance of wing melanization in *Pieris* butterflies (Lepidoptera: Pieridae): Physics, posture, and pattern. *Oecologia* 66:546–553.

———— 1985c. Butterfly engineering. *Sci. Am.* 253:106–113.

———— 1987. Evolution and coadaptation of thermoregulatory behavior and wing pigmentation pattern in pierid butterflies. *Evolution* 41:472–490.

———— 1988. Thermoregulation, flight, and the evolution of wing pattern in pierid butterflies: The topography of adaptive landscapes. *Am. Zool.* 28:899–912.

Kingsolver, J. G., and M. A. R. Koel. 1985. Aerodynamics, thermoregulation, and insect wings: Differential scaling and evolutionary change. *Evolution* 39:488–504.

Kingsolver, J. G., and R. M. Moffat. 1982. Thermoregulation and the determinants of heat transfer in *Colias* butterflies. *Oecologia* 53:27–33.

Kingsolver, J. G., and W. B. Watt. 1983. Thermoregulatory strategies in *Colias* butterflies: Thermal stress and the limits to adaptation in temporally varying environments. *Am. Nat.* 121:32–55.

———— 1984. Mechanistic constraints and optimality models: Thermoregulatory strategies in *Colias* butterflies. *Ecology* 65:1835–1839.

Kingsolver, J. G., and D. C. Wiernasz. 1987. Dissecting correlated characters: Adaptive aspects of phenotypic covariation in melanization patterns of *Pieris* butterflies. *Evolution* 41:491–503.

Krogh, A., and E. Zeuthen. 1941. The mechanism of flight preparation in some insects. *J. Exp. Biol.* 18:1–10.

Kukalova, J. 1968. Permian mayfly nymphs. *Psyche* 75:310–327.

Kukalova-Peck, J. 1978. Origin and evolution of insect wings and their relation to metamorphosis, as documented by the fossil record. *J. Morphol.* 156:53–126.

———— 1985. Ephemeroid wing venation based upon new gigantic Carboniferous mayflies and basic morphology, phylogeny, and metamorphosis of pterygote insects (Insecta, Ephemerida). *Can. J. Zool.* 63:933–955.

———— 1987. New Carboniferous Diplura, Monura, and Thysanura, the hexapod ground plan, and the role of thoracic side lobes in the origin of wings (Insecta). *Can. J. Zool.* 65:2327–2345.

Lewin, R. 1985. On the origin of wings. *Science* 230:428–429.

Longstaff, G. B. 1905a. Notes on the butterflies observed in a tour through India and Ceylon, 1903–4. *Trans. Entomol. Soc. Lond.* 1905:85, 126, 135–36.

———— 1905b. The attitude of *Satyrus semele* at rest. *Entomol. Month. Mag.* 16:44–45.

———— 1906. Some rest-attitudes of butterflies. *Trans. Entomol. Soc. Lond.* 1906:97–118.

Magnus, D. 1958. Experimentelle Untersuchungen zur Bionomie und Ethologie des Kaisermantels, *Argynnis paphila* L. (Lep. Nymph.). *Z. Tierpsychol.* 15:397–426.

Marden, J. 1987. Maximum lift production during takeoff in flying animals. *J. Exp. Biol.* 130:235–258.

Marden, J., and P. Chai. 1989. Effects of palatability and mimicry on butterfly design. Unpublished manuscript.

Masters, A. R., S. B. Malcolm, and L. P. Brower. 1988. Monarch butterfly (*Danaus plexippus*) thermoregulatory behavior and adaptations for overwintering in Mexico. *Ecology* 69:458–467.

Ohsaki, N. 1986. Body temperatures and behavioral thermoregulation strategies of three *Pieris* butterflies in relation to solar radiation. *J. Ethol.* 4:1–9.

Packard, A. S. 1898. *A Text-book of Entomology.* London and New York: Macmillan.

Parker, G. H. 1903. The phototropism of the mourning-cloak butterfly. In *Mark Anniversary Volume: To Edward Lauens Mark*, pp. 453–469. New York: H. Holt Co.

Pictet, A. 1915. A propos des tropismes. Recherches expérimentales sur le comportement des insectes vis-a-vis des facteurs de l'ambiance. *Bull. Soc. Vaud. Sci. Nat.* 50S:423–550.

Pivnick, K. A., and J. N. McNeil. 1986. Sexual differences in thermoregulation of *Thymelicus lineola* adults (Lepidoptera: Hesperiidae). *Ecology* 67:1024–1035.

Pivnick, K. A., and J. N. McNeil. 1987. Diel patterns of activity of *Thymelicus lineola* adults (Lepidoptera: Hesperidae) in relation to weather. *Ecol. Entomol.* 12:197–207.

Polcyn, D. M., and M. A. Chappell. 1986. Analysis of heat transfer in *Vanessa* butterflies: Effects of wing position and orientation to wind and light. *Physiol. Zool.* 59:706–716.

Radl, E. 1903. *Untersuchungen über den Phototropismus der Tiere.* Leipzig. (Not seen by author; quoted in Herter, 1953.)

Rawlins, J. E. 1980. Thermoregulation by the black swallowtail butterfly *Papilio polyxenes* (Lepidoptera: Papilionidae). *Ecology* 61:345–357.

Rawling, J. E., and R. C. Lederhouse. 1978. The influence of environmental factors on roosting in the black swallowtail, *Papilio polyxenes asterius* Stoll (Papilionidae). *J. Lepid. Soc.* 32:145–159.

Riek, E. F., and J. Kukalova-Peck. 1984. A new interpretation of dragonfly wing venation based upon Early Carboniferous fossils from Argentina (Insecta: Odonatoidea) and basic character states in pterygote wings. *Can. J. Zool.* 62:1150–1166.

Robertson, R. M. 1987. Interneurons in the flight system of the cricket *Teleogryllus oceanicus. J. Comp. Physiol.* A160:431–445.

Robertson, R. M., K. G. Pearson, and H. Reichert. 1982. Flight interneurons in the locust and the origin of insect wings. *Science* 217:177–179.

Roland, J. 1982. Melanism and diel activity of alpine *Colias* (Lepidoptera: Pieridae). *Oecologia* (Berlin) 53:214–221.

Shapiro, A. M. 1974. Ecotypic variation in montane butterflies. *Wasmann J. Biol.* 43:267–280.

—— 1977. Phenotypic induction in *Pieris napi* L.: Role of temperature and photoperiod in a coastal California population. *Ecol. Entomol.* 2:217–224.

Srygley, R. B., and P. Chai. 1989. Predation and the elevation of thoracic temperature in brightly-colored, neotropical butterflies. *Am. Nat.* 135:766–787.

Stern, V. M., and R. F. Smith. 1960. Factors affecting egg production and oviposition in populations of *Colias philodice eurytheme* Boisduval (Lepidoptera: Pieridae). *Hilgardia* 29:411–454.

Stone, G. N., J. N. Amos, T. F. Stone, R. I. Knight, H. Gay, and F. Parrott. 1988. Thermal effects on activity patterns and behavioural switching in a concourse of foragers on *Stachytarpheta mutabilis* (Verbenaceae) in Papua New Guinea. *Oecologia* 77:56–63.

Suzuki, N., A. Niizuma, K. Yamashita, M. Watanabe, K. Nozato, A. Ishida, K. Kirstani, and S. Migai. 1985. Studies on ecology and behavior of Japanese black swallowtail butterflies. *Jap. J. Ecol.* 35:21–30.

Tinbergen, N. 1942. The courtship of the Grayling *Eumenis* (= *Satyros*) *semele* L. In *The Animal in Its World: Field Studies*, pp. 197–248. London: Allen & Unwin.

Tongue, A. E. 1909. Resting attitudes of Lepidoptera. *Proc. South Lond. Entomol. Nat. Hist. Soc.* 10:5–8.

Tsuji, J. S. 1980. The in-flight thermal biology of *Colias* butterflies. Undergraduate honors thesis, Stanford University.

Tsuji, J. S., J. G. Kingsolver, and W. B. Watt. 1986. Thermal physiological ecology of *Colias* butterflies in flight. *Oecologia* (Berlin) 69:161–170.

Vielmetter, W. 1954. Die Temperaturregulation des Kaisermantels in der Sonnenstrahlung. *Naturwissenschaften* 41:535–536.

—— 1958. Physiologie des Verhaltens zur Sonnenstrahlung bei dem Tagfalter, *Argynnis paphi* L. 1. Untersuchungen im Freiland. *J. Insect Physiol.* 2:13–37.

Wasserthal, L. T. 1975. The role of butterfly wings in regulation of body temperature. *J. Insect Physiol.* 21:1921–1930.

REFERENCES FOR CHAPTER 2

——— 1983. Haemolymph flows in the wings of pierid butterflies visualized by vital staining (Insecta, Lepidoptera). *Zoomorphology* 103:177–192.

Watt, W. B. 1968. Adaptive significance of pigment polymorphism in *Colias* butterflies. I. Variation of melanin pigment in relation to thermoregulation. *Evolution* 22:437–458.

——— 1969. Adaptive significance of pigment polymorphism in *Colias* butterflies. II. Thermoregulation and photoperiodically controlled melanin variation in *Colias eurytheme*. *Proc. N.A.S.* 63:767–774.

Wickman, P. O. 1985. The influence of temperature on the territorial and mate locating behavior of the small heath butterfly, *Coenonympha pamphilus* (L) Lepidoptera: Satyridae. *Behav. Ecol. Sociobiol.* 16:233–239.

Wigglesworth, V. B. 1976. The evolution of insect flight. In *Insect Flight*, ed. R. C. Rainey. *Symp. R. Entomol. Soc. Lond.* 7:255–269.

Winn, A. F. 1916. Heliotropism in butterflies; or turning toward the sun. *Can. Entomol.* 48:6–9.

Wootton, R. J. 1981. Paleozoic insects. *Ann. Rev. Entomol.* 26:319–344.

3. Dragonflies Now and Then

Bakker, R. T. 1972. Anatomical and ecological evidence of endothermy in dinosaurs. *Nature* 238:81–85.

Bennett, A. F., R. B. Huey, H. John-Alder, and K. A. Nagy. 1984. The parasol tail and thermoregulatory behavior of the Cape ground squirrel *Xerus inauris*. *Physiol. Zool.* 57:57–62.

Brogniart. C. 1894. Recherches pour servir á l'histoire des insects fossiles des temps primaries. *Thèse Fac. Sci. Paris*, no. 821, pp. 1–494.

Carpenter, F. M. 1943. Studies on Carboniferous insects from Commentry, France; Part I. Introduction and families Protagriidae, Meganeuridae, and Campylopteridae. *Bull. Geol. Soc. Am.* 54:537–554.

——— 1947. Lower Permian insects from Oklahoma. Part I. Introduction and the orders Megasecoptera, Protodonata, and Odonata. *Proc. Am. Acad. Arts Sci.* 76:25–54.

Chappell, M. A., and G. A. Bartholomew. 1981. Standard operative temperatures and thermal energetics of the antelope ground squirrel *Ammospermophilus leucurus*. *Physiol. Zool.* 54:81–93.

Church, N. S. 1960. Heat loss and body temperature of flying insects. II. Heat conduction within the body and its loss by radiation and convection. *J. Exp. Biol.* 37:186–212.

Corbet, P. S. 1963. *A Biology of Dragonflies*. Chicago: Quadrangle.

Hankin, E. H. 1921. The soaring flight of dragonflies. *Proc. Camb. Phil. Soc. Biol. Sci.* 20:460–465.

Heinrich, B. 1976. Heat exchange in relation to blood flow between thorax and abdomen in bumblebees. *J. Exp. Biol.* 64:561–585.

——— 1977. Why have some animals evolved to regulate a high body temperature? *Am. Nat.* 111:623–640.

——— 1981. Ecological and evolutionary perspectives. In *Insect Thermoregulation*, ed. B. Heinrich, pp. 235–302. New York: Wiley.

Heinrich, B., and T. M. Casey. 1978. Heat transfer in dragonflies: "Fliers" and "perchers." *J. Exp. Biol.* 74:17–36.

Jurzitza, G. 1967. Über einen reversiblen temperaturabhängigen Farbwechsel bei *Anax imperatur* Leach (Odonata: Aeschnidae). *Deut. Entomol. Z.* (N.F.) 14:387–389.

Kammer, A. E. 1970. A comparative study of motor patterns during preflight warm-up in hawkmoths. *Z. Vergl. Physiol.* 70:45–56.

Kukalova-Peck, J. 1978. Origin and evolution of insect wings and their relation to metamorphosis, as documented by the fossil record. *J. Morphol.* 156:53–126.

Marden, J. H. 1989. Bodybuilding dragonflies: Costs and benefits of maximizing flight muscle. *Physiol. Zool.* 62:505–521.

May, M. 1976a. Thermoregulation and adaptation to temperature in dragonflies (Odonata:Anisoptera). *Ecol. Monogr.* 46:1–32.

——— 1976b. Physiological color change in New World damselfies (Zygoptera). *Odonatologica* 5:165–171.

——— 1976c. Warming rates as a function of body size in periodic endotherms. *J. Comp. Physiol.* 111:55–70.

——— 1977. Thermoregulation and reproductive activity in tropical dragonflies of the genus *Micrathyria*. *Ecology* 58:787–798.

——— 1978. Thermal adaptations of dragonflies. *Odonatologica* 7:27–47.

——— 1979. Energy metabolism of dragonflies (Odonata: Anisoptera) at rest and during endothermic warm-up. *J. Exp. Biol.* 83:79–94.

——— 1981a. Allometric analysis of body and wing dimensions of male Anisoptera. *Odonatologica* 10:279–291.

——— 1981b. Wingstroke frequency of dragonflies (Odonata: Anisoptera) in relation of temperature and body size. *J. Comp. Physiol.* 144:229–240.

——— 1982. Heat exchange and endothermy in Protodonata. *Evolution* 36:1051–1058.

——— 1986. A preliminary investigation of variation in temperature among body regions of *Anax junius* (Drury) (Anisoptera: Aeshnidae). *Odonatologica* 15:119–128.

——— 1987. Body temperature regulation and responses to temperature by male *Tetragoneuria cynosura* (Anisoptera: Corduliidae). *Adv. Odonatol.* 3:103–119.

——— 1990. Thermal adaptations of dragonflies revisited. *Adv. Odonatol.* 5:71–88.

McVey, M. E. 1984. Egg release rates with temperature and body size in libellulid dragonflies (Anisoptera). *Odonatologica* 13:377–385.

Miller, P. L. 1962. Spiracle control in adult dragonflies (Odonata). *J. Exp. Biol.* 39:513–535.

Morgan, K. R., T. S. Shelly, and L. S. Kimsey. 1985. Body temperature regulation, energy metabolism, and foraging in light-seeking and shade-seeking robber flies. *J. Comp. Physiol.* B155:561–570.

O'Farrell, A. F. 1963. Temperature-controlled physiological colour change in some Australian damsel flies (Odonata-Zygoptera). *Austr. J. Sci.* 25:437–438.

——— 1964. On physiological colour change in some Australian Odonata. *J. Entomol. Soc. Austr.* (N.S.W.) 1:1–8.

REFERENCES FOR CHAPTER 3

——— 1968. Physiological colour change and thermal adaptation in some Australian Zygoptera. *Proc. R. Entomol. Soc. Lond.*, ser. C 33:26–29.

Pezalla, V. M. 1979. Behavioral ecology of the dragonfly *Libellula pulchella* Drury (Odonata: Anisoptera). *Am. Midl. Nat.* 102:1–22.

Polcyn, D. M. 1989. The thermal biology of desert dragonflies. Ph.D. dissertation, University of California, Riverside.

Rowe, R. J., and M. J. Winterbourn. 1981. Observations on the body temperature and temperature associated behaviour of three New Zealand dragonflies (Odonata). *Mauri Ora* 9:15–23.

Shelly, T. E. 1982. Comparative foraging behavior of light versus shade-seeking adult damselflies in a lowland neotropical forest (Odonata:Zygoptera). *Physiol. Zool.* 55:335–343.

Singer, F. 1987. A physiological basis of variation in post-copulatory behaviour in a dragonfly *Sympterum obtrusum. Anim. Behav.* 35:1575–1577.

Sternberg, K. 1987. On reversible, temperature-dependent colour change in males of the dragonfly *Aeshna caerulea* (Strom) (Anisoptera: Aeshnidae). *Odonatologica* 16:57–66.

Tracy, C. R., B. J. Tracy, and D. Dobkin. 1979. The role of posturing in behavioral thermoregulation by black dragons (*Hagenius selys;* Odonata). *Physiol. Zool.* 52:565–571.

Tsubaki, Y., and T. Ono. 1987. Effects of age and body size on the male territorial system of the dragonfly *Nannophya pygmaea* Rambur (Odonata: Libellulidae). *Anim. Behav.* 35:518–525.

Veron, J. E. N. 1973. The physiological control of the chromatophores of *Austrolestes annulosus* (Odonata). *J. Insect Physiol.* 19:1689–1703.

——— 1974. The role of physiological color change in the thermoregulation of *Austrolestes annulosus* (Selys) (Odonata). *Austr. J. Zool.* 22:457–469.

——— 1976. Responses of Odonata chromatophores to environmental stimuli. *J. Insect Physiol.* 22:19–30.

Vogt, D., and B. Heinrich. 1983. Thoracic temperature variations in the onset of flight in dragonflies (Odonata: Anisoptera). *Physiol. Zool.* 56:236–241.

Wasserthal, L. T. 1975. The role of butterfly wings in regulation of body temperature. *J. Insect Physiol.* 21:1921–1930.

Wootton, R. J. 1981. Paleozoic insects. *Ann. Rev. Entomol.* 26:319–344.

4. Grasshoppers and Other Orthoptera

Abrams, T. W., and K. G. Pearson. 1982. Effects of temperature on identified central neurons that control jumping in the grasshopper. *J. Neurosci.* 2:1538–1553.

Alcock, J. 1972. Observations on the behaviour of the grasshopper *Taeniopoda eques* (Burmeister), Orthoptera, Acrididae. *Anim. Behav.* 20:237–242.

Alexander, G., and J. R. Hilliard, Jr. 1969. Altitudinal and seasonal distribution of Orthoptera in the Rocky Mountains of northern Colorado. *Ecol. Monogr.* 39:385–431.

Alther, H., H. Sass, and I. Alther. 1977. Relationship between structure and function of antennal chemo-, hygro-, and thermoreceptive sensilla in *Periplaneta americana*. *Cell Tissue Res.* 176:389–405.

Altman, J. S. 1975. Changes in the flight motor pattern during the development of the Australian plague locust, *Chortoicetes terminifera*. *J. Comp. Physiol.* 97:127–142.

Anderson, R. L., and Mutchmor, J. A. 1968. Temperature acclimation and its influence on the electrical activity of the nervous system in three species of cockroaches. *J. Insect Physiol.* 15:243–251.

Anderson, R. V., C. R. Tracy, and Z. Abramsky. 1979. Habitat selection in two species of short-horned grasshoppers. The role of thermal and hydric stresses. *Oecologia* 38:359–374.

Bailey, L. 1954. The respiratory currents in the tracheal system of the adult honey-bee. *J. Exp. Biol.* 31:589–593.

Bailey, W. J., R. J. Cunningham, and L. Lebel. 1990. Song power, spectral distribution and female phonotaxis in the bushcricket *Requena verticalis* (Tettigoniidae: Orthoptera): Active female choice or passive attraction? *Anim. Behav.* 40:33–42.

Bauer, M., and O. von. Helversen. 1987. Separate localization of sound recognizing and sound producing neural mechanisms in a grasshopper. *J. Comp. Physiol.* A161:95–101.

Bentley, D. R., and R. R. Hoy. 1970. Postembryonic development of adult motor patterns in crickets: A neural analysis. *Science* 170:1409–1411.

Bessey, C. A., and E. A. Bessey. 1898. Further notes on thermometer crickets. *Am. Nat.* 32:263–264.

Bodenheimer, R. S. 1929. Studien zur Epidemiologie, Ökologie und Physiologie der afrikanischen Wanderheuschrecke (*Schistocerca gregaria* Forsk). *Z. Angew. Entomol.* 15:1–123.

Brosemer, R. W., W. Vogell, and T. Bücher. 1963. Morphologische und enzymatische Muster bei der Entwicklung indirekter Flugmuskeln von *Locusta migratoria*. *Biochem. Z.* 338:854–910.

Bullock, T. H. 1955. Compensation for temperature in the metabolism and activity of poikilotherms. *Biol. Rev.* 30:311–342.

Buxton, P. A. 1924. Heat, moisture, and animal life in deserts. *Proc. Roy. Soc. Lond.* B96:123–131.

Calhoun, E. H. 1960. Acclimation to cold in insects. *Entomol. Exp. Appl.* 3:27–32.

Carruthers, R. I., T. S. Larkin, H. Firstencel, and Z. Feng. 1992. Influence of thermal ecology of the mycosis of a rangeland grasshopper. *Ecology* 73:190–204.

Chapman, R. R. 1965. The behavior of nymphs of *Schistocerca gregaria* (Farskal) (Orthoptera, Acrididae) in a temperature gradient with special reference to temperature preference. *Behaviour* 24:283–317.

Chappell, M. A. 1983a. Thermal limitations to escape responses in desert grasshoppers. *Anim. Behav.* 31:1088–1093.

———— 1983b. Metabolism and thermoregulation in desert and montane grasshoppers. *Oecologia* (Berlin) 56:126–131.

Dehnel, P. A., and E. Segal. 1956. Acclimation and oxygen consumption to temperature on the American cockroach *(Periplaneta americana)*. *Biol. Bull.* (Woods Hole) 111:53–61.

Doherty, J. A. 1985. Temperature coupling and "trade-off" phenomena in the acoustic communication system of the cricket, *Gryllus bimaculatus* De Geer (Gryllidae). *J. Exp. Biol.* 114:17–35.

Dolbear, A. E. 1897. The cricket as a thermometer. *Am. Nat.* 31:970–971.

Edney, E. B., S. Haynes, and D. Gibo. 1974. Distribution and activity of the desert cockroach *Arenivaga investigata* (Polyphagidae) in relation to microclimate. *Ecology* 55:420–427.

Edwards, G. A., and W. L. Nutting. 1950. The influence of temperature upon the respiration and heart activity of *Thermobia* and *Grylloblatta*. *Psyche* 57:33–44.

Elder, H. Y. 1971. High frequency muscles used in sound production by a katydid. II. Ultrastructure of the singing muscles. *Biol. Bull.* (Woods Hole) 141:434–448.

Ellis, P. E. 1963. An experimental study of feeding, basking, marching and pottering in locust nymphs. *Behavior* 20:282–310.

Farnsworth, E. G. 1972a. Effects of ambient temperature and humidity on internal temperature and wing-beat frequency of *Periplaneta americana*. *J. Insect Physiol.* 18:359–371.

———— 1972b. Effects of ambient temperature, humidity, and age on wing-beat frequency of *Periplaneta* species. *J. Insect Physiol.* 18:827–839.

Fraenkel, D. G. 1929. Untersuchungen über Lebensgewohnheiten, Sinnesphysiologie und Sozialpsychologie der wandernden Larven der afrikanischen Wanderheuschrecke *Schistocerca gregaria* (Forsk). *Biol. Zbl.* 49:657–680.

———— 1930. Die Orientierung von *Schistocerca gregaria* zu strahlender Wärme. *Z. Vergl. Physiol.* 13:300–313.

Gerhardt, H. C. 1978. Temperature coupling in the vocal communication system of the gray treefrog, *Hyla versicolor*. *Science* 199:992–994.

Gillis, J. E., and K. W. Possai. 1983. Thermal niche partitioning in the grasshoppers *Arphia conspersa* and *Trimerotropis suffusa* from a montane habitat in central Colorado. *Ecol. Entomol.* 8:155–161.

Goodman, C. S., and W. J. Heitler. 1977. Isogenic locusts and genetic variability in the effects of temperature on neuronal threshold *J. Comp. Physiol.* 117:183–207.

Gunn, D. L. 1934. The temperature and humidity relations of the cockroach *(Blatella orientalis)*. II. Temperature preferences. *Z. Vergl. Physiol.* 20:617–625.

———— 1942. Body temperature in poikilothermic animals. *Biol. Rev.* 17:293–314.

Hadley, N. F., and D. D. Massion. 1985. Oxygen consumption, water loss and cuticular lipids of high vs. low elevation populations of the grasshopper *Aeropedellus clavatus* (Orthoptera: Acrididae). *Comp. Biochem. Physiol.* 80A:307–311.

Hamilton, A. G. 1936. The relation of humidity and temperature to the development of three species of African locusts—*Locusta migratoria migratorioides* (R. & F.), *Schistocerca gregaria* (Forsk.), *Nomadacris septemfasciata* (Serv.). *Trans. Roy. Entomol. Soc. Lond.* 85:1–60.

———— 1950. Further studies on the relation of humidity and temperature to the development of two species of African locusts—*Locusta migratoria migratorioides* (R. & F.) and *Schistocerca gregaria* (Forsk.). *Trans. R. Entomol. Soc. Lond.* 101:1–58.

Hardman, J. M., and M. K. Mukerji. 1982. A model simulating the population dynamics of the grasshoppers (Acrididae) *Melanoplus sanguinipes* (Fabr.), *M. packardii* (Scudder) and *Camnula pellucida* (Scudder). *Res. Popul. Ecol.* 24:276–301.

Harrison, J. M. 1988. Temperature effects on haemolymph acid-base status *in vivo* and *in vitro* in the two-striped grasshopper *Melanoplus bivittatus. J. Exp. Biol.* 140:421–435.

Heath, J. E., and R. K. Josephson. 1970. Body temperature and singing in the katydid, *Neoconocephalus robustus* (Orthoptera, Tettigoniidae). *Biol Bull.* (Woods Hole) 138:272–285.

Heinrich, B. 1975. Thermoregulation and flight energetics of desert insects. In *Environmental Physiology of Desert Organisms*, ed. N. F. Hadley. Stroudsberg, Penn.: Dowden, Hutchinson and Ross.

——— 1980. Mechanisms of body temperature regulation in honeybees, *Apis mellifera*. II. Regulation of thoracic temperature at high air temperatures. *J. Exp. Biol.* 85:73–87.

Heitler, W. J., C. S. Goodman, and C. H. Frazer-Rowell. 1977. The effects of temperature on the threshold of identified neurons in the locust. *J. Comp. Physiol.* 117:163–182.

Helversen, D. von, and O. von. Helversen. 1981. Korrespondenz zwischen Gesang und auslösendem Schema bei Feldheuschrecken. *Nova Acta Leopold.* 245:449–462.

Herter, K. 1924. Untersuchungen über den Temperatursinn einiger Insekten. *Z. Vergl. Physiol.* 1:122–188.

Hilbert, D. W., and J. A. Logan. 1983. Empirical model of nymphal development for the migratory grasshopper, *Melanoplus sanguinipes* (Orthoptera: Acrididae). *Environ. Entomol.* 12:1–5.

Janiszewski, J. 1984. The temperature of the head, thorax and abdomen of *Periplaneta americana* during rest and flight and high ambient temperatures. *J. Thermal Biol.* 9:177–181.

——— 1985. The effect of head heating on the flight activity of the cockroach. *Experientia* 41:1199–1200.

Janiszewski, J., U. Kosecka-Janiszewska, and D. Otto. 1988. Changes in rate of abdominal ventilatory pumping induced by warming individual ganglia in the male cricket *Gryllus bimaculatus* (De Geer). *J. Therm. Biol.* 13:185–188.

Janiszewski, J., and D. Otto. 1988. Modulation of activity of identified subesophageal neurons in the cricket *Gryllus bimaculatus* by local changes in body temperature. *J. Comp. Physiol.* A162:739–746.

Janiszewski, J., D. Otto, and H. U. Kleindienst. 1987. Descending neurons in the cricket's subesophageal ganglion with activity modulated by localized body cooling. *Naturwissenschaften* 74:500–501.

Jensen, M. 1956. Biology and physics of locust flight. III. The aerodynamics of locust flight. *Phil. Trans. Roy. Soc. Lond.* B239:511–552.

Joern, A. 1982. Importance of behavior and coloration in the control of body temperature by *Brachystola magna* Girard (Orthoptera: Acrididae). *Acrida* 10:117–130.

Josephson, R. K. 1973. Contraction kinetics of the fast muscles used in singing by a katydid. *J. Exp. Biol.* 59:781–801.

———— 1985. The mechanical power output of a tettigoniid wing muscle during singing and flight. *J. Exp. Biol.* 117:357–368.

Josephson, R. K., and H. Y. Elder. 1968. Rapidly contracting muscles used in sound production by a katydid. *Biol. Bull.* (Woods Hole) 135:409.

Josephson, R. K., and R. C. Halverson. 1971. High frequency muscles used in sound production by a katydid. I. Organization of the motor system. *Biol. Bull.* (Woods Hole) 141:411–433.

Kammer, A. E., and S. C. Kinnamon. 1979. Maturation of flight motor pattern without movement in *Manduca sexta*. *J. Comp. Physiol.* 130:29–37.

Kammer, A. E., and M. B. Rheuben. 1976. Adult motor patterns produced by moth pupae during development. *J. Exp. Biol.* 65:65–84.

Kemp, W. P. 1986. Thermoregulation in three rangeland grasshopper species. *Can. Entomol.* 118:335–342.

Kerkut, G. A., and B. J. R. Taylor. 1956. Effect of temperature on the spontaneous activity from the isolated ganglia of the slug, cockroach and crayfish. *Nature* 178:426.

———— 1957. A temperature receptor in the tarsus of the cockroach, *Periplaneta americana*. *J. Exp. Biol.* 34:486–493.

———— 1958. The effect of temperature changes on the activity of poikilotherms. *Behaviour* 13:259–279.

Krogh, A., and T. Weis-Fogh. 1951. The respiratory exchange of the desert locust *(Schistocerca gregaria)* before, during, and after flight. *J. Exp. Biol.* 28:344–357.

Kutsch, W. 1971. The development of the flight motor pattern in the desert locust, *Schistocerca gregaria*. *Z. Vergl. Physiol.* 74:156–168.

———— 1973. The influence of age and culture-temperature on the wing-beat frequency of the migratory locust, *Locusta migratoria*. *J. Insect Physiol.* 19:763–772.

Loftus, R. 1966. Cold receptors in the antenna of *Periplaneta americana*. *Z. Vergl. Physiol.* 52:380–385.

———— 1968. The response of the antennal cold receptors of *Periplaneta americana* to rapid temperature changes and to steady temperature. *Z. Vergl. Physiol.* 59:413–455.

Loher, W., and G. Wiedenmann. 1981. Temperature-dependent changes in circadian patterns of cricket premating behaviour. *Physiol. Entomol.* 6:35–43.

MacKay, W. P. 1982. An altitudinal comparison of oxygen consumption rates in three species of *Pogonomyrmex* harvester ants (Hymenoptera: Formicidae). *Physiol. Zool.* 55:367–377.

Mellanby, K. 1939. Low temperature and insect activity. *Proc. Roy. Soc. Lond.* B127:473–485.

Miles, C. I. 1985. The effects of behaviorally relevant temperatures on mechanosensory neurons of the grasshopper, *Schistocerca americana*. *J. Exp. Biol.* 116:121–139.

Miller, P. L. 1960. Respiration in the desert locust. III. Ventilation and the spiracles during flight. *J. Exp. Biol.* 37:264–278.

Mizisin, A. P., and N. E. Ready. 1986. Growth and development of flight muscle in the locust (*Schistocerca nitens*, Thünberg). *J. Exp. Zool.* 237:45–55.

Morrisey, R., and J. S. Edwards. 1979. Neural function in an alpine grylloblattid: A comparison with the house cricket, *Acheta domesticus*. *Physiol. Entomol.* 4:241–250.

Muchmor, J. A., and A. G. Richards. 1961. Low temperature tolerance in insects in relation to the influence of temperature on muscle apyrase activity. *J. Insect Physiol.* 7:141–158.

Murphy, B. F. Jr., and J. E. Heath. 1983. Temperature sensitivity in the prothoracic ganglion of the cockroach, *Periplaneta americana*, and its relationship to thermoregulation. *J. Exp. Biol.* 105:305–315.

Murrish, D. F., and K. Schmidt-Nielsen. 1970. Exhaled air temperature and water conservation in lizards. *Respir. Physiol.* 10:151–158.

Neville, A. C. 1963. Motor unit distribution of the dorsal longitudinal flight muscles in locusts. *J. Exp. Biol.* 40:123–136.

Neville, A. C., and T. Weis-Fogh. 1963. The effect of temperature on locust flight muscle. *J. Exp. Biol.* 40:111–121.

Novicki, A. 1989a. Rapid postembryonic development of a cricket flight muscle. *J. Exp. Zool.* 250:253–262.

—— 1989b. Control of growth and ultrastructural maturation of a cricket flight muscle. *J. Exp. Zool.* 250:263–272.

Novicki, A., and R. K. Josephson. 1987. Innervation is necessary for the development of fast contraction kinetics of singing muscles in a katydid. *J. Exp. Zool.* 242:309–315.

Otte, D. 1984. *The North American Grasshoppers*. Vol. 2, *Acrididae: Oedipodinae*. Cambridge, Mass.: Harvard University Press.

Parker, J. R. 1930. Some effects of temperature and moisture upon *Melanoplus mexicanus mexicanus* Saussure and *Camnula pellucida* Scudder (Orthoptera). *Mont. Agric. Exp. Stn. Bull.* 223.

Parker, M. A. 1982. Thermoregulation by diurnal movement in the barberpole grasshopper *(Dactylotum bicolor)*. *Am. Midl. Nat.* 107:228–237.

Pepper, J. H., and E. Hastings. 1952. The effects of solar radiation on grasshopper temperature and activities. *Ecology* 33:96–103.

Prange, D. H. 1990. Temperature regulation by respiratory evaporation in grasshoppers. *J. Exp. Biol.* 154:463–474.

Putnam, L. G. 1963. The progress of nymphal development in pest grasshoppers (Acrididae) of western Canada. *Can. Entomol.* 95:1210–1216.

Rainey, R. C. 1958. Some observations on flying locusts and atmospheric turbulence in eastern Africa. *Q. J. Roy. Met. Soc.* 84:334–354.

Ready, N. E. 1986. Development of fast singing muscles in a katydid. *J. Exp. Zool.* 238:43–54.

Rence, B. G., and W. Loher. 1975. Arrhythmically singing crickets: Thermoperiodic re-entrainment after bilobectomy. *Science* 190:385–387.

Roffey, J. 1963. Observations on gliding in the desert locust. *Anim. Behav.* 15:359–366.

Schmidt-Nielsen, K. 1972. *How Animals Work*. Cambridge: Cambridge University Press.

Schwartz, L. M., and J. W. Truman. 1984. Hormonal control of muscle atrophy and degeneration in the moth *Antheraea polyphemus*. *J. Exp. Biol.* 111:13–30.

Shotwell, R. L. 1941. Life histories and habits of some grasshoppers of economic importance on the great plains. *Tech. Bull. U.S. Dept. Agric.* 774.

Skovmand, O., and S. B. Pedersen. 1983. Song recognition and song pattern in a short-horned grasshopper. *J. Comp. Physiol.* 153:393–401.

Stevens, E. D., and R. K. Josephson. 1977. Metabolic rate and body temperature in singing katydids. *Physiol. Zool.* 50:31–42.

Stower, W. J., and J. F. Griffiths. 1966. The body temperature of the desert locust *(Schistocerca gregaria)*. *Entomol. Exp. Appl.* 9:127–178.

Truman, J. W. 1973. Temperature sensitive programming of the silkmoth flight clock: A mechanism for adapting to the seasons. *Science* 182:727–729.

Walker, T. J. 1957. Specificity in the responses of female tree crickets (Orthoptera, Gryllidae, Oecanthinae) to calling songs of males. *Ann. Entomol. Soc. Am.* 50:626–636.

——— 1962. Factors responsible for intra-specific variation in the calling song of crickets. *Evolution* 16:407–428.

——— 1969a. Systematics and acoustic behavior of United States crickets of the genus *Orocharis* (Orthoptera: Gryllidae). *Ann. Entomol. Soc. Am.* 62:752–762.

——— 1969b. Systematics and acoustic behavior of United States crickets of the genus *Cyrtoxipha* (Orthoptera: Gryllidae). *Ann. Entomol. Soc. Am.* 62:945–952.

——— 1975a. Effects of temperature, humidity, and age on stridulatory rates in *Atlanticus* spp. (Orthoptera: Tettigoniidae: Decticinae). *Ann. Entomol. Soc. Am.* 68:607–611.

——— 1975b. Effects of temperature on rates in poikilotherm nervous systems in evidence for calling songs of meadow crickets (Orthoptera: Tettogoniidae: *Orchelimum*) and reanalysis of published data. *J. Comp. Physiol.* 101:57–69.

Waloff, Z. 1963. Field studies on solitary and transient desert locusts in the Red Sea area. *Anti-Locust Bull.* 40:1–93.

Walsh, J. 1986. Return of the locust: A cloud over Africa. *Science* 234:17–19.

Weis-Fogh, T. 1952. Fat combustion and metabolic rate of flying locusts *(Schistocerca gregaria* Forskal). *Phil. Trans. Roy. Soc. Lond.* B237:1–36.

——— 1964a. Functional design of the tracheal system of flying insects as compared with the avian lung. *J. Exp. Biol.* 41:207–227.

——— 1964b. Biology and physics of locust flight. VIII. Lift and metabolic rate of flying locusts. *J. Exp. Biol.* 41:257–271.

——— 1967. Respiration and tracheal ventilation in locusts and other flying insects. *J. Exp. Biol.* 47:561–587.

Whitman, D. W. 1986. Developmental thermal requirements for the grasshopper *Taeniopoda eques* (Orthoptera: Acrididae). *Ann. Entomol. Soc. Am.* 79:711–714.

—— 1987. Thermoregulation and daily activity patterns in a black desert grasshopper, *Taeniopoda eques*. *Anim. Behav.* 35:1814–1826.

—— 1988. Function and evolution of thermoregulation in the desert grasshopper *Taeniopoda eques*. *J. Anim. Ecol.* 57:369–383.

Whitman, D. W., M. S. Blum, and C. G. Jones. 1985. Chemical defense in *Taeniopa eques* (Orthoptera: Acrididae): Role of the metathoracic secretion. *Ann. Entomol. Soc. Am.* 78:451–455.

Whitman, D. W., and L. Orsak. 1985. Biology of *Taeniopoda eques* (Orthoptera: Acrididae) in southeastern Arizona. *Ann. Entomol. Soc. Am.* 78:811–825.

Wigglesworth, V. B., and J. D. Gillett. 1934. The function of the antennae in *Rhodnius prolixus* (Hemiptera) and the mechanism of orientation to the host. *J. Exp. Biol.* 11:120–139.

Wilson, D. M. 1961. The central nervous control of flight in the locust. *J. Exp. Biol.* 38:471–490.

—— 1968. The nervous control of insect flight and related behavior. *Adv. Insect Physiol.* 5:289–338.

Wilson, D. M., and T. Weis-Fogh. 1962. Patterned activity of co-ordinated motor units, studied in flying locusts. *J. Exp. Biol.* 39:643–667.

5. Beetles Large and Small

Barnett, P. S., J. J. A. Heffron, and H. R. Hepburn. 1975. Some thermal characteristics of insect flight; enzyme optima versus intra-thoracic temperature. *S. Afr. J. Sci.* 71:373–374.

Bartholomew, G. A., and T. M. Casey. 1977a. Endothermy during terrestrial activity in large beetles. *Science* 195:882–883.

—— 1977b. Body temperature and oxygen consumption during rest and activity in relation to body size in some tropical beetles. *J. Therm. Biol.* 2:173–176.

Bartholomew, G. A., and B. Heinrich. 1978. Endothermy in African dung beetles during flight, ball making, and ball rolling. *J. Exp. Biol.* 73:65–83.

Bartholomew, G. A., J. R. B. Lighton, and G. N. Louw. 1985. Energetics of locomotion and patterns of respiration in tenebrionid beetles from the Namib Desert. *J. Comp. Physiol.* 155:155–162.

Bolwig, N. 1957. Experiments on the regulation of the body temperature of certain tenebrionid beetles. *J. Ent. Soc. S. Afr.* 20:454–458.

Buxton, P. A. 1924. Heat, moisture and animal life in deserts. *Proc. Roy. Soc. Lond.* B96:123–131.

Chappell, M. A. 1984. Thermoregulation and energetics of the green fig beetle *(Cotinus texana)* during flight and foraging behavior. *Physiol. Zool.* 57:581–589.

Donaldson, J. M. 1981. Population dynamics of adult cetoniinae Coleoptera: Scarabaeidae) and their relationship to metereological conditions. *Phytophylactica* 13:11–21.

Dotterweich, H. 1928. Beiträge zur Nervenphysiologie der Insekten. *Zool. Jb. Abt. Allg. Zool. Physiol. Tiere* 44:399–425.

Dreisig, H. 1980. Daily activity, thermoregulation and water loss in the tiger beetle *Cicindela hybrida*. *Oecologia* (Berlin) 44:376–389.

———— 1981. The rate of predation and its temperature dependence in a tiger beetle, *Cicindela hybrida*. *Oikos* 36:196–202.

———— 1990. Thermoregulatory stilting in tiger beetles, *Cicindela hybrida* L. *J. Arid Environ.* 19:297–302.

Edney, E. B. 1971. The body temperature of tenebrionid beetles in the Namib Desert of southern Africa. *J. Exp. Biol.* 55:253–272.

Ellertson, F. E. 1958. Biology of some Oregon rain beetles, *Plecoma* spp., associated with fruit trees in Wasco and Hood River Counties. Ph.D. thesis, Oregon State University.

Erbeling, L., and W. Paarmann. 1985. Diel activity patterns of the desert carabid beetle *Thermophilium* (=*Anthia*) *sexmaculatum* F. (Coleoptera:Carabidae). *J. Arid Environ.* 8:141–155.

———— 1986. The role of circannual rhythm of thermoregulation in the control of the reproductive cycle of the desert carabid beetle *Thermophilium sexmaculatum* F. In Carabid Beetles: Their Adaptations and Dynamics, ed. P. J. den Boer, M. L. Luff, F. Mossakowski, and F. Weber, pp. 125–146. Stuttgart, New York: Gustav Fisher.

Ganeshaiah, K. N., and V. V. Belavadi. 1986. Habitat segregation in four species of adult tiger beetles (Coleoptera: Cicindelidae). *Ecol. Entomol.* 11:147–154.

Guppy, M., S. Guppy, and J. Hebrard. 1983. Behaviour of the riverine tiger beetle, *Lophyridia dongalensis imperatrix*: Effect of water availability on thermoregulatory strategy. *Entomol. Exp. Appl.* 33:276–282.

Hadley, N. F. 1971. Micrometerology and energy exchange in two desert arthropods. *Ecology* 49:726–734.

Hadley, N. F., T. D. Schultz, and A. Savill. 1988. Spectral reflectances of three tiger beetle subspecies *(Neocicindela perhispida)*: Correlation with habitat substrate. *N. Z. J. Zool.* 15:343–346.

Hamilton, W. J., III. 1971. Competition and thermoregulatory behavior of the Namib Desert tenebrionid genus *Cardiosis*. *Ecology* 52:810–822.

———— 1973. *Life's Color Code*. New York: McGraw.

Hanski, I., and Y. Cambefort, eds. 1991. *Dung Beetle Ecology*. Princeton: Princeton University Press.

Heinrich, B. 1974. Thermoregulation in bumblebees. I. Brood incubation by *Bombus vosnesenskii* queens. *J. Comp. Physiol.* 88:129–140.

Heinrich, B., and G. A. Bartholomew. 1979. Roles of endothermy and size in inter- and intraspecific competition for elephant dung in an African dung beetle, *Scarabaeus laevistriatus*. *Physiol. Zool.* 52:484–496.

Heinrich, B., and M. J. Heinrich. 1983. Heterothermia in foraging workers and drones of the bumblebee *Bombus terricola*. *Physiol. Zool.* 56:563–567.

Heinrich, B., and E. McClain. 1986. "Laziness" and hypothermia as a foraging strategy in flower scarabs (Coleoptera: Scarabaeidae). *Physiol. Zool.* 59:273–282.

Henwood, K. 1975a. Infrared transmittance as an alternative thermal strategy in the desert beetle *Onyymacris plana*. *Science* 189:993–994.

———— 1975b. A field-tested thermoregulation model for two diurnal Namib Desert tenebrionid beetles. *Ecology* 56:1329–1342.

Hölldobler, B. 1972. Behavioral adaptations of beetles to ecological niches in ant colonies. *Verhandlungsbericht Dtsch. Zool. Ges.* 65:137–144.

Holm, E., and E. B. Edney. 1973. Daily activity of Namib Desert arthropods in relation to climate. *Ecology* 54:45–56.

Kenagy, G. J., and R. D. Stevenson. 1982. Role of body temperature in the seasonality of daily activity in tenebrionid beetles of eastern Washington. *Ecology* 63:1491–1503.

Koch, C. 1961. Some aspects of abundant life in the vegetationless sand of the Namib Desert dunes. *J. Sw. Afr. Sci. Soc.* 15:8–34.

––––– 1962. The tenebrionidae of Southern Africa. XXXI. Comprehensive notes on the tenebrionid fauna of the Namib Desert. *Ann. Transv. Mus.* 24:61–106.

Krogh, A., and E. Zeuthen. 1941. The mechanism of flight preparation in some insects. *J. Exp. Biol.* 18:1–10.

Leston, D., J. W. S. Pringle, and D. C. S. White. 1965. Muscular activity during preparation for flight in a beetle. *J. Exp. Biol.* 42:409–414.

Machin, K. E., J. W. S. Pringle, and M. Tamasige. 1962. The physiology of insect fibrillar muscle. IV. The effect of temperature on a beetle flight muscle. *Proc. Roy. Soc. Lond.* B155:493–499.

Marden, J. H. 1987. In pursuit of females: Following and contest behavior by males of a Namib Desert tenebrionid beetle, *Physadesmia globosa*. *Ethology* 75:15–24.

May, M. L., D. L. Pearson, and T. M. Casey. 1986. Oxygen consumption of active and inactive adult tiger beetles. *Physiol. Entomol.* 11:171–179.

McClain, E., C. J. Kok, and A. G. Monard. 1991. Reflective wax blooms on black Namib Desert beetles enhance day activity. *Naturwissenschaften* 78:40–42.

McClain, E., R. L. Praetorius, S. A. Hanrahan, and M. K. Seeley. 1984a. Dynamics of the wax bloom of a seasonal Namib Desert tenebrionid, *Cauricara phalangium* (Coleoptera: Adesmiini). *Oecologia* (Berlin) 63:314–319.

McClain, E., M. J. Savage, and K. Nott. 1984b. Reflectivity of the cuticle of the Namib Desert tenebrionid, *Cauricara phalangium*, with a wax bloom. *S. Afr. J. Sci.* 80:183–184.

McClain, E., M. K. Seeley, N. F. Hadley, and V. Gray. 1985. Wax blooms in tenebrionid beetles of the Namib Desert: Correlates with environment. *Ecology* 66:112–118.

Morgan, K. R. 1985. Body temperature regulation and terrestrial activity in the ectothermic beetle *Cicindela tranquebarica*. *Physiol. Zool.* 58:29–37.

––––– 1987. Temperature regulation, energy metabolism, and mate-searching in rain beetles (*Plecoma* spp.), winter-active, endothermic scarabs (Coleoptera). *J. Exp. Biol.* 128:107–122.

Morgan, K. R., and G. A. Bartholomew. 1982. Homeothermic response to reduced ambient temperature in a scarab beetle. *Science* 216:1409–1411.

Moser, J. C., and W. A. Thompson. 1986. Temperature thresholds related to flight of *Endroctonus frontalis* Zimm. (Col.: Scolytidae) *Agronomie* 6:905–910.

Nicolson, S. W. 1987. Absence of endothermy in flightless dung beetles from southern Africa. *S. Afr. J. Zool.* 22:323–324.

Nicolson, S. W., G. A. Bartholomew, and M. K. Seely. 1984. Ecological correlates of locomotion speed, morphometrics and body temperature in three Namib Desert tenebrionid beetles. *S. Afr. J. Zool.* 19:131–134.

Nicolson, S. W., and G. N. Louw. 1980. Preflight thermogenesis, conductance and thermoregulation in the Protea beetle, *Trichostetha fascicularis* (Scarabaeidae: Cetoniinae). *S. Afr. J. Sci.* 76:124–126.

Oertli, J. J. 1989. Relationship of wing beat frequency and temperature during take-off flight in temperate-zone beetles. *J. Exp. Biol.* 145:321–338.

Paarmann, W., L. Erbeling, and K. Spinnler. 1986. Ant and ant brood preying larvae: An adaptation of carabid beetles to arid environment. In *Carabid Beetles: Their Adaptations and Dynamics*, P. J. den Boer, M. L. Luff, F. Mossakowsi, and F. Weber, pp. 79–90. Stuttgart, New York: Gustav Fisher.

Pearson, D. L., and C. B. Knisley. 1985. Evidence for food as a limiting resource in the life cycle of tiger beetles (Coleoptera: Cicindelidae). *Oikos* 45:161–168.

Pearson, D. L., and E. J. Mury. 1979. Character divergence and convergence among tiger beetles (Coleoptera: Cicindelidae). *Ecology* 60:557–566.

Pearson, D. L., and S. L. Stemberger. 1980. Competition, body size, and the relative energy balance of adult tiger beetles (Coleoptera: Cicindelidae). *Am. Midl. Nat.* 104:373–377.

Roberts, C. S., D. Mitchell, M. K. Seeley, and E. L. McClain. 1991. Beetling the heat: The thermal significance of running in *O. plana*. Unpublished manuscript.

Sato, H., and M. Imamori. 1988. Further observations on the nesting behaviour of a subsocial ball-rolling scarab, *Kheper aegyptiorum*. *Kontyû* (Tokyo) 56:873–878.

Schneider, P. 1980. Contributions to flight physiology in beetles. 4. Body temperature, flight behavior and wing beat frequency. *Zool. Anz.* (Jena) 205:1–19.

Schultz, T. D., and N. F. Hadley. 1987a. Microhabitat segregation and physiological differences in co-occurring tiger beetle species, *Cicindela oregona* and *Cicindela tranquebarica*. *Oecologia* (Berlin) 73:363–370.

——— 1987b. Structural colors of tiger beetles and their role in heat transfer through the integument. *Physiol. Zool.* 60:737–745.

Seely, M. K., and D. Mitchell. 1987. Is the subsurface environment of the Namib Desert dunes a thermal haven for chthonic beetles? *S. Afr. J. Zool.* 22:57–61.

Thiele, H. V. 1977. *Carabid Beetles in Their Environments*. Berlin, Heidelberg, New York: Springer.

Turner, J. S., and A. T. Lombard. 1990. Body color and body temperature in white and black Namib Desert beetles. *J. Arid Environ.* 19:303–315.

Wharton, R. A. 1980. Colouration and diurnal activity patterns in some

Namib Desert Zophosini (Coleoptera: Tenebrionidae). *J. Arid Environ.* 3:309–317.

Young, O. R. 1984. Perching of neotropical dung beetles on leaf surfaces: An example of behavioral thermoregulation? *Biotropica* 16:324–327.

6. *Bumblebees out in the Cold*

Bailey, H. 1954. The respiratory currents in the tracheal system of the adult honeybee. *J. Exp. Biol.* 31:589–593.

Bastian, J., and H. Esch. 1970. The nervous control of the indirect flight muscles of the honeybee. *Z. Vergl. Physiol.* 67:307–324.

Bertsch, A. 1984. Foraging in male bumblebees (*Bombus lucorum* L.): Maximizing energy or minimizing water load? *Oecologia* (Berlin) 62:325–326.

Boettiger, E. G. 1960. Insect flight muscle and their basic physiology. *Ann. Rev. Entomol.* 5:1–15.

Boettiger, E. G., and E. Furshpan. 1954. The response of fibrillar muscle to rapid release and stretch. *Biol. Bull* (Woods Hole) 107:305.

Brower, L. P., J. Z. Brower, and P. W. Westcott. 1960. Experimental studies of mimicry. V. The reactions of toads *(Bufo terrestris)* to bumblebees *(Bombus americanorum)* and their robber fly mimics *(Mallophora bomboides)*, with discussion of aggressive mimicry. *Am. Nat.* 94:343–355.

Buttel-Reepen, H. v. 1903. Die phylogenetische Entstehung des Bienenstaates, sowie Mitteilungen zur Biologie der solitären und sozialen Apiden. *Biol. Zentralbl.* 23:89–108.

Church, N. S. 1960. Heat loss and body temperature of flying insects. I. Heat loss by evaporation of water from the body. II. Heat conduction within the body and its loss by radiation and convection. *J. Exp. Biol.* 37:171–212.

Clark, M. G., D. P. Bloxam, P. C. Holland, and H. A. Lardy. 1973a. Estimation of fructose diphosphatase-phosphofructokinase substrate cycle in the flight muscle of *Bombus affinis. Biochem J.* 134:589–597.

Clark, M. G., C. H. Williams, W. F. Pfeifer, D. P. Bloxam, P. C. Holland, C. A. Taylor, and H. A. Lardy. 1973. Accelerated substrate cycling of fructose-6-phosphate in the muscle of malignant hyperthermic pigs. *Nature* 245:99–101.

Ellington, C. P., K. E. Machin, and T. M. Casey. 1990. Oxygen consumption of bumblebees in forward flight. *Nature* 347:472–473.

Esch, H., and F. Goller. 1991. Neural control of honeybee fibrillar muscle during shivering and flight, *J. Exp. Biol.* 159:419–431.

Esch, H., F. Goller, and B. Heinrich. 1991. How do bees shiver? *Naturwissenschaften* 78:325–328.

Frank, A. 1941. Eigenartige Flugbahnen bei Hummelmännchen. *Z. Vergl. Physiol.* 28:467–484.

Free, J. B., and C. G. Butler. 1959. *Bumblebees.* London: Collins.

Gabritchevsky, E. 1926. Convergence of coloration between American pilose flies and bumblebees *(Bombus). Biol. Bull* (Woods Hole) 51:269–287.

Girard, M. 1869. Études sur la chaleur libre dégagée par les animaux invertebrés et spécialement les insectes. *Ann. Sci. Nat. Zool.* 11:135–274.

Goller, F., and H. Esch. 1990. Comparative study of chill-coma temperatures and muscle potentials in insect flight muscles. *J. Exp. Biol.* 150:221–231.

Greive, H., and B. Surholt. 1990. Dependence of fructose-bis-phosphatase from flight muscles of the bumblebees (*Bombus terrestris* L.) on calcium ions. *Comp. Biochem. Physiol.* 97B:197–200.

Hasselrot, T. B. 1960. Studies on Swedish bumblebees (genus *Bombus* Latr.): Their domestication and biology. *Opuscula Entomol.*, Suppl. 17:1–192.

Heinrich, B. 1972a. Temperature regulation in the bumblebee, *Bombus vagans:* A field study. *Science* 175:185–187.

——— 1972b. Patterns of endothermy in bumblebee queens, drones and workers. *J. Comp. Physiol.* 77:65–79.

——— 1972c. Energetics of temperature regulation and foraging in a bumblebee, *Bombus terricola. J. Comp. Physiol.* 77:49–64.

——— 1972d. Physiology of brood incubation in the bumblebee, *Bombus vosnesenskii. Nature* 239:223–225.

——— 1973. The energetics of the bumblebee. *Sci. Am.* 228:97–102.

——— 1974. Thermoregulation in bumblebees. I. Brood incubation by *Bombus vosnesenskii* queens. *J. Comp. Physiol.* 88:129–140.

——— 1975. Thermoregulation in bumblebees. II. Energetics of warm-up and free flight. *J. Comp. Physiol.* 96:155–166.

——— 1976a. Heat exchange in relation to blood flow between thorax and abdomen in bumblebees. *J. Exp. Biol.* 64:561–585.

——— 1976b. Bumblebee foraging and the economics of sociality. *Am. Sci.* 64:384–395.

——— 1979. *Bumblebee Economics.* Cambridge, Mass.: Harvard University Press.

Heinrich, B., and M. J. E. Heinrich. 1983a. Size and caste in temperature regulation by bumblebees. *Physiol. Zool.* 56:552–562.

——— 1983b. Heterothermia in foraging workers and drones of the bumblebee, *Bombus terricola. Physiol. Zool.* 56:563–567.

Heinrich, B., and A. E. Kammer. 1973. Activation of the fibrillar muscles in the bumblebee during warm-up, stabilization of thoracic temperature and flight. *J. Exp. Biol.* 58:677–688.

Heinrich, B., and C. Pantle. 1975. Thermoregulation in small flies (*Syrphus* sp.): Basking and shivering. *J. Exp. Biol.* 62:599–610.

Heinrich, B., and D. F. Vogt. 1992. Thermoregulation by Arctic vs. temperate bumblebees: Thoracic and abdominal temperature. Unpublished manuscript.

Heran, H. 1952. Untersuchungen über den Temperatursinn der Honigbiene (*Apis mellifica*) unter besonderer Berücksichtigung der Wahrnehmung strahlender Wärme. *Z. Vergl. Physiol.* 34:179–206.

Himmer, A. 1925. Körpertemperaturen an Bienen und anderen Insekten. *Erlanger Jahrb. Bienenkunde* 3:44–115.

——— 1933. Die Nestwärme bei *Bombus agrorum* F. *Biol. Zentralblatt* 53:270–273.

Hochachka, P. W., and G. N. Somero. 1973. *Strategies of Biochemical Adaptation.* Philadelphia, London, Toronto: W. B. Saunders.

Hocking, B., and C. D. Sharplin. 1965. Flower basking in Arctic insects. *Nature* 206:215.

Ikeda, K., and E. G. Boettiger. 1965. Studies on the flight mechanism in insects. II. The innervation and electrical activity of the fibrillar muscles of the bumblebees, *Bombus. J. Insect Physiol.* 11:779–789.

Joos, B., P. A. Young, T. M. Casey. 1991. Wingstroke frequency of foraging bumblebees in relation to morphology and temperature. *Physiol. Zool.* 16:191–200.

Kammer, A. E. 1968. Motor patterns during flight and warm-up in Lepidoptera. *J. Exp. Biol.* 48:89–109.

Kammer, A. E., and B. Heinrich. 1972. Neural control of bumblebee fibrillar muscle during shivering. *J. Comp. Physiol.* 78:337–345.

———— 1974. Metabolic rates related to muscle activity in bumblebees. *J. Exp. Biol.* 61:219–227.

———— 1978. Insect flight metabolism. *Adv. Insect Physiol.* 13:133–228.

Knee, W. J., and J. T. Medler. 1965. The seasonal size increase of bumblebee workers (Hymenoptera: Bombus). *Can. Entomol.* 97:1149–1155.

Krogh, A., and E. Zeuthen. 1941. Mechanisms of flight preparation in some insects. *J. Exp. Biol.* 18:1–10.

Machin, K. E., and J. W. S. Pringle. 1959. The physiology of insect fibrillar muscle. II. Mechanical properties of beetle flight muscle. *Proc. Roy. Soc. Lond.* B151:204–225.

Marsh, R. L., and W. R. Dawson. 1988. Avian adjustments to cold. In *Animal Adaptations to Cold,* ed. L. Wang. New York, Heidelberg, Berlin: Springer-Verlag.

Mulloney, B. 1970. Impulse patterns in the flight motor neurones of *Bombus californicus* and *Oncopeltus fasciatus. J. Exp. Biol.* 52:59–77.

Newport, G. 1837. On the temperature of insects and its connexion with the functions of respiration and circulation in this class of invertebrate animals. *Phil. Trans. Roy. Soc. Lond.* 1837:259–338.

Newsholme, E. A., B. Crabtree, S. J. Higgins, S. D. Thornton, and C. Start. 1972. The activities of fructose diphosphatase in flight muscles from the bumble-bee and the role of this enzyme in heat generation. *Biochem. J.* 128:89–97.

Pekkarinen, A. 1979. Morphometric, colour and enzyme variation in bumblebees (Hymenoptera, Apidae, Bombus) in Fennoscandia and Denmark. *Acta Zool. Fennica* 158:1–60.

Plath, O. E. 1934. *Bumblebees and Their Ways.* New York: Macmillan.

Plowright, R. C., and R. E. Owen. 1980. The evolutionary significance of bumblebee color patterns: A mimetic interpretation. *Evolution* 34:622–637.

Pringle, J. W. S. 1954. The mechanism of the myogenic rhythm of certain insect fibrillar muscles. *J. Physiol.* 124:269–291.

Prŷs-Jones, O. E. 1986. Foraging behavior and activity of substrate cycle enzymes in bumblebees. *Anim. Behav.* 34:609–611.

Richards, K. W. 1973. Biology of *Bombus polaris* Curtis and *B. hyperboreus*

Schönherr at Lake Hazen, Northwest Territories (Hymenoptera: Bombini). *Quest. Entomol.* 9:115–157.

Sladen, F. W. L. 1912. *The Bumble-bee, Its Life History and How to Domesticate It.* London: Macmillan.

Snodgrass, R. E. 1956. *Anatomy of the Honey Bee.* Ithaca, N.Y.: Comstock.

Stiles, E. W. 1979. Evolution of color pattern and pubescence characteristics in male bumblebees: Automimicry vs. thermoregulation. *Evolution* 33:941–957.

Stone, G. N., and P. G. Willmer. 1989. Warm-up rates and body temperatures in bees: The importance of body size, thermal regime and phylogeny. *J. Exp. Biol.* 147:303–328.

Surholt, B., H. Greive, T. Baal, and A. Bertsch. 1990. Non-shivering thermogenesis in asynchronous flight muscles of bumblebees? Comparative studies on males of *Bombus terrestris, Xylocopa sulcatipes* and *Acherontia atropos. Comp. Biochem. Physiol.* 97A:439–499.

Surholt, B., H. Greive, C. Hommel, and A. Bertsch. 1988. Fuel uptake, storage and use in male bumblebees *Bombus terrestris* L. *J. Comp. Physiol.* B158:263–269.

Surholt, B., and E. A. Newsholme. 1981. Maximum activities and properties of glucose 6-phosphate in muscles from vertebrates and invertebrates. *Biochem. J.* 198:621–629.

——— 1983. The rate of substrate cycling between glucose and glucose 6-phosphate in muscle and fat-body of the hawk moth *(Acherontia atropos)* at rest during flight. *Biochem. J.* 210:49–54.

Unwin, D. M., and S. A. Corbet. 1984. Wingbeat frequency, temperature and body size in bees and flies. *Physiol. Entomol.* 9:115–121.

Vogt, F. D., and B. Heinrich. 1992. Thermoregulation in Arctic vs. temperate bumblebees. II. Size and insulation. Unpublished manuscript.

Wolf, T. J., P. Schmidt-Hempel; C. P. Ellington, and R. D. Stevenson. 1989. Physiological correlates of foraging efforts in honey-bees: Oxygen consumption and nectar load. *Funct. Ecol.* 3:417–424.

7. Tropical Bees

Armbruster, W. S., and K. D. McCormick. 1990. Diel foraging patterns of male euglossine bees: Ecological causes and evolutionary response by plants. *Biotropica* 22:160–171.

Baird, J. M. 1986. A field study of thermoregulation in the carpenter bee *Xylocopa virginica virginica* (Hymenoptera:Anthophoridae). *Physiol. Zool.* 59:157–167.

Casey, T. M., M. L. May, and K. R. Morgan. 1985. Flight energetics of euglossine bees in relation to morphology and wing stroke frequency. *J. Exp. Biol.* 116:271–289.

Chappell, M. A. 1982. Temperature regulation of carpenter bees *(Xylocopa californica)* foraging in the Colorado Desert of southern California. *Physiol. Zool.* 55:267–280.

——— 1984. Temperature regulation and energetics of the solitary bee *Centris pallida* during foraging and mate competition. *Physiol. Zool.* 57:215–225.

Dressler, R. L. 1968. Pollination in euglossine bees. *Evolution* 22:202–210.

Ellington, C. P., K. E. Machin, and T. M. Casey. 1990. Oxygen consumption of bumblebees in forward flight. *Nature* 347:472–473.

Gerling, D., P. D. Hurd, and A. Hefetz. 1983. Comparative behavioral biology of two Middle East species of carpenter bees (*Xylocopa* Latreille) (Hymenoptera: Apoidea). *Smithsonian Contributions to Zoology*, no. 369.

Heinrich, B. 1972. Energetics of temperature regulation and foraging in a bumblebee, *Bombus terricola* Kirby. *J. Comp. Physiol.* 77:49–64.

———— 1975. Thermoregulation and flight energetics of desert insects. In *Environmental Physiology of Desert Organisms*, ed. N. F. Hadley, pp. 95–105. Stroudsburg, Penn.: Dowden, Hutchinson and Ross.

———— 1976. Flowering phenologies: Bog, woodland, and disturbed habitats. *Ecology* 57:874–899.

———— 1980. Mechanisms of body temperature regulation in honeybees, *Apis mellifera*. II. Regulation of thoracic temperature at high air temperatures. *J. Exp. Biol.* 85:73–87.

Heinrich, B., and S. L. Buchmann. 1986. Thermoregulatory physiology of the carpenter bee, *Xylocopa varipuncta. J. Comp. Physiol.* B156:557–562.

Heinrich, B., and P. H. Raven. 1972. Energetics and pollination ecology. *Science* 176:597–602.

Inouye, D. W. 1975. Flight temperatures of male euglossine bees (Hymenoptera: Apidae: Euglossini). *J. Kansas Entomol. Soc.* 48:366–370.

Janzen, D. H. 1971. Euglossine bees as long-distance pollinators of tropical plants. *Science* 171:203–205.

Kimsey, L. S. 1980. The behavior of male orchid bees (Apidae, Hymenoptera, Insecta) and the question of leks. *Anim. Behav.* 28:996–1004.

Linsley, E. G. 1958. The ecology of solitary bees. *Hilgardia* 27:543–599.

———— 1960a. Observations on some matinal bees at flowers of *Cucurbita, Ipomaea,* and *Datura* in desert areas of New Mexico and southeastern Arizona. *J. New York Entomol. Soc.* 68:13–20.

———— 1960b. Ethological adaptations of solitary bees for the pollination of desert plants. *Proc. Int. Symp. Pollin. 1st, Copenhagen,* pp. 189–197.

Louw, G. D., and S. W. Nicolson. 1983. Thermal, energetic and nutritional considerations in the foraging and reproduction of the carpenter bee *Xylocopa capitata. J. Entomol. Soc. S. Afr.* 46:227–240.

MacSwain, J. W. 1957. The flight period of *Martinapis luteicornis* (Cockenell) (Hymenoptera: Apoidea). *Pan-Pacific Entomol.* 33:70.

May, M. L., and T. M. Casey. 1983. Thermoregulation and heat exchange in euglossine bees. *Physiol. Zool.* 56:541–551.

Nicolson, S. W., and G. D. Louw. 1982. Simultaneous measurement of evaporative water loss, oxygen consumption and thoracic temperature during flight in a carpenter bee. *J. Exp. Biol.* 222:287–296.

Stone, G. N., and P. G. Willmer. 1989. Warm-up rates and body temperatures in bees: The importance of body size, thermal regime and phylogeny. *J. Exp. Biol.* 147:303–328.

Wille, A. 1958. A comparative study of the dorsal vessel of bees. *Ann. Entomol. Soc. Am.* 51:538–546.

Willmer, G. P. 1988. The role of insect water balance in pollination ecology: *Xylocopa* and *Calotropis. Oecologia* 76:430–438.

8. Hot-Headed Honeybees

Allen, M. D. 1955. Respiration rates of worker honeybees of different ages and at different temperatures. *J. Exp. Biol.* 36:92–101.

Arnhart, L. 1906. Die Bedeutung der Aortenschlangenwindungen des Bienenherzens. *Zool. Anzeiger* 30:721–722.

Bastian, J., and H. Esch. 1970. The nervous control of the flight muscles of the honeybee. *Z. Vergl. Physiol.* 67:307–324.

Boettiger, E. G. 1957. Triggering of the contractile process in insect fibrillar muscle. In *Physiological Triggers,* ed. T. H. Bullock, pp. 103–106. Washington, D.C.: American Physiological Society.

Cahill, K., and S. Lustick. 1976. Oxygen consumption and thermoregulation in *Apis mellifera* workers and drones. *Comp. Biochem. Physiol.* 55A:355–357.

Calder, W. A. 1984. *Size, Function, and Life History.* Cambridge, Mass.: Harvard University Press.

Cena, K., and J. A. Clark. 1972. Effect of solar radiation on temperatures of working honey bees. *Nature* 236:222–223.

Coelho, J. R. 1989. The effect of thorax temperature and body size on flight speed in honey bee drones. *Am. Bee J.* 129:811–1812.

———— 1990. The effect of thorax temperature on force production during tethered flight in honeybee *(Apis mellifera)* drones, workers and queens. *Physiol. Zool.* (MS submitted).

———— 1991. Thermoregulation in honey bee drones *(Apis mellifera).* Unpublished manuscript.

Coelho, J. R., and J. B. Mitton. 1988. Oxygen consumption during hovering is associated with genetic variation of enzymes in honeybees. *Funct. Ecol.* 2:141–146.

Cooper, P., W. M. Schaffer, and S. L. Buchmann. 1985. Temperature regulation of honey bees *(Apis mellifera)* foraging in the Sonoran Desert. *J. Exp. Biol.* 114:1–15.

Duruz, C., and F. Baumann. 1968. Influence de la température sur le potential de repos et le potential récepteur d'une cellule photoréceptrice. *Helv. Physiol. pharmacol. Acta* 26:341–342.

Dyer, F. C., and T. D. Seeley. 1987. Interspecific comparison of endothermy in honeybees *(Apis):* Deviations from the expected size-related patterns. *J. Exp. Biol.* 127:1–26.

Esch, H. 1960. Über die Körpertemperaturen und den Wärmehaushalt von *Apis mellifica. Z. Vergl. Physiol.* 43:305–335.

———— 1964. Über den Zusammenhang zwischen Temperatur, Aktionspotentialen und Thoraxbewegungen bei der Honigbiene *(Apis mellifica* L.) *Z. Vergl. Physiol.* 48:547–551.

———— 1976. Body temperature and flight performance of honey bees in a servomechanically controlled wind tunnel. *J. Comp. Physiol.* 109:254–277.

———— 1988. The effects of temperature on flight muscle potentials in honeybees and cuculiinid winter moths. *J. Exp. Biol.* 135:109–117.

Esch, H., and J. Bastian. 1968. Mechanical and electrical activity in the indirect flight muscles of the honey bee. *Z. Vergl. Physiol.* 58:429–440.

Esch, H., and F. Goller. 1991. Neural control of honeybee fibrillar muscles during shivering and flight. *J. Exp. Biol.* 159:419–431.

Esch, H., W. Nachtigall, and S. N. Kogge. 1975. Correlations between aerodynamic output, electrical activity in the indirect flight muscles and wing positions of bees flying in a servomechanically controlled wind tunnel. *J. Comp. Physiol.* 100:147–159.

Fahrenholz, L. 1986. Die soziale Thermoregulation im Stock der Honigbiene *(Apis mellifera carnica)* und die kalorimetrische Bestimmung der Wärmeproduktion bei Einzeltieren. Ph.D. dissertation, Free University of Berlin.

Feller, P., and W. Nachtigall. 1989. Flight of the honeybee. II. Inner- and surface thorax temperatures and energetic criteria correlated to flight parameters. *J. Comp. Physiol.* B158:719–727.

Free, J. B. 1965. The allocation of duties among worker honeybees. *Zool. Soc. Lond. Symp.* 14:39–50.

Free, J. B., and Y. Spencer-Booth. 1958. Observations of the temperature and food consumption of honeybees *(Apis mellifera)*. *J. Exp. Biol.* 35:930–937.

———— 1960. Chill-coma and cold death temperatures of *Apis mellifera*. *Entomol. Exp. Appl.* 3:222–230.

Freudenstein, K. 1928. Das Herz und das Zirkulationssystem der Honigbiene *(Apis mellifica* L.) *Z. Wiss. Zool.* 132:404–475.

Frisch, K. von. 1967. *Dance Language and Orientation of Bees.* Cambridge, Mass.: Harvard University Press.

Goller, F., and H. Esch. 1990a. Comparative study of chill-coma temperatures and muscle potentials in insect flight muscles. *J. Exp. Biol.* 150:221–231.

———— 1990b. Oxygen consumption and flight muscle activity during warm-up in workers and drones of *Apis mellifera*. *J. Comp. Physiol.* (in press).

———— 1990c. Muscle potentials and temperature acclimation in flight muscles of workers and drones of *Apis mellifera*. *J. Therm. Biol.* 15:307–312.

Harrison, J. M. 1986. Caste-specific changes in honeybee flight activity. *Physiol. Zool.* 59:175–187.

Heinrich, B. 1975. Thermoregulation in bumblebees. II. Energetics of warm-up and free flight. *J. Comp. Physiol.* 96:155–166.

———— 1976. Heat exchange in relation to blood flow between thorax and abdomen in bumblebees. *J. Exp. Biol.* 54:561–585.

———— 1979a. Keeping a cool head: Honeybee thermoregulation. *Science* 205:1269–1271.

———— 1979b. Thermoregulation of African and European honeybees during foraging, attack, and hive exits and returns. *J. Exp. Biol.* 80:217–229.

———— 1979c. *Bumblebee Economics.* Cambridge, Mass.: Harvard University Press.

———— 1980a. Mechanisms of body-temperature regulation in honeybees, *Apis mellifera.* I. Regulation of head temperature. *J. Exp. Biol.* 85:61–72.

———— 1980b. Mechanisms of body-temperature regulation in honeybees, *Apis mellifera.* II. Regulation of thoracic temperature at high air temperatures. *J. Exp. Biol.* 85:73–87.

———— 1981a. Energetics of honeybee swarm thermoregulation. *Science* 212:565–566.

———— 1981b. The mechanisms and energetics of honeybee swarm temperature regulation. *J. Exp. Biol.* 91:25–55.

Heran, H. 1952. Untersuchungen über den Temperatursinn der Honigbiene *(Apis mellifica)* unter besonderer Berücksichtigung der Wahrnehmung strahlender Wärme. *Z. Vergl. Physiol.* 34:179–206.

Herold, R. C. 1965. Development and utrastructural changes of sarcosomes during honeybee flight muscle development. *Dev. Biol.* 12:269–286.

Herold, R. C., and H. Borei. 1963. Cytochrome changes during honeybee flight muscle development. *Dev. Biol.* 8:67–79.

Hersch, M. I., R. M. Crewe, H. R. Hepburn, P. R. Thompson, and N. Savage. 1978. Sequential development of glycolytic competence in the muscles of worker honeybees. *Comp. Biochem. Physiol.* 61B:427–431.

Heusner, A., and M. Roth. 1963. Consommation de l'oxygène de l'abeille à différentes températures. *C. R. Hebd. Séance Acad. Sci., Paris* 256:284–285.

Himmer, A. 1925. Körpertemperaturmessungen an Bienen und anderen Insekten. *Erlanger Jb. Bienenk.* 3:44–115.

———— 1926. Der sozial Wärmehaushalt der Honigbiene. I. Die Wärme im nicht brütenden Wintervolk. *Erlanger Jb. Bienenk.* 4:1–51.

———— 1927. Der soziale Wärmehaushalt der Honigbiene. II. Die Wärme der Bienenbrut. *Erlanger Jb. Bienenk.* 5:1–32.

———— 1932. Die Temperaturverhältnisse bei den sozialen Hymenopteren. *Biol. Rev.* 7:224–253.

Ikeda, K., and E. G. Boettiger. 1965. Studies on the flight mechanism of insects. II. The innervation and electrical activity of the fibrillar muscles of the bumblebee, *Bombus. J. Insect Physiol.* 11:779–789.

Jungmann, R., U. Rothe, and W. Nachtigall. 1989. Flight of the honey bee. II. Thorax surface temperature and thermoregulation during tethered flight. *J. Comp. Physiol.* B158:711–718.

Kammer, A. E., and B. Heinrich. 1974. Metabolic rates related to muscle activity in bumblebees. *J. Exp. Biol.* 61:219–227.

Kosmin, N. P., W. W. Alpatov, and M. S. Resnitschenko. 1932. Zur Kenntnis des Gaswechsels und des Energieverbrauchs der Biene in Beziehung zu deren Aktivität. *Z. Vergl. Physiol.* 17:408–422.

Lensky, Y. 1964a. L'économie de liquides chez les abeilles aux températures élevées. *Insectes Sociaux* (Paris) 11:207–222.

——— 1964b. Résistance des abeilles (*Apis mellifica* L. var. *ligustica*) a des températures élevées. *Insectes Sociaux* (Paris) 11:293–300.

Lindauer, M. 1952. Ein Beitrag der Frage der Arbeitsteilung im Bienenstaat. *Z. Vergl. Physiol.* 34:299–345.

Louw, G., and N. Hadley. 1985. Water economy of the honeybee: A stoichiometric accounting. *J. Exp. Zool.* 235:147–150.

Machin, K. E., J. W. S. Pringle, and M. Tamasige. 1962. The phhsiology of insect fibrillar muscle. IV. The effect of temperature on beetle flight muscle. *Proc. Roy. Soc. Lond.* B155:493–499.

Mardan, M., and P. G. Kevan. 1989. Honeybees and "yellow rain." *Nature* 341:191.

Nachtigall, W., U. Rothe, P. Feller, and R. Jungmann. 1989. Flight of the honeybee. III. Flight metabolic power calculated from gas analysis, thermoregulation and fuel consumption. *J. Comp. Physiol.* B158:729–737.

Neukirch, A. 1982. Dependence of the life span of the honeybee *(Apis mellifica)* upon flight performance and energy consumption. *J. Comp. Physiol.* 146:38–40.

Pearl, R. 1928. *The Rate of Living.* New York: Knopf.

Pirsch, G. B. 1923. Studies on the temperature of individual insects, with special reference to the honey bee. *J. Agric. Res.* 24:257–287.

Pissarew, W. J. 1898. Das Herz der Biene (*Apis mellifera* L.). *Zool. Anzeiger* 21:282–283.

Rösch, G. A. 1925. Untersuchungen über die Arbeitsteilung im Bienenstaat. I. Die Tätigkeiten im normalen Bienenstaate und ihre Beziehungen zum Alter der Arbeiterinnen. *Z. Vergl. Physiol.* 2:571–631.

Rothe, U., and W. Nachtigall. 1989. Flight of the honeybee. IV. Respiratory quotients and metabolic rates during sitting, walking and flying. *J. Comp. Physiol.* 158B:739–749.

Schmaranzer, S. 1983. Thermovision bei trinkenden und tanzenden Honigbienen *(Apis mellifera carnica). Verh. Dtsch. Zool. Ges.* 1983:319.

——— 1984. Körpertemperaturmessungen mittels Thermovision bei Honigbienen während der Futtersuche und des Tanzens. In *International Symposium in Memory of Dr. Franz Saubeer,* 51–53. Univ. für Bodenkultur, Vienna.

Schmaranzer, S., and A. Stabentheiner. 1988. Variability of the thermal behavior of honeybees on a feeding place. *J. Comp. Physiol.* B15:135–141.

Schmidt-Hempel, P., and T. Wolf. 1988. Foraging effort and life span of workers in a social insect. *J. Anim. Ecol.* 57:500–521.

Snodgrass, R. E. 1956. *Anatomy of the Honey Bee.* Ithaca, N.Y.: Comstock Publishing Associates, Cornell University Press.

Sotavalta, O. 1954. On the fuel consumption of the honeybee (*Apis mellifica* L.) in flight experiments. *Ann. Zool. Soc. Vanamo* 16:1–27.

Southwick, E. E. 1985. Bee hair structure and the effect of hair on metabolism at cool temperatures. *Apic. Res.* 24:144–149.

Stabentheiner, A., and S. Schmaranzer. 1986. Thermografie bei Bienen: Körpertemperaturen am Futterplatz und im "Bienenbart." *Verh. Deutsch. Zool. Ges.* 79:417–418.

———— 1987. Thermographic determination of body temperatures in honey bees and hornets: Calibration and applications. *Thermology* 2:563–572.

———— 1988. Flight-related thermobiological investigations of honeybees *(Apis mellifera carnica)*. *Biona* 6:89–102.

Waddington, K. D. 1990. Foraging profits and thoracic temperature of honey bees *(Apis mellifera)*. *J. Comp. Physiol.* B160:325–329.

Witherell, P. C. 1971. Duration of flight and of interflight time in drone honey bees, *Apis mellifera. Ann. Entomol. Soc. Am.* 64:609–612.

Withers, P. C. 1981. The effects of ambient air pressure on oxygen consumption of resting and hovering honeybees. *J. Comp. Physiol.* 141:433–437.

Wolf, T. J., and P. Schmidt-Hempel. 1989. Extra loads and foraging life span in honeybee workers. *J. Anim. Ecol.* 58:943–954.

Wolf, T. J., P. Schmidt-Hempel, C. P. Ellington, and R. D. Stevenson. 1989. Physiological correlates of foraging efforts in honey-bees: Oxygen consumption and nectar load. *Func. Ecol.* 3:417–424.

Zander, E. 1911. *Der Bau der Biene.* Stuttgart: Verlag Eugen Ulmer.

9. The Tolerance of Ants

Bartholomew, G. A., J. R. B. Lighton, and D. H. Feener, Jr. 1988. Energetics of trail running, load carriage, and emigration in the column-raiding army ant *Eciton hamatum. Physiol. Zool.* 61:57–68.

Bernstein, R. A. 1974. Seasonal food abundance and foraging activity in some desert ants. *Am. Nat.* 108:490–498.

Box, T. W. 1960. Notes on the harvester ant, *Pogonomyrmex barbatus* var. *molefacieus,* in southern Texas. *Ecology* 41:381–382.

Briese, D. T., and B. J. MacAuley. 1980. Temporal structure of an ant community in semi-arid Australia. *Austr. J. Ecol.* 5:121–134.

Christian, K. A., and S. R. Morton. 1992. Extreme thermophilia in a central Australian ant, *Melophorus bagoti.* Unpublished manuscript.

Clark, W. H., and P. L. Conners. 1973. A quantitative examination of spring foraging of *Veromessor pergandei* in Northern Death Valley, California (Hymenoptera: Formicidae). *Am. Midl. Nat.* 90:467–474.

Delye, G. 1967. Physiologie et comportement de quelques fourmis (Hym. Formicidae) du Sahara en rapport avec les principaux facteurs du climat. *Insectes Sociaux* (Paris) 14:323–338.

Gamboa, G. J. 1976. Effects of temperature on the surface activity of the desert leaf-cutter ant, *Acromyrmex versicolor versicolor* (Pergande) (Hymenoptera: Formicidae). *Am. Midl. Nat.* 95:485–491.

Greenaway, P. 1981. Temperature limits in trailing activity in the Australian arid-zone meat ant *Iridomyrmex purpureus* form *viridiaeneus. Austr. J. Zool.* 29:621–630.

Hölldobler, B., and R. W. Taylor. 1983. A behavioral study of the primitive ant *Nothomyrmecia macrops* Clark. *Insectes Sociaux* (Paris) 30:384–401.

Hölldobler, B., and E. O. Wilson. 1990. *The Ants.* Cambridge, Mass.: Harvard University Press.

Kay, C. A. R., and W. G. Whitford. 1978. Critical thermal limits of desert

honey ants: Possible ecological implications. *Physiol. Zool.* 51:206–
213.

Leonard, J. G., and J. M. Herbers. 1986. Foraging tempo in two woodland
ant species. *Anim. Behav.* 34:1172–1181.

Lighton, J. R. B., G. A. Bartholomew, and D. H. Feener, Jr. 1987. Energetics of locomotion and load carriage and a model of the energy
cost of foraging of the leaf-cutting ant *Atta colombica* Guer. *Physiol.
Zool.* 60:524–537.

Lubin, Y. D., and J. R. Henschel. 1990. Foraging at the thermal limit:
Burrowing spiders (Seothyra, Eresidae) in the Namib desert dunes.
Oecologia 84:461–467.

Marsh, A. C. 1985a. Microclimatic factors influencing foraging patterns
and success of the thermophilic desert ant, *Ocymyrmex barbiger*. *Insectes Sociaux* (Paris) 32:289–296.

———— 1985b. Thermal responses and temperature tolerance in a diurnal
desert ant, *Ocymyrmex barbiger*. *Physiol. Zool.* 58:629–636.

———— 1987. The foraging ecology of two Namib Desert harvester ant
species. *S. Afr. J. Zool.* 22:130–136.

———— 1988. Activity patterns of some Namib Desert ants. *J. Arid Environ.*
14:61–73.

Mehlhop, P., and N. J. Scott. 1983. Temporal patterns of seed use and
availability in a guild of desert ants. *Ecol. Entomol.* 8:69–85.

Moser, J. C. 1967. Trail of the leaf-cutters. *Nat. Hist.* (New York) 76:32–
35.

Rissing, S. W. 1982. Foraging velocity of seed-harvester ants *Veromessor
pergandei* (Hymenoptera: Formicidae). *Environ. Entomol.* 11:905–
907.

Rogers, L. E. 1974. Foraging activity of the western harvester ant in the
shortgrass plains system. *Environ. Entomol.* 3:420–424.

Sanders, C. J. 1972. Seasonal and daily activity patterns of carpenter ants
(*Camponotus* spp.) in northwestern Ontario (Hymenoptera: Formicidae). *Can. Entomol.* 104:1681–1687.

Schumacher, A., and W. G. Whitford. 1974. The foraging ecology of two
species of Chihuahuan Desert ants: *Formica perpilosa* and *Trachymyr-
mex smithi neomexicanus* (Hymenoptera, Formicidae). *Insectes Sociaux*
(Paris) 21:317–330.

Shapley, H. 1920. Thermokinetics of *Liometopum apiculatum* Mayr. *Proc.
Nat. Acad. Sci.* 6:204–211.

———— 1924. Note on the thermokinetics of dolichoderine ants. *Proc. Nat.
Acad. Sci.* 10:436–439.

Traniello, J. F. A., M. S. Fujita, and R. V. Bowen. 1984. Ant foraging
behavior: Ambient temperature influences prey selection. *Behav.
Ecol. Sociobol.* 15:65–68.

Wehner, R. 1983. Taxonomie, Funktionsmorphologie und Zoogeographie
der saharischen Wüstenameise *Cataglyphus fortis* (Forel 1902) stat.
nov. *Senckenbergiana Biol.* 64:89–132.

———— 1984. Astronavigation in ants. *Ann. Rev. Entomol.* 29:277–298.

———— 1987. Spatial organization of foraging behavior in individually
searching desert ants, *Cataglyphis* (Sahara Desert) and *Ocymyrmex*

(Namib Desert). *Experientia* Suppl. 54:15–42. Basel: Birkhäuser Verlag.

Wehner, R., A. C. Marsh, and S. Wehner. 1992. Sahara ants survive by walking a thermal tightrope. Unpublished manuscript.

Whitford, W. G., and G. Ettershank. 1975. Factors affecting foraging activity in Chihuahuan Desert harvester ants. *Environ. Entomol.* 4:689–696.

10. *Wasps and the Heat of Battle*

Chapman, R. N., C. E. Mickel, J. R. Parker, G. E. Miller, and E. G. Kelly. 1926. Studies on the ecology of sand dune insects. *Ecology* 7:416–426.

Evans, H. E., F. E. Kurczewski, amd J. Alcock. 1980. Observations on the nesting behaviour of seven species of *Crabro* (Hymenoptera, Sphecidae). *J. Nat. Hist.* 14:865–882.

Gibo, D. L., A. Temporale, T. P. Lamarre, B. M. Soutar, and H. E. Dew. 1977. Thermoregulation in colonies of *Vespula arenaria* and *Vespula maculata* (Hymenoptera: Vespidae). III. Heat production in queen nests. *Can. Entomol.* 109:615–620.

Heinrich, B. 1984. Strategies of thermoregulation and foraging in two vespid wasps, *Dolichovespula maculata* and *Vespula vulgaris. J. Comp. Physiol.* B154:175–180.

Heinrich, B., and M. Heinrich. 1983. Size and caste in temperature regulation in bumblebees. *Physiol. Zool.* 56:552–562.

Ishay, J. 1973. Thermoregulation by social wasps in behavior and pheromones. *Trans. N.Y. Acad. Sci.* 35:147–462.

Larsson, F. K. 1989a. Insect mating patterns explained by microclimatic variables. *J. Therm. Biol.* 14:155–157.

——— 1989b. Temperature-induced alternative male mating tactics in a tropical digger wasp. *J. Insect Behav.* 2:849–852.

Larsson, F. K., and J. Tengo. 1989. The effects of temperature and body size on the mating pattern of a gregariously nesting bee, *Colletes cunicularis* (Hymenoptera: Colletidae). *Ecol. Entomol.* 14:279–286.

Makino, S., and S. Yamane. 1980. Heat production by the foundress of *Vespa simillima,* with description of its embryo nest (Hymenoptera: Vespidae). *Insecta Matsumurana* 19:89–101.

Mickel, C. E. 1928. *Biological and Taxonomic Investigations on the Mutillid Wasps.* Bulletin 143, Smithsonian Institution, U.S. National Museum.

O'Neill, K. M., and R. P. O'Neill. 1988. Thermal stress and microhabitat selection in territorial males of the digger wasp *Philanthus psyche* (Hymenoptera: Sphecidae). *J. Therm. Biol.* 13:15–20.

Stabentheiner, A., and S. Schmaranzer. 1987. Thermographic determination of body temperatures in honey bees and hornets: Calibration and applications. *Thermology* 2:563–572.

Steiner, A. 1930. Die Temperaturregulierung im Nest der Feldwespe (*Polistes gallica* var. *biglumis* L.). *Z. Vergl. Physiol.* 11:461–502.

Teräs, I. 1978. The activity of social wasps (Hymenoptera, Vespidae) at low temperatures. *Ann. Entomol. Fenn.* 44:101–104.

Weyrauch, W. 1936. Das Verhalten sozialer Wespen bei Nestüberhitzung. *Z. Vergl. Physiol.* 23:51–63.

Willmer, P. G. 1985a. Thermal ecology, size effects, and the origins of communal behaviour in *Cerceris* wasps. *Behav. Ecol. Sociobiol.* 17:151–160.

—— 1985b. Size effects of the hygrothermal balance and foraging pattern of a sphecid wasp, *Cerceris arenaria. Ecol. Entomol.* 10:469–479.

11. Flies of All Kinds

Alahiotus, S. N. 1983. Heat shock proteins: A new view on temperature compensation. *Comp. Biochem. Physiol.* 75B:379–387.

Asit, B., and C. L. Prosser. 1967. Biochemical changes in tissues of goldfish acclimated to high and low temperatures. I. Protein synthesis. *Comp. Biochem. Physiol.* 21:449–467.

Bartholomew, G. A., and J. R. B. Lighton. 1986. Endothermy and energy metabolism of a giant tropical fly, *Pantopthalmus tabaninus* Thunberg. *J. Comp. Physiol.* B156:461–467.

Byers, G. W. 1969. Evolution of wing reduction in crane flies (Diptera: Tipulidae). *Evolution* 23:346–354.

—— 1983. The crane fly genus *Chionea* in North America. *Univ. Kansas Sci. Bull.* 52:59–195.

Chappell, M. A., and K. R. Morgan. 1987. Temperature regulation, endothermy, resting metabolism, and flight energetics of tachinid flies (*Nowickia* sp.). *Physiol. Zool.* 60:550–559.

Clavel, J. D., and M. F. Clavel. 1969. Influence de la température sur le nombre, le pourcentage d'éclosion et la taille des oeufs fondus par *Drosophila melanogaster. Ann. Soc. Entomol. Fr.* 5:161–177.

Connor, M. E. 1924. Suggestions for developing a campaign to control yellow fever. *Am. J. Trop. Med.* 4:277–307.

Czajka, M., and R. E. Lee, Jr. 1990. A rapid cold-hardening response protecting against cold shock injury in *Drosophila melanogaster. J. Exp. Biol.* 148:245–254.

Digby, P. S. B. 1955. Factors affecting the temperature excess of insects in sunshine. *J. Exp. Biol.* 32:279–298.

Dingley, F., and J. Maynard Smith. 1968. Temperature acclimatization in the absence of protein synthesis in *Drosophila subobscura. J. Insect Physiol.* 14:1185–1194.

Downes, J. A. 1965. Adaptations of insects in the Arctic. *Ann. Rev. Entomol.* 106:257–274.

Edney, E. B., and R. Barrass. 1962. The body temperature of the tse-tse fly, *Glossina morsitans* Westwood (Diptera, Muscidae). *J. Insect Physiol.* 8:469–481.

Frison, T. H. 1935. The stoneflies, or Plecoptera, of Illinois. *Bull. Ill. Nat. Hist. Survey* 20:281–471.

Gerday, C. 1982. Soluble calcium-binding proteins from fish and invertebrate muscle? *Molecular Physiol.* 2:63–87.

Gilbert, F. S. 1984. Thermoregulation and structure of swarms in *Syrphus ribesii* (Syrphidae). *Oikos* 42:249–255.

Hågvar, S. 1971. Field observations on the ecology of a snow insect, *Chionea arancoides* Dalm. (Dipt. Tipulidae). *Norsk Entomol. Tidskr.* 18:33–37.

Haufe, W. O., and L. Burgess. 1956. Development of *Aedes* at Fort Church-ill, Manitoba and predictions of dates of emergence. *Ecology* 37:500–519.

Heinrich, B. 1974. Thermoregulation in endothermic insects. *Science* 185:747–756.

——— 1988. *One Man's Owl.* Princeton, N.J.: Princeton University Press.

Heinrich, B., and G. A. Bartholomew. 1971. An analysis of pre-flight warm-up in the sphinx moth, *Manduca sexta. J. Exp. Biol.* 55:223–239.

Heinrich, B., and C. Pantle. 1975. Thermoregulation in small flies (*Syrphus* sp.): Basking and shivering. *J. Exp. Biol.* 62:595–610.

Hochachka, P. W. 1965. Isoenzymes in metabolic adaptation of a poikilo-therm: Subunit relationships in lactic dehydrogenase of gold-fish. *Arch. Biochem. Biophys.* 111:96–103.

Hocking, B., and C. D. Sharplin. 1965. Flower basking by Arctic insects. *Nature* 206:215.

Hosgood, S. M. W., and P. A. Parsons. 1968. Plymorphism in natural populations of *Drosophila melanogaster* for the ability to withstand temperature shocks. *Experimentation* (Basel) 24:727–728.

Howe, M. A., and M. J. Lelane. 1986. Post-feed buzzing in the tsetse, *Glossina morsitans morsitans*, is an endothermic mechanism. *Physiol. Entomol.* 11:279–286.

Humphrey, W. F., and S. E. Reynolds. 1980. Sound production and endothermy in the horse bot-fly, *Gasterophilus intestinalis. Physiol. Entomol.* 5:235–242.

Jones, J. S., J. A. Coyne, and L. Partridge. 1987. Estimation of the thermal niche of *Drosophila melanogaster* using a temperature-sensitive mu-tation. *Am. Nat.* 130:83–90.

Kevan, P. G. 1972. Heliotropism in some Arctic flowers. *Can. Field Nat.* 86:41–44.

——— 1975. Sun-tracking solar furnaces in High Arctic flowers: Signif-icance for pollination and insects. *Science* 189:723–726.

Kimura, M. T. 1988. Adaptations to temperate climates and evolution of overwintering strategies in the *Drosophila melanogaster* species group. *Evolution* 42:1288–1297.

Kohshima, S. 1984. A novel cold-tolerant insect found in a Himalayan glacier. *Nature* 30:225–227.

Littlewood, S. C. 1966. Temperature threshold for flight of *Trichocera annulata* (Meigen) (Dipt., Trichoceridae). *Entomol. Mon. Mag.* 102:15–18.

Marden, J. H. 1989. Effects of load-lifting constraints on the mating system of a dance fly. *Ecology* 70:496–502.

May, M. L. 1976. Warming rates as a function of body size in periodic endotherms. *J. Comp. Physiol.* 111:55–70.

Maynard Smith, J. 1957. Temperature tolerance and acclimatization in *Drosophila subobscura. J. Exp. Biol.* 34:85–96.

——— 1963. Temperature and rate of aging in poikilotherms. *Nature* 199:400–402.

McAlpine, J. F. 1979. Diptera. In *Canada and Its Insect Fauna*, ed. H. V. Danks, pp. 389–424. Memoirs of the Entomological Society of Canada, 108.

Meats, A. 1973. Rapid acclimation to low temperature in the Queensland fruit fly, *Dacus tryoni*. *J. Insect Physiol.* 19:1903–1911.

Miyan, J. A., and A. W. Ewing. 1985. Is the "click" mechanism of Dipteran flight an artifact of CCl_4 anaesthesia? *J. Exp. Biol.* 116:313–322.

Morgan, K. R., and B. Heinrich. 1987. Temperature regulation in bee- and wasp-mimicking syrphid flies. *J. Exp. Biol.* 133:59–71.

Morgan, K. R., and T. E. Shelly. 1988. Body temperature regulation in desert robber flies (Diptera: Asilidae). *Ecol. Entomol.* 14:419–428.

Morgan, K. R., T. E. Shelly, and L. S. Kimsey. 1985. Body temperature regulation, energy metabolism, and wing loading in light-seeking and shade-seeking robber flies. *J. Comp. Physiol.* B151:561–570.

Morrison, W. W., and R. Milkman. 1978. Modification of heat resistance in *Drosophila* by selection. *Nature* 273:49–50.

Murphy, P. A., J. T. Giesel, and M. N. Manlove. 1983. Temperature effects on life history variation in *Drosophila simulans*. *Evolution* 37:1181–1192.

O'Neill, K. M., W. P. Kemp, and K. A. Johnson. 1990. Behavioural thermoregulation in three species of robber flies (Diptera, Asilidae: *Efferia*). *Anim. Behav.* 39:181–191.

Parsons, P. A. 1978. Boundary conditions for *Drosophila* resource utilization in temperate regions, especially at low temperatures. *Am. Nat.* 112:1063–1074.

Rowe, M. 1989. The own that traded a hoot for a hiss. *Nat. His.* 5:32–33.

Rowley, W. A., and C. L. Graham. 1968. The effect of temperature and relative humidity on the flight performance of female *Aedes aegypti*. *J. Insect Physiol.* 14:1251–1257.

Schnebel, E. M., and J. Grossfield. 1984. Mating-temperature range in *Drosophila*. *Evolution* 38:1296–1307.

Schneiderman, D., and C. M. Williams. 1955. An experimental analysis of the discontinuous respiration of the cecropia moth silkworm. *Biol. Bull.* (Woods Hole) 109:123–143.

Sotavalta, O. 1947. The flight-bee (wing-beat frequency) of insects. *Acta Entomol. Fenn.* 4:1–117.

Stone, A., C. W. Sabrosky, W. W. Wirth, R. I. Foote, and J. R. Colson. 1965. *A Catalogue of the Diptera of America North of Mexico*. USDA Agricultural Handbook, 276.

Sugg, P., J. S. Edwards, and J. Baust. 1983. Phenology and life history of *Belgica antarctica*, an Antarctic midge (Diptera: Chirominidae). *Ecol. Entomol.* 8:105–113.

Thiessen, C. J., and J. A. Mutchmoor. 1967. Some effects of thermal acclimation on muscle apyrase activity and mitochondrial number of *Periplaneta americana* and *Musca domestica*. *J. Insect Physiol.* 13:1837–1842.

Vinogradskaja, O. N. 1942. Body temperature in *Anopheles maculipennis* Messeae Fall. *Zool. Zh.* 21:187–195.

Willmer, P. G. 1982a. Thermoregulatory mechanisms in *Sarcophaga*. *Oecologia* (Berlin) 53:382–385.

———— 1982b. Hygrothermal determinants of insect activity patterns: The Diptera of water-lily leaves. *Ecol. Entomol.* 7:221–231.

Willmer, P. G., and D. M. Unwin. 1981. Field analysis of insect heat budgets: Reflectance, size and heating rates. *Oecologia* (Berlin) 50:250–255.

Yurkiewicz, W. J. 1968. Flight range and energetics of the sheep blowfly during flight at different temperatures. *J. Insect Physiol.* 14:335–339.

Yurkiewicz, W. J., and T. Smyth, Jr. 1966a. Effect of temperature on flight speed of the sheep blowfly. *J. Insect Physiol.* 12:195–226.

———— 1966b. Effects of temperature on oxygen consumption and fuel utilization by the sheep blowfly. *J. Insect Physiol.* 12:403–408.

12. Sweating Cicadas

Bartholomew, G. A., and M. C. Barnhart. 1984. Tracheal gases, respiratory gas exchange, body temperature and flight in some tropical cicadas. *J. Exp. Biol.* 111:131–144.

Chappell, M. 1983. Thermal limitations to escape responses in desert grasshoppers. *Anim. Behav.* 31:1088–1093.

Edney, E. B. 1977. *Water balance in Land Arthropods*. New York: Springer.

Hadley, N. F., E. C. Toolson, and M. Quinlan. 1989. Regional differences in cutilcular permeability in the desert cicada, *Diceroprocta apache:* Implications for evaporative cooling. *J. Exp. Biol.* 141:219–230.

Hadley, N. F., M. C. Quinlan, and M. L. Kennedy. 1991. Evaporative cooling in the desert cicada: Thermal efficiency and water/metabolic costs. *J. Exp. Biol.* 159:269–283.

Heath, J. E. 1967. Temperature responses of the periodical "17 year" cicada, *Magicicada cassini*. *Am. Midl. Nat.* 77:64–76.

Heath, J. E., and P. J. Wilkin. 1970. Temperature responses of the desert cicada, *Diceroprocta apache* (Homoptera, Cicadidae). *Physiol. Zool.* 43:145–154.

Heath, J. E., P. J. Wilkin, and M. S. Heath. 1972. Temperature responses of the cactus dodger, *Cacama valvata* (Homoptera, Cicadidae). *Physiol. Zool.* 45:238–246.

Josephson, R. K., and D. Young. 1979. Body temperature and singing in the bladder cicada, *Cystosoma saundersii*. *J. Exp. Biol.* 80:69–81.

———— 1985. A synchronous insect muscle with an operating frequency greater than 500 Hz. *J. Exp. Biol.* 118:185–208.

Kaser, S. A., and J. Hastings. 1981. Thermal physiology of the cicada, *Tibicen duryi*. *Am. Zool.* 21:1016.

Lloyd, M., and H. J. Dybas. 1966. The periodical cicada problem. I. Population ecology. *Evolution* 20:133–149.

MacNally, R., and D. Young. 1981. Song energetics of the bladder cicada, *Cystosoma saundersii*. *J. Exp. Biol.* 90:185–196.

O'Doherty, R., and J. S. Bale. 1985. Factors affecting the cold hardiness of the peach-potato aphid *Myzus persicae*. *Ann Appl. Biol.* 106:219–228.

Stevens, E. D., and R. K. Josphson. 1077. Metabolic rate and body temperature of singing katydids. *Physiol. Zool.* 50:31–42.

Toolson, E. C. 1984. Interindividual variation in epicuticular hydrocarbon composition and water loss rates of the cicada *Tibicen dealbatus* (Homoptera: Cicadidaea). *Physiol. Zool.* 57:550–556.

——— 1987. Water prolifigacy as an adaptation to hot deserts: Water loss rates and evaporative cooling in the Sonoran Desert cicada, *Diceroprocta apache* (Homoptera: Cicadidae). *Physiol. Zool.* 60:379–385.

——— 1989. Thermobiology of cicadas. Unpublished manuscript.

Toolson, E. C., and N. E. Hadley. 1987. Energy-dependent facilitation of transcuticular water flux contributes to evaporative cooling in the Sonoran Desert cicada, *Diceroprocta apache* (Homoptera: Cicadidae). *J. Exp. Biol.* 131:439–444.

Weis-Fogh, T. 1967. Respiration and tracheal ventilation in locusts and other flying insects. *J. Exp. Biol.* 47:561–587.

13. Warm Caterpillars and Hot Maggots

Ayres, M. P., and S. F. MacLean, Jr. 1987. Molt as a component of insect development: *Galerucella sagittariae* (Chrysomelidae) and *Espirrita autumnata* (Geometridae). *Oikos* 48:273–279.

Brett, J. R. 1971. Energetic responses of salmon to temperature. A study of some thermal relations in the physiology and freshwater ecology of sockeye salmon *(Oncorhynchus nerka). Am. Zool.* 11:99–113.

Buchmann, S. L., and H. G. Spangler. 1992. Social thermoregulation by *Galleria mellonella* L. (Lepidoptera: Pyralidae) larvae and its possible function. Unpublished manuscript.

Capinera, J. L., L. F. Wiener, and P. R. Anamosa. 1980. Behavioral thermoregulation by late-instar range caterpillar larvae *Hemileuca oliviae* Cockerell (Lepidoptera: Saturniidae). *J. Kansas Entomol. Soc.* 53:631–638.

Casey, T. M. 1976. Activity patterns, body temperature and thermal ecology of two desert caterpillars (Lepidoptera: Sphingidae). *Ecology* 56:485–497.

——— 1977. Physiological responses to temperature of caterpillars of desert populations of *Manduca sexta. Comp. Biochem. Physiol.* 57A:485–487.

Casey, T. M., and J. R. Hegel. 1981. Caterpillar setae: Insulation for an ectotherm. *Science* 214:1131–1133.

Casey, T. M., B. Joos, T. D. Fitzerald, M. E. Yurlina, and P. A. Young. 1988. Synchronized group foraging, thermoregulation and growth of eastern tent caterpillars. *Physiol. Zool.* 61:372–377.

Casey, T. M., and R. Knapp. 1987. Caterpillar thermal adaptation: Behavioral differences reflect metabolic thermal sensitivities. *Comp. Biochem. Physiol.* 86A:679–682.

Damman, H. 1987. Leaf quality and enemy avoidance by the larvae of a pyralid moth. *Ecology* 68:88–97.

Downes, J. A. 1965. Adaptations of insects in the Arctic. *Ann. Rev. Entomol.* 106:257–274.

Edwards, D. K. 1964. Activity rhythms in lepidopterous defoliators. II. *Halisidota argentata* Pack (Arctiidae) and *Nephystia phastasmaria* Stkr. (Geometridae). *Can. J. Zool.* 42:939–958.

Eisner, T., E. von Tassel, and J. E. Carrel. 1967. Defense use of a "fecal shield" by a beetle larva. *Science* 158:1471.

Fields, P. G., and J. N. McNeil. 1988. The importance of seasonal variation in hair coloration for thermoregulation of *Ctenucha virginica* larvae (Lepidoptera: Arctiidae). *Physiol. Zool.* 13:165–175.

Fitzgerald, T. D. 1980. An analysis of daily foraging patterns of laboratory colonies of the eastern tent caterpillar, *Malacosoma americanum* (Lepidoptera: Lasiocampidae), recorded photoelectrically. *Can. Entomol.* 112:731–738.

Gerould, J. H. 1921. Blue-green caterpillars: The origin and ecology of a mutuation in hemolymph color in *Colias (Eurymus) philodice*. *J. Exp. Biol.* 34:385–414.

Girard, M. 1865. Note sur la chaleur considérable des larves de la *Galleria cerella. Ann. Soc. Entomol. Fr.* 4:676–677.

——— 1869. Études sur la chaleur libre degagée par les animaux invertébrés et spécialement les insectes. *Ann. Sci. Nat. Zool.*, ser. 5 11:135–274.

Green, G. W. 1955. Temperature relations of ant-lion larvae (Neuroptera: Myemeleontidae). *Can. Entomol.* 87:441–459.

Grossmueller, D. W., and R. C. Lederhouse. 1985. Oviposition site selection: An aid to rapid growth and development in the tiger swollentail butterfly, *Papilio glaucus. Oecologia* (Berlin) 66:68–73.

Haufe, W. O. 1957. Physical environment and behavior of immature stages of *Aedes communis* (Deq.) in subarctic Canada. *Can. Entomol.* 89:120–139.

Heinrich, B. 1971. The effect of leaf geometry on the feeding behavior of the caterpillar of *Manduca sexta* (Sphingidae). *Anim. Behav.* 19:119–124.

——— 1977. Why have some animals evolved to regulate a high body temperature? *Am. Nat.* 111:623–640.

——— 1979. Foraging strategies of caterpillars: Leaf damage and possible avoidance strategies. *Oecologia* (Berlin) 42:325–337.

——— 1992. Avian predators as constraint on caterpillar foraging. In *Ecological and Evolutionary Constraints on Foraging of Caterpillars*, ed. N. E. Stamps and T. M. Casey. New York: Chapman and Hall.

Heinrich, B., and S. L. Collins. 1983. Caterpillar leaf damage, and the game of hide-and-seek with birds. *Ecology* 64:592–602.

Heinrich, B., and M. Heinrich. 1984. The pit-trapping foraging strategy of the ant lion, *Myrmeleon immaculatus* DeGeer (Neuroptera: Myrmeleontidae). *Behav. Ecol. Sociobiol.* 14:151–160.

Hochachka, P. W., and G. N. Somero. 1973. *Strategies of Biochemical Adaptation*. Philadelphia: Saunders.

Hsiao, T., and G. Fraenkel. 1969. Properties of *Leptinotarsa*: A toxic hemolymph protein from the Colorado potato beetle. *Toxicon* 7:119–130.

Huey, R. B. 1982. Temperature, physiology, and ecology of reptiles. In *Biology of Reptilia*, vol. 12, ed. C. Gans and F. H. Pough, pp. 25–91. London: Academic Press.

———— 1992. Physiological consequences of habitat selection. *Am. Nat.* 137 (Suppl.): S91–S115.

Huey, R. B., and P. E. Hertz. 1984. Is a jack-of-all-temperatures a master of none? *Evolution* 38:441–444.

Jones, J. S., J. A. Coyne, and L. Partridge. 1987. Estimation of the thermal niche of *Drosophila melanogaster* using a temperature-sensitive mutation. *Am. Nat.* 130:83–90.

Joos, B., T. M. Casey, T. D. Fitzgerald, and W. A. Buttemer. 1988. Roles of the tent in behavioral thermoregulation of eastern tent caterpillars. *Ecology* 69:2004–2011.

Kavaliers, M. 1981. Rhythmical thermoregulation in larval cranefly (Diptera: Tipulidae). *Can. J. Zool.* 59:555–558.

Kevan, P. G., T. W. Jensen, and J. D. Shorthouse. 1982. Body temperatures and behavioral thermoregulation of High Arctic woolly-bear caterpillars and pupae (*Gynaephora rossii*, Lymantriidae: Lepidoptera) and the importance of sunshine. *Arct. Alp. Res.* 14:125–136.

Kingsolver, J. G. 1979. Thermal and hydric aspects of environmental heterogeneity in the pitcher plant mosquito. *Ecol. Monogr.* 49:357–376.

Knisley, C. B. and D. L. Pearson. 1981. The function of turret building behaviour in the larval tiger beetle, *Cicindela willistoni* (Coleoptera: Cicindelidae). *Ecol. Entomol.* 6:401–410.

Knapp, R., and T. M. Casey. 1986. Thermal ecology, behavior and growth of gypsy moth and eastern tent caterpillars. *Ecology* 67:598–608.

Kukal, O. 1988. Caterpillars on ice: Methuselahs of the insect world, Arctic woolly bears spend most of their long lives in a deep freeze. *Nat. Hist.* 97:36–41.

Kukal, O., and T. E. Dawson. 1989. Temperature and food quality influences feeding behavior, assimilation efficiency and growth rate of Arctic woolly-bear caterpillars. *Oecologia* (Berlin) 79:526–532.

Kukal, O., J. D. Duman, and A. S. Serianni. 1989. Cold-induced mitochondrial degradation and cryoprotectant synthesis in freeze-tolerant Arctic caterpillars. *J. Comp. Physiol.* B158:661–671.

Kukal, O., B. Heinrich, and J. G. Duman. 1988. Behavioural thermoregulation in the freeze-tolerant Arctic caterpillar *Gynaephora groenlandica*. *J. Exp. Biol.* 138:181–193.

Kukal, O., and P. G. Kevan. 1987. The influence of parasitism on the life history of a High Arctic insect, *Gynaephora groenlandica* (Wöcke) (Lepidoptera: Lymantriidae). *Can. J. Zool.* 65:156–163.

Markl, H., and J. Tautz. 1975. The sensitivity of hair receptors in caterpillars of *Barathra brassicae* L. (Lepidoptera, Noctuidae) to particle movement in a sound field. *J. Comp. Physiol.* 99:79–87.

Marsh, A. C. 1987. Thermal responses and temperature tolerances of a desert ant-lion larva. *J. Therm. Biol.* 12:295–300.

May, M. L. 1982. Body temperature and thermoregulation of the Colorado potato beetle, *Leptinotarsa decemlineata*. *Entomol. Exp. Appl.* 31:413–420.

Minnich, D. E. 1925. The reactions of larva of *Vanessa antiopa* L. to sounds. *J. Exp. Zool.* 42:443–469.

Morris, R. F. 1972a. Predation by insects and spiders inhabiting colonial webs of *Hyphantria cunae*. *Can. Entomol.* 104:1197–1207.

REFERENCES FOR CHAPTER 13

———— 1972b. Predation by wasps, birds, and mammals on *Hyphantria cunea. Can. Entomol.* 105:1581–1591.

North, F., and H. Lillywhite. 1980. The function of burrow turrets in a gregariously nesting bee. *Southw. Nat.* 25:373–378.

Porter, K. 1982. Basking behaviour in larvae of the butterfly *Euphydras aurinia. Oikos* 38:308–312.

———— 1989. Sunshine, sex-ratio and behaviour of *Euphydras aurinia* larvae. In *The Biology of Butterflies*, ed. R. I. Vane-Wright and P. R. Ackery, pp. 309–311. Princeton, N.J.: Princeton University Press.

Rawlins, J. E., and R. C. Lederhouse. 1981. Developmental influences of thermal behavior on monarch caterpillars *(Danaus plexippus):* An adaptation for migration (Lepidoptera: Nymphalidae: Danaidae). *J. Kansas Entomol. Soc.* 54:387–408.

Regal, P. J. 1967. Voluntary hypothermia in reptiles. *Science* 155:1551–1553.

Schweitzer, D. 1974. Notes on the biology and distribution of the Cuculiinae (Noctuidae). *J. Lepid. Soc.* 28:5–21.

Seymour, R. S. 1974. Convective and evaporative cooling in the sawfly larvae. *J. Insect Physiol.* 20:2447–2457.

Sherman, P. W., and W. B. Watt. 1973. The thermal ecology of some *Colias* butterfly larvae. *J. Comp. Physiol.* 83:25–40.

Stamp, N. E., and M. D. Bowers. 1990. Variation in food quality and temperature constrain foraging of gregarious caterpillars. *Ecology* 71:1031–1039.

Stewart, P. A. 1969. House sparrows and a field infestation of tobacco hornworm larvae infecting tobacco. *J. Econ. Entomol.* 62:956–957.

Thurston, R., and O. Prachuabmoh. 1971. Predation by birds on tobacco hornworm larvae. *J. Econ. Entomol.* 64:1548–1549.

Wellington, W. G. 1950. Effects of radiation on the temperatures of insectan habitats. *Sci. Agric.* 30:209–234.

Youthed, G. J. 1973. Some adaptations of myrmeleontid (Neuroptera) and rhagionid (Diptera) larvae to life in hot dry sand. Ph.D. thesis, Rhodes University, Grahamstown, South Africa.

14. Fever

Allen, W. W., and R. F. Smith. 1958. Some factors influencing the efficiency of *Apanteles medicaginis* Muesebeck (Hymenoptera: Braconidae) as a parasite of the alfalfa caterpillar, *Colias philodice eurytheme* Boisduval. *Hilgardia* 28:1–42.

Aruga, H., and S. Tanaka. 1968. Effect of high temperature on the resistance of the silkworm, *Bombyx mori* L., to flacherie-virus disease. *J. Seric. Sci. Jpn.* 37:441–444.

Bawden, F. C. 1964. *Plant Viruses and Viral Diseases.* 4th ed. New York: Ronald Press.

Bird, F. T. 1955. Virus diseases of sawflies. *Can. Entomol.* 87:124–127.

Boorstein, S. M., and P. W. Ewald. 1987. Costs and benefits of behavioral fever in *Melanoplus sanguinipes* infected by *Nosema acridophagus. Physiol. Zool.* 60:586–595.

Bronstein, M. S., and W. E. Conner. 1984. Endotoxin-induced behavioral fever in the Madagascar cockroach, *Gramphadorhina protentosa*. *J. Insect Physiol.* 30:327–330.

Brooks, M. A. 1963. Symbiosis and aposymbiosis in arthropods. In *Symboitic Associations*, ed. P. S. Nutman and B. Mosse, pp. 200–231. London: Cambridge University Press.

Cabanac, M., and L. Le Guelte. 1980. Temperature regulation and prostaglandin E_1 fever in scorpions. *J. Physiol.* (London) 303:365–370.

Carruthers, R. I., T. S. Larkin, H. Firstencel, and Z. Feng. 1992. Influence of thermal ecology on the mycosis of rangeland grasshoppers. *Ecology* 73:190–204.

Casterlin, M. E., and W. W. Reynolds. 1977. Behavioral fever in crayfish. *Hydrobiologia* 56:99–101.

———— 1978. Prostaglandin E_1 fever in the crayfish *Cambarus bartoni*. *Pharmacol. Biochem. Behav.* 9:593–595.

Day, M. F., and E. H. Mercer. 1964. Properties of an iridescent virus from the beetle, *Sericesthis pruinosa*. *Austr. J. Biol. Sci.* 17:892–902.

De Lestrange, M. Th. 1954. Action de la température sur le virus responsible de la sensibilit à l'anhydride carbonique chez la Drosophile. *Compt. Rend.* 239:1159–1162.

———— 1963. Contribution à l'étude du virus héréditaire de la Drosophile: action de l'hyperthermie sur le contenu en virus des tissus somatiques de l'hôte. *Ann. Genet.* 6:39–96.

Ewald, P. W. 1980. Evolutionary biology and the treatment of signs and symptoms of infections diseases. *J. Therm. Biol.* 86:169–176.

Force, D. C., and P. S. Messenger. 1964. Duration of development, generation time, and longevity of three hympenopterous parasites of *Therioaphis maculata* reared at various temperatures. *Ann. Entomol. Soc. Am.* 57:405–413.

Girard, M. 1865. Note sur la chaleur considérable des larves de la *Galleria carella*. *Ann. Soc. Entomol. Fr.* 4:676–677.

Guppy, J. C. 1969. Some effects of temperature on immature stages of the armyworm, *Pseudaletia unipuncta* (Lepidoptera: Noctuidae). *Can. Entomol.* 101:1320–1327.

Huger, A. 1954. Experimentelle Eliminierung der Symbioten aus den Myzetomen des Getreide Kapuziners, *Rhizopertha dominca* F. *Naturwissenschaften* 41:170–171.

———— 1956. Experimentelle Untersuchungen über die künstliche Symbiotenelimination bei Vorratsschädlingen: *Rhizopertha dominica* F. (Bostrychidae) und *Oryzaephilus surinamensis* L. *Z. Morphol. Oekol. Tiere* 44:626–701.

Inoue, H., C. Ayuzawa, and A. Kawamura, Jr. 1972. Effects of high temperature on the multiplication of infectious flacherie virus in the silkworm, *Bombyx mori*. *Appl. Entomol. Zool.* 7:155–160.

Inoue, H., and Y. Tanada. 1977. Thermal therapy of the flacherie virus disease in the silkworm, *Bombyx mori*. *J. Invert. Pathol.* 29:63–68.

Kaya, H. K., and Y. Tanada. 1969. Responses to high temperature of the parasite *Apanteles militaris* and of its host, the armyworm, *Pseudaletia unipuncta*. *Ann. Entomol. Soc. Am.* 62:1303–1306.

Kluger, M. J. 1979. *Fever, Its Biology, Evolution and Function.* Princeton, N.J.: Princeton University Press.

Kunkel, L. O. 1936. Heat treatment for the cure of yellows and other virus diseases of peach. *Phytopathology* 26:809–830.

L'Héritier, Ph., and A. Sigot. 1946. Contribution à l'étude du phénomène de la sensibilité au CO_2 chez la Drosophile. Influence du chauffage aux différents stades au développement sur la manifestation de la sensibilité chez l'imago. *Bull. Biol. France et Belgique* 80:171–227.

Louis, C., M. Jourdan, and M. Cabanc. 1986. Behavioral fever and therapy in a rickettsia-infected orthopteran. *Am. J. Physiol.* 250:R991–R995.

Lwoff, A., and M. Lwoff. 1958. L'inhibition du développement du virus poliomyélitique à 39° et le problème du rôle de l'hyperthermie dans l'évolution des infections virales. *Compt. Rend. Acad. Sci. Paris* 246:190–192.

MacLeod, D. M. 1963. Entomophthorales infections. In *Insect Pathology: An Advanced Treatise*, vol. 2, ed. E. A. Steinhaus. New York: Academic.

Maramorosch, K. 1950. Influence of temperature on incubation and transmission of the wound-tumor virus. *Phytopathology* 40:1071–1093.

Mathavan, S., and T. J. Pandian. 1975. Effect of temperature on food utilization in the monarch butterfly *Danaus chrysippus*. *Oikos* 26:60–64.

McClain, E., P. Magnuson, and S. J. Warner. 1988. Behavioral fever in a Namib Desert tenebrionid beetle, *Onymacris plana*. *J. Insect Physiol.* 34:279–284.

McLaughlin, R. E. 1962. The effect of temperature upon larval mortality of the armyworm, *Pseudaletia unipuncta* (Haworth). *J. Insect Physiol.* 4:279–284.

Meeuse, B. J. D. 1966. Production of volatile amines and skatoles at anthesis in some *Arum* lily species. *Plant Physiol.* 41:343–347.

Miyajima, S. 1970. Effects of high temperature on the incidence of infectious flacherie of the silkworm, *Bombyx mori* L. Tokai Branch, Sericultural Society of Japan, *Proceedings* 18:28.

Ono, M., I. Okada, and M. Sasaki. 1987. Heat production by balling in the Japanese honeybee, *Apis cerana japonica*, as a defensive behavior against the hornet, *Vespa simillima xanthoptera* (Hymenoptera: Vespidae). *Experimenta* 43:1031–1032.

Raskin, I., A. Ehmann, W. R. Melander, and B. J. D. Meeuse. 1987. Salicylic acid: A natural inducer of heat production in *Arum* lilies. *Science* 237:1545–1656.

Roberts, D. W., and A. S. Campbell. 1977. Stability of entomopathogenic fungi. *Misc. Publ. Entomol. Soc. Am.* 10:19–76.

Schneider, H. 1956. Morphologische und experimentelle Untersuchungen über die Endosymbiose der Korn- und Reiskäfer (*Calandra granaria* L. und *Calandra oruzae* L.) *Z. Morphol. Oekol. Tiere* 44:555–625.

Sherman, P. W., and W. B. Watt. 1973. The thermal ecology of some *Colias* butterfly larvae. *J. Comp. Physiol.* 83:25–40.

Skubatz, H., T. A. Nelson, A. M. Dong, B. J. D. Meeuse, and A. J. Bendich. 1990. Infrared thermography of *Arum* lily inforescences. *Planta* 182:432–436.

Suzuki, S., R. Kimura, and K. Suzuki. 1963. Restraining effect of high temperature on occurrence of disease by some viruses in the silk-worm, *Bombyx mori* L. Kanto Branch, Sericultural Society of Japan, *Proceedings* 14:65.

Tanada, Y. 1953. Description and characteristics of a granulosis virus of the imported cabbageworm. *Proc. Hawaii. Entomol. Soc.* 15:235–260.

—— 1967. Effects of high temperatures on the resistance of insects to infectious diseases. *J. Seric. Sci. Jpn.* 36:333–339.

Tanada, Y., and G. Y. Chang. 1968. Resistance of the alfalfa caterpillar, *Colias eurytheme*, at high temperatures to a cytoplasmic-polyhedrosis virus and thermal inactivation point of the virus. *J. Invert. Phathol.* 10:79–83.

Tanada, Y. and A. M. Tanabe. 1965. Resistance of *Galleria mellonella* (Linneaus) to the *Tipula* iridescent virus at high temperatures. *J. Invert. Pathol.* 7:184–188.

Thompson, C. G. 1959. Thermal inhibition of certain polyhedrosis virus diseases. *J. Insect Pathol.* 1:189–190.

Van der Walt, E., E. McClain, A. Puren, and N. Savage. 1990. Phylogeny of arthropod immunity: An inducible humoral response in the Kalahari millipede, *Triaenostreptus triodus* (Attems).

Watanabe, H., and Y. Tanada. 1972. Infection of nuclear-polyhedrosis virus in armyworm, *Pseudaletia unipuncta* Haworth (Lepidoptera Noctuidae), reared at a high temperature. *Appl. Entomol. Zool.* 7:43–51.

Yamaguchi, K., Y. Iwashita, and K. Inoue. 1969. On the midgut-nuclear polyhedrosis in the silkworm, *Bombyx mori* L. III. Effects of high temperature treatment in the shape of polyhedron of the infected larvae. *J. Seric. Sci. Jpn.* 38:157–162.

15. Cold Jumpers

Agrell, I. 1941. Zur Ökologie der Collembolen. Untersuchungen im schwedischen Lappland. *Opusc. Entomol.*, Suppl. 3:1–236.

Alexander, R. McN. 1974. The mechanics of jumping by a dog *(Canis familiaris)*. *J. Zool.* 173:549–573.

Barth, R. 1954. O aparelho saltatorio do halticineo *Homophoeta sexnotata* Har. (Coleoptera). *Mem. Inst. Oswaldo Cruz* 52:365–376.

Bennet-Clark, H. C. 1975. The energetics of the jump of the locust *Schistocerca gregaria*. *J. Exp. Biol.* 63:53–83.

Bennet-Clark, H. C., and E. C. A. Lucey. 1967. The jump of the flea: A study of the energetics and a model for the mechanism. *J. Exp. Biol.* 47:59–76.

Boettiger, E. G., and E. Furshpan. 1952. The mechanics of flight movements in Diptera. *Biol. Bull.* (Woods Hole) 102:200–211.

Brown, R. H. J. 1967. Mechanism of locust jumping. *Nature* 214:939.

Camp, C. L., and N. Smith. 1942. Phylogeny and digital ligaments of the horse. *Mem. Univ. Calif.* 13:69–124.

Christian, E. von. 1979. Der Sprung der Collembolen. *Zool. Jb. Abt. Allg. Zool. Physiol. Tiere* 83:457–490.

Dawson, T. J., and C. R. Taylor. 1973. Energetic cost of locomotion in kangaroos. *Nature* 246:313–314.

Edwards, G. A., and W. L. Nutting. 1950. The influence of temperature upon the respiration and heart activity of *Thermobia* and *Grylloblatta*. *Psyche* 57:33–44.

Eisenbeis, G. 1978. Die Thorakal- und Abdominalmuskulatur von Arten der Springschwanz-Gattung *Tomocerus* (Collembola: Tomoceridae). *Entomol. Ger.* 4:55–83.

Eisenbeis, G., and S. Ulmer. 1978. Zur Funktionsmorphologie des Sprung-Apparates der Springschwänze am Beispiel von Arten der Gattung *Tomocerus* (Collembola: Tomoceridae). *Entomol. Gen.* 5:33–55.

Evans, M. E. G. 1972. The jump of the click beetle (Coleoptera, Elateridae)—A preliminary study. *J. Zool.* 167:319–336.

——— 1973. The jump of the click beetle (Coleoptera, Elateridae)—Energetics and mechanics. *J. Zool.* 169:181n194.

Furth, D. G. 1982. The metafemoral spring of flea beetles. *Spixiana*, Suppl. 7:11–27.

——— 1988. The jumping apparatus of flea beetles (Alticidae)—The metafemoral spring. In *Biology of Chrysomelidae*, ed. P. Jolivet, E. Petitpierre, and T. H. Hsiao, pp. 285–297. Dordrecht: Kluwer Academic.

Furth, D. G., and K. Suzuki. 1990a. Comparative morphology of the tibial flexor and extensor tendons in insects. *Syst. Entomol.* 15:433–441.

——— 1990b. The metafemoral extensor and flexor tendons in Coleoptera. *Syst. Entomol.* 15:443–448.

——— 1991. The independent evolution of the metafemoral spring in Coleoptera. Unpublished manuscript.

Furth, D. G., W. Traub, and I. Harpaz. 1983. What makes *Blepharida* jump? A structural study of the metafemoral spring of a flea beetle. *J. Exp. Zool.* 227:43–47.

Gabriel, M. 1985. The development of the locust jumping mechanism: Energy storage and muscle mechanics. *J. Exp. Biol.* 118:327–340.

Godden, D. H. 1969. The neural basis of locust jumping. *Am. Zool.* 9:1139–1140.

Gynther, I. C., and K. G. Pearson. 1986. Intracellular recordings from interneurons and motoneurons during bilateral kicks in the locust: Implications for mechanisms controlling the jump. *J. Exp. Biol.* 122:323–343.

Healey, I. N. 1967. An ecological study of temperatures in a Welsh moorland soil, 1962–63. *J. Anim. Ecol.* 36:425–434.

Heitler, W. J. 1974. The locust jump: Specializations of the metathoracic femoral-tibial joint. *J. Comp. Physiol.* 89:93–104.

Heitler, W. J., and P. Bräunig. 1988. The role of fast extensor motor activity in the locust kick reconsidered. *J. Exp. Biol.* 136:289–309.

Hoyle, G. 1955. Neuromuscular mechanisms of a locust skeletal muscle. *Proc. R. Soc. Lond.* B143:343–367.

Ioff, I. G. 1950. The Alakurt. *Materialy k Poznaniyu Fanny i Flory SSSR Otdel Ektoparazity* 2:4–29.

Ker, R. F. 1977. Some structural and mechanical properties of locust and beetle cuticle. D.Phil. thesis, Oxford University, Oxford, England.

Makings, P. 1973. Activity of the sea-shore bristle-tail (*Petrobius maritimus* Leach) (Thysanura) at low temperatures. *J. Anim. Ecol.* 42:585–598.

Mani, M. S. 1968. *Ecology and Biogeography of High Altitude Insects.* The Hague: W. Junk.

Manton, S. M. 1972. The evolution of arthropod locomotory mechanisms. Part 10. Locomotory habits, morphology and evolution of the hexapod classes. *Zool. J. Linn. Soc.* 51:203–400.

Maulik, S. 1929. On the structure of the hind femur in halticine beetles. *Proc. Zool. Soc. Lond.* 2:305–308.

Neville, A. C. 1965. Energy and economy of insect flight. *Sci. Progress* 53:203–219.

Olroyd, H. 1964. *The Natural History of Flies.* London: Weidenfeld and Nicolson.

Pringle, J. W. S. 1957. *Insect Flight.* Cambridge: Cambridge University Press.

Rothschild, M., and J. Schlein. 1975. The jumping mechanism of *Xenopsylla cheopis.* I. Exoskeletal structures and musculature. *Phil. Trans. R. Soc. Lond.* B271:457–490.

Rothschild, M., Y. Schlein, I. Parker, and K. Sternberg. 1972. Jump of the oriental rat flea *Xenopsylla cheopis* (Roths.). *Nature* 239:45–48.

Rothschild, M., Y. Schlein, K. Parker, C. Neville, and S. Sternberg. 1973. The flying leap of the flea. *Sci. Am.* 229:92–100.

———— 1975. The jumping mechanism of *Xenopsylla cheopis* III. Execution of the jump and activity. *Phil. Trans. R. Soc. Lond.* B271:499–515.

Sander, K. 1957. Bau und Funktion des Sprungapparates von *Pyrilla perpusilla* Walker (Homoptera—fulgoroidae). *Zool. Jb. Abt. Anat. Ontog. Tiere* 25:383–388.

Snodgrass, R. E. 1946. The skeletal anatomy of fleas (Siphonaptera). *Smithson. Miscellaneous. Collns.* 104:1–89.

Usherwood, P. N. R., and H. I. Runion. 1970. Analysis of the mechanical responses of metathoracic extensor tibia muscles of free-walking locusts. *J. Exp. Biol.* 52:39–58.

Weber, H. 1929. Kopf und Thorax von *Psylla mali* Schmidh. (Hemiptera—Homoptera). *Z. Morphol. Oekol. Tiere* 14:59–165.

Weis-Fogh, T. 1960. A rubber-like protein in insect cuticle. *J. Exp. Biol.* 37:889–907.

———— 1961. Power in flapping flight. In *Cell and Organism*, ed. J. A. Ramsey and V. B. Wigglesowrth, pp. 283–300. Cambridge: Cambridge University Press.

Wolkomir, R. 1989. The race to make a "perfect" shoe starts in the laboratory. *Smithsonian* (September 1989):95–104.

16. Social Thermoregulation

Abraham, M., and J. M. Pasteels. 1977. Nest-moving behaviour in the ant *Myrmica rubra. Proceedings of the Eighth International Congress of* the *International Union for the Study of Social Insects,* Wagenigen, Netherlands, p. 286.

Alford, D. V. 1975. *Bumblebees.* London: Davis-Poynter.

Allen, M. D. 1959. Respiration rates of worker honeybees at different ages and temperatures. *J. Exp. Biol.* 36:92–101.

Andrews, E. A. 1927. Ant mounds as to temperature and sunshine. *J. Morphol. Physiol.* 44:608–615.

Belt, T. 1874. *The Naturalist in Nicaragua.* London: John Murray.

Bouillon, A. 1970. Termites of the Ethiopian region. In *Biology of Termites,* vol. 2, ed. K. Krishna and F. M. Weesner. New York: Academic.

Brian, M. V. 1952. The structure of a dense natural ant population. *J. Anim. Ecol.* 21:12–24.

———— 1956. The natural density of *Myrmica rubra* and associated ants in West Scotland. *Insectes Sociaux* 3:474–487.

———— 1973. Temperature choice and its relevance to brood survival and caste determination in the ant *Myrmica rubra* L. *Physiol. Zool.* 46:245–252.

Brian, M. V., and A. D. Brian. 1948. Nest construction by queens of *Vespula sylvestris* Scop. (Hym., Vespidae). *Entomol. Mon. Mag.* 84:193–198.

———— 1951. Insolation and ant populations in the west of Scotland. *Trans. R. Entomol. Soc. Lond.* 102:303–330.

Brückner, D. 1975. Die Abhängigkeit der Temperaturregulierung von der genetischen Variabilität der Honigbiene (*Apis mellifera* L.). *Apidologie* 6:361–380.

———— 1976. Vergleichende Untersuchungen zur Temperaturpraeferenz von ingezüchteten und nichtingezüchtigten Arbeiterinnen der Honigbiene *(Apis mellifera). Apidologie* 7:139–149.

Bruman, F. 1928. Die Luftzirkulation im Bienenstock. *Z. Vergl. Physiol.* 8:366–370.

Büdel, A. 1960. Bienenphysik. In *Biene und Bienenzucht,* ed. A. Büdel and E. Herold. Munich: Ehrenwirth.

———— 1968. Le Microclimat de la ruche. In *Traité de Biologie de l'Abeille,* vol. 4, ed. R. Chauvin. Paris: Masson.

Cahill, K., and S. Lustick. 1976. Oxygen consumption and thermoregulation in *Apis mellifica* workers and drones. *Comp. Biochem. Physiol.* 55A:355–357.

Ceusters, R. 1977. Social homeostasis in colonies of *Formica polyctena* Foerst. (Hymenoptera, Formicidae): Nestform and temperature preferences. *Proceedings of the Eighth International Congress of the International Union for the Study of Social Insects,* Wageningen, Netherlands, pp. 111–112.

Chadwick, P. C. 1931. Ventilation of the hive. *Glean. Bee Cult.* 59:356–358.

Coaton, W. G. H. 1948. *Trinervitermes* species: The snouted harvester termites. *USDA Agric. Bull.* 290:1–24.

Coenen-Stass, D., B. Schaarschmidt, and I. Lamprecht. 1980. Temperature distribution and colorimetric determination of heat production in the nest of the wood ant, *Formica polyctena* (Hymenoptera, Formicidae). *Ecology* 61:238–244.

Cooper, P., W. M. Schaffer, and S. L. Buchmann. 1985. Temperature regulation of honey bees *(Apis mellifera)* foraging in the Sonoran Desert. *J. Exp. Biol.* 114:1–15.

Corkins, C. L. 1930. The metabolism of the honey bee colony during winter. *Bull. Wyo. Arranger. Exp. Sta.* 175:1–54.

———— 1932. The temperature relationship of the honeybee cluster under controlled temperature conditions. *J. Econ. Entomol.* 25:820–825.

Corkins, C. L., and C. S. Gilbert. 1932. The metabolism of honey-bees in winter. *Bull. Wyo. Agr. Exp. Sta.* 187:1–30.

Cowles, R. B. 1930. The life history of *Varanus niloticus* (Lin.) as observed in Natal, South Africa. *J. Entomol. Zool.* 22:1–31.

Cumber, R. A. 1949. The biology of bumble-bees with special references to the production of the worker caste. *Trans. Roy. Entomol. Soc. Lond.* 100:1–45.

Darchen, R. 1973. La thermorégulation et l'écologie de quelques espèces d'abeilles sociales d'Afrique (Apidae, Trigonini et *Apis mellifica* var. *adansonii*). *Apidologie* 4:341–370.

Délye, G. 1967. Physiologie et comportement de quelques fourmis (Hym. Formicidae) du Sahara en rapport avec les principaux facteurs du climat. *Insectes Sociaux* 14:323–338.

Dreyer, W. A. 1942. Further observations on the occurrence and size of ant mounds with reference to their age. *Ecology* 23:486–490.

Dunham, W. E. 1929. The influence of external temperature on the hive temperatures during the summer. *J. Econ. Entomol.* 22:798–801.

———— 1931. A colony of bees exposed to high external temperatures. *J. Econ. Entomol.* 24:606–611.

———— 1933. Hive temperatures during the summer. *Glean. Bee Cult.* 61:527–529.

Ebner, R. 1926. Einige Beobachtungen an Termitenbauten. *Denkschr. Akad. Wiss. Wien, Math.-Nat. Kl.* 100:75–76.

Elton, C. 1932. Orientation of the nests of *Formica turncorum* in north Norway. *J. Anim. Ecol.* 1:192–193.

Emerson, A. E. 1938. Termite nests, a study of the phylogeny of behaviour. *Ecol. Monogr.* 8:247–284.

———— 1956. Regenerative behaviour and social homeostasis of termites. *Ecology* 37:245–258.

Esch, H. 1960. Über die Körpertemperaturen und den Wärmehaushalt von *Apis mellifica*. *Z. Vergl. Physiol.* 43:305–335.

Etterschank, G. 1971. Some aspects of the ecology and the nest microclimatology of the meat ant, *Iridomyrmex purpureus* (Sm.). *Proc. Roy. Soc. Victoria* 84:137–152.

Fabritius, M. 1976. Experimentelle Untersuchung des Wärmeverhaltens der Hornissen *(Vespa crabro)*. Dissertation, Frankfurt.

Fahrenholz, L. 1986. Die soziale Thermoregulation im Stock der Honigbiene *(Apis mellifera carnica)* und die kalorimetrische Bestimmung der Wärmeproduktion bei Einzeltieren. Diplomaarbeit, Freie Universität, Berlin.

Farrar, C. L. 1943. An interpretation of the problems of wintering the honeybee colony. *Glean. Bee Cult.* 71:513–518.

Fielde, A. M. 1904. Observations on ants in their relation to temperature and to submergence. *Biol. Bull.* (Woods Hole) 7:170–174.

Forel, A. 1874. *Les Mourmis de la Suisse.* Zurich: Société Helvétique des Sciences Naturelles.

Franks, N. R. 1989. Thermoregulation in army ant bivouacs. *Physiol. Zool.* 14:397–404.

Free, J. B., and C. G. Butler. 1959. *Bumblebees.* London: Collins.

Free, J. B., and J. Simpson. 1963. The respiratory metabolism of honey-bee colonies at low temperatures. *Ent. Exp. Appl.* 6:234–238.

Free, J. B., and H. Y. Spencer-Booth. 1958. Observations on the temperature regulation and food consumption of honey-bee *(Apis mellifica).* *J. Exp. Biol.* 35:930–937.

——— 1960. Chill-coma and cold death temperatures of *Apis mellifica.* *Entomol. Exp. Appl.* 3:222–230.

Frisch, K. von. 1974. *Animal Architecture.* New York: Harcourt Brace Jovanovich.

Fye, R. E., and J. T. Medler. 1954. Temperature studies in bumblebee domiciles. *J. Econ. Entomol.* 47:847–852.

Gates, B. N. 1914. The temperature of the bee colony. *Bull. U.S. Dept. Agric.* 96:1–29.

Gaul, A. T. 1952a. Additions to vespine biology. IX. Temperature regulation in the colonly. *Bull. Brooklyn Entomol. Soc.* 47:79–82.

——— 1952b. Metabolic cycles and the flight of vespine wasps. *J. N.Y. Entomol Soc.* 60:21–24.

Gay, F. J., and J. H. Calaby. 1970. Termites from the Australian region. In *Biology of Termites,* vol. 2, ed. K. Krishna and F. M. Weesner. New York: Academic.

Geyer, J. W. 1951. A comparison between the temperatures in a termite supplementary fungus garden and in the soil at equal depth. *J. Entomol. Soc. S. Africa* 14:36–43.

Gibo, D. L., R. M. Yarascavitch, and H. E. Dew. 1974a. Thermoregulation in colonies of *Vespula arenaria* and *Vespula maculata* (Hymenoptera: Vespidae) under normal conditions and under cold stress. *Can. Entomol.* 106:873–879.

Gibo, D. L., H. E. Dew and A. S. Hajduk. 1974b. Thermoregulation in colonies of *Vespula arenaria* and *Vespula maculata* (Hymenoptera: Vespidae). II. The relation between colony biomass and calorie production. *Can. Entomol.* 106:873–879.

Gibo, D. L., A. Temporale, T. P. Lamarre, B. M. Soutar, and H. E. Dew. 1977. Thermoregulation in colonies of *Vespula arenaria* and *Vespula maculata* (Hymenoptera: Vespidae). III. Heat production in queen nests. *Can. Entomol.* 109:615–620.

Grassé, P. P. 1944. Recherches sur la biologie des termites champignonnistes (Macrotermitinae). *Ann. Sci. Nat. Zool.* 6:97–171; 7(1945):115–146.

——— 1949. Ordre des Isoptères ou Termites. In *Traité de Zoologie,* vol. 9, ed. P. P. Grassé. Paris: Masson.

Grassé, P. P., and C. Noirot. 1958. Le comportement des termites à l'égard de l'air libre. L'atmosphère des termitières et son renouvellement. *Ann. Sci. Nat. Zool.* 20:1–28.

Greaves, T. 1964. Temperature studies of termite colonies in living trees. *Aust. J. Zool.* 12:250–262.

———— 1967. Experiments to determine the populations of tree-dwelling colonies of termites *Coptotermes acinaciformis* (Froggat) and *C. frenchi* (Hill). Division of Entomology Technical Paper No. 7, Commonwealth Scientific and Industrial Research Organization, Australia.

Grigg, G. C. 1973. Some consequences of the shape and orientation of "magnetic" termite mounds. *Aust. J. Zool.* 21:231–237.

Grigg, G. C., and A. J. Underwood. 1977. An analysis of the orientation of "magnetic" termite mounds. *Aust. J. Zool.* 25:87–94.

Grigg, G., P. Jacklyn, and L. Taplin. 1988. The effect of buried magnets on colonies of *Amitermes* spp. building magnetic mounds in northern Australia. *Physiol. Zool.* 13:285–289.

Harkness, R. D., and R. Wehner. 1977. *Cataglyphis. Endeavour,* new ser. 1:115–121.

Hasselrot, T. B. 1960. Studies on Swedish bumblebees (genus *Bombus* Latr.). *Opusc. Entomol.,* Suppl. 17:1–200.

Hazelhoff, E. H. 1954. Ventilation in a bee-hive during summer. *Physiologia Comp. Oecol.* 3:343–364.

Heimann, M. 1963. Zum Wärmehaushalt der kleinen roten Waldameise (*Formica polyctena* Foerst.). *Waldhygiene* 5:1–21.

Heinrich, B. 1974a. Pheromone-induced brooding behavior in *Bombus vosnesenskii* and *B. edwardsii* (Hymenoptera: Bombidae). *J. Kans. Entomol. Soc.* 47:396–404.

———— 1974b. Thermoregulation in bumblebees. I. Brood incubation by *Bombus vosnesenskii* queens. *J. Comp. Physiol.* 88:129–140.

———— 197a. *Bumblebee Economics.* Cambridge, Mass.: Harvard University Press.

———— 1979b. Thermoregulation of African and European honeybees during foraging, attack, and hive exits and returns. *J. Exp. Biol.* 80:217–229.

———— 1981a. Energetics of honeybee swarm thermoregulation. *Science* 212:565–566.

———— 1981b. The mechanisms and energetics of honeybee swarm temperature regulation. *J. Exp. Biol.* 91:25–55.

———— 1981c. The regulation of temperature in the honeybee swarm. *Sci. Am.* 244:146–160.

———— 1985. The social physiology of temperature regulation in honeybees. In *Fortschritte der Zoologie,* vol. 31, ed. B. Hölldobler and M. Lindauer. Stuttgart, New York: F. Fisher.

Hepburn, H. R., E. Armstrong, and S. Kurstjens. 1983. The ductility of native beeswax is optimally related to honeybee colony temperature. *S. Afr. J. Sci.* 79:416–417.

Heran, J. 1952. Untersuchungen über den Temperatursinn der Honigbiene unter besonderer Berücksichtigung der Wahrnehmung strahlender Wärme. *Z. Vergl. Physiol.* 34:179–206.

Hess, W. R. 1926. Die Temperaturregulierung im Bienenvolk. *Z. Vergl. Physiol.* 4:465–487.

Hill, G. F. 1942. *Termites (Isoptera) from the Australian Region.* Melbourne: Commonwealth Scientific and Industrial Research Organization.

Himmer, A. 1926. Der soziale Wärmehaushalt der Honigbiene. I. Die

Wärme im nichtbrütenden Wintervolk. *Erlanger Jahrb. Bienenk.* 4:1–51.

———. 1927. Ein Beitrag zur Kenntnis des Wärmehaushalts im Nestbau sozialer Hautflügler. *Z. Vergl. Physiol.* 5:375–389.

———. 1931. Über die Wärme im Hornissennest. *Z. Vergl. Physiol.* 13:748–761.

———. 1932. Die Temperaturverhältnisse bei den sozialen Hymenopteren. *Biol. Rev.* 7:224–253.

———. 1933. Die Nestwärme bei *Bombus agrorum* F. *Biol. Zentralbl.* 53:270–276.

Holdaway, F. G., and F. J. Gay. 1948. Temperature studies of the habitat of *Eutermes exitiosus* with special reference to the temperature within the mound. *Aust. J. Sci. Res.* B1:464–493.

Hölldobler, B., and E. O. Wilson. 1990. *The Ants.* Cambridge, Mass.: Harvard University Press.

Horstmann, K., and H. Schmid. 1986. Temperature regulation in nests of the wood ant, *Formica polyctena* (Hymenoptera Formicidae). *Entomol. Gen.* 11:229–236.

Hubbard, M. D., and W. G. Cunningham. 1977. Orientation of mounds in the ant *Solenopsis invicta* (Hymenoptera, Formicidae, Myrmicinae). *Insectes Sociaux* 24:3–8.

Huber, P. 1810. *Recherches sur les Moeurs des Fourmis Indigènes.* Paris: Paschoud.

Hunter, J. 1792. Observations on bees. *Phil. Trans. Roy. Soc. Lond.* 82:128–196.

Ishay, J. 1972. Thermoregulatory pheromones in wasps. *Experientia* 28:1185–1187.

———. 1973. Thermoregulation by social wasps: Behavior and pheromones. *Trans. N.Y. Acad. Sci.* 35:447–462.

Ishay, J. and R. Ikan. 1968a. Food exchange between adults and larvae in *Vespa orientalis* F. *Anim. Behav.* 16:298–303.

———. 1968b. Gluconeogensis in the oriental hornet, *Vespa orientalis* F. *Ecology* 49:169–171.

Ishay, J., and F. Ruttner. 1971. Thermoregulation im Hornissennest. *Z. Vergl. Physiol.* 72:423–434.

Ishay, J., H. Bytinski-Salz, and A. Shulov. 1967. Contributions to the bionomics of the oriental hornet (*Vespa orientalis* Fab.). *Isr. J. Entomol.* 2:45–106.

Jackson, W. B. 1957. Microclimate patterns in the army ant bivouac. *Ecology* 38:276–285.

Janet, C. 1895. Études sur les Fourmis, les Guêpes et les Abeilles. Neuvième note. Sur *Vespa crabro* L. *Mém. Soc. Zool. Fr.* 8:1–140.

Jordan, R. 1936. Beobachtung der Arbeitsteilung im Hummelstaate *(Bombus muscorum).* *Arch. Bienenk.* 17:81–91.

Josens, G. 1971. Variations thermiques dans les nids de *Trinervitermes geminatus* Wassermann, en relation avec le milieu extérieur, dans la savane de Lamto (Côte d'Ivoire). *Insectes Sociaux* 18:1–14.

Kato, M. 1939. The diurnal rhythm of temperature in the mound of an ant, *Formica truncorum truncorum* var. *Yessenni* Forel, widely distributed at Mt. Hakkoda. *Sci. Rep. Tohoku Univ. Sendai* 14:53–64.

Kiechle, H. 1961. Die soziale Regulation der Wassersammeltätigkeit im

Bienenstaat und deren physiiologische Grundlage. *Z. Vergl. Physiol.* 45:154–192.

Kneitz, G. 1964. Untersuchungen zum Aufbau und zur Erhaltung des Nestwärmehaushaltes bei *Formica polyctena* Foerst. (Hym. Formicidae). Dissertation, Würzburg.

———— 1970. Saisonale Veränderungen des Nestwärmehaushaltes bei Waldameisen in Abhängigkeit von der Konstitution und dem Verhalten der Arbeiterinnen also Beispiel vorteilhafter Anpassung eines Insektenstaates an das Jahreszeitenklima. *Verh. Dtsch. Zool. Ges.* 64:318–322.

Koeniger, N. 1975. Experimentelle Untersuchungen über das Wärmen der Brut bei *Vespa crabro* und *Apis mellifica. Verh. Dtsch. Zool. Ges.,* 148.

———— 1978. Das Wärmen der Brut bei der Honigbiene (*Apis mellifera* L.). *Apidologie* 9:305–320.

Kondoh, M. 1968. Bioeconomic studies on the colony of an ant species, *Formica japonica* Motschulsky, I. Nest structure and seasonal change of the colony members. *Jap. J. Ecol.* 18:124–133.

Kronenberg, F. 1979. Characteristics of colonial thermoregulation in honey bees. Ph.D. thesis, Stanford University, Stanford, Calif.

Kronenberg, F. and H. C. Heller, 1982. Colonial thermoregulation in honeybees *(Apis mellifera). J. Comp. Physiol.* 148:65–76.

Kukal, O., and D. L. Pattie. 1988. Colonization of snow bunting, *Plectrophenax nivalis,* nests by bumblebees, *Bombus polaris,* in the High Arctic. *Can. Entomol.* 102:544.

Lacher, V. 1964. Elektrophysiologische Untersuchungen en einzelnen Rezeptoren für Geruch, Luftfeuchtigkeit und Temperatur auf den Antenne der Arbeitsbiene und der Drohne. *Z. Vergl. Physiol.* 48:587–623.

Lange, R. 1959. Experimentelle Untersuchungen über den Nestbau der Waldameisen: Nesthügel und Volkstärke. *Entomophaga* 4:47–55.

Lee, K. E., and T. G. Wood. 1971. *Termites and Soils* New York: Academic.

Lensky, Y. 1964. Comportement d'une colonie d'abeilles a des températures extremes. *J. Insect Physiol.* 10:1–12.

Lindauer, M. 1954. Temperaturregulierung und Wasserhaushalt im Bienenstaat. *Z. Vergl. Physiol.* 36:391–432.

Lindauer, M., and W. E. Kerr. 1958. Die gegenseitige Verständigung bei den stachellosen Bienen. *Z. Vergl. Physiol.* 41:405–434.

Linder, C. 1908. Observations sur les Fourmilières-Boussoles. *Bull. Soc. Vaudoise Sci. Nat.,* ser. 5 44:303–310.

Lindhard, E. 1912. Humlebien som Husdyr. Spredt Traek af nogle danske Humlebiarters Biologi. *Tidsskr. Plavl.* 19:335–352.

Loos, R. 1964. A sensitive anemometer and its use for the measurement of air currents in the nests of *Macrotermes natalensis* (Haviland). In *Études sur les Termites Africains,* ed. A. Bouillon. Paris: Masson.

Lüscher, M. 1951. Significance of "fungus gardens" in termite nests. *Nature* 167:34–35.

———— 1955. Der Sauerstoffverbrauch bei Termiten und die Ventilation des Nestes bei *Macrotermes natalensis* (Haviland). *Acta Trop.* 12:289–307.

REFERENCES FOR CHAPTER 16

———— 1961. Air-conditioned termite nests. *Sci. Am.* 205:138–145.

MacKay, W. P., and E. MacKay. 1985. Temperature modification of the nest of *Pogonomyrmex montanus* (Hymenoptera: Formicidae). *Southw. Nat.* 30:307–309.

Makino, S., and S. Yamane. 1980. Heat production by the foundress of *Vespa simillima*, with description of its embryo nest (Hymenoptera: Vespidae). *Insecta Matsumurana* 19:89–101.

Mardan, M., and P. K. Kevan. 1989. Honeybees and "yellow rain." *Nature* 341:191.

Martin, M. A., and J. S. Martin. 1978. Cellulose digestion in the midgut of the fungus-growing termite *Macrotermes natalensis:* The role of acquired digestive enzymes. *Science* 199:1453–1455.

Maschwitz, U. 1966. Das Speichelsekret der Wespenlarven und seine biologische Bedeutung. *Z. Vergl. Physiol.* 53:228–252.

Meudec, M. 1977. Le comportement de transport du couvain lors d'une perturbation du nid chez *Tapinoma erraticum* (Dolichoderinae). Rôle de l'individu. *Insectes Sociaux* 24:345–353.

Meyer, W. 1956. Arbeitsteilung im Bienenschwarm. *Insectes Sociaux* 3:303–323.

Michener, C. D. 1974. *The Social Behavior of the Bees: A Comparative Study.* Cambridge, Mass.: Harvard University Press.

Michener, C. D., and A. Wille. 1961. The bionomics of a primitively social bee, *Lasioglossum inconspicuum. Univ. Kans. Sci. Bull.* 42:1123–1202.

Michener, C. D., R. B. Lange, J. J. Bigarella, and R. Salamuni. 1958. Factors influencing the distribution of bees' nests in earth banks. *Ecology* 29:207–217.

Möglich, M. 1978. Social organization of nest emigration in *Leptothorax* (Hym., Form). *Insectes Sociaux* 25:205–225.

Morimoto, R. 1960. Experimental study on the trophallactic behavior in *Polistes* (Hymenoptera, Vespidae). *Acta Hymenoptera* 1:99–103.

Nagy, K. A., and J. N. Stallone. 1976. Temperature maintenance and CO_2 concentration in a swarm cluster of honeybees, *Apis mellifera. Comp. Biochem. Physiol.* 55A:169–171.

Neuhaus, W., and R. Wohlgemuth. 1960. Über das Fächeln der Bienen und dessen Verhältnis zum Fliegen. *Z. Vergl. Physiol.* 43:615–641.

Nielsen, E. T. 1938. Temperatures in a nest of *Bombus hypnorum* L. *Vidensk. Medd. Dan. Naturhist. Foren. Khobenhavn* 102:1–6.

Noirot, C. 1970. The nests of termites. In *Biology of Termites*, vol. 2, ed. K. Krishna and F. M. Weesner. New York: Academic.

Nolan, W. J. 1925. The brood-rearing cycle of the honeybee. *Bull. U.S. Dept. Agric.* 1349:1–56.

Nye, P. H. 1955. Some soil-forming processes in the humid tropics. IV. The action of the soil fauna. *J. Soil Sci.* 6:73–83.

Ofer, J. 1970. *Polyrachis simplex*, the weaver ant of Israel. *Insectes Sociaux* 17:49–82.

Omholt, S. W. 1987. Thermoregulation in the winter cluster of the honeybee, *Apis mellifera. J. Theor. Biol.* 128:219–231.

Phillips, E. F., and G. S. Demuth. 1914. The temperature of the honeybee cluster in winter. *Bull. U.S. Dept. Agric.* 93:1–16.

Plowright, R. C. 1977. Nest architecture and the biosystematics of bumblebees. *Proceedings of the Eighth International Congress of the International Union for the Study of Social Insects,* Wageningen, Netherlands, pp. 183–185.

Pontin, A. J. 1960. Field experiments on colony foundation by *Lasius niger* (L.) and *L. flavus* (F.) (Hym., Formicidae). *Insectes Sociaux* 7:227–230.

———— 1963. Further considerations of competition and the ecology of the ants *Lasius flavus* (F.) and *L. niger* (L.) *J. Anim. Ecol.* 32:565–574.

Postner, M. 1951. Biologisch-Ökologische Untersuchungen an Hummeln und ihren Nestern. *Veröff. Mus.* (Bremen) 1:46–86.

Raignier, A. 1948. L'économie thermique d'une colonie polycalique de la fourmi des Bois (*Formica rufa polyctena* Foerst). *Cellule* 51:281–368.

Réaumur, R. A. F. de. 1742. *Mémoires pour Servir à l'Histoire des Insectes,* vol. 6. Paris: Royale.

Ribbands, C. R. 1953. *The Behaviour and Social Life of Honeybees.* London: Bee Research Association.

Richards, K. W. 1973. Biology of *Bombus polaris* Curtis and *B. hypereus* Schönherr at Lake Hazen, Northwest Territories (Hymenoptera: Bombini). *Quest. Entomol.* 9:115–157.

Ritter, W. 1978. Der Einfluss der Brut auf die Änderung der Wärmebildung in Bienenvölkern *(Apis mellifera carnica). Verh. Dtsch. Zool. Ges.,* p. 220.

Ritter, W., and N. Koeniger. 1977. Influence of the brood on the thermoregulation of honeybee colonies. *Proceedings of the Eighth International Congress of the International Union for the Study of Social Insects,* Wageningen, Netherlands, pp. 283–284.

Roland, C. 1969. Rôle de l'involucre et du nourissement au sucre dans la régulation thermique à l'intérieur d'un nid de Vespides. *C.R. Hebd. Seances Acad. Sci.* 269:914–916.

Rosengren, R., W. Fortelius, K. Lindström, and A. Luther. 1987. Phenology and causation of nest heating and thermoregulation in red wood ants of the *Formica rufa* group studied in coniferous forest habitats in southern Finland. *Ann. Zool. Fennici* 24:147–155.

Roth, M. 1965. La production de chaleur chez *Apis mellifica* L. *Ann. Abeille* 8:5–77.

Roubik, D. W., and F. J. A. Peralta. 1983. Thermodynamics in nests of two *Melipona* species in Brazil. *Acta Amazonica* 13:453–466.

Ruelle, J. E. 1964. L'architecture du nid de *Macrotermes natalensis* et son sens fonctionel. In *Études sur les Termites Africains,* ed. A. Bouillon. Paris: Masson.

Ruttner, F. 1968. Les Races d'Abeilles. In *Traitée de Biologie de l'Abeille,* vol. 1, ed. R. Chauvin. Paris: Masson.

Sakagami, S. F. 1971. Ethosoziologischer Vergleich zwischen Honigbienen und Stachellosen Bienen. *Z. Tierpsychol.* 28:337–350.

Sakagami, S. F., and K. Hayashida. 1961. Biology of the primitive social bee *Halictus duplex* Dalle Torre. III. Activities in spring solitary phase. *J. Fac. Sci. Hokkaido Univ.,* ser. 6, Zool. 14:639–682.

Sakagami, S. F., and C. D. Michener. 1962. *The Nest Architecture of the Sweat Bees.* Lawrence: University of Kansas.

Sanders, C. J. 1972. Seasonal and daily activity patterns of carpenter ants (*Camponotus* spp.) in Northwestern Ontario (Hymenoptera: Formicidae). *Can. Entomol.* 104:1681–1687.

Sands, W. A. 1969. The association of termites and fungi. In *Biology of Termites*, vol. 1, ed. K. Krishna and F. M. Weesner. New York: Academic.

Scherba, G. 1958. Reproduction, nest orientation and population structure of an aggregation of mound nests of *Formica ulkei* Emery. *Insectes Sociaux* 5:201–213.

——— 1962. Mound temperatures of the ant *Formica ulkei* Emery. *Am. Midl. Nat.* 67:373–385.

Schneirla, T. C., R. Z. Brown, and F. C. Brown. 1954. The bivouac or temporary nest as an adaptive factor in certain terrestrial species of army ants. *Ecol. Monogr.* 24:269–296.

Seeley, T. D. 1977. Measurement of nest cavity volume by the honey bee (*Apis mellifera*). *Behav. Ecol. Sociobiol.* 2:201–227.

Seeley, T. D., and B. Heinrich. 1981. Regulation in the nests of social insects. In *Insect Thermoregulation,* ed. B. Heinrich. New York: Wiley.

Seeley, T. D., and R. A. Morse. 1976. Nest site selection by the honey bee, *Apis mellifera. Insectes Sociaux* 25:323–337.

Seeley, T. D., and P. K. Visscher. 1985. Survival of honeybees in cold climates. The critical timing of colony growth and reproduction. *Ecol. Entomol.* 10:81–88.

Simpson, J. 1961. Nest climate regulation in honey bee colonies. *Science* 133:1327–1333.

Skaife, S. H. 1955. *Dwellers in Darkness.* London: Longmans Green.

Sladen, F. W. L. 1912. *The Bumble-bee, Its Life-History and How to Domesticate It, with Descriptions of All the British Species of* Bombus *and* Psithyrus. London: Macmillan.

Snyder, T. E. 1926. Preventing damage by termites or white ants. *U.S. Dept. Agric. Farmers' Bull.* 1472:1–21.

Southwick, E. E. 1982. Metabolic energy of intact honeybee colonies. *Comp. Biochem. Physiol.* 71A:277–281.

——— 1983. The honey bee cluster as a homeothermic superorganism. *Comp. Biochem. Physiol.* 75A:641–645.

——— 1984. Metabolismus von Honigbienen und Einfluss der Traubengrösse bei niedrigen Temperaturen. *Apidologie* 15:267–269.

——— 1985. Allometric relations, metabolism and heat conductance in clusters of honey bees at cool temperatures. *J. Comp. Physiol.* B156:143–149.

——— 1987. Cooperative metabolism in honey bees: An alternative to antifreeze and hibernation. *J. Therm. Biol.* 12:155–158.

——— 1988. Thermoregulation in honey-bee colonies. In *Africanized Honey Bees and Bee Mites.* ed. G. R. Needham, R. E. Page Jr., M. Delfinado-Baker, and C. E. Bowman. New York: Halsted Press.

——— 1990. The colony as a thermoregulating superorganism. In *The Behavior and Physiology of Bees,* ed. L. J. Goodman and R. C. Fisher. Wallingford, U.K.: CAB International.

Southwick, E. E., and G. Heldmaier. 1987. Temperature control in honey bee colonies. *Bioscience* 37:395–399.

Southwick, E. E., and R. F. A. Moritz. 1987. Social control of air ventilation in colonies of honey bees, *Apis mellifera. J. Insect Physiol.* 33:623–626.

Southwick, E. E., and J. N. Mugaas. 1971. A hypothetical homeotherm: The honeybee hive. *Comp. Biochem. Physiol.* 40A:935–944.

Southwick, E. E., D. W. Roubik, and J. M. Williams. 1990. Comparative energy balance in groups of Africanized and European honey bees: Ecological implications. *Comp. Biochem. Physiol.* 97A:1–7.

Steiner, A. 1924. Über den sozialen Wärmehaushalt der Waldameise (*Formica rufa* var. *rufo-pratensis* For.). *Z. Vergl. Physiol.* 2:23–56.

———— 1926. Temperaturmessungen in den Nestern der Waldameise (*Formica rufa* var. *rufa-pratensis*) und der Wegameise *(Lasius niger)* während des Winters. *Mitt. Naturforsch. Ges. Bern,* pp. 1–19.

———— 1929. Temperaturuntersuchungen in Ameisennestern mit Erdkuppeln, im Nest von *Formica exsecta* Nyl. und in Nestern unter Steinen. *Z. Vergl. Physiol.* 9:1–66.

———— 1930. Die Temperaturregulation im Nest der Feldwespe (*Polistes gallica* var. *biglumis* L.). *Z. Vergl. Physiol.* 11:461–502.

———— 1932. Die Arbeitsteilung der Feldwespe *Polistes dubia* K. *Z. Vergl. Physiol.* 17:101–151.

Stuart, A. M. 1977. A polyethic and homeostatic response to a simple stimulus in a tropical termite. *Proceedings of the Eighth International Congress of the International Union for the Study of Social Insects,* Wageningen, Netherlands, pp. 149–151.

Stussi, T. 1972a. Réaction de thermogenèse au froid chez la guêpe ouvrière et autres Hyménoptères sociaux. *C.R. Hebd. Seances Acad. Sci.* 274:2687–2689.

Stussi, T. 1972b. L'heterothermie de l'abeille. *Arch. Sci. Physiol.* 26:131–159.

Sudd, J. H., J. M. Douglas, T. Gaynard, D. M. Murray, and J. M. Stockdale. 1977. The distribution of wood-ants (*Formica lugubris* Zetterstedt) in a northern English forest. *Ecol. Entomol.* 2:301–313.

Tschinkel, W. R. 1987. Seasonal life history and nest architecture of a winter-active ant, *Prenolepis imparis. Insectes Sociaux* 34:143–164.

Underwood, B. A. 1990. Seasonal nesting cycle and migration patterns of the Himalayan honey bee *Apis laboriosa. Natl. Geogr. Res.* 6:276–290.

Vanderplank, F. L. 1960. The bionomics and ecology of the red tree ant *Oecophylla* sp., and its relationship to the coconut bug *Pseudotheraptus wayi* (Brown) (Coreidae). *J. Anim. Ecol.* 29:15–33.

Veith, H. J., and N. Koeniger. 1978. Identifizierung von *cis*-9 Pentacosen als Auslöser für das Wärmen der Brut bei der Hornisse. *Naturwissenschaften* 65:263.

Villa, J. D. 1987. Africanized and European colony conditions at different elevations in Colombia. *Am. Bee J.* 127:53–57.

Vogt, F. D. 1986a. Thermoregulation in bumblebee colonies. I. Thermoregulatory versus brood-maintenance behaviors during acute changes in ambient temperature. *Physiol. Zool.* 59:55–59.

———— 1986b. Thermoregulation in bumblebee colonies. II. Behavioral and demographic variation throughout the colony cycle. *Physiol. Zool.* 59:60–68.

Waloff, N., and R. E. Blacklith. 1962. The growth and distribution of the mounds of *Lasius flavus* (Fabricius) (Hym. Formicidae) in Silwood Park, Berkshire. *J. Anim. Ecol.* 31:421–437.

Wasmann, E. 1915. *Das Gesellschaftsleben der Ameisen.* Münster: Aschendorf.

Weir, J. S. 1973. Air flow, evaporation and mineral accumulation in mounds of *Macrotermes subhyalinus* (Rambur). *J. Anim. Ecol.* 42:509–520.

Wellenstein, G. 1928. Beiträge zur Biologie der roten Waldameise (*Formica rufa* L.) mit besonderer Berücksichtigung klimatischer und förstlicher Verhältnisse. *Z. Angew. Entomol.* 14:1–68.

——— 1967. Zur Frage der Standortansprüche hügelbauender Waldameisen (*F. rufa*-Gruppe). *Z. Angew. Zool.* 54:139–166.

Weyrauch, W. 1936. Das Verhalten sozialer Wespen bei Nestüberhitzung. *Z. Vergl. Physiol.* 23:51–63.

Wille, A. 1976. Las abejas jicótes del género *Melipona* (Apidae: Meliponini) de Costa Rica. *Rev. Biol. Trop.* 24:123–147.

Wilson, E. O. 1959. Some ecological characteristics of ants in New Guinea rain forest. *Ecology* 40:437–447.

——— 1971. *The Insect Societies.* Cambridge, Mass.: Harvard University Press.

Wohlgemuth, R. 1957. Die Temperaturregulation des Bienenvolkes unter regeltheoretischen Gesichtspunkten. *Z. Vergl. Physiol.* 40:119–161.

Wójtowski, F. 1963. Studies on heat and water economy in bumble-bee nests. *Zool. Poloniae* 13:19–36.

Woodworth, C. E. 1936. Effect of reduced temperature and pressure on honeybee respiration. *J. Econ. Entomol.* 29:1128–1138.

Zahn, M. 1958. Temperatursinn, Wärmehaushalt und Bauweise der Roten Waldameise. *Zool. Beiträge,* new ser. 3:127–194.

Zucchi, R., and S. F. Sakagami. 1972. Capacidade termo-reguladora em *Trigona spinipes* e em algumas outras espécies de abelhas sem ferrão. In *Homenagem à Warwick E. Kerr.* Brazil: Rio Claro.

Acknowledgments

Many friends and colleagues have contributed to the preparation of this book. I thank the following for making available previously unpublished information: R. L. Anderson, W. S. Armbruster, S. L. Buchmann, R. I Carruthers, R. Carter, P. Chai, K. A. Christian, J. R. Coelho, H. Esch, D. G. Furth, F. Goller, R. B. Huey, B. Lyon, J. H. Marden, A. C. Marsh, E. L. McClain, K. D. McCormack, D. Mitchell, S. R. Morton, C. S. Roberts, M. K. Seeley, A. Shmida, H. G. Spangler, K. Suzuki, F. D. Vogt, and R. and S. Wehner.

I am also indebted to those who read drafts of various chapters and provided useful comments and criticisms: Steve Buchmann, Timothy Casey, Harald Esch, David G. Furth, William J. Hamilton III, Ray B. Huey, Robert K. Josephson, Jarmila Kukalova-Peck, Justin Malloy, James Marden, Michael May, Elizabeth McClain, Kenneth Morgan, Timothy Otter, R. D. Stevenson, Lutz Wasserthal, and Rudiger Wehner.

Ross Bell aided me in numerous aspects of insect identification and taxonomy. I am particularly grateful to Ray B. Huey for much thoughtful advice and many helpful criticisms, and I thank Freeman Dyson for drawing my attention to an important work on ants. Erika Geiger's help in typing the manuscript was invaluable. Financial support from the National Science Foundation has made possible most of my own research and that of many others described in this book and is gratefully acknowledged. Finally, an Alexander von Humboldt Senior Scientist Award from the Federal Republic of Germany gave me the necessary impetus and time to start writing while at the Philipps University in Marburg, and a Lady Davis Fellowship to the Hebrew University of Jerusalem helped me to finish it. I thank my two hosts, Andreas Bertsch and Avishai Shmida.

Howard Boyer at Harvard University Press gave invaluable encouragement and moral support to keep on with the project, which was necessary on several occasions when my endurance started to flag. Meticulous editing by Kate Schmit saved me from numerous slips and helped me say what I wanted to say; I am grateful for her invaluable help in the many technical details that would have got me down. Finally, I thank my daughter, Erica, for contributing her precious time and considerable energies to work on the index.

My thanks also to the following publishers for permitting me to use or adapt artwork that originally appeared elsewhere:
Figures 1.4, 1.5, and 1.6 From A. E. Kammer and M. B. Rheuben, "Adult motor patterns produced by moth pupae during development," *Journal of Experimental Biology* 65(1976):65–84, with permission from the Company of Biologists Ltd., Cambridge, England.

Figure 1.7 From B. Heinrich and G. A. Bartholomew, "An analysis of pre-flight warm-up in the sphinx moth, *Manduca sexta*," *Journal of Experimental Biology* 55(1971):223–239, with permission from the Company of Biologists Ltd., Cambridge, England.

Figures 1.15, 1.16, and 1.17B From B. Heinrich, "Thoracic temperature stabilization by blood circulation in a free-flying moth," *Science* 168(1970):580–582, and "Nervous control of the heart during thoracic temperature regulation in a sphinx moth," *Science* 169(1970):606–607, copyright 1970 by the AAAS.

Figure 1.18 From B. Heinrich, "Temperature regulation of the sphinx moth, *Manduca sexta*. II. Regulation of heat loss by control of blood circulation," *Journal of Experimental Biology* 54(1971):153–166, with permission from the Company of Biologists Ltd., Cambridge, England.

Figure 1.19 From J. I. Hegel and T. M. Casey, "Thermoregulation and control of heat temperature in the sphinx moth, *Manduca sexta*," *Journal of Experimental Biology* 101(1982):1–15, with permission from the Company of Biologists Ltd., Cambridge, England.

Figure 1.20 From T. M Casey, "Flight energetics of sphinx moths: Heat production and heat loss in *Hyles lineata* during free flight," *Journal of Experimental Biology* 64(1976):545–560, with permission from the Company of Biologists Ltd., Cambridge, England.

Figure 1.23 From G. A. Bartholomew and B. H. Heinrich, "A field study of flight temperatures in moths in relation to body weight and wing loading," *Journal of Experimental Biology* 58 (1973):123–135, with permission from the Company of Biologists Ltd., Cambridge, England.

Figure 1.24 From T. M. Casey and B. A. Joos, "Morphometrics, conductance, thoracic temperature, and flight energetics of noctuid and geometrid moths," *Physiological Zoology* 56(1983):160–173, copyright 1983 University of Chicago Press.

Figures 1.30, 1.31B, 1.32, 1.33, and 1.34 From B. Heinrich, "Thermoregulation by winter-flying endothermic moths," *Journal of Experimental Biology* 127(1987):313–332, with permission from the Company of Biologists Ltd., Cambridge, England.

Figure 1.35 From B. Heinrich and T. P. Mommsen, "Flight of winter moths near 0 degrees C," *Science* 228(1985):177–179, copyright 1985 by the AAAS.

Figures 2.8, 2.9, 2.19 From B. Heinrich, "Comparative thermoregulation of four montane butterflies of different mass," *Physiological Zoology* 59(1986):616–626, copyright 1986 University of Chicago Press.

Figures 2.15, 2.16, 2.17 From B. Heinrich, "Thermoregulation and flight activity of the satyr, *Coenonympha inornata* (Lepidoptera:Satyrida)," *Ecology* 67(1986):593–597, with permission from the Ecological Society of America.

Figure 2.20 From J. Kukalova-Peck, "Ephemeroid wing venation based upon new gigantic Carboniferous mayflies and basic morphology, phylogeny, and metamorphosis of pterygote insects (Insecta, Ephemerida)," *Canadian Journal of Zoology* 63(1985):933–955, with permission from the National Research Council of Canada.

Figure 2.22 From J. P. C. Dumont and R. M. Robertson, "Neural circuits: An evolutionary perspective," *Science* 233(1986):849–853, copyright 1986 by the AAAS.

Figures 3.4, 3.5, 3.6, 3.8, 3.9, and 3.11 From B. Heinrich and T. M. Casey, "Heat transfer in dragonflies: 'Fliers' and 'perchers,'" *Journal of Experimental Biology* 74(1978):17–36, with permission from the Company of Biologists Ltd., Cambridge, England.

Figures 4.3 and 4.4 From M. A. Chappell, "Metabolism and thermoregulation in desert and montane grasshoppers," *Oecologia* (Berlin) 56(1983):126–131, with permission from Springer-Verlag, Heidelberg.

Figure 4.7   From D. W. Whitman, "Thermoregulation and daily activity patterns in a black desert grasshopper, *Taeniopoda eques*," *Animal Behavior* 35(1987):1814–1826, with permission from Academic Press Inc., London.

Figure 4.8   From D. W. Whitman, "Function and evolution of thermoregulation in the desert grasshopper *Taeniopoda eques*," *Journal of Animal Ecology* 57(1988):369–383.

Figure 4.9   From M. Bauer and O. von Helversen, "Separate localization of sound recognizing and sound producing neural mechanisms in a grasshopper," *Journal of Comparative Physiology* B161(1987):95–101, with permission from Springer-Verlag, Heidelberg.

Figure 4.18   From B. F. Murphy, Jr., and J. E. Heath, "Temperature sensitivity in the prothoracic ganglion of the cockroach, *Periplaneta americana*, and its relationship to thermoregulation," *Journal of Experimental Biology* 105(1983):305–315, with permission from the Company of Biologists Ltd., Cambridge, England.

Figures 5.7, 5.10, and 5.11   From K. R. Morgan, "Body temperature regulation and terrestrial activity in the ectothermic beetle *Cicindela tranquebarica*," *Physiological Zoology* 58(1985):29–37, copyright 1985 University of Chicago Press.

Figure 5.8   From H. Dreisig, "Daily activity, thermoregulation and water loss in the tiger beetle *Cicindela hybrida*," *Oecologia* (Berlin) 44(1980):376–389, with permission from Springer-Verlag, Heidelberg.

Figures 5.14, 5.23, and 5.24   From G. A. Bartholomew and B. Heinrich, "Endothermy in African dung beetles during flight, ball making, and ball rolling," *Journal of Experimental Biology* 73(1978):65–83, with permission from the Company of Biologists Ltd., Cambridge, England.

Figure 5.16   From B. Heinrich and G. A. Bartholomew, "Roles of endothermy and size in inter- and intraspecific competition for elephant dung in an African dung beetle, *Scarabaeus laevistriatus*," *Physiological Zoology* 52(1979):484–496, copyright 1979 University of Chicago Press.

Figures 5.18 and 5.19   From B. Heinrich and E. McClain, "'Laziness' and hypothermia as a foraging strategy in flower scarabs (Coleoptera: Scarabaeidae)," *Physiological Zoology* 59(1986):273–282, copyright 1986 University of Chicago Press.

Figure 5.21   From K. R. Morgan, "Temperature regulation, energy metabolism, and mate-searching in rain beetles (*Plecoma* spp.), winter-active, endothermic scarabs (Coleoptera)," *Journal of Experimental Biology* 128(1987)107–122, with permission from the Company of Biologists Ltd., Cambridge, England.

Figures 6.8 and 6.9   From H. Esch, F. Goller, and B. Heinrich, "How do bees shiver?" *Naturwissenschaften* 78(1991):325–328, with permission from Springer-Verlag, Heidelberg.

Figures 6.8 and 8.4   From H. Esch and F. Goller, "Neural control of honeybee fibrillar muscle during shivering and flight," *Journal of Experimental Biology* 159(1991):419–431, with permission from the Company of Biologists Ltd., Cambridge, England.

Figure 6.15   From B. Heinrich and M. J. E. Heinrich, "Size and caste in temperature regulation by bumblebees," *Physiological Zoology* 56(1983):552–562, copyright 1983 University of Chicago Press.

Figure 7.1   From M. L. May and T. M. Casey, "Thermoregulation and heat exchange in euglossine bees," *Physiological Zoology* 56(1983):541–551, copyright 1983 University of Chicago Press.

Figures 7.3 and 7.4   From M. A. Chappell, "Temperature regulation and energetics of the solitary bee *Centris pallida* during foraging and mate competition," *Physiological Zoology* 57(1984):215–225, copyright 1984 University of Chicago Press.

Figures 7.5, 7.6, 7.8, and 7.9   From B. Heinrich and S. L. Buchmann, "Thermoregulatory physiology of the carpenter bee, *Xylocopa varipuncta*," *Journal of*

Comparative Physiology B156(1986):557–562, with permission from Springer-Verlag, Heidelberg.

Figures 8.8 and 8.11 From B. Heinrich, "Mechanisms of body-temperature regulation in honeybees, *Apis mellifera*. II. Regulation of thoracic temperature at high air temperatures," *Journal of Experimental Biology* 85(1980):73–87, with permission from the Company of Biologists Ltd., Cambridge, England.

Figure 8.9 From B. Heinrich, "Keeping a cool head: Honeybee thermoregulation," *Science* 205(1979):1269–1271, copyright 1979 by the AAAS.

Figure 8.12 From P. Cooper, W. M. Schaffer, and S. L. Buchmann, "Temperature regulation of honey bees *(Apis mellifera)* foraging in the Sonoran Desert," *Journal of Experimental Biology* 114(1985):1–15, with permission from the Company of Biologists Ltd., Cambridge, England.

Figures 9.4, 9.5, and 9.6 From A. C. Marsh, "Thermal responses and temperature tolerance in a diurnal desert ant, *Ocymyrmex barbiger*," *Physiological Zoology* 58(1985):629–636, copyright 1985 University of Chicago Press.

Figures 10.3 and 10.6 From B. Heinrich, "Strategies of thermoregulation and foraging in two vespid wasps, *Dolichovespula maculata* and *Vespula vulgaris*," *Journal of Comparative Physiology* B154(1984):175–180, with permission from Springer-Verlag, Heidelberg.

Figure 11.9 From M. A. Chappell, and K. R. Morgan, "Temperature regulation, endothermy, resting metabolism, and flight energetics of tachinid flies (*Nowickia* sp.)," *Physiological Zoology* 60(1987):550–559, copyright (1987) University of Chicago Press.

Figures 11.7 and 11.8 From K. R. Morgan and B. Heinrich, "Temperature regulation in bee- and wasp-mimicking syrphid flies," *Journal of Experimental Biology* 133(1987):59–71, with permission from the Company of Biologists Ltd., Cambridge, England.

Figure 11.10 From G. A. Bartholomew and J. R. B. Lighton, "Endothermy and energy Metabolism of a giant tropical fly, *Pantopthalmus tabaninus* Thunberg," *Journal of Comparative Physiology* B156(1986):461–467, with permission from Springer-Verlag, Heidelberg.

Figures 11.11 and 11.13 From K. R. Morgan, T. E. Shelly, and L. S. Kimsey, "Body temperature regulation, energy metabolism, and wing loading in light-seeking and shade-seeking robber flies," *Journal of Comparative Physiology* B151(1985):561–570, with permission from Springer-Verlag, Heidelberg.

Figure 11.16 From K. R. Morgan and T. E. Shelly, "Body temperature regulation in desert robber flies (Diptera: Asilidae)," *Economic Entomology* 14(1988):419–428, with permission from the Entomological Society of America.

Figure 12.3 From E. C. Toolson, "Water profligacy as an adaptation to hot deserts; Water loss rates and evaporative cooling in the Sonoran Desert cicada, *Diceroprocta apache* (Homoptera: Cicadidae)," *Physiological Zoology* 70(1987):379–385, copyright (1987) University of Chicago Press.

Figure 12.4 From E. C. Toolson and N. E. Hadley, "Energy-dependent facilitation of transcuticular water flux contributes to evaporative cooling in the Sonoran Desert cicada, *Diceroprocta apache* (Homoptera: Cicadidae)," *Journal of Experimental Biology* 131(1987):439–444, with permission from the Company of Biologists Ltd., Cambridge, England.

Figure 12.5 From N. F. Hadley, M. C. Quinlan, and J. L. Kennedy, "Evaporative cooling in the desert cicada: Thermal efficiency and water / metabolic costs," *Journal of Experimental Biology* 159(1991):269–283, with permission from the Company of Biologists Ltd., Cambridge, England.

Figure 13.3 From T. M. Casey, "Activity patterns, body temperature and thermal ecology of two desert caterpillars (Lepidoptera: Sphingidae)," *Ecology* 56(1976):485–497, with permission from the Ecological Society of America.

Figures 13.7 and 13.8 From R. Knapp and T. M. Casey, "Thermal ecology, behavior and growth of gypsy moth and eastern tent caterpillars," *Ecology* 67(1986):598–608, with permission from the Ecological Society of America.

Figure 13.12 From O. Kukal, B. Heinrich, and J. G. Duman, "Behavioural thermoregulation in the freeze-tolerant Arctic caterpillar *Gynaephora groenlandica*," *Journal of Experimental Biology* 138(1988):181–193, with permission from the Company of Biologists Ltd., Cambridge, England.

Figure 13.18 From A. C. Marsh, "Thermal responses and temperature tolerances of a desert ant-lion larva," *Journal of Thermal Biology* 12(1987):295–300, with permission from Pergamon Press PLC.

Figure 14.3 From E. McClain, P. Magnuson, and S. J. Warner, "Behavioral fever in a Namib Desert tenebrionid beetle, *Onymacris plana*," *Journal of Insect Physiology* 34(1988):279–284, with permission from Pergamon Press PLC.

Figures 16.12, 16.13, 16.14, 16.15, 16.16, and 16.18 From B. Heinrich, "The mechanisms and energetics of honeybee swarm temperature regulation," *Journal of Experimental Biology* 91(1981):25–55, with permission from the Company of Biologists Ltd., Cambridge, England.

Figure 16.19 From E. E. Southwick, "Cooperative metabolism in honey bees: An alternative to antifreeze and hibernation," *Journal of Thermal Biology* 12(1987):156, with permission from Pergamon Press PLC.

Index of Authors Cited

260, 261, 262, 263,
264, 265, 266, 267,
268, 269, 270, 271,
272, 273, 274, 280,
284, 285, 286, 287,
288, 289, 295, 298,
300, 301, 302, 303,
304, 305, 306, 307,
310, 315, 316, 319,
320, 335, 336, 337,
338, 339, 340, 348,
351, 352, 353, 354,
355, 356, 357, 359,
360, 361, 364, 367,
385, 386, 388, 389,
395, 397, 398, 404,
405, 408, 447, 454,
465, 466, 467, 474,
476, 477, 478, 479,
480, 481, 482, 484,
485, 490
Heinrich, M. J. E., 212,
214, 240, 246, 248,
250, 251, 258, 259,
340
Heitler, W. J., 182, 184,
436, 437, 438
Heldmair, G., 479, 484
Heller, H. C., 490
Henschel, J. R., 331
Henwood, K., 193, 195,
196, 198, 199
Hepburn, H. R., 214,
472
Heran, H., 266, 313,
476, 479, 489, 494
Herbers, J. M., 333
Herold, R. C., 314
Hersch, M. I., 314
Herter, K., 79, 179, 180
Hertz, P. E., 385
Hess, W. R., 476, 490,
492
Heusner, A., 295
Hidaka, T., 79
Hilbert, D. W., 159
Hill, G. F., 500, 501
Hilliard, J. R., Jr., 159
Himmer, A., 229, 260,
295, 313, 319, 459,
463, 465, 472, 476,
483, 490
Hochachka, P. W., 3, 7,
239, 348, 385
Hoffman, J. A., 9, 103

Holdaway, F. G., 500,
501, 503, 505
Hölldobler, B., 201, 323,
326, 329, 332, 447,
450
Holm, E., 192, 196,
197, 200
Hosgood, S. M. W., 344
Howe, M. A., 359
Hoy, R. R., 187
Hoyle, G., 436
Hsiao, T., 408
Hubbard, M. D., 450
Huey, R. B., 7, 9, 383,
385, 389
Huger, A., 415
Humphrey, W. F., 359,
360
Hunter, J., 472
Hurd, P. D., 287
Huxley, J. S., 3

Ikan, R., 462, 463
Ikeda, K., 231, 233,
235, 236, 237, 240,
295
Imamori, M., 208
Ioff, I. F., 443
Inoue, H., 412, 413,
414, 415
Inouye, D. W., 278
Ishay, J., 335, 459, 460,
462, 463
Iwashita, Y., 414

Jacklyn, P., 502
Jackson, W. B., 454
Jacobs, M. E., 30
Janet, C., 459, 462
Janiszewski, J., 179,
182, 183
Janzen, D., 277, 382
Jensen, M., 145, 396,
399
Johnson, K. A., 364
Jones, J. S., 167
Jones, R. E., 76, 345,
408
Joos, B., 50, 53, 54, 59,
253, 392
Jordan, R., 470
Josens, G., 500
Josephson, R. K., 51,
151, 172, 173, 175,

176, 177, 178, 187,
373, 374
Jourdan, M., 417
Jungermann, R., 300
Jwrzitza, G., 131

Kalmus, H., 50
Kammer, A. E., 20, 21,
22, 23, 24, 30, 44, 57,
58, 81, 88, 89, 138,
188, 230, 233, 234,
235, 236, 239, 247,
263, 295, 300
Kaser, S. A., 375, 376
Kato, M., 450
Kavaliers, M., 408
Kawamura, A., Jr., 413,
414
Kay, C. A. R., 325, 326,
330
Kaya, H. K., 415
Kemp, W. P., 155, 168,
364
Kenagy, G. J., 196, 199
Kennedy, M. L., 376,
377
Ker, R. F., 423, 425,
426
Kerkut, G. A., 181, 182,
183, 185, 186, 445
Kerr, W. E., 470
Kevan, P., 91, 306, 349,
396, 397, 398, 399,
496
Kiechle, H., 495
Kimsey, L. S., 278, 362,
364
Kimsey, Y., 135
Kimura, M. T., 7, 344
Kimura, R., 413
Kingsolver, J. G., 84,
86, 92, 93, 96, 106,
107, 383, 408
Kinnamon, S. C., 22,
188
Kleindienst, H. U., 183
Kluger, M. J., 411, 412,
416
Knapp, R., 386, 390,
391, 392, 395, 396
Knee, W. J., 258
Kneitz, G., 450, 451
Knisley, C. B., 201, 404,
405
Koch, C., 192, 193, 195

General Index

Conditional strategy, 212
Contest competition, 209–212, 218
Convective cooling, 78, 81–83, 87–88,
    92–97, 128, 131, 197, 203, 248,
    286–288, 316–317, 328, 340, 399–
    400, 402
Counter-currents. *See* Blood circula-
    tion

Dancing, 312–313
Developmental rate, 159–161, 167–
    168
Digestion, 160–162, 394, 398

Ectotherms, 9
Efficiency, 50–51, 177, 246, 385
Elytra (as sun shield), 193, 195
Endotherms, 8, 9, 292–293, 400
Energy: units of measure, 10–13; effi
    ciency, 50–51; storage, 422–444
Environmental temperatures, 10, 520
Enzymes: temperature adaptation, 71–
    72, 214, 386; futile cycling, 238–
    246; endothermy, 314
Evaporative cooling, 46, 153, 273–
    274, 285, 305–307, 359, 361, 375–
    380, 402, 457–458, 463, 495–496
Exercise, 187–188, 314, 373

Fanning, 457, 463, 468, 470, 492–
    495
Fecal shields, 404
Fecundity, 76
Feeding rate, 383
Fibrillar flight muscles, 176, 230–231,
    296, 300
Flight, 30–31, 78, 118; evolution, 2,
    90, 105, 104–113; neural activa-
    tion, 21, 298–300; motor patterns,
    23, 298–300; heat production and
    endothermy, 32–33, 71, 89, 97,
    215–217, 219, 252, 263, 282, 298,
    351, 366, 370, 374–375; time, 61;
    lift, 71–73, 300; acceleration, 90;
    muscle mass, 90, 349, 357; palata-
    bility, 91; duration, 94, 99, 128,
    282; speed, 288, 338, 348; power,
    298–300, 348; wing-beat frequency,
    299
Flight mill, 31, 148–149, 292, 298–
    299, 348
Flower basking, 349
Flyer (dragonflies), 118–127
Foraging, 289, 308–310, 326, 335,
    339

Freezing, 62, 398
Fur. *See* Insulation; Pubescence
Futile cycle, 239

Gliding, 148. *See also* Soaring
Gobetting, 306
Growth rates, 168, 383, 415, 419

Hawk-dove strategy, 212
Head temperature: warm-up, 25, 120,
    286; flight, 44–45, 65–66, 286–
    287; thermal preferences, 170, 182
Heart activity, 38–39, 43–44, 121–125
Heat loss: flight, 33–46, 285–286, 340
Heat therapy, 411–412
Heterothermy, 8
Homeothermy, 8, 224
Hypothermia, 214, 251

Incubation, 260–264, 335, 337, 459–
    462, 465, 490
Insulation, 40–41, 46, 56, 63, 65, 77,
    218, 253–255, 289–290, 335, 341,
    355, 357

Larvae: thermoregulation, 146, 160,
    186; as honeypots, 462; develop-
    ment defects, 463, 472
Leks, 352, 360
Lethal temperatures, 192, 282, 420
Lift, 137, 149
Limits of measurements, 13
Load, 253
Locust plagues, 144
Longevity, 311, 318, 332

Maggots, 401
Mass (body): competition, 210; body
    temperature, 220–221, 287; cooling
    rate, 227, 257–258; ecology, 259–
    260, 340; evolution, 260, 340
Maxithermy, 199–200, 326–328, 331,
    372, 387–389
Metabolic water, 151–152
Metafemoral spring, 423–426
Microhabitat selection, 62, 70, 91,
    198, 283, 327, 330, 352, 364, 366,
    386–387, 408
Migration, 492
Mitochondria, 178
Moltings, 394
Mounds: location, 449; architecture,
    450–451, 491–504; Wärmeträger-
    theorie, 451; compost heap, 452;
    metabolic heating, 452–454; brood

placement, 455–456; airducts, 502–504

Muscle: force, 23; mass, 90, 127; efficiency, 149–151; twitches, 149–150; speed, 178–179

Myogenic flight muscles, 176, 230–231

Nests: location, 448–449, 457, 464, 496, 500; mortality, 448–449; insulation, 459, 465, 468; incubation, 459–462, 465, 490; architecture, 471, 499–504

Neurogenic, 176, 300, 373–374

Non-shivering thermogenesis, 238–246

Obelisk (posture), 129–131, 328

Operative temperature, 520

Ovary, 264–265

Pace, 316–318, 332, 491

Palatability, 91

Panting, 152–153

Parasitism, 395–397, 415

Percher (dragonflies), 127–131

Phase theory, 144

Pheromones, 462, 465, 477, 490

Phototaxis, 146

Pile. See Insulation; Pubescence

Pits, 404–406

Pleurosternal muscle, 238

Postflight cooling, 46, 50

Posture, 146–147, 154–156, 165

Power, 149–151, 177–178, 253, 295, 298–300, 428

Predator avoidance, 90, 162, 327, 372, 385, 387, 392, 395, 420

Prey capture, 339

Profit (energy), 250

Prostaglandins, 416

Protodonates, 138–141

Pubescence, 253–255, 301, 335

Pyrogen, 418

Q_{10} effect, 293, 320

Release mechanisms (of jumps), 425

Resilin, 440–441, 444

Rest: metabolic rate, 70, 158–159, 214, 320, 479, 482

Running: convective cooling, 197; metabolic rate, 197; endothermy, 197, 205, 209, 210, 222–223; pace, 324, 333

Sarcoplasmic reticulum, 59–60, 176, 178, 348, 374

Scramble competition, 209, 218, 250

Setae, 394–395

Set-points, 306, 316

Sexual signaling, 200. See also Color

Shivering. See Warm-up

Singing, 370, 373. See also Calling rate

Soaring, 89, 95, 126–127

Sound production, 168–179

Spathes, 411

Specific heat, 11

Stilting, 147, 154, 156, 165, 193, 198, 203, 328

Stridulation, 170–178

Super-cooling, 62

Superorganism, 476, 489

Swarms, 319–320; temperatures, 474; composition, 479; ventilation, 482; metabolism, 484

Temperature tolerance, 134–135, 158, 162, 164, 167, 346, 374, 384, 386, 405

Tempo. See Pace

Tetanus, 233, 243

Thermal imaging, 14, 66, 310, 313, 346, 364–365

Thermal refuges, 330

Thermal refugia: tents, 392–394, 399; turrets, 404–405

Thermal respiting, 330

Thermal sensors, 42–43, 133–134, 147–148, 179, 184, 306, 316, 447, 489–490

Thermal tolerances, 327–328, 330, 331, 370

Thermoconformity, 220

Thermocouples, 14–15

Thermophilic, 326–328, 331, 370–372

Thermoregulation: definitions, 8, 9, 13, 26, 28, 40; errors, 14; controversy, 28–30; evolution, 57–61, 104–113, 117, 138–141, 176, 199, 219, 382, 392; behavioral, 63, 76–103, 280–283, 417–418; ontogeny, 187–188

Thermosiphon, 503–505

Thermostat, 30–31, 489

Thermovision. See Thermal imaging

Thoracic temperature: muscular activity, 19–21; wing-beat frequency, 26; flight, 29–30, 31, 32, 40, 52–56, 58, 63–64, 92, 279, 282, 284; set-points, 31; errors, 92

Lightning Source UK Ltd.
Milton Keynes UK
UKOW05f0432291215

265468UK00006B/59/P